CATALYTIC AIR POLLUTION CONTROL

CATALYTIC AIR POLLUTION CONTROL
Commercial Technology

THIRD EDITION

Ronald M. Heck
Robert J. Farrauto
with Suresh T. Gulati

A JOHN WILEY & SONS, INC., PUBLICATION

Library of Congress Cataloging-in-Publication Data:
Heck, Ronald M., 1943-
    Catalytic air pollution control : commercial technology / Ronald M. Heck, Robert J. Farrauto, Suresh T. Gulati. – 3rd ed.
        p.   cm.
    Includes bibliographical references and index.
    ISBN 978-0-470-27503-0 (cloth)
1. Air–Purification–Equipment and supplies.   2. Catalysts.   3. Automobiles–Catalytic converters.   4. Trucks-Catalytic converters.   I. Farrauto, Robert J., 1941-   II. Gulati, Suresh T.   III. Title.
    TD889.H43   2009
    629.25′28–dc22

                                                                    2008032182

Printed in the United States of America

10 9 8 7 6 5 4 3 2 1

To my wife, Barbara, whose friendship, support, understanding (especially on lost weekends), humor and selflessness made this endeavor much easier. I'm glad I will have more time for her as this project is finished. Unk for always being there for support; and to Merc and Dutch who were overseeing it all.

Ronald M. Heck

To my wife Olga (Olechka) who has given me love, understanding, focus and a new vision of the wonders of life; my loving daughters Jill Marie and Maryellen and their husbands Glenn and Tom. To my grandchildren Nicholas, Matthew, Kevin, Jillian, Owen and Brendan who represent everything that is beautiful in life.

To the memory of my loving parents who gave me a sense of values that has propelled me to help others.

Robert J. Farrauto

To my wife Teresa whose encouragement and support helped complete this project, my sons Raj and Prem for their genuine support, and my darling daughter Sonya for her "how can I help you, dad?" attitude throughout this project.

Suresh T. Gulati

Contents

Preface

Environmental quality is certainly a worldwide concern. Air pollution knows no boundaries, and reducing it is of the utmost importance. Countries are now establishing environmental regulations that must be met by mobile as well as by stationary pollution sources. Exhausts containing volatile organic compounds, carbon monoxide, nitric oxides, ozone, and so forth all can be converted to harmless nonpollutants at reasonable temperatures and with cost-effective systems using heterogeneous catalysts. The use of the right catalyst system converts pollutants to nonpollutants at low-energy requirements and at higher rates, resulting in cost-effective pollution control. The use of catalytic systems for pollution abatement was virtually nonexistent before 1976, but now it is a multibillion-dollar worldwide business that addresses gasoline- and diesel-fueled automobiles and trucks, destruction of volatile organic compounds from stationary sources such as chemical processing plants, reduction of nitric oxides from power plants and stationary engines, decomposition of ozone in high-flying commercial aircraft, pollution from small engines, and so on. The utilization of catalysts for future pollution abatement applications promises to grow at a strong pace over the next decade.

This book is designed to be a stand-alone introductory reference or textbook on the commercially available catalytic systems used today for reducing harmful emissions for both mobile and stationary sources. It is like no other book currently available, because it describes modern catalytic air pollution abatement techniques from a practical point of view. The subjects are discussed in clear and succinct language, with emphasis placed on the real-world catalytic system performance. It is intended to serve as a bridge for academic and industrial catalysis.

Part I has been expanded to include more catalyst fundamentals to give readers a more detailed understanding of kinetics, characterization, and deactivation modes for environmental catalysts. Additionally questions have been added at the end of each chapter to challenge the reader's knowledge of the material presented. In both of these regards, the book now can serve as an introductory text with special emphasis on the applied aspects of environmental catalysis normally not covered in fundamental textbooks. It is especially useful for newcomers as well as for experienced catalyst scientists and engineers.

Part II discusses the application of catalytic systems for mobile source emission control, the automobile catalytic converter, diesel oxidation catalysts, diesel particulate filters (the newest major application of environmental catalysis), and the decomposition of ozone in high-flying aircraft. The sections on both automotive and diesel have been substantially expanded to reflect the opportunities for catalysis with the new emission standards particularly in diesel.

Part III describes the stationary application of catalysts, including volatile organic compounds, reduction of nitric oxides, and oxidation of gaseous carbon monoxide and hydrocarbons. A new section has been added to small-engine applications as this is a new developing field of catalytic applications.

Finally, Part IV presents new and emerging applications that, if developed, will dramatically change current catalytic technology for environmental control. For this reason, materials for the hydrogen economy including fuel cells have been updated.

* * *

Dr. Ron Heck is president of RMH Consulting where he specializes in consultation on environmental catalysis for auto, diesel, and stationary source; general catalysis; fuel cells; reaction engineering; combustion technology; chemical engineering; and expert witness. He retired as a principal scientist and research manager from Engelhard Corporation in 2003 where he was responsible for developing new catalyst technology for Engelhard Corporation's worldwide customers in environmental catalysis. He was responsible for developing the PremAir® family of catalysts technologies for removing pollutants from the ambient air and the close-coupled catalyst technology that allowed commercial development of ultra-low-emission systems. Ron was with Engelhard for 31 years and worked on development of catalytic processes for Engelhard in SCR NO_x, NSCR NO_x, automotive catalyst, diesel catalyst, PremAir® catalyst systems, hydrogenation technology, ozone abatement, volatile organic compound abatement, ammonia oxidation, chemical feedstock purification, and chemical synthesis.

Ron is a member of American Men and Women of Science and Who's Who in Technology Today. He is a recipient of the Forest R. McFarland Award from the Society of Automotive Engineers for outstanding contributions to this professional society. Ron is an SAE Fellow in recognition of engineering creativity and contributions to the profession and the public at large. With Dr. Gulati, Ron teaches a 2-day course on automotive emission control catalysis and diesel emission control catalysis organized by the Society of Automobile Engineers.

He is the co-author of the book with Dr. Farrauto entitled *Catalytic Air Pollution Control: Commercial Technology*. Ron was a co-editor of the "News-Brief" section of *Applied Catalysis B: Environmental* and was a member of the Scientific Advisory Board (SAB) for environmental studies for the Strategic Environmental Research & Development Program (SERDP).

Ron and his former research team from Engelhard received the 2004 Thomas Alva Edison Patent Award from the R&D Council of New Jersey for the invention of close-coupled catalyst technology for ultra-low-emission gasoline vehicles.

Ron has been involved in over 80 publications in commercial applications of catalysts and holds 36 U.S. patents on catalytic processes.

Ron received his B.S. in chemical engineering and his Ph.D. from the University of Maryland and his M.A. in theology from the College of Saint Elizabeth.

Dr. Farrauto is a research fellow at the Corporate Research Laboratories of BASF Catalysts LLC (formerly Engelhard) in Iselin, New Jersey. He has worked extensively in the development of catalysts for the environmental, chemical, and alternative energy industries. His major responsibilities have included the development of advanced automobile emission control catalysts for passenger cars. He was technical leader of the Engelhard team that developed diesel oxidation catalysts now commercialized in the United States and Europe for trucks and passenger cars. Currently he is managing a research team investigating hydrogen production for fuel cells to be used for stationary, portable, and vehicular applications.

In addition to the third edition of *Catalytic Air Pollution Control: Commercial Technology*, he is the co author of *Fundamentals of Industrial Catalytic Processes*, second edition, published by Wiley. Dr. Farrauto is the author of 85 publications, and 50 U.S. patents, and he served as the North and South American Editor of *Applied Catalysis B: Environmental*. He is the recipient of the 2008 Ciapetta Lectureship Award sponsored by the North American Catalysis Society and in 2005 received the Catalysis and Reaction Engineering Practice Award in sponsored by the American Institute of Chemical Engineering. He is the recipient of the Canadian Catalysis Foundation (Year 2000–2001) Cross Canada Lectureship Award and in 2001 received the Henry Albert Award for excellence in precious metal catalysis sponsored by the International Precious Metal Institutes. He is an adjunct professor in the Earth and Environmental Engineering Department of Columbia University in New York City where he teaches catalysis courses and supervises graduate students.

He received his B.S. in chemistry from Manhattan College in New York City and a Ph.D. from Rensselaer Polytechnic Institute in Troy, New York.

Dr. Suresh Gulati is a former corporate fellow of Corning Inc. and currently is a consultant to Corning. He has spent 33 years helping Corning develop and optimize ceramic catalyst supports and particulate filters for gasoline- and diesel-powered vehicles. Prior to retiring in 2000, he applied his mechanical engineering background to ensure long-term reliability of glass and ceramic products like space windows, CRT, fiber optics, and liquid crystal displays. He has published extensively in refereed journals, holds 15 U.S. patents, and continues to give talks at conferences organized by the SAE, ASME, and ACerS.

He is a fellow of these professional societies. In addition, he is a recipient of two SAE awards: the Lloyd L. Withrow Distinguished Speaker Award in 2000 and the Forest R. McFarland Award in 2003. Both Drs. Gulati and Heck teach a 2-day course on catalytic converters that is organized by the Society of Automobile Engineers.

Dr. Gulati holds a B.S. from the University of Bombay, an M.S. from the Illinois Institute of Technology, and a Ph.D. from the University of Colorado—all in mechanical engineering.

Acknowledgments

The authors want to acknowledge several people who have helped in preparing and reviewing various aspects of the manuscript.

Ron Heck wants to acknowledge the many outside agencies and companies which provided updates on their technology, in particular MECA (Joe Kubsh), EPA (Joe McDonald), Umicore (Bill Staron and Janette Tomes), NGK (Paul Busch), Johnson Matthey (Andy Walker), Caterpillar (Herbert DaCosta), ORNL (Bruce Bunting and Jim Parks), Donaldson Company (Ted Angelo), Ford (Christine Lambert), Cormetech (George Wensell), BASF (Mike Galligan), and Emitec (Klaus Mueller-Hauss).

Robert Farrauto wants to acknowledge the many Engelhard (now BASF Catalysts LLC) scientists and engineers, past and present, who have pioneered in the development of the technologies described in this book. A special thanks to Dr. J. G. Cohn who gave special input to earlier editions. Ron and Bob want to express their deep appreciation for the leadership in environmental catalysis provided by Dr. John Steger. All the authors wish to thank the BASF Materials Characterization team, including Bob Geise, Joanne St. Amanda, Sharon Goresh, John Motylewski, Patricia Nelson, Tom Gegan, James Drozd, Nancy Brungard, George Munzing, Gail Hodges, Scott Hedrick, Beth Nartowicz, and Xinsheng Liu, who `have provided many of the photographs of catalysts used in environmental technology. They are grateful to Maurica Fedors and Arda Argulian who have provided literature and patent searches. Bob would like to express his gratitude to the graduate students at Columbia University whose questions and feedback provided an important contribution to this edition.

Suresh Gulati wants to thank Drs. Pronob Bardhan and Joseph Antos of the Science and Technology Division of Corning Incorporated for their thorough review and valuable suggestions, Linda Newell for preparing figures, and Julie Berman for preparing revisions for the manuscript. Also, special acknowledgment is given to BASF (formerly Engelhard Corporation) and Corning Incorporated for their pioneering development of ceramic substrates and catalysts.

Acknowledgments, First Edition

The authors want to acknowledge several people who have helped in preparing and reviewing various aspects of the manuscript. First and foremost they express their deepest gratitude to Dr. J. Gunther Cohn, who critically reviewed the entire book and offered many helpful suggestions. Special appreciation is extended to Drs. Jennifer Feeley, Jim Chen, John Hochmuth, Michel Deeba, Barry Speronello, Jordan Lampert, Harold Rabinowitz, and Michael Spencer, all of Engelhard R&D, who recommended many changes that have been incorporated into the final manuscript. The authors are indebted to Marisa Fedors of the Technical Information Center of Engelhard for her efforts in searching the literature. Thanks to Denise Lenci, Donna Gallagher, Michel Stryjewski, Jon Lederman, and Terry Lomuntad of Engelhard Corporation's Communications Department who helped with typing and recommending appropriate photos. The efforts in preparing figures by Dave Antonucci, Ray Tisch, and O. J. Natale of Engelhard are greatly appreciated. The authors are grateful to Drs. Jerry Spivey, John Armor, and Professor Scott Cowley for their many helpful suggestions.

Finally, they would like to acknowledge Engelhard Corporation and many of its employees, past and present, who pioneered in the development of processes and catalysts for treating environmental problems, and who will continue to do so in the future.

Acknowledgments, Second Edition

The authors want to acknowledge several people who have helped in preparing and reviewing various aspects of the manuscript, including Maurica Fedors, Arda Agulian, and Jane Szeg of the Engelhard Technical Information Center for considerable help in literature searching; Bob Ianniello, Joanne St. Amand, Sharon Goresh, John Motylewski, Patricia Nelson, Zeneida Gutierrez, George Munzing, Nancy Brungard, and Tom Gegan who contributed important characterization data; and Bob Womelsdorf, Mike Durilla, Jim Fu, Jim Chen, Rudy Lechelt, Rosto Brezny, and Nick Bayachek for help with documentation.

Robert Farrauto wants to acknowledge the entire fuel processing team at Engelhard with special thanks to Dr. Wolfgang Ruettinger and Dr. Larry Shore who provided proofreading and valuable suggestions. Special thanks to Dr. Ed Wolynic, Dr. Terry Poles, and Dr. Bruce Robertson who have given strong commercial leadership and inspiration to move forward in fuel cell technology.

Ron Heck wants to acknowledge the entire PremAir® team who made new technology possible with special thanks to Dr. Jeff Hoke who provided proofreading and valuable suggestions. Special thanks to Dr. Steger who has been a support in communicating this subject to the world.

Ron Heck and Robert Farrauto want to acknowledge the staff at Engelhard who have pioneered in emission control, and who will continue to advance the technology.

Suresh Gulati wants to acknowledge George Beall, Lou Manfredo, and Daniel Ricoult of Corning S & T, who recommended many changes. The efforts in preparing figures by Nancy Foster of Corning and parts of the manuscript by Virginia Doud, formerly of Corning, are greatly appreciated. Finally, special acknowledgment is made to Engelhard Corporation and Corning Incorporated, who pioneered in the development of substrates and catalysts.

PART I
Fundamentals

1 Catalyst Fundamentals

1.1 INTRODUCTION

Chemical reactions occur by breaking chemical bonds of reactants and by forming new bonds and new compounds. Breaking stable bonds requires the absorption of energy, whereas making new bonds results in the liberation of energy. The combination of these energies results in either an exothermic reaction in which the conversion of reactants to products liberates energy or an endothermic process in which the conversion process requires energy. In the former case, the energy of the product is lower than that of the reactants, with the difference being the heat liberated. In the latter case, the product energy is greater by the amount that must be added to conserve the total energy of the system. Under the same reaction conditions, the heat of reaction (ΔH) being a thermodynamic function does not depend on the path or the rate by which reactants are converted to products. Similarly, the free energy of reaction (ΔG) of the reaction is not dependent on the reaction path because it too is a thermodynamic state function. This will be emphasized once we discuss catalytic reactions. The rate of reaction is determined by the slowest step in a conversion process independent of the energy content of the reactants or products.

1.2 CATALYZED VERSUS NONCATALYZED REACTIONS

A few decades ago, chlorofluorocarbons (i.e., CF_2Cl_2), emitted primarily from refrigerants, were found to catalyze the destruction of the ozone (O_3) layer in the stratosphere necessary to protect us from harmful ultraviolet (UV) radiation and its skin cancer consequences. Fortunately alternative chemicals are now used, and this problem is no longer of great concern. It does, however, serve as an excellent example of a homogeneous gas phase catalytic reaction. First let us consider the very slow noncatalytic reaction between gaseous O_3 and O atoms produced by dissociation of O_2 by solar radiation in the upper atmosphere:

$$O_3 + O \rightarrow 2O_2 \tag{1.1}$$

Catalytic Air Pollution Control: Commercial Technology, by Ronald M. Heck and Robert J. Farrauto, with Suresh T. Gulati.
Copyright © 2009 John Wiley & Sons, Inc.

Chlorine atoms, produced by solar radiation of chlorofluorohydrocarbons, catalyze the decomposition of ozone by reacting with it to form ClO and O_2 (1.2). The ClO then reacts with the O atoms regenerating Cl and producing more O_2 (1.3).

$$Cl + O_3 \rightarrow ClO + O_2 \tag{1.2}$$

$$ClO + O \rightarrow Cl + O_2 \tag{1.3}$$

Adding both reactions results in Eq. (1.1) and completes the catalytic cycle since the Cl and ClO are both consumed and regenerated in the two reactions. Thus, Cl is a homogeneous catalyst for the destruction of O_3. The uncatalyzed reaction is very slow, and its reaction profile can be described kinetically by the Arrhenius profile in which reactants convert to products by surmounting the noncatalytic activation energy barrier (E_{NC}) as shown in Figure 1.1. The rate constant k of the reaction is inversely related to the exponential of the activation energy, where T is the absolute temperature, R is the universal gas constant, and k_o is the preexponential constant. The Arrhenius equation (1.4) indicates that the rate constant k decreases the higher the activation energy (E).

$$k = k_o \mathrm{Exp} \left(-E/RT\right) \tag{1.4}$$

Since the catalyzed reaction has a lower activation energy (E_C), its reaction rate is greater. The barrier was lowered by the Cl catalyst providing a chemical shortcut to products. Although the rate is greater for the catalyzed reaction, the enthalpy (ΔH) and free energy (ΔG) are not changed. Similarly the equilibrium constant for both catalyzed and noncatalyzed reactions is not changed

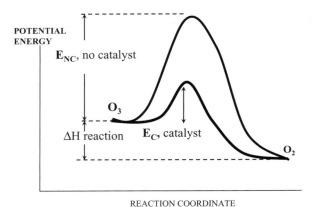

FIGURE 1.1. Catalyzed and uncatalyzed reaction energy paths for O_3 decomposition to O_2. Activation energy for catalyzed reaction E_C is lower, and the reaction is faster than the noncatalyzed E_{NC}.

since both operate under the same reaction conditions in the stratosphere. The catalyst can only influence the rate of which reactants are converted to products in accordance to the equilibrium constant and cannot make thermodynamically unfavorable reactions occur. In industrial practice, reactions conditions, such as temperature and pressure, are varied to bring the free energy to a desirable value to permit the reaction to occur.

Now we will consider the conversion of carbon monoxide (CO), a known human poison, to CO_2, a reaction of great importance to the quality of air we breathe daily. The overall rate of the noncatalytic reaction is controlled by the dissociation of the O_2 molecule to O atoms (rate-limiting step), which rapidly react with CO forming CO_2. The temperature required to initiate the dissociation of O_2 is greater than $700\,^\circ C$, and once provided, the reaction rapidly goes to completion with a net liberation of energy (the heat of reaction is exothermic). The requirement to bring about the O_2 dissociation and ultimately the conversion of CO to CO_2 has an activation energy (E_{NC}). Reaction occurs when a sufficient number of molecules (O_2) possess the energy necessary (as determined by the Boltzmann distribution) to surmount the activation energy barrier (E_{NC}) shown in Figure 1.2a). The rate of reaction is expressed in accordance with the Arrhenius equation (1.4). Typically the activation energy for the noncatalytic or thermal conversion of CO to CO_2 is about 40 Kcal/mole.

Let us now discuss the effect of passing the same gaseous reactants, CO and O_2, through a reactor containing a solid catalyst. Since the process is now carried out in two separate phases, the term *heterogeneous catalytic reaction* is used. In the presence of a catalyst such as Pt, the O_2 and CO molecules adsorb on separate sites in a process called chemisorption in which a chemical partial bond is formed between reactants and the catalyst surface. Dissociation of chemisorbed O_2 molecules to chemisorbed O atoms is rapid, occurring essentially at room temperature. Highly reactive adsorbed O atoms react with chemisorbed CO on adjacent Pt sites producing CO_2, which desorbs from the Pt site, completing the reaction and freeing the catalytic site for another cycle. Thus, the activation energy for the Pt catalyzed reaction (E_c), shown in Figure 1.2b), is considerably smaller than that for the noncatalyzed reaction, enhancing the conversion kinetics. Typically the activation energy for Pt catalyzed CO to CO_2 is less than about 20 Kcal/mole. Figure 1.3 shows the initial lightoff of a conversion versus temperature plot for the catalyzed reaction occurring around $100\,^\circ C$. The noncatalyzed reaction has a considerably higher lightoff temperature (around $700\,^\circ C$) because of its higher activation energy. More input energy is necessary to provide the molecules the necessary energy to surmount the activation barrier so lightoff occurs at higher temperatures. It should be noted, however, that the noncatalyzed reaction has a greater sensitivity to temperature, (slope of plot). Thus, the reaction with the higher the activation energy has the greater sensitivity to temperature, making it increase to a greater extent with temperature than that with a lower activation energy. This is a serious problem for highly exothermic reactions, such as CO and hydrocarbon oxidation, where noncatalyzed free radical reactions, with large

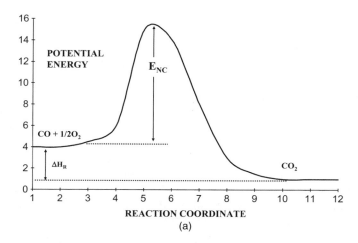

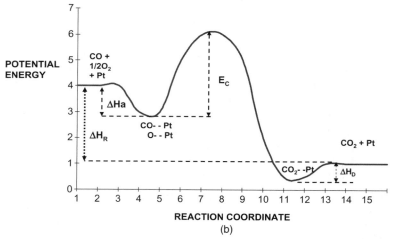

FIGURE 1.2. Activation energy diagram for a) thermal reaction of CO and O_2 and b) the same reaction in the presence of Pt. Activation energy for the noncatalyzed reaction is E_{NC}. The Pt catalyzed reaction activation energy is designated E_c. Note that the heat of reaction ΔH_R is the same for both reactions. ΔH_a = heat of adsorption; ΔH_D = heat of desorption.

activation energies, can lead to undesirable products. Thus, the temperature must be carefully controlled within the reactor.

Equations relating reaction rates to activation energies will be discussed in considerable detail in Chapter 4, but for now, it is sufficient to understand that an inverse relationship exists between the activation energy and the reaction rate.

The environmental significance of catalyzed reactions is now apparent; a reaction can be carried out at much lower temperatures consistent with startup conditions in an automobile converter. Kinetic rate studies indicate that the

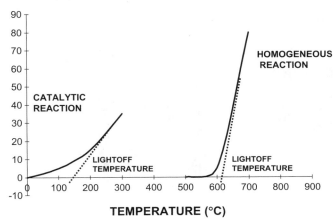

CO CONVERSION (%)

TEMPERATURE (°C)

FIGURE 1.3. Conversion of CO versus temperature for a noncatalyzed (homogeneous) and a catalyzed reaction.

rate-limiting step is the reaction of chemisorbed CO with chemisorbed O atoms on adjacent Pt sites. The reaction occurs around 100 °C far below the 700 °C required for the noncatalytic process described above. Thus, the catalyst provides a new reaction pathway in which the rate-limiting step is altered from one of high-temperature dissociation of O_2 to that of the reaction between two adsorbed moieties on adjacent Pt sites at a significantly lower temperature. This shows the great importance of catalysis in enhancing rates of reaction, allowing them to occur at moderate temperatures as indicated in this example. A lower operating temperature translates into energy savings, less expensive reactor materials of construction, and preferred product distributions with greater rates of production with smaller size reactors. For this reason, catalysts are commonly used in many industrial applications ranging from petroleum processing, chemical and energy synthesis, to environmental emission control (Bartholomew and Farrauto 2006).

Inspection of Figure 1.2b) indicates an energy decrease associated with the adsorption of CO and O_2 on the Pt surface (ΔH_a) because of its exothermic nature. This is a consequence of the decreased entropy (ΔS) when the molecules are confined in an adsorbed state with the commensurate loss in a degree of freedom. Because ΔG_a must be negative and $-T\Delta S_a$ is positive, ΔH_a must be negative in accordance with $\Delta G_a = \Delta H_a - T\Delta S_a$. Desorption is always endothermic.

The catalytic reaction is usually carried out in a fixed-bed reactor similar to that shown in Figure 1.4. Here we introduce the monolithic support upon the walls of which is deposited the catalysts called a catalyzed washcoat. This is the common support used in environmental applications for reasons to be given in later chapters. The proper flow of inlet reactant gas is established by

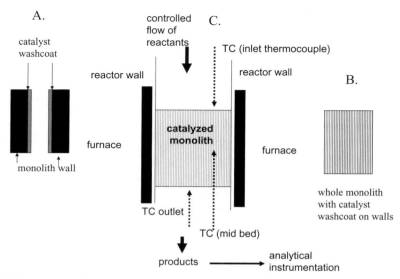

FIGURE 1.4. Fixed bed reactor containing a catalyzed monolith for measurement of conversion versus temperature. A) Single channel of catalyzed monolith. B) Monolith structure with walls coated with catalyzed washcoat. C) Catalyzed washcoat on a monolith in the reactor surrounded by a furnace.

individually controlled mass flow meters. The mixed reactants are continuously preheated, and the conversion of reactants and the appearance of products are observed using analytical instrumentation such as a gas chromatograph or other suitable equipment specific for reactants and products such as a CO and CO_2 analyzer, hydrocarbon analyzer, and so on. The inlet, mid-bed, and outlet temperatures are commonly measured using thermocouples.

1.2.1 Adsorption and Kinetic Models for CO Oxidation on Pt: Langmuir–Hinshelwood Kinetics

1.2.1.1 Langmuir Isotherm. The most widely accepted kinetic model for the CO oxidation reaction on Pt is based on the Langmuir isotherm from which is derived Langmuir–Hinshelwood (LH) kinetics (Hinshelwood 1940; Hougen and Watson 1943). The Langmuir isotherm is based on the key assumption that all sites on the adsorbent surface are of equal energies. It also assumes the rate-limiting step is the surface reaction between adsorbed species with all others fast and in equilibrium. Despite this ideal view, its application yields reasonable predictions.

Consider the strong adsorption of CO in equilibrium with the surface of Pt.

$$CO + Pt \leftrightarrow CO \text{ - - - } Pt \qquad (1.5)$$

The rate of forward (CO adsorption) is given by

$$(\text{Rate})_{fCO} = k_{fCO} P_{CO}(1 - \theta_{CO}) \qquad (1.6)$$

where k_{fCO} = the forward rate constant of CO adsorption on Pt, P_{CO} = the partial pressure of CO, and θ_{CO} the fraction of the surface of Pt covered by CO. The term $(1 - \theta_{CO})$ is the fractional number of sites available for additional CO adsorption on the Pt surface. The isotherm also assumes each site is occupied by only one adsorbate molecule and full coverage is a monolayer.

The rate of reverse (CO desorption) is as follows:

$$(\text{Rate})_{dCO} = k_{dCO} \theta_{CO} \qquad (1.7)$$

At equilibrium, the forward and desorption rates are equal and the ratio of the forward rate to reverse rate is

$$k_{fCO}/k_{dCO} = K_{CO} \qquad (1.8)$$

$$k_{fCO} P_{CO}(1 - \theta_{CO}) = k_{dCO} \theta_{CO} \qquad (1.9)$$

$$\theta_{CO} = K_{CO} P_{CO}/(1 + K_{CO} P_{CO}) \qquad (1.10)$$

Plotting Eq. (1.10) generates Figure 1.5.
When P_{CO} is large,

$$(1 + K_{CO} P_{CO}) \sim K_{CO} P_{CO} \qquad (1.11)$$

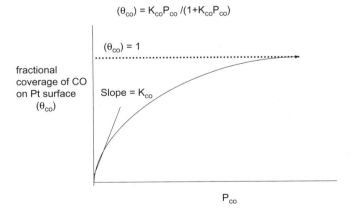

FIGURE 1.5. Adsorption isotherm (θ_{CO}) for CO on Pt for large, moderate, and low partial pressures of CO. The slope at low partial pressures of CO equals the adsorption equilibrium constant K_{CO}.

Eq. (1.10) reduces to $\theta_{CO} = 1$ representing monolayer coverage of the surface by CO.

When CO is small,

$$(1 + K_{CO}P_{CO}) \sim 1 \tag{1.12}$$

and Eq. (1.10) reduces to $\theta_{CO} = K_{CO}P_{CO}$ with the slope equal to K_{CO}.

At moderate values of P_{CO}, Eq. (1.10) applies as written with the curve becoming more shallow as P_{CO} increases.

The isotherm for dissociative chemisorption of O_2 on Pt is similarly generated, where θ_O refers to the fractional coverage by O atoms consistent with the stiochiomety for CO oxidation of one O atom for each CO.

$$O_2 + Pt \leftrightarrow 2O \text{ --- } Pt \tag{1.13}$$

The rate of the forward (adsorption of O_2 on Pt) is as follows:

$$(Rate)_{fO2} = k_{fO2}P_{O2}(1 - \theta_O)^2 \tag{1.14}$$

The rate of the reverse reaction (O desorption from Pt) is as follows:

$$(Rate)_{dO2} = k_{dO2}\theta_O^2 \tag{1.15}$$

The square term for both the forward and reverse rates is from the lower probability that two adjacent Pt sites will be available to accommodate two oxygen atoms resulting from the dissociative chemisorption of O_2 on Pt. Similarly two adsorbed O atoms on Pt must be adjacent for desorption and recombination of diatomic O_2 to occur. Also the adsorption equilibrium constant (K_{O2}) for O_2 on Pt is k_{fO2}/k_{dO2}.

Equating forward and reverse rates,

$$k_{fO2}P_{O2}(1 - \theta_O)^2 = k_{dO2}\theta_O^2 \tag{1.16}$$

$$\theta_O = K_{O2}^{1/2}P_{O2}^{1/2} / (1 + K_{O2}P_{O2})^{1/2} \tag{1.17}$$

Plotting the fractional coverage of oxygen atoms versus $P_{O2}^{1/2}$ generates a similar plot as in Figure 1.5, but the slope at low $P_{O2} = K_{O2}^{1/2}$. At high P_{O2} $\theta_O = 1$.

1.2.1.2 Langmuir–Hinshelwood Kinetics for CO Oxidation on Pt.

The adsorption isotherms for CO and O_2 were considered separately in Section 1.2.1.1, but for the oxidation of CO by O_2, it is necessary to consider both gases present with each competing for the same sites on Pt. We will use k as the rate constant for the oxidation of CO.

The net rate of reaction for CO oxidation will be

$$[\text{Rate}]_{CO} = k\theta_{CO}\theta_{O} \tag{1.18}$$

Coverage for both θ_{CO} and θ_{O} must be modified to include competitive adsorption on Pt sites.

To account for the sites occupied by O atoms, the rate for forward adsorption of CO is written as follows:

$$(\text{Rate})_{fCO} = k_{fCO}P_{CO}(1 - \theta_{CO} - \theta_{O}) \tag{1.19}$$

The desorption rate for CO only depends on the sites occupied by CO [as shown in Eq. (1.7) and reprinted below]:

$$(\text{Rate})_{dCO} = k_{dCO}\theta_{CO}$$

Equating adsorption (1.19) and desorption (1.7) rates at equilibrium and recognizing that $k_{fCO} / k_{dCO} = K_{CO}$

$$K_{CO}P_{CO} = \theta_{CO}/(1 - \theta_{CO} - \theta_{O}) \tag{1.20}$$

For the rate of adsorption and desorption of O_2, we obtain

$$K_{O2}^{1/2}P_{O2}^{1/2} = \theta_{O}/(1 - \theta_{CO} - \theta_{O}) \tag{1.21}$$

Simplification is achieved by dividing Eq. (1.20) by Eq. (1.21):

$$\theta_{O} = \theta_{CO}K_{O2}^{1/2}P_{O2}^{1/2}/K_{CO}P_{CO} \tag{1.22}$$

Substitute Eq. (1.22) into Eq. (1.20):

$$\theta_{CO} = K_{CO}P_{CO}/(1 + K_{CO}P_{CO} + K_{O2}^{1/2}P_{O2}^{1/2}) \tag{1.23}$$

Substitute Eq. (1.23) into Eq. (1.22):

$$\theta_{O} = K_{O2}^{1/2}P_{O2}^{1/2}/(1 + K_{CO}P_{CO} + K_{O2}^{1/2}P_{O2}^{1/2}) \tag{1.24}$$

Now we have an expression for θ_{CO} and one for θ_{O}, so substituting these terms into [as shown in Eq. (1.18) and reprinted below]:

$$[\text{Rate}]_{CO} = k\,\theta_{CO}\theta_{O}$$

$$[\text{Rate}]_{CO} = k\,K_{CO}P_{CO}K_{O2}^{1/2}P_{O2}^{1/2}/(1 + K_{CO}P_{CO} + K_{O2}^{1/2}P_{O2}^{1/2})^{2} \tag{1.25}$$

For low P_{CO}, Eq. (1.25) reduces to Eq. (1.26):

$$[Rate]_{CO} = k\ K_{CO}P_{CO}K_{O2}^{1/2}P_{O2}^{1/2}\big/\big(1+K_{O2}^{1/2}P_{O2}^{1/2}\big)^2 \tag{1.26}$$

This shows a direct relationship between the rate and the P_{CO} when P_{O2} is constant

For large P_{CO}, Eq. (1.25) reduces to Eq. (1.27):

$$[Rate]_{CO} = k\ K_{O2}^{1/2}P_{O2}^{1/2}/(P_{CO}K_{CO}) \tag{1.27}$$

This shows that at high P_{CO}, the reaction rate is inhibited by CO and its rate of oxidation decreases. Thus, a maximum in rate exists when

$$\theta_O = \theta_{CO}$$

This is shown graphically in Figure 1.6.

Applying this model was useful in designing an optimum system for the first gasoline oxidation catalyst for the automobile converter. When the CO was high during the cold-start portion of the driving cycle, the addition of extra O_2 (from air) decreased the P_{CO} more than P_{O2} and the rate of the reaction for CO oxidation increased. Thus, understanding kinetics and the rate expressions helped design a workable system to meet regulations.

From a fundamental point of view, it should be noted that the assumption of uniform energy sites on the catalyst in the Langmuir isotherm is not correct.

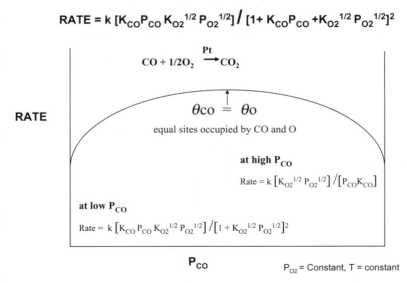

FIGURE 1.6. LH kinetics applied to increasing P_{CO} at constant P_{O2}. Maximum rate was achieved when an equal number of CO molecules and O atoms is adsorbed ($\theta_O = \theta_{CO}$) on adjacent Pt sites.

A heterogeneous catalytic surface consists of a distribution of strong, moderate, and weak sites upon which the reactant molecules adsorb. Naturally with increasing temperature, only the stronger sites retain adsorbed molecules and the fractional coverage decreases. This action results in a change in the overall energy of the adsorbed states on the activation energy profile of Figure 1.2b). Fundamentally, with increasing temperature, this causes a small change in the activation energy but for all intents and purposes can be ignored when making activation energy measurements. Measurements of activation energies will be discussed later.

1.3   CATALYTIC COMPONENTS

Precious metals (often referred to as the Pt group metals or PGMs) are located in group VIIIB of the periodic table and consist of Ru, Rh, Pd, Ir, Pt, and Os. For environmental emission control applications, Pt, Pd, and Rh dominate as the main catalytic components individually or in combinations; however, they also are commonly used in processing petroleum to transportation fuels and for the synthesis of many important chemicals (Bartholomew and Farrauto 2006). Ironically the PGMs are also referred to as the noble metals for their resistance to oxidation, various poisons, and high temperatures; yet they are some of the most catalytically active elements in nature because of their ability to chemisorb and convert adsorbed species with high rates. They are rare and expensive, and thus when no longer performing satisfactorily, they are recycled, purified, and reused. They are primarily mined in South Africa and Russia with small deposits in Canada and the United States.

Other group VIIIB metals and their oxides such as Fe, Co, and Ni are also catalysts as are Cu and Ag (group IB) , V (group VB), and Cr and Mo (group VIB) for industrial applications (Bartholomew and Farrauto 2006). Some of these base metals have modest catalytic activities relative to Pt, Pd, and Rh but are much less expensive and, in certain cases, more selective. It is common practice in commercial applications, especially in automotive catalysts, to pursue base metal materials as replacements for precious metals; however, because of their lower activity, vulnerability to poisons, and lack of hydrothermal stability in the severe environment of an automobile exhaust, this quest has not been successful.

It should be understood that often the active catalytic component is not always present in its native elemental state but may be present as an oxide. For the oxidation of many hydrocarbons, Pd is most catalytically active as PdO, whereas for the hydrogenation reactions, Pd metal is most active. Vanadium pentoxide (V_2O_5) is an active catalyst for oxidizing SO_2 to SO_3 in the manufacture of sulfuric acid. In contrast, Ni metal is active for hydrogenation reactions.

1.4 SELECTIVITY

The catalyst also affects the selectivity or rate of desired product formation by preferentially lowering the activation energy for a particular step in the reaction sequence and increasing the rate at which this step proceeds. Selectivity is an issue for many reactions in which multiple products can occur in parallel. Different catalysts and/or reaction conditions can preferentially enhance the rate at which reactants are converted to desired products even though many other paths are thermodynamically more or less favorable. Selectivity is the rate of one reaction compared with the rate of parallel paths the reactants may take to other products. Industry describes a process based on the % selectivity to a particular product. The most desirable catalyst and process conditions will have the highest selectivity.

Consider the comparison between the products produced when Pt is used as a catalyst as opposed to oxides of vanadium for the oxidation of ethylene, a hydrocarbon component in automobile exhaust:

$$C_2H_4 + 3O_2 \xrightarrow{\ Pt\ } 2CO_2 + 2H_2O \tag{1.28}$$

V_2O_5 catalyzes the formation of the aldehyde because this pathway with this catalyst has the lowest activation energy compared with the complete combustion to CO_2 and H_2O:

$$C_2H_4 + 1/2O_2 \xrightarrow{\ V_2O_5\ } CH_3CH{=}O \tag{1.29}$$

For Pt the reaction products are exclusively CO_2 and H_2O; thus, selectivity is essentially 100%, making it a good catalyst for pollution abatement. However, for V_2O_5, the selectivity is about 80–90% toward the aldehyde, with the balance of 20–10% being CO_2 and H_2O. Clearly V_2O_5 would not be desirable for conversion of hydrocarbons such as ethylene to harmless CO_2 and H_2O, but it is used commercially for selective partial oxidation reactions to desirable chemicals. So it is the function of the catalyst with optimum reaction conditions to reduce the activation energy of the path that will yield the most desirable product. The ability to (1) enhance reaction rates and (2) direct reactants to specific products makes catalysis extremely important in the environmental, petroleum, and chemical industries.

An important reaction in the automotive catalytic converter is the reduction of NO by H_2 during a specific driving mode that will be discussed later in the automobile converter chapter. Two parallel reaction pathways are possible: one desirable leading to N_2 formation and the other undesirable producing toxic NH_3:

$$2H_2 + 2NO \xrightarrow{\ Rh\ } N_2 + 2H_2O \tag{1.30}$$

$$5H_2 + 2NO \xrightarrow{\ Pt\ } 2NH_3 + 2H_2O \tag{1.31}$$

Clearly Rh is more selective and dominates the NO-to-N_2 pathway with a rate considerably higher than that undesired pathway leading to NH_3 formation when Pt is used.

The reaction conditions also have a pronounced effect on product distribution depending on the activation energies for all possible reactions. For example, NO (a component of acid rain and a contributor to ozone formation) emitted from automobile engines and power plant exhausts can be reduced using a V_2O_5-containing catalyst with high selectivity provided the temperature is maintained between 250 °C and 400 °C:

$$2NH_3 + NO + O_2 \xrightarrow{250-400°C} 3/2\,N_2 + 3H_2O \qquad (1.32)$$

$$2NH_3 + NO + 5/2\,O_2 \xrightarrow{>400°C} 3NO + 3H_2O \qquad (1.33)$$

Ammonia also decomposes to N_2 above 400 °C and, thus, is not available to reduce the NO:

$$2NH_3 + 3/2\,O_2 \xrightarrow{>400°C} N_2 + 3H_2O \qquad (1.34)$$

The desired reaction is favored below 400 °C since it has the lowest activation energy of the other two reactions. Once the temperature exceeds 400 °C, the reactions with the higher activation energy (greater temperature sensitivity) become favored and mixed products form.

1.5 PROMOTERS AND THEIR EFFECT ON ACTIVITY AND SELECTIVITY

The main catalytic component dominates the activity and the selectivity, but oxides or metal promoters, which may or may not be catalytic for the reaction of interest, can promote the activity or enhance one reaction over another. There are many examples in the industrial literature for a wide variety of petroleum and chemical processes (Bartholomew and Farrauto 2006), but we will focus on environmental applications. The addition of CeO_2 to a precious metal catalyst such as Pt or Pd in the automotive catalyst promotes the oxidation of hydrocarbons and carbon monoxide decreasing the minimum temperature needed to initiate catalytic oxidation during the cold-start portion of the automobile cycle requirement. It has other functions as well, which will be described in the catalytic converter chapter. Using TiO_2 as a carrier for Pt enhances the rate of oxidation reactions for some and finds use in low-temperature applications (Bollinger and Vannice 1996; Grisel and Nieuwenhuys 2001; Bond et al. 2006). One can speculate that the presence of TiO_2 as a carrier for metals promotes the oxidation rate by providing sites for O_2 to dissociate and adsorb, minimizing competition with CO for metal. The addition of a small amount of iron oxide as a promoter to a Pt-containing catalyst enhances the kinetics of the oxidation of CO in the presence of large excesses of H_2. This purification technique, called preferential oxidation or PROX, is used to produce fuel cell quality H_2 with less than 10 vppm of CO. In the presence of a Pt catalyst promoted with oxides of Fe, the O_2 added reacts selectively with the CO without oxidizing appreciable amounts of the H_2. The O_2 dissociatively chemisorbs on the iron oxide and reacts with the CO adsorbed on Pt. In the absence of the promoter oxide, the reaction rate is considerably slower and seems to obey LH-type kinetics where the CO and O_2 compete for Pt sites. Thus, a competitive reaction is altered to one that is noncompetitive and the kinetics are significantly enhanced (Liu et al. 2002; Korotkikh and Farrauto 2000).

One may also consider a mechanism in which the lattice oxygen associated with the TiO_2 or CeO_2 contributes the O atom directly to the species to be oxidized and is replenished by gas phase O_2. This mechanism is called Mars–van Krevlen and has been applied to reactions involving reducible oxide carriers as well as to base metal oxides that are active for both CO and hydrocarbon oxidation reactions (Mars and van Krevelen 1954). It has also been suggested for the oxidation of methane on Pd/Al_2O_3 (Hurtado et al. 2004; Avgouropoulos et al. 2002).

One more example will be instructive in demonstrating the important role of small amounts of promoters in altering catalyst selectivity. In the generation of fuel cell quality H_2, the gas stream is enriched in H_2 by promoting the oxidation of CO by water in the water gas shift reaction using a Pt-containing catalyst (1.35). For a Pt-only catalyst, appreciable amounts of undesired methane are formed during the reaction at 300 °C. The addition of 5–10% ZnO to the catalyst suppresses the methanation reaction (1.36) and avoids the consumption of H_2 and the large exotherm associated with the reaction (Korotkikh et al. 2003):

$$CO + H_2O \rightarrow H_2 + CO_2 \qquad (1.35)$$

$$CO + 3H_2 \rightarrow CH_4 + H_2O \qquad (1.36)$$

1.6 DISPERSED MODEL FOR CATALYTIC COMPONENT ON CARRIER: PT ON AL₂O₃

In many industrial reactions, the number of reactant molecules converted to products in a given time is directly related to the number of catalytic sites available to the reactants. It is, therefore, common practice to maximize the number of active sites by dispersing the catalytic components onto a surface. Maximizing the surface area of the catalytic components, such as Pt, Fe, Ni, Rh, Pd, CuO, PdO, CoO, and so forth, increases the number of sites upon which chemisorption and catalytic reaction can occur. It is common practice to disperse the catalytic components on a high-surface-area carrier, such as Al_2O_3,

SiO_2, TiO_2, SiO_2-Al_2O_3, zeolites, CeO_2, and so on. In some cases, but not for environmental applications, nonoxides such as high-surface-area carbons are also used as carriers. The carriers themselves usually are not catalytically active for the specific reaction in question, but they do play a major role in promoting the activity and selectivity as well as in maintaining the overall stability and durability of the finished catalyst.

The preparation and properties of these materials and their influence on catalytic reactions will be discussed in Chapter 2, but, for now, Al_2O_3 (the most commonly used carrier in catalysis especially for environmental applications) will be used to develop a model of a heterogeneous catalyst. Figure 1.7 is a drawing of a few select pores of a high-surface-area Al_2O_3.

The drawing shown has 20 and 100 Å (2–10 nanometers) pores into which Pt, or any other catalytic component, has been deposited by solution impregnation. The Pt particles or crystallites are represented as dots. When the Al_2O_3 is bonded to a monolithic honeycomb support, which will be described in Chapter 2, it is called a *washcoat*. The internal surface of the Al_2O_3 is rich in surface OH^- groups (not shown), depending on the type of Al_2O_3 and its thermal history. These OH^- species that cover the entire internal surface and are part of the walls of each pore represent sites upon which one can chemically or physically bond a catalytic substance. The physical surface area of the Al_2O_3 is the sum of all internal areas of the oxide from all the walls of each and every pore. It is upon these internal walls and at the OH^- sites that the catalytic components are bound. The catalytic surface area is the sum of all the areas of the active catalytic components in this example, Pt. The smaller the individual size of the crystallites of the active catalytic material (higher catalytic surface area), the more sites are available for the reactants to interact. As a rough approximation, one assumes the higher the catalytic surface area, the higher the rate of reaction for a process controlled by kinetics. This is often

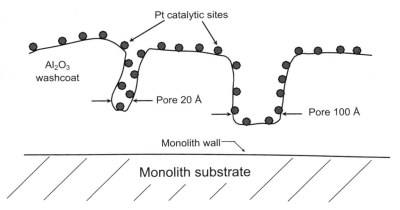

FIGURE 1.7. Conceptual, highly idealized model for catalytic sites dispersed on a high-surface-area Al_2O_3 carrier bonded to a monolith support.

the case, but there are exceptions in which a particular reaction is said to be structurally sensitive and the rate is a maximum when interacting with a catalytic crystal size of a specific size range.

The tiny Pt-containing particles shown in Figure 1.7 are dispersed throughout the porous Al_2O_3 carrier network and generate a high-Pt surface area. As it is shown, every Pt atom is available to the reactants. This model is highly idealized since the Pt sites typically vary in size and usually are not 100% dispersed as shown. This procedure maximizes the catalytic area but also introduces other physical processes such as mass transfer of the reactants to the catalytic sites. Each of these processes has a rate influenced by the hydrodynamics of the fluid flow, the pore size and structure of the carrier, and the molecular dimensions of the diffusing molecule (Bartholomew and Farrauto 2006; Morbidelli et al. 2001). These are discussed below.

1.7 CHEMICAL AND PHYSICAL STEPS IN HETEROGENEOUS CATALYSIS

To maximize reaction rates, it is essential to ensure the accessibility of all reactants to the active catalytic component sites dispersed within the internal pore network of the carrier. Consider a reaction in which CO and O_2 molecules are flowing through a bed of a heterogeneous catalyst. To be converted to CO_2, the following physical and chemical steps must occur:

1. CO and O_2 must make contact with the outer surface of the carrier (or washcoat in the case of a monolithic-supported catalyst) containing the catalytic sites. To do so, they must diffuse through a stagnant thin layer of gas or boundary layer in close contact with the catalyzed carrier. Bulk molecular diffusion rates vary approximately with $T^{3/2}$ and typically have "apparent" activation energies, $E_1 = 2$–4 Kcal/mole.

   The term "apparent" activation energy is used here to distinguish the physical phenomena of diffusion from the truly activated chemical processes that occur at the catalytic site. Diffusion reactions are physical phenomena and thus are not activated processes. The term "apparent" activation energy is a convenient term used to give a figure of merit for reaction sensitivity to temperature.

2. Since the bulk of the catalytic components are internally dispersed, most CO and O_2 molecules must diffuse through the porous network toward the active catalytic sites. The "apparent" activation energy for pore diffusion E_2 is approximately $1/2$ that of a chemical reaction or about 6–10 Kcal/mole.

3. Once molecule CO and O_2 arrive at the catalytic site, O_2 dissociates quickly and chemisorption of both O and CO occurs on adjacent catalytic sites. The kinetics generally follow exponential dependence on tem-

perature; i.e., exp $(-E_3/RT)$, where E_3 is the activation energy, which for chemisorption is typically greater than 10 Kcal/mole.

4. An activated complex forms between adsorbed CO and adsorbed O with an energy equal to that at the peak of the activation energy profile since this is the rate-limiting step. At this point, the activated complex has sufficient energy to convert to CO_2, which remains adsorbed on the catalytic site. Kinetics also follow exponential dependence on temperature, i.e., $(-E_4/RT)$, with activation energies typically greater than 10 Kcal/mole.

5. CO_2 desorbs from the site obeying exponential kinetics, i.e., $exp(-E_5/RT)$, with activation energies typically greater than 10 Kcal/mole.

6. The desorbed CO_2 diffuses through the porous network toward the outer surface with an "apparent" activation energy and kinetics similar to step 2.

7. CO_2 must diffuse through the stagnant layer and, finally, into the bulk gas. Reaction rates follow $T^{3/2}$ dependence. "Apparent" activation energies are also similar to step 1 are less than 2–4 Kcal/mole.

Steps 1 and 7 represent bulk mass transfer, which is a function of the specific molecules, the dynamics of the flow conditions, and the geometric surface area (outside or external area) of the catalyst/carrier. Pore diffusion, as illustrated in steps 2 and 6, depends primarily on the size and shape of both the pore and the diffusing reactants and product. Steps 3–5 are related to the chemical interactions of reactants and products (i.e., CO, O_2, and CO_2) at the catalytic site(s).

Any of the seven steps listed above can be rate limiting and control the overall rate of reaction. Let us take, for example, the conversion of any reactant to product using a heterogeneous catalyst. This is shown graphically in Figure 1.8.

Chemically controlled reactions determine the overall reaction rate initially, but their high sensitivity to temperature (high activation energy) relative to those controlled by diffusion is apparent by its steep rise in conversion with temperature. Pore diffusion then becomes rate limiting as the temperature increases, but eventually the least temperature-sensitive bulk mass transfer process becomes rate limiting. This is also demonstrated in Figure 1.9, which depicts the three relative rates of reaction. It shows that of the three rate-limiting phenomena, bulk mass transfer (BMT) is the fastest process at low temperatures but has a shallow dependence on temperature because of its lower "apparent" activation energy. Pore diffusion has a lower reaction rate than BMT because of its higher "apparent" activation energy, but its temperature dependence is greater. The highest temperature dependence occurs for a reaction controlled by chemical kinetics, but because of its higher activation energy, the rate is low at low temperatures relative to those controlled by diffusion (Morbidelli et al. 2001).

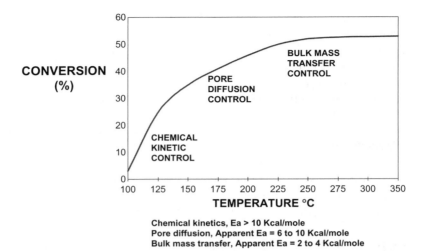

Chemical kinetics, Ea > 10 Kcal/mole
Pore diffusion, Apparent Ea = 6 to 10 Kcal/mole
Bulk mass transfer, Apparent Ea = 2 to 4 Kcal/mole

FIGURE 1.8. Conversion versus temperature profile illustrating regions for chemical kinetics, pore diffusion, and bulk mass transfer control.

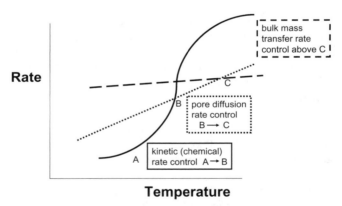

FIGURE 1.9. Relative rates of bulk mass transfer, pore diffusion, and chemical kinetics as a function of temperature. Chemical kinetics controls the rate between temperatures A and B. Pore diffusion controls from B to C, whereas bulk mass transfer controls at temperatures greater than C.

1.7.1 Reactant Concentration Gradients within the Catalyzed Washcoat

In the chemical kinetic control region, the reaction of chemisorbed CO with chemisorbed O is slow relative to diffusion and, thus, is rate limiting. As the temperature is further increased, control of the overall rate will shift to pore diffusion. Here the surface reaction between CO and O is faster than the rate gaseous CO and O_2 can be supplied to the sites and a concentration gradient exists decreasing within the washcoat. This is referred to as intraparticle diffusion in which the catalytic components deep within the washcoat are not

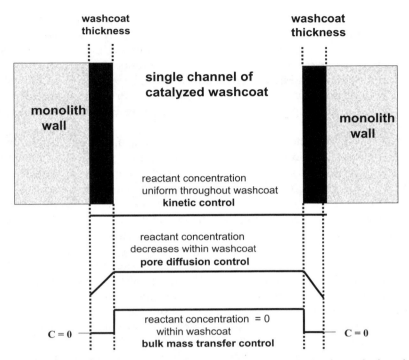

FIGURE 1.10. Reactant concentration gradients within a single channel of washcoat for three regimes controlling the rate of reaction.

being completely used or have an effectiveness factor less than 1. The effectiveness factor is the ratio of the actual rate versus the theoretical maximum rate and can be thought of as a measure of the utilization of the catalytic component(s). At higher temperatures, the rate of diffusion of the CO and O_2 from the bulk gas to the external surface of the washcoat is slow relative to the other processes and the rate becomes controlled by bulk mass transfer. In this regime, the CO and O_2 are converted to CO_2 as soon as they arrive at the external surface of the washcoat. The concentration of reactant and product is essentially zero at the external interface of the washcoat and the bulk fluid. The effectiveness factor is close to zero. Figure 1.10 graphically shows the relative gradients in concentration for reactants for the three rate-controlling processes in a catalyzed washcoat.

1.8   PRACTICAL SIGNIFICANCE OF KNOWING THE RATE-LIMITING STEP

The efficiency with which a catalyst functions in a process depends on what controls the overall reaction rate. If the kinetics of a process are measured and found to be in a regime where chemical kinetics are rate controlling, the catalyst should be made with as high a catalytic surface area as possible. This is accomplished by increasing the catalytic component loading and/or dispersion so that every catalytic site is available to the reactants. Furthermore, the catalytic components should be dispersed uniformly throughout the interior of the carrier with an effectiveness factor approaching 1. Process parameters, such as an increase in temperature, promote a reaction controlled by chemical kinetics.

When it is known that a process will have significant pore diffusion limitations, the carrier should be selected with large pores and locate the catalytic components as close to the surface as possible to improve the effectiveness factor. To enhance the transport rate, one can decrease the thickness of the washcoat to decrease the diffusion path of reactants and products. A temperature increase will have some effect on enhancing the rate but to a lesser extent than for those reactions controlled by chemical kinetics.

The rate of mass transfer is enhanced by increasing turbulence in the bulk gas and by increasing the geometric surface area (i.e., external area) of the washcoated monolithic catalyst. This can be accomplished by selecting a monolith with a high geometric surface area or density of channels per unit area. Clearly increasing the catalytic components surface area, the loading of the catalytic components, or the size of the pores will have no effect on enhancing the rate of mass transfer since these catalyst properties do not participate in the rate-limiting step. Also, temperature will have virtually no impact on the BMT rate.

REFERENCES

Avgouropoulos, G., Ioannides, T., Papadopoulos, C., Hocevar, S., and Matralis, H. "A comparative study of Pt/Al₂O₃, Au/Ce-Fe₂O₃ and CuO-CeO₂ catalysts for the selective oxidation of carbon monoxide in excess hydrogen," *Catalysis Today* 75: 157 (2002).

Bartholomew, C., and Farrauto, R. *Fundamentals of Industrial Catalytic Processes*, Second Edition, Wiley and Sons, Hoboken, NJ (2006).

Bollinger, M., and Vannice, M. A. "A kinetics and Drift study of low temperature carbon monoxide oxidation over Au-TiO2 catalysts," *APCAT B: Environmental* 8: 417 (1996).

Bond, G., Lous, C., and Thompson, D. *Catalysis by Gold* Imperial College Press, London, England (2006).

Grisel, R., and Nieuwenhuys, B. "A comparative study of the oxidation of CO and CH₄ over Au/MOₓ/Al₂O₃ catalysts," *Catalysis Today* 64: 69 (2001).

Hinshelwood, C. N. *The Kinetics of Chemical Change*, Oxford, Clarendon Press, London, England (1940).

Hougen, O., and Watson, K. "Solid catalysts and reaction rates general principles," *Industrial Engineering Chemistry* 35: 529 (1943).

Hurtado, P., Ordonez, S., Sastre, H., and Diez, F. "Development of a kinetic model for the oxidation of methane over Pd/Al$_2$O$_3$ at dry and wet conditions," *Applied Catalysis B: Environmental* 51: 229–238 (2004).

Korotkikh, O., and Farrauto, R. "Selective catalytic oxidation of CO in H$_2$: Fuel cell applications," *Catalysis Today* 62: 2 (2000).

Korotkikh, O., Ruettinger, W., and Farrauto, R. "Suppression of methanation activity by water gas shift reaction catalyst," U.S. patent 6, 562, 315 (2003).

Liu, X., Korotkikh, O., and Farrauto, R. "Selective catalytic oxidation of CO in H$_2$: A structural of Fe oxide promoted Pt/alumina catalyst," *Applied Catalysis B: Environmental* 226: 293 (2002).

Mars, P., and Van Krevelen, D. "Oxidations carried out by means of vanadium oxide catalysts," *Chemical Engineering Series* 3: 41 (1954).

Morbidelli, M., Garvriilidis, A., and Varma, A. *Catalyst Design: Optimal Distribution of Catalyst in Pellets, Reactors and Membranes*, Cambridge University Press, Cambridge, England (2001).

CHAPTER 1 QUESTIONS

1. Give everyday examples of how you encounter rate-limiting steps using analogies to intrinsic, pore diffusion, and mass transfer-controlled reactions.

2. Give examples of how in your everyday life you make decisions regarding selectivity.

3. List three examples for each of the petroleum and chemical products produced by catalytic processes.

4. Distinguish a homogeneous catalyst and process from a heterogeneous catalyst and process.

5. Why is it important to maximize the number of active sites in a heterogeneous catalyst?

6. What is the practical value in preparing a catalyst or in adjusting the process conditions by knowing the activation energy of a reaction?

7. How would you change the process conditions by knowing that in a hydrocarbon oxidation reaction, the hydrocarbon has a very large inhibition effect as in the equation:

$$\text{Rate} \sim k[(\text{Kads}_{HC}\ P_{HC})(\text{Kads}_{O2}\ P_{O2})]/(\text{Kads}_{HC}\ P_{HC})$$

8.
   a. How does the presence of a catalyst change the thermodynamic equilibrium constant of a reaction?

   b. Two reactions are thermodynamically feasible (both have negative free energies), but one is much more negative than the other. Can a catalyst direct the reactants to the least favorable?

9. What are the benefits of using a catalyst for abating emissions?

2 The Preparation of Catalytic Materials: Carriers, Active Components, and Monolithic Substrates

2.1 INTRODUCTION

The most common catalysts used in the environmental industry are metals or metal oxides, such as Pt, Pd, Rh, V_2O_5, and so on, that are dispersed on a high-surface-area carrier such as Al_2O_3, SiO_2, TiO_2, or crystalline alumina-silicates or zeolites. Once catalyzed, they are most often deposited onto the walls of monolithic supports (see Section 2.5).

It is important to understand that most industrial catalysts are prepared, activated, and conditioned by general procedures, but many critical details of the preparations are maintained as trade secrets to protect the proprietary nature of the suppliers' products. Although many patents are held by suppliers for various preparation procedures, there is no guarantee that what is disclosed or claimed is actually practiced. This is essential for each supplier to maintain a unique advantage in its product over the competitor's products. Consequently, what is described below are the more general, commonly used procedures.

2.2 CARRIERS

A carrier is usually a high-surface-area inorganic material containing a complex pore structure through which catalytic materials are deposited. At one time it was thought to provide only a surface to disperse the catalytic substance to maximize the catalytic surface area. However, it is now clear that it can and often does play a critical role in maintaining the activity, selectivity, and durability of the finished catalyst. By far the most common carriers are the high-surface-area inorganic oxides, most of which are described below.

Catalytic Air Pollution Control: Commercial Technology, by Ronald M. Heck and Robert J. Farrauto, with Suresh T. Gulati.
Copyright © 2009 John Wiley & Sons, Inc.

2.2.1 Al₂O₃

Alumina is by far the most commonly used carrier used in commercial environmental applications. Many different sources of alumina have varying surface areas, pore size distributions, surface acidic properties, composition of trace components, and crystal structures (Johnson 1990; Wefers and Misra 1987). Its properties depend on its preparation, purity, and thermal history. Various crystalline alumina hydrates are produced by precipitation from either acid or basic solutions. It is an amphoteric oxide soluble at pH levels above about 12 and below about 6. Within this broad pH range, it forms several different crystalline hydrates. For example, at a pH of 11, it forms a trihydrate species ($Al_2O_3 \cdot 3H_2O$) called bayerite, whereas at a pH of 9, it forms pseudo-boehmite, which is a monohydrate crystal ($Al_2O_3 \cdot H_2O$). On the more acidic side, such as a pH of 6, the precipitate lacks any definite long-range crystal structure and is classified as amorphous. The high surface area is created by heat treating or calcining in air, typically at about 500 °C where a network formed from Al_2O_3 particles 20–50 Å in diameter bonds together forming polymer-type chains. A scanning electron micrograph of a specially prepared gamma γ-Al_2O_3 crystal, magnified 80,000 times, is shown in Figure 2.1a) with a surface area of about 150 m²/g. One can see that the structure is composed of primary Al_2O_3 particles agglomerated forming highly porous networks.

Once precipitated, it is thoroughly washed to remove impurities from the precursor salts. If an acidic solution of Al^{+3} is neutralized with NaOH, the Na^+ should be removed by washing. Drying is usually performed at about 110 °C to remove excess H_2O and other salts containing volatile species such as NH_3 present from the precursor salts. Calcinations at different temperatures determine the final crystal structure, which in turn determines its chemical and physical properties.

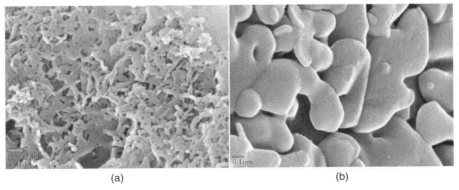

(a)   (b)

FIGURE 2.1. a) Scanning electron microscope (SEM) image of gamma (γ)-Al_2O_3 (80,000 magnification). b) SEM of alpha (α)-Al_2O_3 (80,000 magnification).

- The monohydrate (boehmite) and trihydrate (bayerite) alumina structures change as a function of the temperature (°C) in air (Wefers and Misra 1987).

$$\text{boehmite (monohydrate)} \rightarrow \text{(gamma) } \gamma\text{-Al}_2\text{O}_3 \text{ (500–800)}$$
$$\rightarrow \text{(delta) } \delta\text{-Al}_2\text{O}_3 \text{ (800–1000)}$$
$$\rightarrow \text{(theta) } \theta\text{-Al}_2\text{O}_3 \text{ (1000–1100)}$$
$$\rightarrow \text{(alpha) } \alpha\text{-Al}_2\text{O}_3 \text{ (>1100)} \tag{2.1}$$

$$\text{bayerite (trihydrate)} \rightarrow \text{(eta) } \eta\text{-Al}_2\text{O}_3 \text{ (300–800)}$$
$$\rightarrow \text{(theta) } \theta\text{-Al}_2\text{O}_3 \text{ (800–1150)}$$
$$\rightarrow \text{(alpha) } \alpha\text{-Al}_2\text{O}_3 \text{ (>1100)} \tag{2.2}$$

As the temperature is increased, between the ranges given above for a given Al_2O_3 crystal structure, there is a gradual dehydration, which causes an irreversible loss in the internal porosity or internal surface area (the material is being densified) and a loss in its surface OH⁻ or Bronsted acid sites. Continued heating causes a complete transformation to another crystal structure with a continuing loss in internal surface area and surface hydroxyl groups. For example, boehmite loses the bulk of its water below about 300 °C, during and after which it begins to sinter or lose internal surface area. At roughly 500 °C, it converts to γ-Al$_2$O$_3$, which typically has an internal surface area of 100–200 m^2/g. Continuous heating causes additional sintering and/or phase changes and loss of surface hydroxyl groups up to about 1,100 °C, where it converts to the lowest internal surface area structure (1–5 m^2/g). This is called alpha α-Al$_2$O$_3$. An SEM shown in Figure 2.1b) magnified by 80,000 clearly shows that its morphology is much more densely packed than the γ-Al$_2$O$_3$ shown in Figure 2.1a). Paralleling the structural changes that occur during heat treatment, the surface becomes progressively more dehydrated or more hydrophobic. These transformations obey time-temperature relationships and depend on the environment in which the material is exposed. The presence of steam, for example, greatly accelerates these transformations, so transformation temperatures can vary significantly.

The irreversible phase transitions of Al_2O_3 result in the collapse of the internal or physical surface area of the carrier, which occludes the active catalytic species within its pore structure, resulting in a loss of accessibility by the reactants. This is a primary source of thermally induced catalyst deactivation that will be discussed in Chapter 5. There is an advantage, however, in that the appearance of the different phases and irreversible changes in physical surface area allow the catalyst's thermal history to be accessed. This is important when studying the causes of catalyst deactivation in commercial installations because of excessive temperature exposure.

The presence of certain elements in the Al_2O_3 can have a profound influence on its physical surface-area retention after exposure to high temperatures. Small amounts of Na_2O present in the Al_2O_3 can enhance (or catalyze) the sintering of the Al_2O_3 and thus act as a flux (Wefers and Misra 1987). The rate of γ-Al_2O_3 sintering (ultimately to α-Al_2O_3) is enhanced by increasing amounts of Na_2O present.

In contrast, certain elements can act as "negative catalysts" and actually reduce the sintering rate. The presence of a few percent stabilizer such as La_2O_3 can greatly retard the sintering of the γ-Al_2O_3. This was of enormous importance in the development of high-temperature, durable catalytic converters for the automobile.

Many studies have been offering mechanisms explaining the stabilization effects of CeO_2 (Wan and Dettling 1986), La_2O_3 (Kato et al. 1987), BaO (Machida et al. 1988), and SiO_2 (Beguin et al. 1991). It is generally accepted that a solid solution of the stabilizing ion in the Al_2O_3 structure decreases the mobility of the Al and O ions, resulting in a reduction in the rate of sintering and/or phase transition.

2.2.2 SiO_2

The inertness of SiO_2 toward reacting with sulfur oxide (SO_x) compounds in exhaust streams makes it a suitable catalyst carrier. In contrast, Al_2O_3 is highly reactive with SO_3 and forms compounds that alter the internal surface of the carrier. This results in catalyst deactivation.

Alkaline solutions of silicate (pH > 12) can be neutralized with acid, resulting in the formation of silicic acid. This can then polymerize, forming a high-surface-area network with interconnecting pores of varying sizes:

$$SiO_4^{-2} \xrightarrow{\text{H}^+} [Si(OH)_4]_x \rightarrow SiO_2 \cdot H_2O \qquad (2.3)$$

Similar to Al_2O_3, it is then washed, dried, and calcined. High-surface-area SiO_2 materials can be 300–400 m^2/g. They have a small amount of chemically held water, giving rise to some surface acidic hydroxyl groups.

2.2.3 TiO_2

Because of its inertness to sulfate formation and its surface properties, TiO_2 is a preferred carrier for vanadia in selective catalytic reduction of NO_x from stationary sources where sulfur is often present especially in power plant exhausts. There are two crystal structures of importance: anatase and rutile. Catalytically, the anatase form is the most important in that it has the highest surface area (50–80 m^2/g) and is thermally stable up to about 500 °C. The rutile structure has a low surface area (<10 m^2/g), and it forms about 550 °C, resulting in occlusion of the vanadia and, ultimately, deactivation.

Anatase is formed by precipitation from titanate solutions or by the decomposition of organo-titanates followed by washing, drying, and mild calcination. Aqueous slurries are prepared, into which a honeycomb is dipped for washcoating. After drying, the washcoated monolith is calcined in air at 300–500 °C.

2.2.4   Zeolites

Natural occurring or synthetic alumina-silicate materials with well-defined crystalline structures and pore size are called *zeolites*. The Al_2O_3 and SiO_2 are bound in a tetrahedral structure with each Al and Si cation bonded to four oxygen anions. In turn, each O^{-2} is bonded to either a Si^{+4} or Al^{+3} in an arrangement similar to that shown below:

$$
\begin{array}{ccccc}
O & O & O & O & O \\
| & | & | & | & | \\
O-Si-O-Al-O-Si-O-Si-O-Al-O \\
| & | & | & | & | \\
O & O^- & O & O & O^- \\
  & H^+ & & & Na^+
\end{array}
$$

To maintain charge neutrality, an extra Na^+ or H^+ must be bonded to the AlO^-, giving rise to an exchangeable cation site. These sites are acidic when the cation is H^+, and they remain on the internal or external surface, accessible by reactant molecules. The pore structure dimensions of zeolites are between about 3 (0.3 nm) and 8 Å (0.8 nm), which fall into the range of molecular sizes. Any molecule with a larger cross-sectional area is prevented from entering the channel of the zeolite cage. It is for this reason that zeolites are often referred to as *molecular sieves*.

Treatment in dilute acid converts this structure to the H^+ exchanged state. In the pore of a zeolite, the surface is composed of AlO^-H^+ or AlO^-M^+ (for the case of the metal-exchanged site) that provides the active sites for the desired catalytic reactions.

Zeolites are of great significance because of their well-defined crystalline structures and surface properties. Modifications to the preparation alter the SiO_2/Al_2O_3, which in turn has a profound influence on the number of surface acid or exchangeable cation sites that are the origins of the active sites. As the SiO_2/Al_2O_3 ratio increases, there are fewer Al cations within the framework and, thus, fewer H^+ for exchange. One might be tempted to increase the Al content in the zeolite to increase the active sites; however, thermal stability decreases as the SiO_2/Al_2O_3 increases. Thus, zeolites offer a great variety of properties as carriers for various catalytic metal ions dependent on the reaction of interest and the conditions in which it must function.

The zeolite mordenite typically has a SiO_2/Al_2O_3 ratio of 5–10 with its most important aperture (or pore size) about 6.6 Å (0.66 nm) in size. Its aperture has only one dimension into which molecules can pass. It was one of the first zeolites used as a catalyst in environmental applications. Its acidic properties provided the active sites for the selective NO_x conversion with NH_3 in power plants in a technology called the selective catalytic reduction (SCR). Synthetic zeolites such as mordenite are generally prepared from aqueous solutions of alkali salts of aluminum and silicon and sometimes an organic amine, called a *template*, which aids in establishing a particular crystalline structure. Reaction is carried out usually in autoclaves at temperatures between 150 °C and 180 °C. Its structure is shown in Figure 2.2a). The zeolite is drawn as a stick figure with an O present in the midpoint of each stick that connects two Si or a Si and Al depending on the concentration of each. It is the O ion associated with the Al cation that provides either the H^+ or the metal cation. Mordenite has 12 member rings. Another zeolite, ZSM-5, has also been used in SCR. It has a SiO_2/Al_2O_3 of about 10 with one aperture 5.5 Å (0.55 nm) in diameter. Unlike mordenite, it is a three-dimensional structure with entry from all three sides. Each ring has 10 members. This is shown in Figure 2.2b). Beta zeolite (Figure 2.2c) also has 12 member rings and is three dimensional. But it has two apertures, one about 6.6 Å (0.66 nm) and the other about 5.6 Å (0.56 nm). Beta has been incorporated into diesel oxidation catalysts as a hydrocarbon trap and is

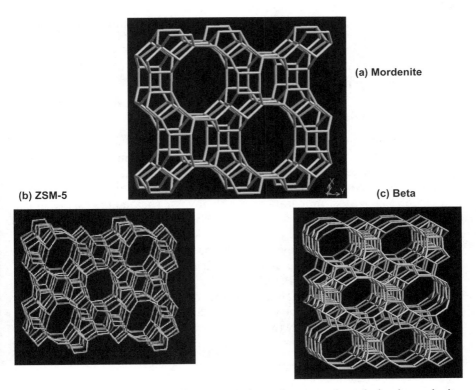

(a) Mordenite

(b) ZSM-5

(c) Beta

FIGURE 2.2. Three zeolites of importance in environmental catalysis: a) mordenite, b) ZSM-5, and c) beta. Figures copied from http://topaz.ethz.ch/IZA-SC/SearchRef. htm.

a carrier for Cu and/or Co for NO_x reduction with SCR. Both SCR and diesel oxidation catalysts will be discussed in their respective chapters.

Zeolite technology is extensively used in the chemical, petroleum, and gas separation industries. The reader is referred to a recent summary (Bartholomew and Farrauto 2006).

2.2.5 CeO_2

High-surface-area ceria (CeO_2) in combination with varying amounts of other metal oxides such as ZrO_2 has become an important oxygen storage component in the automotive catalytic converters. Its ability to store and release oxygen rapidly and reversibly allows it to moderate the conversion when the engine operates near the stoichiometric air-to-fuel point in the three-way automobile catalytic converter. It also contributes to the steam-reforming reaction during fuel-rich operation. Finally it serves as a carrier for some catalytic components promoting the activity for a variety of reactions, including the three-way automotive catalytic converter. It will be more fully described in the chapter describing the automotive three-way catalytic converter.

2.3 MAKING THE FINISHED CATALYST

2.3.1 Impregnation

The most common commercial procedure for dispersing the catalytic species within the carrier is by impregnating an aqueous solution containing a salt (precursor) of the catalytic element or elements (Komiyama 1985; Thomas and Brundrett 1980; Worstell 1992; Stiles 1983; Trimm 1980; LePage 1997; Schuth and Unger 1997). Most preparations simply involve soaking the carrier in the solution and allowing capillary and electrostatic forces to distribute the salt over the internal surface of the porous network. The salt generating the cations or anions containing the catalytic element are chosen to be compatible with the surface charge of the carrier to obtain efficient adsorption or, in some cases, ion exchange. For example, $Pt(NH_3)_2^{+2}$ salts can ion exchange with the H^+ present on the hydroxyl containing surfaces of Al_2O_3. Anions such as $PtCl_4^{-2}$ will be electrostatically attracted to the H^+ sites. The isoelectric point of the carrier (the charge assumed by the carrier surface), which is dependent on pH, is useful in making decisions regarding salts and pH conditions for the preparation.

2.3.1.1 Incipient Wetness or Capillary Impregnation. The maximum water uptake by the carrier is referred to as the *water pore volume*. This is determined by slowly adding water to a carrier until it is saturated, as evident by the beading of the excess H_2O. The precursor salt is then dissolved in an amount of water equal to the water pore volume. Once dried, the carrier pore structure is certain to contain the precise amount of catalytic species.

2.3.1.2 Electrostatic Adsorption. It is customary to use a precursor salt that generates a charge opposite to that of the carrier, which is determined by the pH at which the carrier has a point of zero charge (PZC). In weakly alkaline solutions, the surface charge on Al_2O_3 or SiO_2 is generally negative (with respect to the PZC) so any cation should preferentially adsorb uniformly over the entire surface. Cations such as Pd^{+2}, $Pt(NH_3)_2^{+2}$, and others derived from nitrates or oxalate salts are commonly used, whereas anions are generated from chloride precursor salts (e.g., $PdCl_2^{-2}$ from Na_2PdCl_4). The literature contains more complete discussion of surface phenomena and its importance to catalyst preparation (Regalbutto 2006; Park and Regalbutto 1995).

2.3.1.3 Ion Exchange. This method has the advantage of producing a highly dispersed catalytic component within the carrier. Assuming a carrier has a well-defined exchange capacity, a cation salt containing the catalytic species can exchange with the surface carrier cation. Ion exchange is most commonly used for zeolite catalysts.

It is common practice to first treat the acid form of the zeolite (H^+Z^- or simply HZ) with an aqueous solution containing NH_4^+ (NH_4NO_3) to form the ammonium-exchanged zeolite (NH_4Z). This can then be treated with a salt solution containing a catalytic cation forming the metal-exchanged zeolite (MZ):

$$HZ + NH_4^+ \rightarrow NH_4Z + H^+ \tag{2.4}$$

$$NH_4Z + M^+ \rightarrow MZ + NH_4^+ \tag{2.5}$$

The finished exchanged zeolite is washed and dried, and an aqueous slurry is prepared for coating onto the walls of the monolith.

2.3.2 Fixing the Catalytic Species

After impregnation, it is often desirable to fix the catalytic species so subsequent processing steps such as washing, drying, and high-temperature calcination will not cause significant movement or agglomeration of the well-dispersed catalytic species.

The pH of the solution is adjusted to precipitate the catalytic species in the pores of the carrier. For example, by presoaking Al_2O_3 in a solution of NH_4OH, the addition of an acidic Pd salt, such as $Pd(NO_3)_2$, will precipitate hydrated PdO on the surfaces of the pores within the carrier. In some select cases, H_2S gas can be used as a precipitating agent. After all the preparation processing steps, the catalyst is treated at high temperature in air to decompose and drive off the sulphur. This method has been used in the fixing of Rh onto Al_2O_3 in automobile exhaust catalysts:

$$Rh_2O_3 + H_2S \rightarrow Rh_2S_3 + H^+ \tag{2.6}$$

$$Rh_2S_3 + O_2 \xrightarrow{600°C} Rh + SO_2 \tag{2.7}$$

(Note: (2.6) and (2.7) not balanced for simplicity)

An alternative approach is the addition of reducing agents to precipitate catalytic species as metals within the pore structure of the carrier:

$$HCOOH + Pd^{+2} \rightarrow Pd + 2H^{+1} + CO_2 \tag{2.8}$$

This method is particularly effective for the precious metals because they are easily reduced to their metallic states. The advantage of the reducing agents mentioned above is that, upon subsequent heat treatment, they leave no residue.

2.3.3 Drying

Excess water and other volatile species are removed during forced air drying at about 110 °C.

2.3.4 Calcination

It is most common to calcine the catalyst in forced air to about 400–500 °C to remove all traces of decomposable salts used to prepare the catalyst.

2.4 NOMENCLATURE FOR DISPERSED CATALYSTS

The combination of the catalytic species supported on a carrier is presented by stating the amount and specific catalytic material, followed by the name of the carrier separated by a slash, i.e., 0.5% Pt/SiO_2, 1% Pd/Al_2O_3, 3% V_2O_5/TiO_2, and so on. It must be clearly understood that this only describes the general composition of the catalyst and does not describe the nature or the chemistry of the active sites responsible for the particular catalytic reaction. These are often not known in real processes.

2.5 MONOLITHIC MATERIALS AS CATALYST SUBSTRATES

Monolithic or honeycomb materials offer several advantages over more traditional pellet-shaped catalysts and, thus, are now the supports of choice for almost all environmental applications. It is a unitary structure composed of

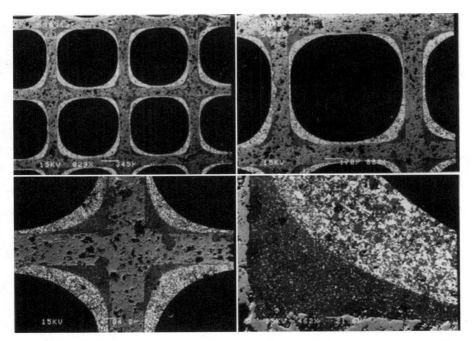

FIGURE 2.3. Optical micrographs of double-layered washcoated ceramic monoliths.

inorganic oxides or metals in the structure of a honeycomb with equally sized and parallel channels, which may be square, sinusoidal, triangular, hexagonal, round, and so on. Monoliths will be described in great detail later, but for now, it is sufficient to say monolithic structures are available as ceramic and metal with different channel dimensions and shapes.

Commercial ceramic monoliths have large pores and low surface areas (i.e., $0.3\,m^2/g$), so it is necessary to deposit the high-surface-area carrier often but not always containing the catalytic components onto the channel walls. The catalyzed coating is composed of a high surface carrier such as Al_2O_3 impregnated with a catalytic components. This is referred to as the *catalyzed washcoat*. Figure 2.3 shows various magnifications of a 400-cells-per-square-inch (cpsi) ceramic monolith with a double washcoat commonly used in automotive applications.

The washcoat can be seen deposited over the entire wall, but it is concentrated at the corners of the square-channel ceramic monolith. The thickness of the "fillet" depends primarily on the geometry of the channel and on the coating method. The pollutant-containing gases enter the channels uniformly and diffuse to and through the washcoat pore structure to the catalytic sites where they are converted catalytically. The amount of geometric surface area, upon which the washcoat is deposited, is determined by the number and diameter of the channels. There is a limit as to how much washcoat can be deposited, since too much results in a decrease of the effective channel diameter, thereby increasing the pressure drop to an unacceptable level.

Monoliths offer several engineering design advantages that have led to their widespread use in environmental applications. However, one of the most important is low-pressure drop associated with high flow rates. The monolith has a large open frontal area and with straight parallel channels offers less resistance to flow than a pellet-type catalyst. Low-pressure drop translates to lower compressor costs for stationary applications and to greater power savings for mobile sources. Other advantages are excellent attrition resistance, good mechanical and thermal shock properties, ability to make compact reactors, freedom in reactor orientation, and so on.

A more detailed description of monoliths will be presented in Chapters 7 and 9 on automotive and diesel substrates.

2.5.1 Ceramic Monoliths

Synthetic cordierite, $2MgO \cdot 2Al_2O_3 \cdot 5SiO_2$, is by far the most commonly used ceramic for monolithic catalyst support applications. The raw materials such as kaolin, talc, alumina, aluminum hydroxide, and silica are blended into a paste and extruded and calcined. For sizes up to about 11 in (27.94 cm) in diameter and 7 in (17.78 cm) in length, cell densities from about 9 to 1200 cpsi can be made. The conversion desired, the physical space available for the reactor, and engineering constraints, such as pressure drop (Kolb et al. 1993; Lachman and Williams 1992), are considered when designing the monolith size.

Some physical properties of selective ceramic monolith structures are given in Table 2.1.

An increase in cell density from 100 to 300 cpsi significantly increases the geometric area from 398 to 660 ft²/ft³ (157 to 260 cm²/cm³) but decreases the channel diameter from 0.083 to 0.046 in (0.21 to 0.12 cm). The wall of the ceramic drops in thickness from 0.017 to 0.012 in (0.04 to 0.03 cm). The increase in cell density does cause an increase in pressure drop at a given flow rate. For example, with a flow rate of 300 standard cubic feet per minute (SCFM) (8.4×10^6 cm³/min) through a monolith of 1 ft² (929 cm²) by 1 in thick (2.54 cm) the pressure drop for a 100 cpsi is about 0.1 in (0.254 cm) of water compared with about 0.3 for a 300-cpsi monolith. It should be understood that the pressure drop values are for the monolith without the catalyzed washcoat. Applying a washcoat will increase the pressure drop as a function of its thickness and surface roughness.

2.5.1.1 Thermal Shock Resistance.

By nature of its low thermal expansion coefficient ($10 \times 10^{-7}/°C$), cordierite undergoes little dimensional change when cycled over a wide temperature range. Thus, it resists cracking from thermal shock. Other materials such as mullite, zirconyl mullite, and alpha alumina have higher melting points but from five to ten times the thermal expansion coefficients, making these much less resistant to thermal shock. This

TABLE 2.1. Physical Properties of Ceramic Monoliths.

Cell density (cpsi)	Hydraulic channel diameter (inches)	Open frontal area (%)	Geometric surface area (ft^2/ft^3)	Pressuer drop (inches of water)	Ceramic wall thickness (inches)
64	0.099	70	340	0.075	0.019
100	0.083	69	398	0.095	0.017
200	0.059	72	576	0.210	0.012
300	0.046	65	660	0.300	0.012
400	0.044	71	852	—	0.006

Pressure drop for 300 standard cubic feet per minute of gas flow through an uncoated monolith 12 inches × 12 in × 1 in (thick).

is a critical parameter for automotive applications where operational changes create large axial temperature gradients within the honeycomb during normal driving.

The washcoat influences the thermal shock resistance of the monolith (especially during rapid temperature changes) because it expands to a larger extent than the monolith. These differences must be factored into the design of the finished catalyst to maintain a good adherent coating without substantially weakening the monolith. The particle size of the carrier and the thickness of the washcoat are two key parameters that must be optimized.

2.5.1.2   Mechanical Strength. Monoliths are made with axial strengths of over 3,000 pounds per square inch (psi). They must be resistant to both axial and mechanical perturbations experienced in automotive, truck, and aircraft applications. The high mechanical integrity is derived from the physical and chemical properties of the raw materials and from the final processing after extrusion.

2.5.1.3   Melting Point. The melting point of cordierite is over 1,300 °C—far greater than temperatures expected for modern environmental applications. The materials are also resistant to the harsh environment of high-temperature steam, sulfur oxides, and oil additive constituents, which are present in the exhaust of many applications. New materials, such as silicon carbide and aluminum titanates, have also been developed that expand the temperature range capabilities well beyond cordierite for applications like diesel particulate filters where temperatures may exceed 1,500 °C during soot regeneration.

2.5.1.4   Catalyst Compatibility. Automotive ceramic monoliths have well-designed pore structures (of about 3–4 microns) that allow good chemical and mechanical bonding to the washcoat. The chemical components in the ceramic are immobilized, so little migration from the monolith into the catalyzed washcoat occurs.

2.5.2   Metal Monoliths

Monoliths made of high-temperature-resistant, aluminum-containing steels are becoming increasingly popular as catalyst supports, mainly because they can be prepared with thinner walls than a ceramic. This offers the potential for higher cell densities with lower pressure drop (Emitec 1992). The wall thickness of a 400-cpsi metal substrate used for automotive applications is about 25% lower than its ceramic counterpart, i.e., 0.0015–0.002 in (0.004–0.005 cm) compared with 0.006–0.008 in (0.015–0.02 cm), respectively. The open frontal area of the metal is typically about 90% versus 70% for the ceramic with the same cell density. Its thermal conductivity is also considerably higher (about 15–20 times) than the ceramic, resulting in faster heat-up. This property is particularly important for oxidizing hydrocarbons and carbon monoxide emissions when a vehicle is cold. Metal substrates also offer some advantages for installation of the converter in that they can be directly welded into the exhaust system.

A common design is that of corrugated sheets of metal welded or wrapped together into a monolithic structure. In most cases, the washcoat is deposited onto a roughened metallic surface of the already fabricated monolith by a dipping or controlled vacuum deposition. There are some new designs in which the washcoat is first deposited onto a flat, roughened surface prior to final wrapping into the monolith shape.

Adhesion of the oxide-based washcoat to the metallic surface and corrosion of the steel in high-temperature steam environments were early problems that prevented their widespread use in all but some specialized automotive applications. Surface pretreatment of the metal has improved the adherence problems, and new corrosion-resistant steels are allowing metals to slowly penetrate the automotive markets. They are currently used extensively for low-temperature applications such as NO_x in power plants, O_3 abatement in airplanes, CO and VOC abatement, and abatement of oil-based emissions from restaurants. Being electrically conductive, they have found use in electrically heated catalytic converters for rapid conversion of emissions during startup. They are finding greater use in high-performance vehicles where response time is critical during acceleration. Here the low-pressure drop of the metal monolith is its most desirable property. They are usually more expensive than their ceramic counterpart.

2.6   PREPARING MONOLITHIC CATALYSTS

The catalyzed carrier is made into an acidified aqueous slurry with a solids content from 30% to 50%. The mixture is ball milled for at least 2 hrs to reduce the particle size (typically 5–20 microns) and to generate the proper rheology for the subsequent monolith dipping operation.

The preparation of the finished catalyst involves dipping the monolith into the slurry. The monolith generally has some wall porosity or surface roughness to ensure adhesion of the catalyzed washcoat. The excess slurry is air blown to clear the channels and dried at about 110 °C. The final step is calcination, which bonds the catalyzed washcoat securely to the monolith walls and decomposes and volatilizes the excess preparation components. Calcinations are performed in air at temperatures between 300 °C and 500 °C. Great care must be taken to avoid rapid heat-up since H_2O trapped in the micropores can build up sufficient pressure to crack the monolith. Furthermore, exothermic reactions caused by decomposing salts can cause localized high temperatures within the catalyst material that can accelerate sintering.

An alternative approach is to first coat the monolithic honeycomb with the uncatalyzed carrier, followed by drying and calcining. It is then dipped it into a solution containing the catalytic salts. This method relies on the electrostatic adsorption of the salts to the carrier surface. The supported catalyst is then dried and calcined to its final state.

A recent method uses a vacuum on one end of the monolith, and the slurry is sucked up to a given axial length. It is then turned up to 180 degrees where the uncoated portion of the monolith is similarly coated. This technology allows zone coating where the front of the bed may contain a different formulation than the back.

Some manufacturers that use metal substrates for lower temperature nonautomotive application precoat them with the washcoat prior to wrapping or forming the metal into the monolithic structure. It is common to pretreat the surface of the metal to generate roughness. This process ensures good bonding to the washcoat. The metal is then calcined to produce a stable bond between the surface and the washcoat. The major advantage is coating uniformity, with no corners containing high localized amounts of washcoat.

2.7   CATALYTIC MONOLITHS

Technology has been developed that allows the vanadia and titania plus additives such as silica to be extruded directly into a low-cell-density honeycomb (Lachman and Williams 1992). Organic additives such as polyvinyl alcohol are sometimes added as plasticizers to aid in the extrusion process. The final material is calcined at temperatures high enough to burn out the organics but low enough to prevent the vanadia and titania from sintering or degrading chemically.

A homogeneous extruded catalyst, in the form of a monolith having a moderately high surface area, is now available for special applications. Since the entire monolith is a catalyst, a higher concentration of active component is present than would be for a similar washcoated honeycomb. Furthermore, the absence of a washcoat eliminates a coating manufacturing step. The monoliths are calcined in their final preparation step to about 600 °C to retain the surface area. This process produces a substantially weaker material, requiring thick walls and low cell densities (<100 cpsi).

The extrusion technology is similar to that used for cordierite substrates. For some cases, a paste of TiO_2 powder is first extruded, calcined at 500 °C, impregnated with ammonium vanadate/oxalic acid, and calcined to the finished product. The first applications are with V_2O_5/TiO_2 and zeolites for selective catalytic reduction of NO_x for stationary pollution abatement applications.

The major disadvantage to date of these monoliths is the inability to make high-cell-density material of sufficient strength to maintain mechanical integrity in operation.

2.8   CATALYZED MONOLITH NOMENCLATURE

The common nomenclature is to state the washcoat loading in grams per cubic inch (g/in^3) and in grams of catalytic component (especially for precious metals) per cubic foot (g/ft^3) of monolith. The monolith or honeycomb volume is calculated based on its cross-sectional area and length. It does not take into account the number of channels per square inch. Consequently, the cell density is always stated when describing the finished catalyst dimensions. This is also important when a space velocity is stated.

2.9   PRECIOUS METAL RECOVERY FROM MONOLITHIC CATALYSTS

Almost all commercially available environmental catalysts contain expensive precious metals that can be recovered and reused after the catalyst has lost its effectiveness. The hydrometallurgical procedure simply involves crushing the spent catalyst and treating it with acid to dissolve only the ceramic components. This leaves an insoluble precious-metal-rich residue that is then further purified by chemical procedures (Harris 1993; Mishra 1993).

The pyrometallurgical method uses smelting in which the ceramic floats to the top as a slag. The highly dense precious metals alloy sink and alloy to an added metal (i.e., Cu or Fe) at the bottom of the smelter. Here they are chemically removed and further purified for reuse.

REFERENCES

Auerbach, S. M., Carrado, K. A., and Dutta, P. K., Editors. *Handbook of Zeolite Science and Technology*. Dekker, New York (2003).

Bartholomew, C., and Farrauto, R. J. *Fundamentals of Industrial Catalytic Processes*. Second Edition. Wiley and Sons, New York (2006).

Beguin, B., Garbowski, E., and Primet, M. "Stabilization of alumina toward thermal sintering by silicon addition," *Journal of Catalysis* 127: 595–604 (1991).

Emitec Product Literature, *The New Generation of Metallic Catalytic Converter Substrates*. Lohmar, Germany (1992).

Harris, G. "A review of precious metal refining," *Precious Metals 1993*, Editor. Mishra, R., pp 351–374, Seventeenth International Precious Metals Conference, Newport, RI (1993).

Johnson, M. J. "Surface area stability of aluminas," *Journal of Catalysis* 123: 245–259 (1990).

Kato, A., Yamashita, H., Kawagoshi, H., and Matsuda, S. "Preparation of lanthanum beta alumina with high surface area by co-precipitation," *Communications of the American Ceramic Society* 70(7): C157–161 (1987).

Kolb, W. B., Papadimitriou, A. A., Cerro, R., Leavitt, D. D. and Summers, J. "The ins and outs of coating monolithic structures," *Journal of Chemical Engineering Progress* 89(2): 61–67 (1993).

Komiyama, M. "Design and preparation of impregnated catalysts," *Catalysis Reviews: Science and Engineering* 27(2): 342–372 (1985).

Lachman, I., and Williams, J. "Extruded monolithic catalyst supports," *Catalysis Today* 14: 317–329 (1992).

LePage, J. F. "Preparation of solid catalysts," pp 49–72 in *Handbook of Heterogeneous Catalysis*. Editors Ertl, G., Knozinger, H., and Weitkamp, J., VCH Weinheim, Germany (1997).

Machida, M., Eguchi, K., and Arai, H. "Preparation and characterization of large surface area $BaO \cdot 6Al_2O_3$," *Bulletin of the Chemical Society of Japan* 61: 3659–3665 (1988).

Mishra, R. "A review of platinum group metals recovery from automobile catalytic converters," *Precious Metals 1993*. Editor Mishra, R., pp 449–474, Seventeenth International Precious Metal Conference, Newport, RI (1993).

Park, J., and Regalbutto, J. "A simple accurate determination of oxide PZC and the strong buffering effect of oxide surfaces at incipient wetness," *Journal of Colloid and Interfacial Science* 175: 239–252 (1995).

Regalbutto, J., Editor. *Catalyst Preparation*. Francis and Taylor, New York (2006).

Schuth, F., and Unger, K. "Precipitation and co-precipitation," pp 72–86 in *Handbook of Heterogeneous Catalysis*. Editors Ertl, G., Knozinger, H., and Weitkamp, J., VCH Weinheim, Germany (1997).

Stiles, A. *Catalyst Manufacture: Laboratory and Commercial Preparations*. Dekker, New York (1983).

Thomas, A., and Brundrett, C. "Catalyst development: lab to commercial scale," *Chemical Engineering Progress* 76(6): 41–45 (1980).

Trimm, D. *Design of Industrial Catalysts*. Elsevier Scientific Amsterdam, The Netherlands (1980).

Wan, C., and Dettling, J. "High temperature catalyst and compositions for internal combustion engines," US. Patent 4,624,940 (1986).

Wefers, K., and Misra, C. *Oxides and Hydroxides of Aluminum*. Alcoa Laboratories, East Saint Louis, IL (1987).

Worstell, J.H. "Succeed at catalyst upgrading," *Chemical Engineering Progress* 88(6): 33–39 (1992).

CHAPTER 2 QUESTIONS

1. Using a search engine such as "Google", find the following zeolites and their physical and chemical properties such as aperture size, Si-to-Al ratio, degree of acidity (related to Al), and number of tetrahedral sites per ring. What is the significance of some of these properties?

   a. Faujasite (FAU):

   b. Mordenite (MOR):

   c. ZSM-5 (MFI):

2. Compare ceramic and metal monoliths as supports for catalyzed washcoats. When would you use each? Compare porous particulates and monoliths. State the advantages and disadvantages.

3. What are the performance consequences of changing the following monolith properties?

   a. Smaller channel diameter.

   b. Thinner monolith wall thickness.

   c. Decreasing cells/in^2 with increased channel diameter?

3 Catalyst Characterization

3.1 INTRODUCTION

The characterization of a heterogeneous catalyst is the quantitative measure of its physical and chemical properties assumed to be responsible for its performance in a given reaction. These measurements have value in the preparation and optimization of a catalyst and, even more importantly, in elucidating mechanisms of deactivation and subsequent catalyst design to minimize such deactivation. Physical properties such as the pore size, surface area, and morphology of the carrier, as well as the geometry and strength of the monolithic catalyst support, must be well defined for the given end-use application. Similarly, determining the composition, structure, and nature of the carrier and the active catalytic components and their changes during the catalysis process is a critical goal in characterization. Several reference sources can provide additional detail regarding the equipment and procedures (ASTM Committee D-32 on Catalysis 1988; Anderson and Dawson 1976; Delannay 1984; Deviney and Gland 1985; Farrauto and Hobson 1992; Bartholomew and Farrauto 2006; Knozinger 2003).

The importance of the physical and chemical properties can be appreciated when we restate the fundamental steps involved in a heterogeneous catalytic process. (These were stated in more detail in Chapter 1.) The example used refers to a catalyzed carrier supported on a monolith.

1. *Bulk mass transport of the reactants from the bulk fluid to the outside surface of the catalyst.* This is strongly influenced by the hydraulic diameter of the honeycomb channel, the geometric surface area of the catalyst, and the flow turbulence.

2. *Diffusion or transport of reactant(s) to active sites through the pore structure of the catalyst.* This depends on the pore size and washcoat thickness of the catalyst. A well-optimized catalyst will have a sufficiently large pore size and proper washcoat thickness to permit easy access of reactant and product molecules to and from the active sites, respectively.

3. *Chemisorption of reactants(s) onto the catalytic active sites.* This is dependent on the number and nature of the active site(s) and on the chemistry of the adsorbing molecule(s).

4. *Chemical conversion of the chemisorbed species to products.* This step depends on the number and nature of the activated complexes formed by the chemisorbed molecule and the catalytic site.

5. *Desorption of products(s) from active sites.* This step depends on the number of adsorbed species and on the strength of their bonds to the active surface.

6. *Diffusion or transport of product(s) through the pore structure.* This step is affected mostly by the size and shape of the diffusing molecule and by the pore and thickness of the washcoat.

7. *Bulk diffusion of the products from the outside surface of the washcoat to the bulk fluid.* The process is influenced by the hydraulic diameter of the monolith channel and by the turbulence of the flow.

During the conversion process, several physical and chemical properties have to be well defined to produce an optimized catalytic system. The most important properties for environmental monolithic catalysts are discussed below.

3.2   PHYSICAL PROPERTIES OF CATALYSTS

3.2.1   Surface Area and Pore Size

Surface area, pore size, pore size distribution, pore structure, and pore volume of the carrier are among the most fundamentally important properties in catalysis because the active sites are present or dispersed throughout the internal surface through which reactants and products are transported. The size and number of pores determine the internal surface area. It is usually advantageous to have a high surface area (large number of small pores) to maximize the dispersion of catalytic components. However, if the pore size is too small, diffusional resistance becomes a problem.

3.2.2   Surface Area and Pore Size Measurements

A standardized procedure for determining the internal surface area of a porous material (ASTM D4641-87 1988) with surface areas greater than 1 or $2\,m^2/g$ is based on the adsorption of N_2 at liquid N_2 temperature onto the internal surfaces of the carrier.

Each adsorbed molecule occupies an area of the surface comparable with its cross-sectional area ($16.2\,\text{Å}^2$). By measuring the number of N_2 molecules adsorbed at monolayer coverage, one can calculate the internal surface area. This plot is shown in Figure 3.1a). The adsorption of N_2 first rapidly rises with pressure and then flattens in the general region where monolayer coverage is

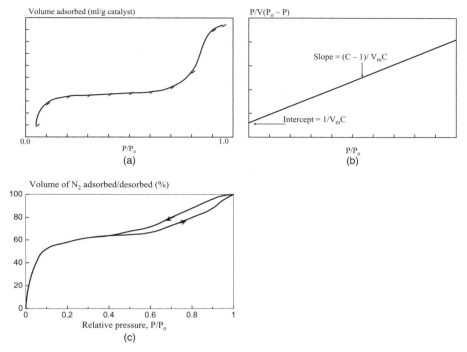

FIGURE 3.1. a) Adsorption isotherm for nitrogen for BET surface area measurement. b) Linear plot of BET equation for surface area measurement. c) Nitrogen adsorption/desorption isotherm for pore size measurement.

occurring. At high relative N_2 partial pressures, coverage beyond a monolayer occurs, as does condensation of liquid N_2 in the pores, giving rise to the large increase in volume adsorbed. The Brunauer, Emmett, and Teller (BET) equation (3.1) describes the relationship between the volume adsorbed at a given partial pressure and the volume adsorbed at monolayer coverage:

$$P/V(P_o - P) = 1/V_m C + (C-1)P/CP_o V_m \tag{3.1}$$

where

P = partial pressure of N_2

P_o = saturation pressure at the experimental temperature

V = volume adsorbed at P

V_m = volume adsorbed at monolayer coverage

C = constant

This equation can be stated in the linear form $Y = sx + b$, where P/P_o is x and the term on the left is Y as plotted in Figure 3.1b). The intercept b is equal to $1/V_m C$, and the slope s is $(C-1)/V_m C$. The most reliable results are obtained at relative pressures (P/P_o) between 0.05 and 0.3.

The same equipment can be used to determine the pore size distribution of porous materials with diameters less than 100 Å (10 nm), except that high relative pressures are used to condense N_2 in the catalyst pores. The procedure involves measuring the volume adsorbed in either the ascending or the descending branch of the BET plot at relative pressures close to 1.0. Figure 3.1c) shows the volume of N_2 adsorbed and desorbed as the pressure is increased and decreased, respectively. Capillary condensation occurs in the pores in accordance with the Kelvin equation [Eq. (3.2)].

$$\ln(P/P_o) = -2\sigma V \cos\theta / rRT \qquad (3.2)$$

where

σ = surface tension of liquid nitrogen

θ = contact angle

V = molar volume of liquid nitrogen

r = radius of the pore

R = gas constant

T = absolute temperature

P = measured pressure

P_o = saturation pressure

Hysteresis between the adsorption and desorption isotherms in Figure 3.1c) at relative pressures 0.6–0.9 is observed with carriers having a significant volume of mesopores (diameters between 20 (2 nm) and 500 Å (50 nm)). It results because pores, which fill at a given pressure during adsorption, require a lower pressure to empty during desorption. The form of the Kelvin equation given above describes the desorption isotherm, and it is the preferred one for calculations of the pore size distribution.

3.2.3   Pore Size by Mercury Intrusion

For materials with pore diameters greater than about 30 Å (3 nm), the mercury intrusion method is preferred (ASTM D4284-83 1988). The penetration of mercury into the pores of a material is a function of applied pressure. At low pressures, mercury penetrates the large pores, whereas at higher pressures, the smaller pores are progressively filled. Due to the nonwetting nature of mercury on oxide carriers, penetration is met with resistance. The Washburn equation [see Eq. (3.3)] relates the pore diameter d with the applied pressure P:

$$d = -4\gamma \cos\theta / P \qquad (3.3)$$

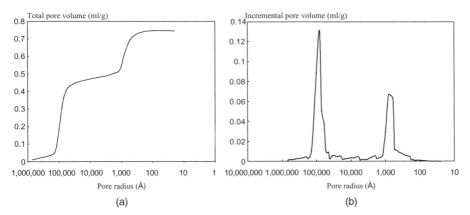

FIGURE 3.2. a) Mercury penetration as a function of pore size or catalyst. b) Differential porosimetry for a porous catalyst.

The wetting or contact angle θ between mercury and the solid is usually $130°$, and the surface tension of the mercury γ is 0.48 N/m. Pressure is expressed in atmospheres and d in nanometers (10 Å). This technique is satisfactory for pores down to a 30 Å (3 nm) diameter; however, this is a function of the instrument capability. The maximum diameters measured are usually 10^6 Å. Typical pore size distribution data are shown in Figure 3.2a), where the integral penetration of mercury into the pores is plotted as a function of applied pressure. The differential curve [Figure 3.2b)] clearly shows a bimodal pore distribution with mean pore diameters at $10,000$ (1000 nm) and 1000 Å (100 nm). A nitrogen desorption isotherm is required to obtain an accurate measure in the region below 100 Å (10 nm) (see Section 3.2.2).

3.2.4 Particle Size Distribution of the Carrier

The size of the agglomerated particles that make up the carrier must be compatible with the surface roughness and macropore structure of the honeycomb surface. Slightly acidified aqueous mixtures of the carrier powders are ball milled to produce the slurry from which the honeycomb will be coated. The rheology of the slurry depends on the solids content, the particle size and surface properties of the carrier material, and the acidity of the slurry.

Sieves of various mesh sizes have been standardized, and thus, one can determine particle size ranges by noting the percentage of material, usually based on weight, that passes through one mesh size but is retained on the next finer screen (ASTM D4513-85 1988). Sieves are stacked with the coarsest on top and the underlying screens progressively finer. A precise weight of catalyst material is added to the top screen. The stack of sieves is vibrated, allowing the finer particles to pass through coarser screens until retained by those screens finer in opening than the particle size of the material of interest. Each fraction is then weighed, and a distribution is determined.

This method is reliable only for particles larger than about 40 µm. Below this size, sieving is slow and charging effects influence measured values. Sophisticated instrumentation is available for measuring the distribution of finer particles. Methods include electronic counting, light scattering, image analysis, sedimentation, centrifugation, and volume exclusion. An example of electronic counting is described in ASTM D4438-85 (1988). In this method, the dried powder from the slurry is suspended in an electrolyte and is pumped through a tube containing a small orifice. An electric current passes through this tube, and as individual particles pass through the orifice, a fraction of the current is interrupted. This fractional change in current is proportional to the particle size. The magnitude of the change in current flow is subdivided over the range of sizes limited by the size of the orifice. The particle diameters are calculated on the equivalent sphere of the excluded volume, and assuming constant density for the particles, the results are commonly recorded as weight percentage as a function of incremental particle size.

More recently, laser techniques have been found effective (ASTM D4464-85 1988). A He–Ne laser beam is passed through an aqueous suspension of particles, and it is diffracted in proportion to the radius of the particle. A distribution for an Al_2O_3 carrier is illustrated in Figure 3.3. The values on the left axis indicate the cumulative % of particles, the size of which is displayed

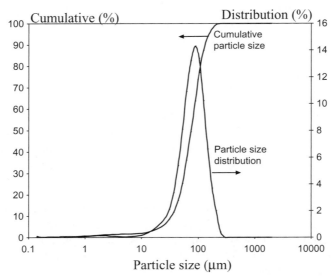

FIGURE 3.3. Particle size measurement using laser light scattering analysis.

on the x-axis. The right axis shows the % distribution for a given particle size.

3.2.5 Washcoat Thickness

Optical microscopy is the method used most frequently to obtain washcoat thickness directly. A portion of washcoated monolith is mounted in epoxy and sliced to obtain a cross section. The contrast between washcoat and monolith is sufficient to permit optical thickness measurements. A typical cross section of a washcoat on a ceramic auto exhaust monolith is shown in Figure 2.3 (see Chapter 2).

3.2.6 Mechanical Strength of a Monolith

Monoliths, particularly when used in a stacked mode (e.g., in stationary pollution abatement), must resist crushing axially. For automobile and truck applications, and ozone abatement in aircraft, resistance to vibration and radial strength is important during the process of housing the catalytic system in a steel container.

The strength measurement for monoliths is simple. A single unit representative of the lot is placed between parallel plates of a device capable of exerting compressive stress, and the force necessary to crush the material is noted. The monolith can be placed within the plates so as to measure axial or radial crush strength. A sufficient number of units must be tested to obtain proper statistics.

3.2.7 Adhesion

Washcoated monolithic catalysts are subjected to high gaseous flow rates and to rapid temperature fluctuations. As a result, adhesion loss is a likely cause for concern. The most common technique is to subject the washcoated monolith to a jet of gas that simulates the linear velocities anticipated in use and to note the weight change of the catalyst due to erosion. One can also collect the attrited particles and correlate their weight with the weight of washcoat initially bound to the monolith.

Monolithic materials are frequently subjected to substantial thermal gradients (thermal shocks) in start-up and shutdown, as in auto exhaust emission control. Frequent thermal shocks cause the washcoat to delaminate due to the expansion difference between it and the monolith. This delamination is most pronounced when the monolith is metallic because of its greater thermal expansion compared with an oxide washcoat.

One can evaluate this phenomenon by cycling the catalyzed monolith between temperature extremes anticipated in service and by noting weight losses. Catalyzed monolith can be mounted on a rotating "carousel" that moves in and out of streams of heated gas.

3.3   CHEMICAL AND PHYSICAL MORPHOLOGY STRUCTURES OF CATALYTIC MATERIALS

3.3.1   Elemental Analysis

The proper combination of chemical elements is essential in catalysts for optimum performance. More often than not, small amounts of promoter oxides intentionally added (often less than 0.1%) can influence activity, selectivity, and life.

Catalyst suppliers must meet many manufacturing specifications that apply to the proper chemical analysis of a catalyst. Carriers are derived from raw materials, which contain various impurities. Some impurities can be detrimental to catalytic performance and, therefore, must be removed. Impurities such as alkali and alkaline earth compounds, if used in excess, act as fluxes, causing sintering or loss of surface area in Al_2O_3. When added in the proper amount, the same impurities can enhance stability against sintering or, in some cases, improve selectivity. Hence, it is not just the presence of impurities but also the manner in which they have been introduced that is important. Obviously, one has greater control when starting with a relatively pure material to which predetermined amounts of promoters can be added.

The quantitative procedures used to analyze catalysts are no different than those for any other chemical material. Special procedures are often needed to dissolve catalysts in preparation for analysis—particularly refractory materials such as certain noble metals and ceramics. Refer to ASTM D4642-86 for procedures for analyzing Pt supported on Al_2O_3.

3.3.2   Thermal Gravimetric Analysis (TGA) and Differential Thermal Analysis (DTA)

Thermal gravimetric analysis measures the weight change of a material upon heating in a highly sensitive balance. The energy change (heat adsorbed or liberated) associated with any transformation is measured with the differential thermal analysis mode. It is common practice for both to be measured simultaneously in a combined TGA/DTA unit. A few milligrams of catalyst are loaded into a quartz pan suspended in the microbalance. A controlled gas flow and temperature ramp is initiated, and a profile of weight and energy change versus temperature is recorded. Frequently, TGA units are equipped with a mass spectrometer so the off-gases from the catalyst can be measured as a function of temperature. Catalyst salt components are impregnated, usually from aqueous solution, into a porous carrier. After deposition onto the carrier, the combination is calcined in air to a temperature sufficient to decompose the salt:

$$Ba(CH_3COO)_2 + 4O_2 \rightarrow BaO + 4CO_2 + 3H_2O \qquad (3.4)$$

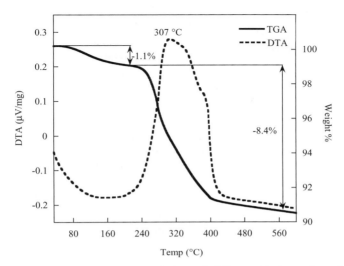

FIGURE 3.4. Thermal gravimetric analysis and differential thermal analysis of the decomposition of barium acetate on ceria.

This is shown in Figure 3.4 where barium acetate is decomposed after it has been impregnated into a high-surface-area carrier such as cerium oxide.

The solid line shows the weight change associated with heating the ceria after impregnation with the acetate salt of barium. A small weight loss (measured on the right axis) is observed between 80 °C and 240 °C due to the evaporation of H_2O. The dotted line (measured on the left axis) shows the DTA profile with a negative slope consistent with the endothermic evaporation of H_2O. Decomposition of the acetate begins around 250 °C where the weight change (solid line) is observed to decrease sharply until about 400 °C where it plateaus. Associated with this exothermic decomposition event (positive slope) is the heat liberated in the DTA profile showing a maximum at 307 °C. The ceria was preconditioned to be stable during this procedure. This simple test allows production to establish the optimized conditions for drying and calcining the catalyst during its preparation. This particular material, BaO/CeO_2, is being considered for adsorbing or trapping NO_x in a lean-burn engine exhaust, which will be discussed in the section on controlling emissions from diesel engines.

This technique is also used to establish the temperature and environmental conditions under which a metal will oxidize (temperature-programmed oxidation or TPO) or a metal oxide is reduced in an environment such as H_2 or CO (temperature-programmed reduction or TPR).

In Chapter 5 on catalyst deactivation, other examples of how these useful tools can be used to understand catalyst deactivation and regeneration modes will be described.

3.3.3   The Morphology of Catalytic Materials by Scanning Electron Microscopy (SEM)

In Chapter 2, the reader was introduced to a variety of carrier materials upon which the catalytic components are dispersed to maximize the number of sites available for reactants to chemisorb. Clearly, for environmental applications, γ-Al_2O_3 is the most common material. It has a surface area of over $200\,m^2/g$ with a highly porous structure as shown in the SEM of Figure 2.1a). Its pores range from about 20 to $200\,\text{Å}$ (2–20 nm). As it experiences elevated temperatures, it slowly transforms into other lower porosity and more crystalline intermediary structures terminating with alpha α-Al_2O_3, which has the lowest surface area and is the most crystalline as shown in Figure 2.1b). The structural transformations of γ-Al_2O_3 will be discussed in more detail in Chapter 5 "Catalyst Deactivation" because they represent one of the most commonly occurring deactivation modes for supported catalysts.

3.3.4   Location and Analysis of Species within the Catalyst by Electron Microscopy

Environmental monolithic catalysts contain a variety of catalytic metal and metal oxide promoters in addition to the Al_2O_3 carrier. Each has a specific function and must have its proper place within the washcoat. Some components must be in intimate contact with each other to contribute to electronic promoting effects, but they could poison other components and cause deactivation. Such is the case with certain metals that form undesirable alloys or compounds. For this reason, some catalyst manufacturers use multiple washcoat layers of specific compositions to maximize effectiveness while minimizing the poisoning effects. This process can be observed when the electron microscope is used in the backscatter mode (wavelength-dispersive analysis or WDS) as shown in Figures 3.5a–d). In the scanning mode, the electron beam focused on the sample is scanned by a set of deflection coils. Backscattered electrons or secondary electrons emitted from the sample are detected. As the electron beam passes over the surface of the sample, variations in composition and topology produce variations in the intensity of the secondary electrons. The raster of the electron beam is synchronized with that of a cathode ray tube, and the detected signal then produces an image on the tube. Spot or area analysis is also possible when an electron microscope is equipped with an energy dispersive analyzer (EDX) or WDS. The bombardment of a sample with electrons generates X rays characteristic of the elements present. Thus, the EDX can determine the composition of any portion of the sample, whereas the WDS permits the mapping of the location of species present. This is particularly important when foreign matter is present either from a contamination problem in manufacturing or by poisoning during a catalytic process. The electron microprobe is another form of electron microscopy that is extremely important for metal location studies requiring high resolution (Delannay 1984;

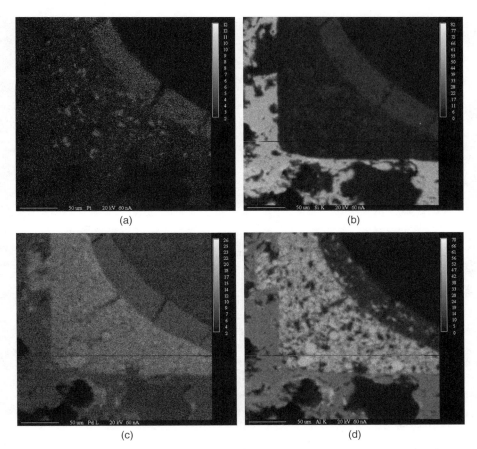

FIGURE 3.5. a) Wavelength dispersive analysis of Pt in a double-layered washcoat. b) Wavelength dispersive analysis of SiO_2 in a double-layered washcoat. c) Wavelength dispersive analysis of Pd in a double-layered washcoat. d) Wavelength dispersive analysis of Al_2O_3 in a double-layered washcoat.

Bartholomew and Farrauto 2006). It is similar to the scanning electron microscope; however, its primary function is to detect characteristic X rays produced by the electron beam interaction with the specimen. The X-ray emissions can be used to determine the elemental composition of the specimen quantitatively and to detect the location of a particular element within the morphology or topological structure of the specimen.

Examples of the use of wavelength dispersive analysis is shown in a series of micrographs shown as Figures 3.5a)–d) for a double-coated layered washcoat. The top coat is composed of Pt dispersed on SiO_2 [Figures 3.5a) and 3.5b)], and the Pd is dispersed on Al_2O_3 as the bottom coat [Figures 3.5c) and d)].

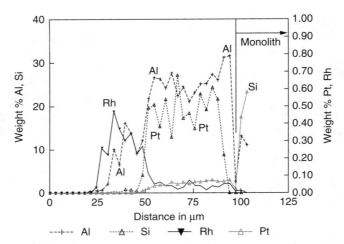

FIGURE 3.6. Electron microprobe showing a two-washcoat-layer monolith catalyst. The top layer is Rh on Al_2O_3, and the bottom layer is Pt on Al_2O_3.

Figure 3.6 shows a microprobe line profile of a ceramic monolith upon which has been deposited two different washcoat layers. The top layer from roughly 25 to 50 microns contains Rh on Al_2O_3, whereas the second or bottom layer from 50 to 100 microns contains Pt on Al_2O_3. The total thickness of the two layers is 100 microns as evidenced by the decrease and the absence of Pt at depths greater than 100 microns and by the appearance of Si, which is a component of the monolith substance.

3.3.5 Structural Analysis by X-Ray Diffraction

Provided that a material is sufficiently crystalline to diffract X rays and is present in an amount greater than 1%, X-ray diffraction (XRD) can be used for qualitative and quantitative analysis. Crystal structures possess planes made by repetitive arrangements of atoms, which are capable of diffracting X rays. The angles of diffraction differ for the various planes within the crystal. Thus, every compound or element has its own somewhat unique diffraction pattern. Comparing the patterns allows differentiation of various structures.

3.3.6 Structural Analysis of Al_2O_3 Carriers

Figure 3.7 shows the XRD patterns of two Al_2O_3 structures: amorphous γ-Al_2O_3 and α-Al_2O_3. The gamma (γ) is the high-surface-area, lower temperature structure, whereas the alpha (α) is produced at higher temperatures and has low surface area (Farrauto and Hobson 1992; Anderson and Dawson 1976; Bartholomew and Farrauto 2006). Below crystallite sizes of 50 Å, a well-defined X-ray pattern will not be obtained. Materials with crystallites smaller than this

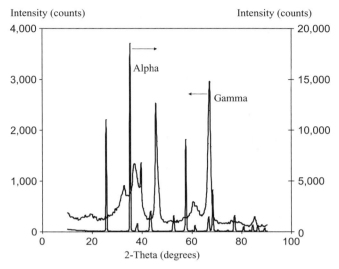

FIGURE 3.7. X-ray diffraction patterns of γ- and α-Al_2O_3.

are more precisely called amorphous since they possess no long-range order to diffract X rays. Structures in this class, which are common for freshly prepared catalysts, must be characterized by other techniques such as those listed below.

3.3.7 Dispersion or Crystallite Size of Catalytic Species

3.3.7.1 Chemisorption. One of the most frustrating facts facing the catalytic scientist is that often when a structure has a definite XRD pattern and can be well characterized, it usually has less-than-optimum activity. This is because most catalytic reactions are favored by either amorphous materials or extremely small crystallites. Small crystals can agglomerate or grow and produce large crystallites that diffract X rays and thus generate easily read patterns; however, the atoms of the small crystals are buried within the larger crystal, making them inaccessible to reactant molecules. Frequently, the purpose of the preparation technique is to disperse the catalytic components in such a way as to maximize their availability to reactants.

$$\% \text{ Dispersion} = \frac{\text{Number of catalytc sites on the surface}}{\text{Theoretical number of sites present as if atoms}} \times 100$$

(3.5)

When this is done effectively, only small crystals are present and the diffraction of X rays is minimized, because little long-range structure exists. As the crystals get smaller and smaller, the XRD peaks get broader and broader and eventually are undetectable above the background. However, it is these "X-ray-amorphous" species that are often the most active for a given catalytic reaction.

Standardized techniques exist (ASTM D3908-82 1988; Anderson and Dawson 1976; Bartholomew and Farrauto 2006) for obtaining information regarding the distribution and number of catalytic components dispersed within or on the carrier. Selective chemisorption can be used to measure the accessible catalytic component on the surface by noting the amount of gas adsorbed per unit weight of catalyst. The stoichiometry of the chemisorption process must be known to estimate the available catalytic surface area. One assumes that the catalytic surface area is proportional to the number of active sites. A gas that will selectively chemisorb only on to the metal and not the support is used under predetermined conditions. Hydrogen and carbon monoxide are most commonly used as selective adsorbates for many supported metals. There are reports in the literature of instances in which gases such as NO and O_2 have been used to measure catalytic areas of metal oxides; however, due to difficulty in interpretation, they are of limited use.

The measurements are usually carried out in a static vacuum system similar to that used for BET surface area measurements. The pressure of gas above the sample is increased, and the amount adsorbed is measured at equilibrium as shown in Figure 3.8a). When there is no further adsorption with increasing pressure (shown in the flat portion of the figure), the catalytic surface is saturated with a monolayer of adsorbate. Noting the number of molecules of gas adsorbed and knowing its stoichiometry with the surface site (i.e., 1 H per metal site), one can determine the catalytic surface area by multiplying molecules adsorbed by cross-sectional area of the site and dividing by the weight of catalyst used in the measurement. For example, the cross-sectional area of Pt is 8.9 Å^2 and Ni 6.5 Å^2.

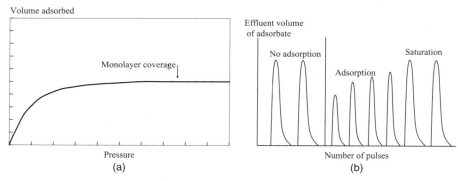

FIGURE 3.8. a) Chemisorption isotherm for determining the surface area of the catalytic component. b) Pulse chemisorption profiles for the dynamic chemisorption method.

The static vacuum technique is time consuming, so alternative methods have been devised. For years, a dynamic pulse technique has been used in which a pulse of adsorbate such as H_2 or CO is injected into a stream of inert gas and passed through a bed of catalyst. Gas adsorption is measured by comparing the amount injected with the amount passing through the bed unadsorbed.

As shown from left to right in Figure 3.8b), the first two pulses are used for calibration and bypass the catalyst sample. The second set of pulses that pass through the catalyst are first diminished due to adsorption. Once saturation or monolayer coverage is reached, no further adsorption from the gas phase occurs. The amount adsorbed is found by the difference in areas under the peaks compared with those under the calibration pulses. The major difference between dynamic and static methods is that the former measures only that which is strongly adsorbed, whereas the latter, performed under equilibrium conditions, measures strong and weakly chemisorbed species. Thus, static techniques usually give better dispersion results.

3.3.7.2 Transmission Electron Microscopy. In transmission electron microscopy (TEM), a thin sample, usually prepared by a microtome, is subjected to a beam of electrons. The dark spots on the positive of the detecting film correspond to dense areas in the sample that inhibit electron transmission. These dark spots form the outline of metal particles or crystallites, and hence, their sizes can be determined.

Figure 3.9 shows a TEM of sintered Pt, about 500 Å (50 nm) in size, dispersed on TiO_2. Crystallites of Pt about 100 Å (10 nm) in size, dispersed on CeO_2, are shown in the TEM of Figure 3.10. Using image analysis, a size

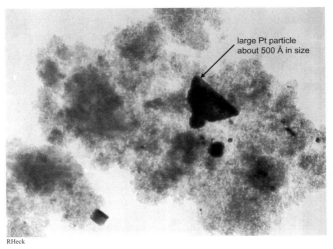

large Pt particle about 500 Å in size

RHeck

FIGURE 3.9. Transmission electron micrograph of Pt on TiO_2.

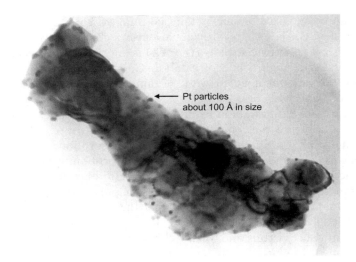

FIGURE 3.10. Transmission electron micrograph of Pt on CeO_2.

distribution and average crystallite size can be calculated. Assuming a spherical shape for the crystallites, the percent dispersion (i.e., ratio of surface atoms to total atoms in the crystallite ×100) can be calculated. It should be understood that this technique measures only a small fraction of the catalyst, so obtaining data representative of the entire sample is difficult. However, it does give a direct measure of the catalytic components.

3.3.8 X-Ray Diffraction

The larger the crystals of a given component, the sharper the peaks on the XRD pattern for each crystal plane. The Scherrer equation relates the breadth B at half-peak height of an XRD line due to a specific crystalline plane to the size of the crystallites L:

$$B = k\lambda/L\cos\theta \tag{3.6}$$

where

λ = X-ray wavelength
θ = diffraction angle
k = constant usually equal to 1

As the crystallite size increases, the line breadth B decreases (Bartholomew and Farrauto 2006). Figure 3.11 shows the sharp XRD pattern of CeO_2 treated at 1,500 °C. The CeO_2 treated at 800 °C has a broad profile indicative of a much

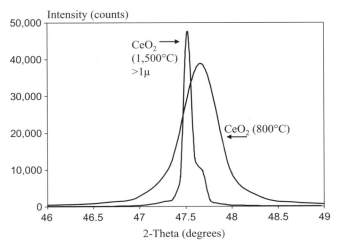

FIGURE 3.11. X-ray diffraction profile for different crystallite sizes of CeO_2.

smaller crystallite size. Cerium oxide is a constituent in automobile catalytic converters, and its crystal size impacts its ability to function as an oxygen storage component (see Chapter 7, "Automotive Catalysis").

3.3.9   Surface Composition of Catalysts by X-ray Photoelectron Spectroscopy

The composition of the catalyst surface, as opposed to its bulk, is of critical importance since this is where the reactants and products interact. It is on these surfaces that the active sites exist and where chemisorption, chemical reaction, and desorption take place. Furthermore, poisons deposit in the layers of the surface of the catalyst and, thus, knowing its concentration will give valuable insight into activity and deactivation. Techniques such as XRD and electron microscopy measure the structure and/or chemical composition of catalysts extending below the catalytic surface. The composition of the surface is usually different from that of the bulk, and thus, its analysis must be carried out by techniques specific to the surface.

The tools available for surface composition characterization (Bartholmew and Farrauto 2006; Deviney and Gland 1985) are X-ray photoelectron spectroscopy (XPS), Auger spectroscopy (AES), ion scattering spectroscopy (ISS), and secondary ion mass spectroscopy (SIMS). XPS is used more widely than the others for studying the surface composition and oxidation states of industrial catalysts, and thus, its application will be discussed in limited detail.

The acronym "XPS" refers to the technique of bombarding the surface with X-ray photons to produce the emission of characteristic electrons. These are measured as a function of electron energy. Because of the low energy of the

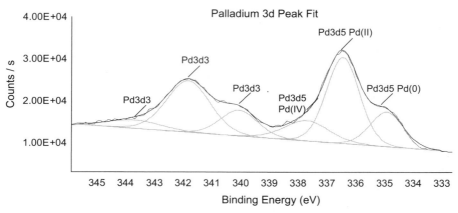

FIGURE 3.12. An XPS spectrum of various oxidation states of Palladium on Al_2O_3 is shown.

characteristic electrons, the depth to which the analysis is made is only about 40 Å (4 nm). The composition of this thin layer as a function of depth can be determined by removing or sputtering away top layers and by analyzing the underlying surfaces. Several important catalytic properties have been studied by this technique, including the oxidation state of the active species, the interaction of a metal with an oxide carrier, and the nature of chemisorbed poisons and other impurities.

If a small amount of a gas phase impurity, i.e., S, Cl, P, and so on, deposits on the surface of the catalyst, its concentration is not likely to be detected by bulk chemical analysis. XPS allows only the top monolayer to be analyzed and, thus, allows insight into its effect on the performance. Another example is the oxidation state of the surface catalytic component. Palladium can exist in three oxidation states, all of which have different activities toward specific reactions. For example, Pd metal is the active component in hydrogenation reactions, whereas higher oxidation states of Pd are active for hydrocarbon oxidations. XPS allows us to determine the respective oxidation states. In Figure 3.12, the Pd is found in three oxidation states: 25% as Pd (o), 56% as Pd (+2), and 19% as Pd (+4).

3.3.10 The Bonding Environment of Metal Oxides by Nuclear Magnetic Resonance (NMR)

Zeolites have long been of great importance in the chemical and petroleum industries for their unique pores size and acidity, both of which have a strong influence on catalytic activity. They are now finding greater use in environmental applications as adsorbents for hydrocarbons during cold exhaust conditions and as carriers for metal cations that generate high activities for special reactions such as the selective NO_x reduction.

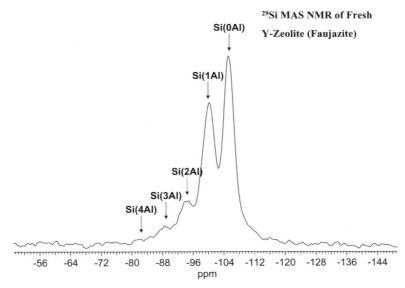

FIGURE 3.13. NMR profile of a Y faujasite with varying amounts of Si–O–Al bridges.

The zeolites of greatest interest in the environmental catalysis are composed of varying ratios of SiO_2 and Al_2O_3. Si(+4) and Al(+3) are bound together through an oxygen bridge forming a tetrahedral structure. NMR allows one to determine the number of Si–O–Al bridges. In Figure 3.13, an NMR profile of a Y faujasite zeolite is shown.

The first major peak occurs at −106-ppm units indicating no bridges Si(OAl), whereas the second occurs at −100-ppm units indicating the number of single Si–O–Al bridges Si(1Al). The peaks at progressively lower ppm units show fewer bridges Si(2Al), Si(3Al), Si(4Al) as is evident by their low intensity. Thus, the distribution of tetrahedral Si–O–Al sites that represent the active sites can be determined. The higher the Si/Al ratio in the zeolite, the more tetrahedral bridges are expected to exist. When most zeolites undergo high-temperature exposure, they deactivate and the tetrahedral Si–O–Al bridge is broken. One can follow deactivation by the decrease in peak intensities. The importance of this will become clear when deactivation of zeolites is discussed in Chapter 5.

Atomic nuclei spin in much the same ways as electrons. The movement of an electric charge generates a magnetic field. The silicon isotope (^{29}Si) with an uneven number of protons and neutrons is present in significant amounts in all compounds, including zeolites. In such a case, a dipole (separation of charge) exists that can interact with an external magnetic field. The dipole can exist in different spin states, and at a particular frequency of the imposed magnetic field, it can be elevated to a higher spin state during "resonance."

Upon relaxation of the external field, the dipole decays to a lower energy state and emits energy that is characteristic of that material. The frequency in which "resonance" occurs is influenced by the elements to which the isotope is bonded. So the number of Si–O–Al bridges will alter the frequency of the resonant energy of ^{29}Si, creating the profile shown in Figure 3.13.

3.4 TECHNIQUES FOR FUNDAMENTAL STUDIES

This chapter has attempted to show the methodology commonly used for the characterization of real commercial catalysts. Many techniques are designed to obtain more fundamental characterizations that are beyond the scope of this book. The reader should review external sources (Bartholomew and Farrauto 2006).

All of the techniques described in this chapter are performed after the catalyst is removed from the reactor; that is, *ex situ*. It is very desirable, but also extremely difficult, to measure catalyst properties *in situ* during an actual process. Once removed from a catalytic environment, the nature of the adsorbed surface species will be changed, and what is measured will significantly differ from the actual catalytic surface. Nevertheless, many of its other properties (e.g., surface area, pore size, crystalline structure, crystallite size, and chemical composition) are essential data in determining which factors influenced its performance. Infrared DRIFT spectroscopy allows the nature of the adsorbed species to be observed before, during, and after catalytic reaction. In Figure 3.14, CO adsorption is studied on the surface of a automobile catalyst

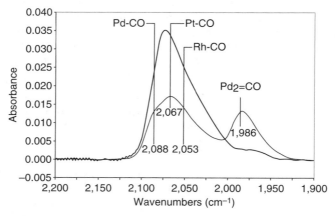

FIGURE 3.14. DRIFT spectra of CO chemisorbed on different precious metal particles of catalysts prepared in different ways. The CO chemisorption followed by FT-IR measurements was performed at room temperature after the catalysts were treated at 400°C for 1 hr. with a 7% H_2 in Ar gas.

containing Pt, Pd, and Rh prior to the addition of O_2. The nature of the bonding of the CO to each element is shown. The addition of O_2 will remove the CO from the most active sites.

Catalytic scientists have created an exciting new approach toward the study of fundamental catalytic reactions. So-called operand spectroscopy permits the characterization of the surface of a catalyst during real catalysis (Banares 2005).

REFERENCES

American Society for Testing and Materials (ASTM): Committee D-32 on Catalysis, Second Edition, Philadelphia, PA (1988).

ASTM: D3663-85, "Standard test method for surface area of catalysts," pp 3–6 (1988).

ASTM: D4641-87, "Standard test method for determination of nitrogen adsorption/ desorption isotherm for pore size measurements," pp 41–44 (1988).

ASTM: D4284-83, "Standard test method for determination of pore volume distribution by mercury porosimetry," pp 26–29 (1988).

ASTM: D4513-85, "Standard test method of particle size of catalytic materials by sieving," pp 71–72 (1988).

ASTM: D4438-85, "Standard test method for distribution of catalytic materials by electronic counting," pp 63–65 (1988).

ASTM: D4464-85, "Standard test method for particle size by laser light scattering," pp 66–67 (1988).

ASTM: D4642-86, "Standard test method for chemical analysis of Pt on Al_2O_3 catalysts," pp 14–17 (1988).

ASTM: D3908-82, "Standard test method for hydrogen chemisorption of supported Pt on Al_2O_3 catalyst by volumetric method," pp 13–16 (1988).

Anderson, R., and Dawson, P. *Experimental Methods in Catalytic Research*. Academic Press, New York (1976).

Banares, M. "Operando methodology: Combination of in-situ spectroscopy and simultaneous activity measurements under catalytic reaction conditions," *Catalysis Today* 100: 71 (2005).

Bartholomew, C., and Farrauto, R. J. *Fundamentals of Industrial Catalytic Processes*. Chapter 3. Second Edition. Wiley and Sons, New York (2006).

Delannay, F. *Characterization of Heterogeneous Catalysts*. Dekker, New York (1984).

Deviney, M. L., and Gland, J. *Catalyst Characterization Science: Surface and Solid State Chemistry, ACS Symposium Series No. 288*. American Chemical Society, Washington, DC (1985).

Farrauto, R. J., and Hobson, M. C. "Catalyst characterization," in *Encyclopedia of Physical Science and Technology*. Vol 2. Editor Meyers, R. A. pp 735–761. Academic Press, New York (1992).

Knozinger, H. "Catalyst characterization" in *Encyclopedia of Catalysis*. Vol. 2, pp 142–182. Editor Horvath, I, Wiley-Interscience, Hoboken, NJ (2003).

CHAPTER 3 QUESTIONS

1. Why are BET surface areas and Hg measured pore sizes important in heterogeneous catalysis?
2. What do X-ray diffraction patterns tell us about the components of the catalyst?
3. How may a TGA/DTA be used in
    a. Catalyst preparation?
    b. Activation of a catalyst?
    c. Burn-off of hydrocarbons oils deposited on the catalyst?
4. How is NMR used in determining the thermal history of zeolites?
5. What can we learn from CO chemisorption measurements of metal-supported catalysts?

4 Monolithic Reactors for Environmental Catalysis

4.1 INTRODUCTION

For pollution abatement applications, it is common to use a monolithic honeycomb-washcoated supported catalyst to minimize the pressure drop associated with high flow rates. The honeycomb is usually inside a steel housing and is physically fixed in the exhaust. This allows the process effluent gases to pass uniformly through the channels of the honeycomb. The incoming pollutant-laden stream is often hot from the upstream process, or via a burner in the process line or a heat exchanger, it can be preheated to a sufficient temperature to initiate the catalytic reactions.

4.2 CHEMICAL KINETIC CONTROL

At the entrance to the honeycomb bed, it is common for the reaction to be controlled by chemical kinetics rather than by diffusion to or within the catalyst pore structure. For example, in the automobile catalytic converter, the incoming gases are relatively cold when the vehicle has been dormant for an extended period of time, so the reaction rate is governed by the chemical kinetics. In addition to the sensible heat from the gas stream, the catalyst surface heats up due to the heat generated by the catalytic reactions (if exothermic). When sufficiently hot, the rate will be determined by mass transfer. Many chemical and petroleum processes, where selectivity is important, operate under conditions that give maximum yield of the desired product and may be specifically designed to operate under chemical or pore diffusion control. Knowledge of all the rate-controlling steps throughout the entire process is essential in designing the catalyst and the reactor.

In the laboratory, when screening a large number of catalyst candidates, it is important to measure activity at low conversion levels to ensure the catalyst is evaluated in the intrinsic or chemical rate-controlling regime. The optimized

Catalytic Air Pollution Control: Commercial Technology, by Ronald M. Heck and Robert J. Farrauto, with Suresh T. Gulati.
Copyright © 2009 John Wiley & Sons, Inc.

intrinsic activity of the catalyst is essential even for those processes that will operate under mass transfer control. Mass transfer control occurs only when the intrinsic activity is higher than transport effects. At low conversion levels, heating effects due to exothermic or endothermic reactions are small, so the temperature within the bed is essentially equal to that of the bulk gas. Since monoliths are made up of a series of parallel channels, there is essentially no external heat transfer, and the reactor operates almost adiabatically. Therefore, once significant conversions of greater than about 20% occur, the catalyst surface and gas within the bed quickly heat up. Good laboratory practice is to maintain all conversions below about 20% for kinetic measurements. For highly exothermic reactions, i.e., $\Delta H > 50$ Kcal/mole, measurements should be made at conversions no greater than 10%. Once the proper catalyst material is identified and mass transfer effects are understood, the reactor design work can be initiated. Whenever possible, it is important to understand under what conditions the catalyst will function in the "real world" so physical and chemical properties of the catalyst can be modified.

Regardless of the field conditions in which the catalyst will function, it is always necessary to determine the chemical kinetics of a reaction. The discussion below is designed to give some simple and useful methods, and it should serve as a refresher from earlier physical chemistry or reaction engineering studies.

The net rate ($Rate_{net}$) of a reaction is equal to $Rate_{forward} - Rate_{reverse}$. At equilibrium, the forward and reverse rates are equal so the net rate is zero. Thus, to measure only the forward reaction, you need to be far removed from equilibrium so the reverse reaction is negligible. By only feeding in reactants and operating at conversions <10%, the reverse reaction can usually be ignored provided you are still far from equilibrium. To test this, determine the expected final concentration of reactants and products at <10% conversion and fit into the usual equilibrium expression of product concentrations divided by reactant concentrations. This value (K') should be no more than 0.1 or 10% of the concentrations at equilibrium (K_e) at the temperature of the kinetic measurement.

$$aA + bB = cC + dD$$

Ratio in kinetic measurement: $K' = [C]^c [D]^d / [A]^a [B]^b$

$$K'/K_e < 0.1$$

A material balance across any reactor gives the following equation assuming a one-dimensional, plug-flow, steady-state operation:

$$\frac{d(vC)}{dz} = -r \tag{4.1}$$

where

v = velocity (cm/s)
C = molar concentration (g-moles/cc)
z = length (cm)
r = rate of reaction (g-moles/cc-s)

When the conversion or the reactant concentration is low, the reactor is considered isothermal; hence,

$$v\frac{dC}{dz} = -r \tag{4.2}$$

Assume the oxidation of ethane to CO_2 and H_2O in a large excess of O_2 in a fixed bed of catalyst:

$$C_2H_6 + 7/2O_2 \rightarrow 2CO_2 + 3H_2O \tag{4.3}$$

When a large excess of any one reagent is present, the reaction rate is insensitive to the small changes it undergoes in the process. In this case, given the large excess of O_2, we can assume the rate is independent of O_2. This particular reaction is known to obey first-order kinetics (pseudo-zero order in O_2), so the reaction rate is expressed as follows:

$$v\frac{dC_{(C_2H_6)}}{dz} = -k'C_{(C_2H_6)} \tag{4.4}$$

where k' = the apparent rate constant.
    Integrating this equation between the reactor inlet (i) and outlet (o) gives

$$\ln\frac{C_o}{C_i} = -\frac{k'z}{v} = -k't \tag{4.5}$$

where t is the actual residence time (seconds) the C_2H_6 spends in the catalyst bed.
    In heterogeneous fixed-bed catalysis, the residence time is defined in terms of the volumetric flow and volume of catalyst, both of which can be easily measured. Dividing the volumetric gaseous flow rate at STP (standard temperature and pressure) by the volume of catalyst plus void volume, one obtains the space velocity (VHSV) term used for monoliths. Its reciprocal is the residence time "t" at STP. For monolith reactors, the space velocity is calculated on the basis of its outside physical dimensions, i.e., diameter and length for a cylinder of honeycomb. This is not fundamentally correct, since it does not take into account the cell density or catalyst loading, but it is conventionally used and widely acceptable.

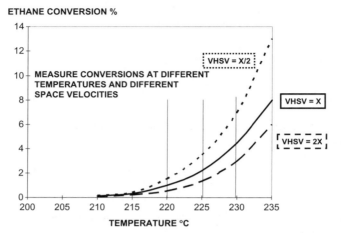

FIGURE 4.1. Ethane conversion versus temperature at different space velocities. Experiment performed to determine the rate constant at various temperatures.

For VHSV = volume flow rate of feed / physical volume of catalyst,

$$\text{VHSV} = 1/\text{time} \tag{4.6}$$

The rate expression then becomes

$$\ln \frac{C_o}{C_i} = -\frac{k'}{\text{VHSV}} \tag{4.7}$$

By varying the space velocity, the change in conversion, as measured by the change in $C_{C_2H_6}$, can be determined. The slope of the plot yields the k' of the reaction at STP conditions. The actual rate constant can be obtained by using the residence time (via space velocity) at actual temperature and pressures.

In Figure 4.1, the conversion at any temperature is increased with decreasing space velocity, simply because the residence time (SV^{-1}) is increased.

4.3 THE ARRHENIUS EQUATION AND REACTION PARAMETERS

We can derive a simple expression to aid us in determining which step is rate determining by use of the Arrhenius equation (Broadbelt 2003). By taking the natural logarithm of the Arrhenius equation (1.4), a linear expression results [see Eq. (4.8)] in which one can generate a series of straight lines whose slopes are directly related to the activation energy of the rate-controlling step, as shown in Figure 4.2:

$$\ln k = \ln k_o - E/RT \tag{4.8}$$

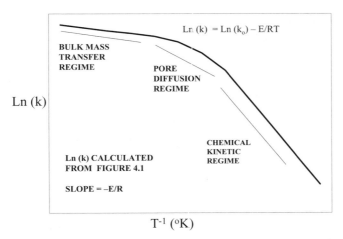

$$\text{Ln} (k) = \text{Ln} (k_o) - E/RT$$

FIGURE 4.2. Arrhenius plot for determining activation energies.

The plot of ln k versus 1/T shows three distinctly different slopes each related to a possible rate-limiting step. The largest slope is for a reaction controlled by chemical kinetics, the intermediate slope for pore diffusion, and the smallest slope for bulk mass transfer control.

When the reaction is controlled by one of the chemical steps, mass transfer of reactants to the active sites is fast. The concentration of reactants within the catalyst/carrier is essentially uniform. With pore diffusion control, the concentration of reactants decreases from the outer periphery of the catalytic surface toward the center. Finally, with bulk mass transfer control, the concentration of reactants approaches zero at the boundary layer near the outside surface of the catalyst.

When the rate constant is experimentally determined at several temperatures, one can generate sufficient data to apply the Arrhenius equation and subsequent plot (Figure 4.2) for the calculation of the activation energy.

Knowing that the rate is the moles converted per volume per unit time and is proportional to the change in reactant species concentration, one can make a plot of reactant concentration against the reciprocal of space velocity (i.e., time) at a specific temperature to generate the reaction rate as shown in Figure 4.3. It is important to understand that kinetic parameters for reactions controlled by chemical kinetics are always measured with small changes in concentration and far from equilibrium conditions. This allows one to ignore the reverse reaction and mass transfer contributions. It also minimizes heating effects that result from highly endothermic or exothermic reactions. Thus, measuring the rate as shown in Figure 4.3, the space velocity should be varied to limit changes in concentration to no more than 20%.

There is a very useful and easy method to determine the activation energy. Consider the general reaction:

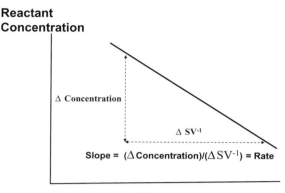

FIGURE 4.3. Concentration of reactant versus reciprocal of space velocity, where the slope = $(\Delta\text{Concentration})/(\Delta SV^{-1})$ = −Reaction Rate (moles/volume-time). Concentration change should be no more than 20%.

$$xA + yB = zC \tag{4.9}$$

where k_F is the forward reaction rate constant. The empirical power rate law is

$$\text{Rate} = -k_F C_A^x C_B^y \tag{4.10}$$

Taking the Ln of the equation gives

$$-\text{Ln}[\text{Rate}] = \text{Ln}\,k_F + x\,(\text{Ln}\,C_A) + y\,(\text{Ln}\,C_B) \tag{4.11}$$

Applying the Arrhenius expression for Ln k_F:

$$-\text{Ln}[\text{Rate}] = -E/RT + x\,(\text{Ln}\,C_A) + y\,(\text{Ln}\,C_B) \tag{4.12}$$

Experimentally one measures the variation in rate with temperatures but adjusts the conditions for small conversions of reactants so all concentration terms are relatively constant, and therefore, the terms x $(\text{Ln}C_A)$ + y $(\text{Ln}C_B)$ are constant. A plot of Ln [Rate] against $1/T$ generates a slope = $-(E/R)$.

One can also experimentally determine the reaction orders x and y for A and B, respectively in Eq. (4.12). In determining the reaction order x for C_A, the overall reaction must be far from equilibrium with low concentrations of product present so the reverse reaction can be neglected and no product inhibition will occur. The percent (%) change in concentration of the reactant, whose order is being measured $[C_A]$, should be small to maintain the reaction in the kinetic regime. All other reactant concentrations $[C_B]$ must be relatively large so little change in their concentration values will occur. This way Ln [C_B] can be treated as constant throughout the experiment and adsorbed into the constant k_F (forward reaction rate constant) giving a new constant k_F''. The slope of a plot of Ln [Rate]$_A$ against Ln [C_A] gives a straight line with the slope equal to the reaction order x.

Consider that C_i–C_o is the change in concentration of reactant A:

$$-Ln[Rate\ C_A] = -Ln[(C_i - C_o)(\Delta SV)] = Ln\,k_F'' + x\,Ln[C_A] \qquad (4.13)$$

It should be understood that this reaction order measured corresponds to the specific concentration of reactant in which it is measured. For other concentration ranges, the reaction order may be different and, thus, should be measured in the new range. This is especially true as the reaction proceeds down the monolith channel.

Reaction orders reflect the coefficients of reactants and products in the rate-limiting step of a chemical reaction. Often power law rate expressions show fractional orders and not the expected whole numbers. This is because there are often more complicated terms in real rate-limiting steps such is the case for the Langmuir–Hinshelwood kinetics [see, for example, Eq. (1.25)], where depending on the conditions of the experiment, there can be positive and negative terms for the same reactant.

4.4   BULK MASS TRANSFER

When experiments are conducted with an extremely active catalyst or at high temperatures, diffusional effects are introduced, and the intrinsic kinetics of the catalytic material is not determined accurately. This is because mass transfer effects dominate the overall reaction kinetics. The apparent activation energy will decrease as pore diffusion and bulk mass transfer become more significant.

Stationary environmental abatement processes are designed to operate in the bulk mass transfer regime where maximum conversion of the pollutant to the nontoxic product is desired. When a reaction is bulk mass transfer controlled, the chemical kinetics is rapid relative to transport of reactants to the outer surface of the catalyzed washcoat. For such processes, the concentration of reactant is zero at the external surface of the catalyzed washcoat.

Diffusion processes have small temperature dependency (low activation energies). For this reason, rates controlled by bulk mass transfer can easily be differentiated from chemical-controlled reactions that have a high degree of dependence on temperature (high activation energies).

An important benefit of designing the pollution abatement catalytic system in the bulk mass transfer regime is that the physical size and other geometric parameters of the honeycomb for a required conversion can be obtained using fundamental parameters of mass transfer.

Referring back to Eq. (4.2), when the reaction rate is bulk mass transfer controlled, the following expression is obtained:

$$v \frac{dC}{dz} = -K_g aC \tag{4.14}$$

where

K_g = mass transfer coefficient (cm/s)
a = geometric surface area per unit volume (cm^2/cm^3)
C = reactant gas phase concentration (g-moles/cc)

Again integrating, this becomes

$$\ln \frac{C_o}{C_i} = -K_g at \tag{4.15}$$

For stationary abatement applications, the temperature rise across the catalyst configuration is assumed small because the concentration of pollutants is not usually more than about 1,000 ppm. For such a condition, a simplistic mass transfer model can be derived from Eq. (4.14):

$$\text{Fractional Conversion} = 1 - \exp(-K_g at) \tag{4.16}$$

Using the following dimensionless numbers:

$$N_{Sh} = K_g d_{ch}/D \qquad \text{Sherwood number} \tag{4.17}$$

$$N_{Sc} = D\mu/\rho \qquad \text{Schmidt number} \tag{4.18}$$

$$N_{Re} = (W/A\varepsilon)d_{ch}/\mu \qquad \text{channel Reynolds number} \tag{4.19}$$

where

D = diffusivity of the pollutant in air (cm^2/s)
L = honeycomb length (cm)
W = total mass flow rate to honeycomb catalyst (g/s)
A = frontal area of honeycomb (cm^2)
d_{ch} = hydraulic diameter of the honeycomb channel (cm)
ρ = gas density at operating conditions (g/cm^3)
μ = gas viscosity at operating conditions (g/s-cm)
ε = void fraction of honeycomb, dimensionless
      an equation can be derived involving these mass transfer parameters.

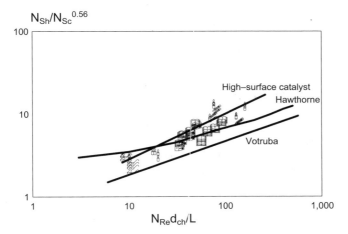

FIGURE 4.4. Mass transfer correlation versus reduced Reynold's number. Reprinted from *Encyclopedia of Environmental Control Technology,* Vol. 1, copyright © 1989 by Gulf Publishing Co. (Bonacci et al. 1989).

$$\text{Fractional Conversion} = 1 - \exp - \frac{(N_{Sh}a/\varepsilon L)}{(N_{Sc}N_{Re})} \qquad (4.20)$$

The conversion can be estimated from this model by determining the Sherwood number, the Schmidt number, and the Reynolds number. The honeycomb size parameters are estimated and specified by a, L, d_{ch}, and ε. The physical parameters, such as the geometric surface area of a honeycomb with different channel densities and channel diameters, are shown presented in Table 2.1 (see Chapter 2). Diffusivities D are calculated from standard formulas.

The density and viscosity of the bulk gas stream can be obtained from any physical property manual. Correlations for the Sherwood number have been developed for catalyzed honeycombs as depicted in Figure 4.4. Knowing the size of the monolith that is acceptable in a confined space, one can estimate the conversion. Also, using this Eq. (4.20), various design options can be considered, such as honeycomb cell density, reactor diameter, and length.

Example: Calculation for Mass Transfer Conversion

As an example, consider the removal of propane (C_3H_8) in a stream of air at 300 °C and atmospheric pressure with the following given parameters:

Temperature, T	=300 °C
Pressure, P	=1 atm
Flow Rate, W	=1,000 lb/hr (126 g/s)

Diameter of monolith, D   =6 in (15.24 cm)
Length of monolith, L   =6 in (15.24 cm)
Area of monolith, A   =28.27 in^2 (182.4 cm^2)
Monolith geometry   =100 cpsi (15.5 cells/cm^2)
C_3H_8 feed fraction, x   =1,000 vppm

These are only sample conditions, and many combinations of variables are possible, depending on the constraints of the actual physical application. However, this will serve as an illustration of the calculation approach.

Calculate the predicted mass transfer conversion using Eq. (4.20) and Figure 4.4. Equation (4.20) is as follows:

$$\text{Fractional Conversion} = 1 - \exp - \frac{(N_{Sh} a/\varepsilon L)}{(N_{Sc} N_{Re})} \tag{4.21}$$

The geometric properties of the monolith to be used are found in the literature as follows (Lachman and McNally 1985):

Channel diameter, d_{ch} =0.083 in (0.21 cm)
Void fraction, ε =0.69
Surface to volume ratio, a   =33 in^2/in^3 (13 cm^2/cm^3)

The following dimensionless groups need to be determined to use Figure 4.4 to determine the N_{Sh} number:

$$\text{Reynolds number, } N_{Re} = (W/A\varepsilon) d_{ch}/\mu \tag{4.22}$$

$$\text{Schmidt number, } N_{Sc} = \mu/(\rho D) \tag{4.23}$$

From the literature, the physical and chemical properties can be determined. Using Hodgman's (1960) *Handbook of Chemistry and Physics*, the density (ρ) and viscosity (μ) can be found:

$$\rho @ 300\,°C = 6.16 \times 10^{-4}\,\text{g/cc}$$

$$\mu @ 300\,°C = 297 \times 10^{-6}\,\text{g/s-cm}$$

Therefore using Eq. (4.16), the N_{Re} is

$$N_{Re} = \frac{(126/s)(0.21\,\text{cm})(0.69 \times 182.4\,\text{cm}^2)}{(297 \times 10^{-6}\,\text{g/s-cm})}$$

$$N_{Re} = 707.9 \tag{4.24}$$

Furthermore, to use Figure 4.4, the following term must be determined:

$$N_{Re}d_{ch}/L = (707.9)(0.21\,cm)/(15.24\,cm) = 9.75 \qquad (4.25)$$

To calculate the N_{Sc} number using Eq. (4.23), the viscosity (μ), the density (ρ), and an estimate of the binary mass diffusivity (D) are needed. Again, use Hodgman's *Handbook of Chemistry and Physics* to find the viscosity and density.

Because the mixture is dilute and at low pressure, the following equation can be used to calculate the diffusity (D) for a binary system (Bird et al. 1970):

$$D_{a\text{-}b} = 0.0018583 \frac{\sqrt{T^3(1/M_A + 1/M_B)}}{P\sigma_{AB}^2 \Omega_{D,AB}} \qquad (4.26)$$

where

M   = molecular weight of species, A = air; B = C_3H_8 (g/g-mole)

P   = total pressure (atm)

σ_{AB} = collision diameter for binary system (A)

T   = absolute operating temperature (°K)

$\Omega_{D,AB}$ = collision integral for binary system, dimensionless

So the unknown in Eq. (4.26) is the collision integral $\Omega_{D,AB}$ and the collision diameter σ_{AB}. These are determined as follows. Using Table B.1 (Intermolecular Force Parameters and Critical Properties) from Bird et al. (1970), the following parameters for each component are found:

$$\text{for air } M_A = 28.97, \sigma_A = 3.617\,A, \varepsilon_A/k = 97.0\,K$$

$$\text{for } C_3H_8\ M_B = 44.09, \sigma_B = 5.061\,A, \varepsilon_B/k = 254\,K$$

The terms σ and ε/k are Leonard–Jones parameters for the single components.

Now for the binary system, the same parameters need to be calculated using the following equations:

$$\sigma_{AB} = 1/2\,(\sigma_A + \sigma_B) \qquad (4.27)$$

$$\sigma_{AB} = 1/2\,(3.617 + 5.061) = 4.339\,A \qquad (4.28)$$

$$\varepsilon_{AB}/k = \sqrt{(\varepsilon_A/k)(\varepsilon_B/k)} \qquad (4.29)$$

$$\varepsilon_{AB}/k = \sqrt{(97)(254)} = 156.96\,K \qquad (4.30)$$

The term ε_{AB} is needed for the calculation of the following parameter:

$$kT/\varepsilon_{AB} = 573\,K/156.96\,K = 3.65 \qquad (4.31)$$

Using this calculated kT/ε_{AB} and Table B.2 (Functions for Prediction of Transport Properties of Gases at Low Densities) of Bird et al. (1960), the collision integral $\Omega_{AB} = 0.903$.

Substituting all the above information into Eq. (4.27) gives the following:

$$D = 0.0018583 \frac{\sqrt{(573)^3(1/28.97 + 1/44.09)}}{(1)(4.339)^2(0.903)} = 0.359 \, cm^2/s \qquad (4.32)$$

Therefore, the Schmidt number is calculated using Eq. (4.23) as follows:

$$N_{Sc} = (297 \times 10^{-6} \, g/s\text{-}cm)/(6.16 \times 10^{-4} \, g/cc)(0.359 \, cm^2/s) = 1.34 \quad (4.33)$$

Using Figure 4.4, and the calculation of $(N_{Re}d_{ch})/L = 9.75$ from Eq. (4.25), the value of $N_{Sh}/N_{Sc}^{0.56} = 3.8$ is found from the graph. Using the $N_{Sc} = 1.34$ [from Eq. (4.28)], the $N_{Sh} = 4.4$ is calculated.

With the above information, the mass transfer-controlled fractional conversion can now be calculated from Eq. (4.20) as follows:

$$\text{Fractional Conversion} = 1 - e^{-\frac{(4.4)(13 \, cm^2/cm^3/0.69)(15.24 \, cm)}{(1.34)(707.9)}}$$

$$\text{Fractional Conversion} = 1 - e^{-1.332}$$

$$\text{Fractional Conversion} = 0.736 \text{ or } 73.6\% \qquad (4.34)$$

Therefore, the mass transfer-controlled percent conversion of C_3H_8 is calculated to be 73.6%, and this is the maximum conversion that can be achieved in the honeycomb catalyst.

4.5   REACTOR BED PRESSURE DROP

When the pollutant-containing process gas stream enters the channels of the honeycomb, its flow contracts within the restrictive channel diameter. A pressure drop (usually denoted as ΔP) develops along the monolith length. Because this restricts flow for an automobile, it represents a loss in power. For many stationary applications, it increases the load on the compressor. Additionally, the presence of a washcoat on the surface of the honeycomb channel creates friction and adds further resistance to the gaseous flow. In all applications, there are maximum permitted pressure drops as part of the overall system engineering and economics.

The basic equation for pressure drop can be derived from the energy balance and results in the following expression:

$$-\frac{1}{\rho}\frac{dP}{dL} = \frac{2fv^2}{g_c d_{ch}} \qquad (4.35)$$

where

P = total pressure (atm)
f = friction factor, dimensionless
d_{ch} = honeycomb channel diameter (cm)
g_c = gravitational constant (980.665 cm/s^2)
L = length (cm)
v = velocity in channel at operating conditions as in (4.36) (cm/s)
ρ = gas density at operating conditions (g/cm^3)

Correlations have been developed for catalyzed honeycombs relating the apparent Fanning Friction factor f to the channel Reynolds number as shown in Figure 4.5.

The velocity (v) in the channel is calculated using the mass flow rate (W), density (ρ), void fraction (ε), and cross-sectional area (A) of the honeycomb as follows:

$$v = W/(\rho A \varepsilon) \tag{4.36}$$

ε is the percent open frontal area of the honeycomb as listed in Table 2.2 (see Chapter 2):

Equation (4.35) simplifies to

$$\Delta P = \frac{2fL\rho v_{ch}^2}{g_c d_{ch}} \tag{4.37}$$

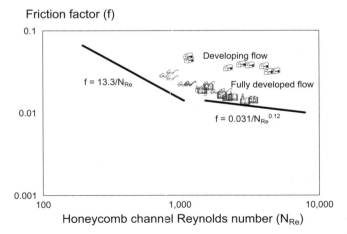

FIGURE 4.5. Friction factor correlation. Reprinted from *Encyclopedia of Environmental Control Technology,* Vol. 1, copyright © 1989 by Gulf Publishing Co. (Bonacci et al. 1989).

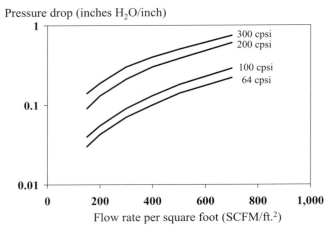

Pressure drop (inches H_2O/inch)

FIGURE 4.6. Monolithic pressure drop versus flow rate. Reprinted from *Encyclopedia of Environmental Control Technology,* Vol. 1, copyright © 1989 by Gulf Publishing Co. (Bonacci et al. 1989).

Note that using Figure 4.5 for friction factor (f) with Eq. (4.37), one finds that in laminar flow ($N_{Re} < 1000$) the ΔP is proportional to v, and in turbulent flow ΔP is proportional to v^2.

Using the fundamental friction factor and the equation for pressure drop, design curves can be generated to show the interaction of various honeycomb cell densities for minimizing the pressure drop. Figure 4.6 shows a series of pressure drop curves for various monolith cell densities in air. The information is then used with the mass transfer calculations to select the optimum honeycomb geometry (volume, cross-sectional area, length, cpsi, and so on) for a given application.

4.6   SUMMARY

What has been developed in this chapter is a generalized approach for determining kinetics and mass transfer conversion and pressure drop for any application of monoliths. More specific application of these principles will be illustrated in Chapters 6 and 8 for automotive catalysis and diesel catalysis. Since every application of the basic principles appears different when used as a specific case, assumption, and author's nomenclature, they are still derived from the approach set forth in this chapter.

REFERENCES

Bird, R., Stewart, W., and Lightfoot, E. *Transport Phenomena.* Wiley and Sons, New York. (1970).

Bonacci, J., Farrauto, R., and Heck, R. "Catalytic incineration of hazardous waste," in *Encyclopedia of Environmental Control Technology Volume I: Thermal Treatment of Hazardous Wastes*. Editor Cheremisinoff, P. N. Gulf Publishing Co. Houston, TX (1989).

Broadbelt, L. "Kinetics of catalyzed reactions-heterogeneous," pp 472–490 in *Encyclopedia of Catalysis*. Editor Horvath, I. Wiley-Interscience, Hoboken, NJ (2003).

Hodgman C. D., Editor. *Handbook of Chemistry and Physics*. The Chemical Rubber Publishing Co. Cleveland, OH (1960).

Lachman, I., and McNally, R. "Monolithic honeycomb supports for catalysis," *Chemical Engineering Progress* 84(1): 29–31 (1985).

CHAPTER 4   QUESTIONS

1. What temperature is required for a catalyst with an activation energy of 15,000 cal/mol to operate with the same volume as one with an activation energy of $E = 12,000$ cal/mol (operation temperature is 227 °C) with the same conversion? Assume k_o are the same for simplicity.

2. You are employed as an environmental engineer for a major chemical company that uses alcohols as liquid carriers for pigments to be used for decorating. Currently, to meet local environmental regulations, the harmful emissions (1,500 vppm) resulting from spraying and drying the coatings are thermally incinerated. However, this requires a thermal burner that is expensive, consumes large amounts of fuel, and generates some NO_x (NO, NO_2, and/or N_2O). The plant manager, knowing you had a course in catalysis, is seeking an alternative to thermal incineration. You are asked to evaluate some fixed-bed catalysts that will abate the alcohol emissions. You contact a catalyst company, and they suggest you use a monolithic support with the benefits of a lower low pressure drop than a particulate bed of catalyst. They offer two catalysts: an inexpensive one with an activation energy (E) of 21,200 cal/mol (21.2 Kcal/mol) and a second more expensive one with an activation energy of 12,000 cal/mol (12 Kcal/mol). Compare the volumes needed for each assuming you want to achieve 90% conversion at 227 °C. Assume both catalysts have the same k_o (pre-exponential function).

3. Compare two catalysts with $E = 12$ and 15 Kcal/mol.

4. What must the activation energy (E) be for a catalyst to operate at 275 °C (548 °K) where one with an $E = 30$ Kcal/mol operates at 550 °C (823 °K). Assume both will give the same conversion and both will have the same k and k_o. (This is not likely since k_o includes the number of active sites and other more fundamental parameters, including collision frequencies, that govern conversion for each catalyst.)

5. Determine the activation energy for a fixed-bed catalyst where the reaction conditions are adjusted so % conversion for each temperature is low and about the same for the following three rates:

Rate $(400°K) = 10$

Rate $(410°K) = 20$

Rate $(420°K) = 40$

6. Why is it useful to know the reaction orders for the reactants and products? Consider the empirical equation where A and B are reactants and C and D products and a,b,c, and d are reaction orders:

Rate Net $= [A]^a [B]^b [C]^c [D]^d$

5   Catalyst Deactivation

5.1   INTRODUCTION

One of the major sources of catalyst deactivation in environmental applications occurs due to high-temperature exposure. This is especially true in the automobile catalytic converter where temperatures close to $1,000\,°C$ can occur. Other sources of deactivation such as poisoning can occur due to exhaust or process contaminants adsorbing onto or blocking active catalytic sites (Bartholomew 2003; Forzatti et al. 1984; Bartholomew and Farrauto 2006; Trimm 1997). It is essential to understand the modes of poisoning in order to develop resistant materials and methods of regeneration when possible. Since environmental catalytic applications use washcoated monoliths, deactivation by attrition and erosion of the washcoat from the support must also be considered.

Throughout the application sections of this book, reference will be made to common sources of deactivation specific to a particular application, methods of regeneration, and design considerations to minimize these effects. A convenient tool for studying deactivation and regeneration is the model reaction. Here one takes a representative model compound and conditions under which the catalyst is expected to function and develop a conversion versus temperature curve as shown in Figure 1.3. Shifts in the profile provide great insight into the mechanism of deactivation. This theme will be developed in more detail throughout this chapter.

This chapter will serve as a convenient working guide for determining the mechanism of deactivation and diagnostic tools to monitor these changes. It will be particularly useful when attempting to perform a technical service on a catalyst that is not performing in accordance with specifications.

5.2   THERMALLY INDUCED DEACTIVATION

It is generally the objective of the catalyst manufacturer to maximize the accessibility of the reactants to the active sites by depositing the catalytic components on a carrier. A perfectly dispersed (100% dispersion) catalyst is one in which every atom (or molecule) of an active component is available to

Catalytic Air Pollution Control: Commercial Technology, by Ronald M. Heck and Robert J. Farrauto, with Suresh T. Gulati.
Copyright © 2009 John Wiley & Sons, Inc.

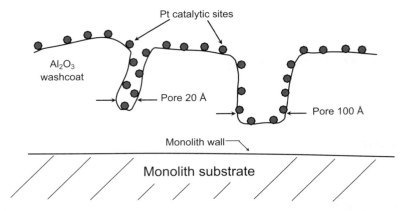

FIGURE 5.1. Idealized cartoon of perfectly dispersed Pt on a high-surface-area γ-Al$_2$O$_3$ washcoat bound to a monolith.

the reactants. This is shown as a cartoon in Figure 5.1 where the black dots represent the catalytic component (indicated as Pt here) dispersed on a high-surface-area γ-Al$_2$O$_3$ that is bonded to a monolith. It should be understood that the surface of the Al$_2$O$_3$ contains hydroxide sites, not shown for convenience, which are in direct contact with the catalytic components and play an important role in immobilizing them. The carrier can be thought of as an inorganic sponge possessing small pores that are varying in size and shape into which reactants can flow and interact with catalytic components dispersed on the surface. The products then pass through the porous network out to the bulk fluid.

Some catalysts are made in this highly active state but are highly unstable, and thermal effects cause crystal growth resulting in a loss of catalytic surface area. Additionally, the carrier with a large internal surface network of pores tends to undergo sintering with a consequent loss in internal surface area. Not uncommon are reactions of the catalytically active species with the carrier resulting in the formation of a less catalytically active species. All of these processes are influenced by the nature of the catalytic species, the carrier, and the process gas environment, but mostly by high temperatures.

5.2.1 Sintering of the Catalytic Component

It is common for a highly dispersed catalytic species to undergo growth to more well-defined crystals as a consequence of their high surface-to-volume ratio. As this process proceeds, the crystals grow larger decreasing the surface-to-volume ratio with fewer catalytic atoms or molecules on the surface of the crystal available to the reactants. In other words, many active sites are buried within the crystal, and with fewer sites participating in the reaction, a decline in performance frequently is noted.

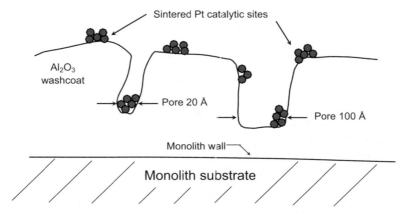

FIGURE 5.2. Conceptual diagram of sintering of the catalytic component on a carrier.

This phenomena is represented in Figure 5.2 by a simple model in which individual active catalytic components are designated as dots on a carrier.

Initially they are well dispersed but undergo coalescence or crystal growth induced thermally. As the catalytic components undergo coalescence, the number of surface sites decreases and the reaction rate decreases.

The driving force for catalytic sintering can be explained by the high surface-to-volume energy ratio possessed by small crystallites in the nanometer range. Thermodynamically, this is an unstable state, and crystal growth, or sintering, occurs to minimize the free energy.

The extent of sintering of the catalytic component can be measured by selective chemisorption techniques in which a thermally aged catalyst adsorbs much less adsorbate than when it was fresh. The growth in crystal structures can also be observed by the X-ray diffraction (XRD) pattern; however, for most environmental catalysts, the catalytic component is present in such a dilute concentration, or crystals below about 35 Å (3.5 nm), that XRD cannot usually detect the crystals. High-resolution transmission electron microscopy permits the most direct method of observation of the growth process but requires a great number of micrographs for statistical significance and frequently is not considered a routine test in commercial practice. An excellent example of a sintered precious metal catalyst is shown in the transmission electron microscope (TEM) image of Figure 5.3. Here the Pt is observed to have grown from 20 to 200 Å (2 nm to 20 nm).

Catalytic scientists have found preparation and compositional additions to decrease the rate of the sintering process by using the proper carriers and/or stabilizers, but with time, all materials reach some quasi-steady state as they seek to approach minimum free energy as dictated by thermodynamics.

Automotive Catalyst Aging Effects

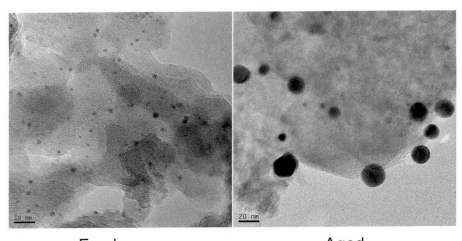

Fresh   Aged

FIGURE 5.3. TEM of fresh and sintered Pt on Al_2O_3 in an automobile catalytic converter application.

The loss of performance due to Pt, Pd, and Rh sintering (Klimish et al. 1975) is so important for automotive emission control catalysts that considerable research in learning to stabilize high dispersions has been conducted. The addition of certain stabilizers to the catalyst formulation has been a fruitful approach. Certain rare earth oxides, such as CeO_2 and La_2O_3 (Oudet et al. 1989), have been effective in reducing the sintering rates of Pt in the automobile exhaust catalytic converter. Once again the precise mechanism of stabilization is not clearly understood, but obviously the stabilizers fix the catalytic components to the surface minimizing mobility and crystal growth. In some cases, the sintering can occur simply by migration over the surface as a function of the wetting angle of the metal or metal oxide to the carrier surface. Catalytic components sinter more readily on SiO_2, which has a lower concentration of hydroxide species on its surface compared with γ-Al_2O_3.

In the idealized conversion versus temperature plot of Figure 5.4, the conversion of the sintered catalyst is displaced to higher temperatures relative to the fresh catalysts since fewer sites are available. It should be noted that the activation energy, as evidenced by the parallel profile, is unchanged since the kinetic control is still the operative rate-limiting step.

5.2.2 Carrier Sintering

Within a given structure, such as γ-Al_2O_3, the loss of surface area is associated with a loss of surface hydroxyl groups and with a gradual loss of the internal

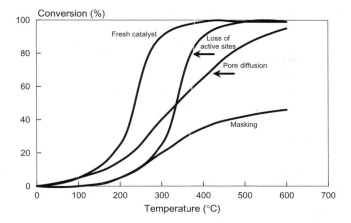

FIGURE 5.4. Idealized conversion versus temperature for various aging phenomena.

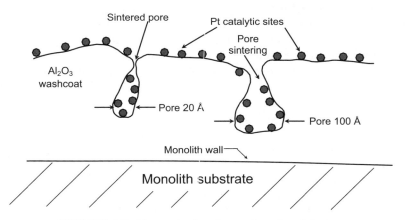

FIGURE 5.5. Conceptual cartoon of carrier sintering.

pore structure network, as shown in the simple cartoon of Figure 5.5. The surface hydroxyl groups are not shown, but the sintering process results in the expulsion of a surface H_2O molecule and the bonding of adjacent Al sites to the remaining O ion.

As the sintering phenomena are occurring, the pore openings are getting progressively smaller, introducing greater pore diffusion resistance. Thus, a chemically controlled reaction may gradually become limited by pore diffusion. The occurrence of this phenomena is determined by a progressive decrease in the activation energy of the catalytic reaction. In the plot of conversion versus temperature of Figure 5.4, the slope of the curve becomes progressively lower. In the extreme case, the pores completely close, making the catalytic species within that pore inaccessible to the reactants. The 2θ peaks in the X-ray diffraction pattern become sharper relative to the less sintered material.

The second mechanism for carrier sintering involves an irreversible conversion to a new crystal structure. Each conversion results in a decrease in the porosity and surface area of the Al_2O_3. At about 800 °C, γ-Al_2O_3 converts to δ-Al_2O_3, with a surface area about 50–60% that of γ, with a more crystalline structure as determined by X-ray diffraction. At about 1,000 °C, δ-Al_2O_3 converts to θ-Al_2O_3 with a surface area of about 75–95% that of γ. The X-ray diffraction peaks become even sharper and more crystalline. The final high-temperature stable structure at about 1,100 °C is α-Al_2O_3, which is the most crystalline of all Al_2O_3 structures with virtually no porosity or surface area (i.e., 1–2% γ). It is highly compact as was shown in Figure 2.1b in Chapter 2. In a similar manner, TiO_2, which is used as a carrier for supporting V_2O_5 that is used as a catalyst for selective reduction of NO_x with NH_3, is irreversibly converted from its high-surface-area anatase structure (~80 m^2/g) to its rutile structure with a surface area <10 m^2/g at about 550 °C. Figure 5.6 shows the morphology change associated with this transformation. The X-ray diffraction pattern has different 2θ values with sharper and more well-defined peaks than anatase.

Conversion from a high-surface-area structure to a low one usually results in the encapsulation of the catalytic components, as symbolically shown in Figure 5.5, making them inaccessible, which leads to a more deactivated catalyst.

In the conversion versus temperature profile of Figure 5.4, carrier sintering results in a shift to higher temperatures since some active sites are no longer accessible to the reactants. When the pores get smaller, the rate-limiting step changes from kinetic control to pore diffusion and the slope becomes lower since the activation energy for pore diffusion is lower than for kinetic control.

The occurrence of either mechanism is primarily detected by N_2 surface area measurements, i.e., Brunauer, Emmett, and Teller (BET), XRD, and to a lesser extent, pore size distribution. X-ray diffraction is the preferred method because it directly shows a sharpening of the diffraction patterns of the various crystal planes as sintering occurs. Conversion from one phase to another will generate a completely different XRD pattern and will allow a semiquantitative estimate of crystal sizes. The X-ray diffraction patterns for

(a)

(b)

FIGURE 5.6. Scanning electron microscope (SEM) image of TiO_2 in a) a high-surface-area anatase structure and b) a low-surface-area rutile structure.

high-surface-area and low-surface-area structures were shown in Figure 3.7. The BET surface area measurement is also commonly used; however, other mechanisms, i.e., masking or fouling, can lead to a decline in apparent surface area without sintering having occurred. Similarly, pore size measurements are useful, but they do not present a complete picture since pore blockage due to masking also cause an apparent change in the pore size distribution.

The presence of specific amounts of stabilizers, such as BaO, La_2O_3, SiO_2, and ZrO_2 (see Chapter 2), can retard the rate of sintering in certain carriers (Wan and Dettling 1986). They are believed to form solid solutions with the carrier surface decreasing their surface reactivity that leads to sintering.

When a zeolite experiences a high temperature, the Si–O–Al bridges undergo a process called de-alumination, where the Al is extracted from the framework of the tetrahedral structure. Since the Al is the origin for catalytic sites such as H^+ or metal cations, they also leave the framework and the electronic environment with the pore of the zeolite is altered. Naturally this changes the activity/selectivity of the catalyst. The Al-containing species form new structures with penta and octahedral coordination sites, none of which have well-defined pore sizes like the zeolite from which they are derived. Thus, the crystallinity of the zeolite is decreased. This is shown in Figure 5.7a), where the fresh zeolite loses Si–O–Al bridges, i.e., Si(3Al), Si(2Al), and Si(Al) after high-temperature exposure. The formation of penta and octahedral sites is shown in Figure 5.7b) as a function of various prolonged thermal exposures.

5.2.3   Catalytic Species–Carrier Interactions

The reaction of the active catalytic component with the carrier can be a source of deactivation if the product is less active than the initial dispersed species. For example, Rh_2O_3 reacts with a high-surface-area γ-Al_2O_3 forming an inactive compound during high-temperature lean conditions in the automobile exhaust. This is a particularly important mechanism for deactivation of NO_x reduction activity. The reaction believed to be occurring is conceptually shown below in Eq. (5.1):

$$Rh_2O_3 + Al_2O_3 \xrightarrow[\text{air}]{800°C} Rh_2Al_2O_4 \tag{5.1}$$

The conversion versus temperature plot (Figure 5.4) shifts to higher temperatures due to the loss of active sites. The slope may or may not change depending on the catalytic activity and activation energy of the new structure formed. This undesirable reaction has led to the development of carriers such as SiO_2, ZrO_2, TiO_2, and their combinations that are less reactive with Rh_2O_3 than with Al_2O_3. The interaction problem can be solved by these alternative carriers, but often they are not as stable against sintering; thus, there is a trade-off in performance that must be factored into the new catalyst formulations.

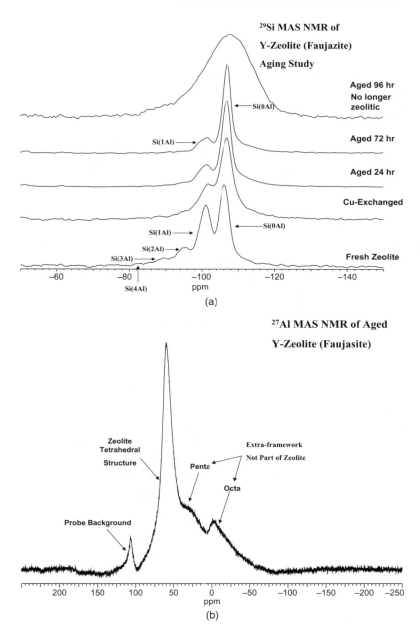

FIGURE 5.7. Nuclear magnetic resonance (NMR) profile of a thermally a) aged zeolite showing the loss of the Si–O–Al bridges. Si(3Al), Si(2Al), and Si(Al) are observed to decrease in intensity with progressively more severe thermal aging. b) Growth of penta and octahedral coordination sites in a thermally deactivated zeolite.

Rh also has an undesirable reaction with the oxygen storage component (CeO_2-containing compounds), which is an important component in the three-way catalytic converter. Separating these components by incorporating them in different washcoat layers on the monolith is one approach toward eliminating deactivation.

Some interactions are beneficial to the catalyst activity. For example, both Pt and Pd have greatly improved activities when deposited on CeO_2-containing carriers compared with more traditional Al_2O_3 carriers.

Reactions between components that form new compounds can be monitored by XRD provided the amount present is large enough to be XRD visible. The Rh_2O_3/Al_2O_3 compound formed cannot be easily studied by XRD due to the relatively small amount of Rh present. Selective chemisorption may be considered the most useful technique, but its measurement requires a reduction of the Rh to the metallic state that decomposes the Rh–Al compound and consequently regenerates the Rh. Temperature-programmed reduction in which the amount and temperature at which H_2 decomposes the Rh_2O_3–Al_2O_3 complex is a convenient technique for studying these interactions. One measures the amount of H_2 consumed in the reduction of the Rh ion in the compound. Less convenient techniques such as NMR or X-ray photoelectron spectroscopy (XPS) may also be useful in elucidating this type of deactivation mode.

5.3 POISONING

A common cause of catalyst deactivation results from contaminants present in the feedstock or in the process equipment depositing onto the catalyst surface. There are two basic mechanisms by which poisoning occurs: (1) selective poisoning, in which an undesirable contaminant directly reacts with the active site or the carrier, rendering it less or completely inactive; and (2) nonselective poisoning such as deposition of fouling agents onto or into the catalyst carrier, masking sites, and pores, resulting in a loss in performance due to a decrease in accessibility of reactants to active sites.

5.3.1 Selective Poisoning

This is a discriminating process, by which a poison directly reacts with an active site, decreasing its activity or selectivity for a given reaction, as the cartoon shows in Figure 5.8.

Some poisons chemically react with the catalytic component such as Pb, Hg, Cd, and so on, which react directly with Pt, Pd, or Rh forming a catalytically inactive alloy. Such poisoning leads to permanent deactivation. Some poisons merely adsorb (chemisorb) onto sites, i.e., SO_2 onto a metal site (i.e., Pd), and block that site from further reaction (Lampert et al. 1997). These mechanisms sometimes are reversible (Chen et al. 1992) in that heat treatment, washing,

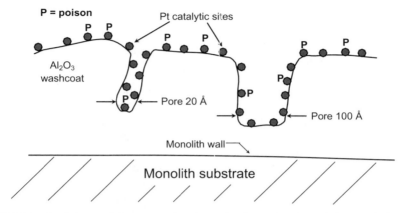

FIGURE 5.8. Conceptual cartoon showing selective poisoning of the catalytic sites.

or simply removing the poison from the process stream often desorbs the poison from the catalytic site and restores its catalytic activity. When active sites are directly poisoned, there is a shift to higher temperatures but with no change in the slope since the remaining sites can function as before with no change in activation energy. The conversion versus temperature diagram will look similar to that for catalytic sintering with a loss in active sites and a parallel shift to higher temperatures.

When the carrier reacts with a constituent in the gas stream to form a new compound, as in the case of $Al_2(SO_4)_3$, pores are generally partially blocked resulting in increased diffusional resistance. This will cause a decrease in the activation energy, and thus, the conversion versus temperature curve will shift to higher temperatures with a lower slope as shown in Figure 5.4.

An interesting example contrasting the poisoning behavior of $Pd/\gamma\text{-}Al_2O_3$ and $Pt/\gamma\text{-}Al_2O_3$ has been reported in which SO_2 chemisorbs onto Pd causing deactivation for methane oxidation (Lampert et al. 1997). Some of the adsorbed SO_2 is converted to SO_3, which spills over forming $Al_2(SO_4)_3$. Using nonsulfating carriers such as ZrO_2 or SiO_2 leads to a faster rate of deactivation since no reservoir is available for spillover. With the Pt catalyst, the SO_2 is readily converted to SO_3, which rapidly desorbs and reacts with the Al_2O_3 forming $Al_2(SO_4)_3$ that slowly causes pore plugging. By using nonsulfating supports, the Pt catalyst can be made resistant to deactivation.

Bulk chemical analysis, surface analysis by XPS, and loss of chemisorption catalytic surface area are commonly used procedures to characterize this type of deactivation. Chemisorption methods are acceptable provided the pretreatment does not remove the poison. X-ray diffraction is useful only when large amounts of a new compound are formed. X-ray photoelectron spectroscopy more precisely measures the adsorbed species at the catalyst surface and, for this reason, has become an important tool in characterization of poisoned catalysts. The high vacuum treatment necessary for this measurement, however, can greatly alter the surface, so some interpretation and judgement is essential.

5.3.2 Nonselective Poisoning or Masking

In commercial environmental processes, it is common for aerosoles or high-molecular-weight material from upstream to deposit physically onto the surface of the washcoat. This mechanism of deactivation is referred to as fouling or masking. The reactor scale (i.e., Fe, Ni, Cr, etc.) from corrosion, silica, and alumina containing dusts, phosphorous from lubricating oils, and so on are frequently found on catalysts (Hegedus and Baron 1978; Heck et al. 1992). A cartoon sketch of these phenomena is shown in Figure 5.9.

These mechanisms are nondiscriminating in that the species deposit physically on the outer surface of the catalyst. The conversion versus temperature curve will show a substantial stepwise drop in the maximum conversion due to the loss in geometric area that impacts the bulk mass transfer area (see Figure 5.4). Those poisons that penetrate into the pore network will coat the inside of the pores leading to enhanced pore diffusion resistance, and the slope of the conversion versus temperature plot will be decreased due to the decrease in activation energy. Methods for regeneration will be discussed in Chapter 11.

Surface chemical analysis by SEM or XPS is a commonly used method for detecting the nature of the masking agent. Accompanying masking, there usually is a decline in surface area due to pore blockage.

Electron microscopy is helpful in seeing deposits on the surface. Figure 5.10 shows a fresh a) and aged b) surface of Pt/Al_2O_3 in which massive amounts of Si have physically deposited causing deactivation. The catalyst was used to abate hydrocarbon and carbon monoxide emissions from a silicone wire coating operation.

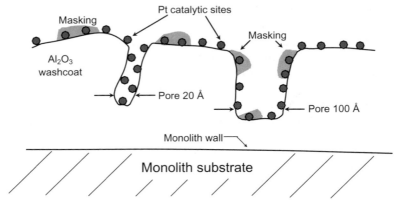

FIGURE 5.9. Conceptual cartoon showing masking or fouling of a catalyst washcoat.

Fresh catalyst

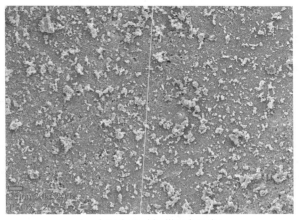

Aged catalyst with silica deposits

FIGURE 5.10. SEM of Pt/Al$_2$O$_3$ surfaces: (top) fresh and (bottom) aged with Si deposits (Magnification = 200×).

An XPS spectrum of a poisoned Pt/Al$_2$O$_3$ catalyst used to abate emissions from a sulfur-containing, fuel-rich burning engine shows a heavy concentration of poisons, as shown in Figures 5.11. The surface has large concentrations of carbon (6.6%), Fe (2.7%), and S (2.4%). Also shown are the fresh components of the catalysts Al at 35.6%, the Pt at 0.2%, and O at 52.5%. The Fe was found to be derived from reactor scale corrosion.

A special type of masking or fouling is coking in which a carbon-rich, hydrogen-deficient material is formed on and in the pore network. This mechanism is not often a problem in environmental applications since sufficient oxygen is usually present and the temperature is sufficiently high to oxidize any coke precursor materials.

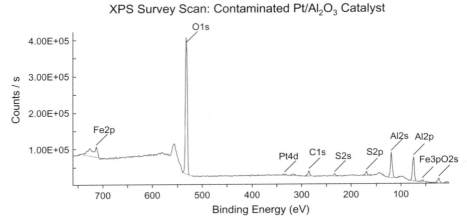

FIGURE 5.11. XPS spectrum of the surface of a contaminated Pt on Al_2O_3 catalyst.

Selective and nonselective poisons deposit on catalysts and deactivate by different mechanisms. Nonselective poisoning, such as fouling, is a physical phenomena, and contaminants usually accumulate at the inlet of the bed or in the case of the monolith about 1–3 cm axially downstream from the inlet face. Furthermore, they tend to concentrate in the top of the washcoat usually no more than about 30 microns deep from the gas–washcoat interface. Selective poisons, such as sulfur oxides, are more discriminating and will usually deposit on the specific sites; thus, they may penetrate down the axis of the monolith and deep inside the washcoat.

These two mechanisms can be clearly observed in Figure 5.12. Aerosols of phosphorous, calcium, and zinc, originating from the detergent package in lubricating oils, have deposited within about 30 microns depth of the washcoat thickness. (The units on the x-axis require multiplication by 3.) The sulfur, existing in the gas phase as SO_2/SO_3, forms $Al_2(SO_4)_3$ and is present more uniformly throughout the depths of the washcoat. The signal for sulfur ends at about 3×45 microns, and the appearance of the Si identifies the monolith–washcoat interface at about 3×50 microns.

Thermal analysis [thermal gravimetric analysis (TGA)/differential thermal analysis (DTA)] is an excellent tool to establish conditions for regenerating poisoned catalysts. By measuring the weight and energy change of a representative sample of a poisoned catalyst in a microbalance, the temperature needed for coke burn-off and regeneration can be determined. This is shown in Figure 5.13 for a Rh-containing, hydrocarbon, steam-reforming catalyst used for hydrogen generation for fuel cell applications. The first minor weight loss below 200 °C is due to volatile components. Note that the weight loss is endothermic, which indicates volatility. At about 425 °C, the coke begins to combust. The weight loss is accompanied by the exothermic heat of combustion, which peaks at 556 °C.

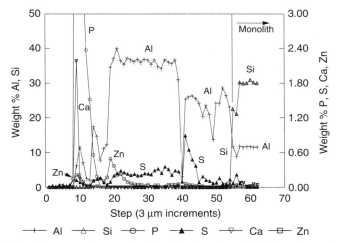

FIGURE 5.12. Electron microprobe showing the deposition location of the poisons within the washcoat of a monolith catalyst. The X-ray beam is scanned perpendicular to the axial direction through the thickness of the washcoat.

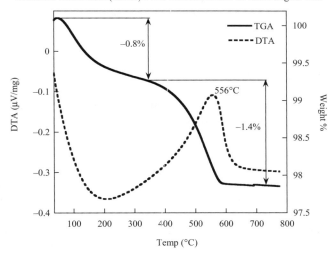

FIGURE 5.13. TGA/DTA in the air of coke burn-off from a catalyst.

In Figure 5.14, a TGA/DTA profile for endothermic desorption of sulfur from a Pd-containing/Al_2O_3 diesel oxidation catalyst is shown. Desulfation from the metal is complete around 800 °C, after which the $Al_2(SO_4)_3$ begins to decompose. The high temperature for its decomposition indicates it is a stable compound and, therefore, is not preferred for a high sulfur-containing gas

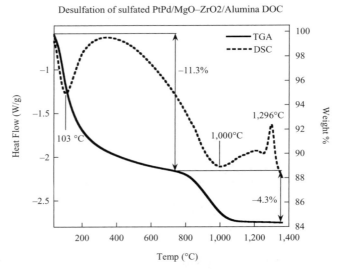

FIGURE 5.14. TGA/DTA profile for desulfation of Pd on Al_2O_3 catalyst.

stream. The $Al_2(SO_4)_3$ is a large-volume compound that blocks pores and causes enhanced diffusional resistance for the reactants, decreasing activity and reducing its life.

5.4 WASHCOAT LOSS

Loss of the catalytic washcoat due to attrition or erosion is a serious source of irreversible deactivation. Washcoat loss from a monolithic environmental catalyst could likely be a problem since the gases are flowing at high linear velocities and process and temperature changes occur rapidly. The thermal expansion differences between the washcoat and the monolith, especially metal substrates, lead to a loss of bonding and to a loss of washcoat. Special surface roughening methods are being used for metal monoliths to provide a more receptive host for generating adherent washcoats. Occasionally proprietary binders containing SiO_2 and/or Al_2O_3 are added to the washcoat formulation to improve the chemical bond between the washcoat and the substrate to ensure an adherent washcoat.

Washcoat loss is observed by preparing a cross section of the honeycomb catalyst and scanning the wall of the channel with either an optical or scanning electron microscope. The loss of catalytic washcoat material is irreversible and results in a shift in the conversion versus temperature curve (Figure 5.4) to higher temperatures. It also decreases the bulk mass transfer conversion achievable, which is directly proportional to the geometric surface area of the washcoat.

5.5 GENERAL COMMENTS ON DEACTIVATION DIAGNOSTICS IN MONOLITHIC CATALYSTS FOR ENVIRONMENTAL APPLICATIONS

5.5.1 Duty Cycles

Catalysts must be prepared with the knowledge of the expected duty cycle to ensure life expectancy. Modern three-way automobile catalytic converters are designed for a lifetime of 150,000 miles to meet U.S. standards. This means it is absolutely necessary to understand the conditions under which the catalyst will be exposed in the hands of the driving public. Will the vehicle be maintained in accordance with the manufacturer's recommendations? Will the vehicle be driven on highways and/or the city? Will excessively high speeds be common? What is the terrain to be driven? What quality of fuel and air will be used for combustion? These are just a few of the conditions that have to be factored into the catalyst and design of the control strategy. Given all of these, what are the most likely causes for anticipated deactivation? In the mid-1970s, it was clear that the presence of lead in the gasoline, added as an octane booster, was the most serious source of deactivation by alloying with the active precious metal components. Lead was subsequently regulated for health reasons and removed by the U.S. federal government in 1975; now it is universally accepted that lead be removed from all gasoline to be used for all catalytically equipped vehicles. The dominant cause of deactivation then became factors such as thermal stresses and poisoning by sulfur present in the fuel. To meet the ever increasing emission standards, governmental regulations were enacted that decreased sulfur content in gasoline and diesel fuel. Today we find the major sources of deactivation are high-temperature sintering of the carrier, the oxygen storage component, and the catalytic metals. Poisoning both selectively and nonselectively by additives in oil such as P, Zn, and Ca is still an issue. Catalyst companies have designed new materials and stabilizers for the oxides present and sintering inhibitors for the catalytic components. The problem of poisoning by oil additives is still a problem. Characterization methods, such as the electron microprobe, provide the researcher the knowledge of where the poisons will be accumulating and thus allow new designs to minimize the negative effects. Washcoat materials, with the proper stabilized pore sizes, can act as filters when properly positioned within the washcoat, protecting the catalytic components from the poisons.

5.5.2 Monitoring Catalytic Activity in a Monolith

It is necessary to know the composition of reactants entering the reactor and the products being produced at its exit. What happens inside the monolith is also critical because catalysts, like many products, are designed to expect some loss of performance with use. So even if no apparent loss in performance is observed at the outlet, it is likely that deactivation is occurring along the length

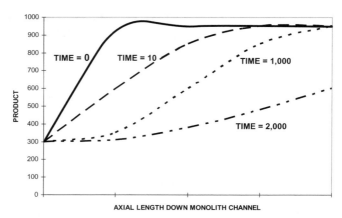

FIGURE 5.15. Conversion proceeding axially down the channel of a monolith with poisoning. Units of time are arbitrary units.

and that performance is being compensated by the catalyst downstream from the major reaction front. This can be best illustrated by reference to Figure 5.15, which shows the product-generated front moving down the bed as deactivation proceeds from inlet to outlet.

The generic illustration is for any catalytic reaction where time is in arbitrary units with no real significance other than to demonstrate the progress of deactivation. At TIME = 0, maximum product formation occurs near the inlet of the axial length. If the catalyst in the front of the bed is being subjected to selective poisoning, the reaction front will begin to move downstream. The rate at which the profile moves downstream increases as the concentration of poisons increases. As poisoning continues, the reaction profile is shifted downstream to a new location at TIME = 100 and TIME = 1,000. Still the outlet concentration of products will not change given a false sense of catalyst stability. Finally, when the bed becomes sufficiently poisoned, breakthrough occurs at TIME = 2,000, indicating that the remaining catalyst cannot sustain the degree of desired conversion. This phenomenon is monitored by inserting analysis probes at various axial locations within the monolith channels so changes can be observed and, if possible, corrective action can be taken before the system undergoes complete failure.

For an exothermic reaction along a catalyzed monolith, typical of environmental pollution abatement applications, the change in temperature with length down the channel ($\Delta T/\Delta L$) is initially very high at the inlet and quickly decreases as the pollutant concentration is reduced to close to zero. This is shown in Figure 5.16 for TIME = 0 where time is arbitrary. The high peak temperature causes catalyst sintering, resulting in a loss in activity, and the maximum moves down the channel (TIME = 100). Effectively the exotherm is distributed more uniformly, reducing the peak temperature. With time the

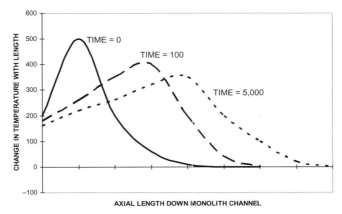

FIGURE 5.16. Temperature profiles ($\Delta T/\Delta L$) for an exothermic reaction down the axial length of a catalyzed–monolith channel caused by sintering.

peak temperature is further reduced and deactivation slows (TIME = 5,000) to essentially zero provided no other mechanisms are operative.

In any high-temperature process, sintering of the catalytic components and the carrier occurs in the early stages but then equilibrates to a given crystal size. Any temperature upsets can cause additional sintering and loss of activity.

REFERENCES

Bartholomew, C. "Catalyst deactivation/regeneration," pp 182–316 in *Encyclopedia of Catalysis*. Editor Horvath, I. Wiley-Interscience, Hoboken, NJ (2003).

Bartholomew, C., and Farrauto, R. *Fundamentals of Industrial Catalytic Processes*. Second Edition. Wiley, Hoboken, NJ (2006).

Chen, J., Heck, R. M., and Farrauto, R. J. "Deactivation, regeneration and poison resistant catalysts: Commercial experience in stationary pollution abatement," *Catalysis Today* 11(4): 517–546 (1992).

Forzatti, P., Buzzi-Ferraris, G., Morbidelli, M., and Carra, S. "Deactivation of catalysts," *Chemical and Kinetic Aspects. International Chemical Engineering* 24(1): 60–73 (1984).

Heck, R., Farrauto, R., and Lee, H. "Commercial development and experience with catalytic ozone abatement in jet aircraft," *Catalysis Today* 13: 43 (1992).

Hegedus, L., and Baron, L. "Phosphorus accumulation in automotive catalyst," *Journal of Catalysis* 54: 115 (1978).

Klimish, R., Summers, J., and Schlatter, J. "The chemical degradation of automobile emission control catalyst," p 103. *ACS series 143*, Editor McEvoy, J. American Chemical Society, Washington, D.C. (1975).

Lampert, J., Kazi, S., and Farrauto, R. "Palladium catalyst performance for methane emissions abatement for lean burn natural gas engines," *Applied Catalysis B: Environmental* 14: 211 (1997).

Oudet, F., Vejux, A., and Courtine, P. "Evolution during thermal treatment of pure and lanthanum doped Pt/Al_2O_3 and $Pt-Rh/Al_2O_3$ automotive exhaust catalysts," *Applied Catalysis* 50: 79–86 (1989).

Trimm, D. "Deactivation and regeneration," pp 1263–1282, in *Handbook of Heterogeneous Catalysis*. VCH, Weinheim, Germany (1997).

Wan, C., and Dettling, J. "High temperature catalyst compositions for internal combustion engines," U.S. patent 4,624,940 (1986).

CHAPTER 5 QUESTIONS

1. How can you use TGA/DTA for catalysis preparation, deactivation, and regeneration?

2. A Pt/Al_2O_3/monolith catalyst is returned from the field because the customer is complaining about poor performance. You extract a sample for laboratory testing to determine deactivation modes. Suggest the causes of deactivation for each set of results:

   a. The conversion/temperature profile is shifted to the right of the fresh catalyst, but the slope is the same. The chemisorption of CO is decreased by 75%, but the BET surface area is unchanged. The Pt crystallite size was examined by microscopy (TEM), and no change in size of the Pt seems to occur.

   b. A Pd/TiO_2/monolithic catalyst used for combusting solvent hydrocarbons emitted from an automobile paint factory shows a small decrease in BET surface area, and the XRD shows that the TiO_2 washcoat has sharper peaks. The CO chemisorption is the same, but the slope of the conversion/temperature profile has decreased. What are the possible causes of deactivation?

3. A Pd/Al_2O_3/monolith ozone decomposition catalyst is removed from the air intake of a Boeing 777 wide-body commercial jet after 10,000 hr of operation. In the lab, it is noted that the conversion/temperature profile is shifted to the right with a reduction in slope. It never reaches its maximum fresh activity. What is the deactivation mode?

4. A customer complains that the catalyst is deactivating in the process stream, but when you remove it and test it in the lab with a model gas stream containing the main components to be converted, there is no sign of deactivation. What are some possible explanations besides the customer being crazy?

5. A $2\% Pt/Al_2O_3$ catalyst, as a tablet, is to be used to abate emissions from a coal-fired power plant. What characterization methods would be used to determine the various modes of deactivation? Also sketch the conver-

sion/temperature profile from a sample of deactivated catalysts using a model reaction in the lab. In some cases, more than one deactivation mode may be possible. Consider each:

   a. The data recorded overnight indicated the catalyst saw an excessively high temperature due to a failed heat exchanger upstream.

   b. A compressor fails upstream, and oil leaks into the exit gas to be abated.

   c. What process and catalyst changes would you suggest to minimize the possible deactivation modes?

   d. Secondary air is necessary to maintain a lean (excess air) environment for the catalyst. If the filter on the injected air stream gets clogged with dust, what might happen to the catalyst?

   e. How might you recover the activity for d?

6. Consider a highly exothermic hydrocarbon oxidation reaction proceeding down the length of a Pd/TiO_2/monolith. The feed-stream contains a high concentration of sulfur dioxide, which is known to poison only the Pd sites selectively, reducing its activity to zero. Assuming you have a series of thermocouple probes inserted at different axial lengths within the monolith, how would the temperature profile (conversion) change with time?

7. a. An inexpensive motorcycle operates fuel rich with a two-cycle engine in which the oil is mixed directly with the fuel. Its exhaust, therefore, contains significant amounts of inorganic components (ash) from the oil such as the oxides of P, Zn, and Ca added as lubricants. This exhaust contains a high concentration of CO, and additional air must be added. A Pt/ZrO_2/monolith is located in the engine exhaust to abate the CO. The ash nonselectively deposits a few millimeters within the inlet of the monolithic catalyst. Assuming you have a series of thermocouple probes inserted at different axial lengths within the monolith, how would the temperature profile (conversion) change with time?

   b. Can you suggest a solution to preventing ash from decreasing the activity of the catalyst?

8. A Cu-exchanged zeolite is used as a selective nitric oxide (NO_x) reduction catalyst to elemental nitrogen in a desulfurized natural gas-fueled power plant operating in an air-rich mode. Ammonia is injected as the selective reductant. The outlet NO_x levels are very low until there is a large temperature spike in the catalyst caused by a leak of some natural gas and its oxidation in the catalyst bed. A large breakthrough of NO_x appears and never decreases. The catalyst is replaced with a new one, but the plant manager wants to understand what changes occurred in the catalyst. What characterization techniques would you use to determine the cause?

PART II
Mobile Sources

6 Automotive Catalyst

6.1 EMISSIONS AND REGULATIONS

6.1.1 Origins of Emissions

The development of the four-cycle, spark-ignited combustion engine permitted the controlled combustion of gasoline that provides the power to operate the automobile. Gasoline, which contains a mixture of paraffins and aromatic hydrocarbons, is combusted with controlled amounts of air producing complete combustion products of CO_2 and H_2O:

$$\text{Hydrocarbons in gasoline} + O_2 \rightarrow CO_2 + H_2O + \text{Heat} \qquad (6.1)$$

and also some incomplete combustion products of CO and unburned hydrocarbons (UHCs). The CO levels range from 1 to 2 vol. %, whereas the unburned hydrocarbons are from 500 to 1,000 vppm. During the combustion process, very high temperatures are reached due to diffusion burning of the gasoline droplets, resulting in thermal fixation of the nitrogen in the air to form NO_x, which is a combination of NO, NO_2, and N_2O (Zeldovich 1946). Levels of NO_x are in the 100- to 3,000-vppm ranges. The exhaust also contains approximately 0.3 moles of H_2 per mole of CO. The quantity of pollutants varies with many of the operating conditions of the engine but is influenced predominantly by the air-to-fuel ratio in the combustion cylinder. Figure 6.1 shows the engine emissions from a spark-ignited gasoline engine as a function of the air-to-fuel ratio (Kummer 1980).

The term used to describe the air-to-fuel (A/F) weight ratio is the lambda (λ) ratio that is defined as the actual air-to-fuel ratio divided by the air-to-fuel ratio at the stoichiometric point or $\lambda = (\text{A/F})_{\text{actual}} / (\text{A/F})_{\text{stoichiometric}}$. The stoichiometric point $\lambda = 1$ is the precise amount of air required to oxidize all of the fuel and for gasoline is approximately 14.6 (wt/wt). For A/F rich (insufficient air) $\lambda < 1$, and for A/F lean (excess air) $\lambda > 1$.

When the engine is operated rich of stoichiometric, the CO and HC emissions are highest, whereas the NO_x emissions are depressed. This is because complete burning of the gasoline is prevented by the deficiency in O_2. The

Catalytic Air Pollution Control: Commercial Technology, by Ronald M. Heck and Robert J. Farrauto, with Suresh T. Gulati.
Copyright © 2009 John Wiley & Sons, Inc.

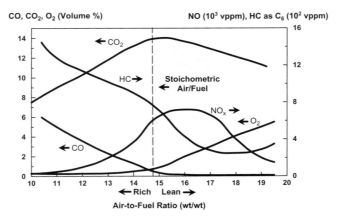

FIGURE 6.1. Spark-ignited gasoline engine emissions as a function of the air-to-fuel ratio. Copyright © 1980, reproduced with kind permission from Elsevier Sciences Ltd. (Kummer 1980).

level of NO_x is reduced because the adiabatic flame temperature is reduced. On the lean side of stoichiometric, the CO and HC are reduced since nearly complete combustion dominates. Again, the NO_x is reduced since the operating temperature is decreased. Just lean of stoichiometric operation, the NO_x is a maximum, since the adiabatic flame temperature is the highest. At stoichiometric, the adiabatic flame temperature is lowered slightly because of the heat of vaporization of the liquid fuel gasoline. The actual operating region of combustion for the spark-ignited engine is defined by the lean and rich flame stability, beyond which the combustion is too unstable (Searles 1989).

Within the region of operation of the spark-ignited engine, a significant amount of CO, HC, and NO_x is emitted to the atmosphere. The consequences of these emissions have been well documented (Viala 1993), but briefly, CO is a direct poison to humans, whereas HC and NO_x undergo photochemical reactions in the sunlight leading to the generation of smog and ozone.

6.1.2 Regulations in the United States

The need to control engine emissions was recognized as early as 1909 (Frankel 1909). The necessity to control automobile emissions in the United States came in 1970 when the U.S. Congress passed the Clean Air Act. The requirements under the Clean Air Act were changing as the technology was being evaluated. As a point of reference, the 1975–1976 Federal (49 states) requirements were 1.5 g/mile HC, 15.0 g/mile CO, and 3.1 g/mile NO_x (Hightower 1974). The Environmental Protection Agency (EPA) established a test procedure (FTP) simulating the average driving conditions in the United States in which CO, HC, and NO_x would be measured. The FTP cycle was conducted on a vehicle dynamometer and included measurements from the automobile during three conditions: (1) cold start, after the engine was idle for 8 hours; (2) hot start; and (3) a combination of urban and highway driving conditions. Separate bags would collect the emissions from all three modes, and a weighing factor would be applied for calculating the total emissions. Complete details on the FTP test procedure are contained later in this chapter. Typical precontrolled vehicle emissions in the total FTP cycle were 83–90 g/mile of CO, 13–16 g/mile of HC, and 3.5–7.0 g/mile of NO_x (Hydrocarbon Processing 1971). Several changes in engine design and control technology were implemented to lower the engine out emissions; however, the catalyst was still required to obtain greater than 90% conversion of CO and HC by 1976 and to maintain performance for 50,000 miles.

Amendments in the early 1990s to the Clean Air Act set up more stringent requirements for automotive emissions (Calvert et al. 1993). The catalysts have been required to last 100,000 miles for new automobiles after 1996. Furthermore, these amendments (which were contingent on tier 2 standards to be set by the Environmental Protection Agency), reduced nonmethane hydrocarbon (NMHC) emissions to a maximum of 0.125 g/mile starting in 2004 (down from 0.41 g/mile in 1991), carbon monoxide to 1.7 g/mile (down from 3.4 g in 1991), and nitrogen oxides to 0.2 g/mile (down from 1.0 g). California in this same time frame continued to set even more stringent regulations: NMHC emissions had to be reduced to 0.075 g/mile by 2000 for 96% of all passenger cars. By 2003, 10% of these had to have emissions no greater than 0.04 g/mile and 10% could emit no NMHCs at all.

The current summary of the California emission standards for passenger cars is given below. LEV is the abbreviation for low-emission vehicle, whereas T is transitional, U is ultra, and S is super. ZEV stands for zero emission vehicle and P is partial. The NMOG is nonmethane organics. As of 2000, the regulations are as follows:

Category	Durability Basis (miles)	NMOG (g/mile)	CO (g/mile)	NO_x (g/mile)
TLEV	50,000	0.125	3.4	0.4
	120,000	0.156	4.2	0.6
LEV	50,000	0.075	3.4	0.05
	120,000	0.09	4.2	0.07
ULEV	50,000	0.04	1.7	0.05
	120,000	0.055	2.1	0.07
SULEV	120,000	0.010	1.0	0.02
PZEV	150,000	0.010	1.0	0.02
ZEV	-0-	-0-	-0-	-0-

Engine manufacturers have explored a wide variety of technologies to meet the requirements of the Clean Air Act. Catalysis has proven to be the most effective passive system. Currently, the major worldwide suppliers of automotive catalysts are BASF (formerly Engelhard), Johnson Matthey, Umicore, and Delphi (now part of Umicore). As the automobile engine has become more sophisticated, the control devices and combustion modifications have proven to be compatible with catalyst technology, to the point where today's engineering design incorporates the emission control unit and strategy for each vehicle. The ZEV vehicles will probably be battery operated. The California Air Resources Board (CARB) is also implementing the PZEV vehicle, which is basically a SULEV vehicle with zero evaporative emissions. Current federal EPA regulations involve tier 1 and NLEV standards with the tier 2 standards effective from the 2004 to the 2009 model year (ICT 2007). There has been some controversy involving the tier 2 standards, and the EPA is moving forward to full enforcement (EPA 2007). Merging of the CARB and EPA standards is likely in the future. There are also gasoline fuel standards in the United States, with the main regulation being 30 ppm of sulfur fuel. The 2007 global emission regulations are available on the International Catalyst Technologies website for further reference and comparison with U.S. and CARB regulations (ICT 2007).

The year 2010 will approach 800 million passenger cars in use worldwide with an annual worldwide production of new cars approaching 100 million. In addition, about 40% more passenger vehicles are represented by trucks. The majority of these vehicles (automobiles and trucks) use a spark-ignited gasoline engine to provide power, and this has become the most frequent form of transportation. Gasoline blend remains a mixture of paraffins and aromatic hydrocarbons that combust in air at a very high efficiency.

However, oxygenates, mostly alcohols derived from crops such as corn or sugar cane, are now added to the fuel to decrease fossil fuel usage. Ethanol has a high octane rating and, thus, replaces aromatics. Octane is a measure of the ability of the fuel to resist precombustion during the compression stroke in the spark-ignition engine. Various blends of ethanol (10% to 20%) are currently in use in the United States, and mixtures of 85% ethanol and 15% gasoline (E85) are now available to decrease the use of gasoline.

6.2   THE CATALYTIC REACTIONS FOR POLLUTION ABATEMENT

The basic operation of the catalyst is to perform the following reactions in the exhaust of the automobile:

Oxidation of CO and HC to CO_2 and H_2O:

$$C_yH_n + (1+n/4)O_2 \rightarrow yCO_2 + n/2\,H_2O \tag{6.2}$$

$$CO + 1/2\,O_2 \rightarrow CO_2 \tag{6.3}$$

$$CO + H_2O \rightarrow CO_2 + H_2 \tag{6.4}$$

Reduction of NO/NO_2 to N_2

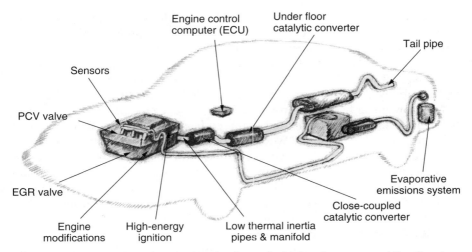

FIGURE 6.2. Location of a catalyst in the underbody of an automobile. Courtesy Engelhard Corporation (now BASF Catalysts).

$$NO\,(or\,NO_2) + CO \rightarrow 1/2\,N_2 + CO_2 \qquad (6.5)$$

$$NO\,(or\,NO_2) + H_2 \rightarrow 1/2\,N_2 + H_2O \qquad (6.6)$$

$$(2 + n/2)\,NO\,(or\,NO_2) + C_yH_n \rightarrow (1 + n/4)\,N_2 + yCO_2 + n/2\,H_2O \qquad (6.7)$$

Hydrogen may also be present in the exhaust, and this reacts rapidly essentially at room temperature and is not usually depicted in the exhaust emissions.

The underbody location of the catalytic converter in the automobile is shown pictorially in Figure 6.2.

When a driver first starts the automobile, both the engine and the catalyst are cold. As the exhaust gradually warms, it reaches a temperature high enough to initiate the catalytic reactions. This is referred to as the *lightoff temperature*, and the rate of reaction is kinetically controlled; that is, it depends on the chemistry of the catalyst since all the transport reactions are fast. Typically, the CO reaction begins first, followed by the HC and NO_x reaction. When the vehicle exhaust is hot, the chemical reaction rates are fast, and pore diffusion and/or bulk mass transfer controls the overall conversion of the exhaust pollutants.

6.3 THE PHYSICAL STRUCTURE OF THE CATALYTIC CONVERTER

Both high-surface-area γ-Al_2O_3 beaded (or particulate) and washcoated monolithic catalysts have been used for passenger vehicles from the onset of automotive emissions controls. In the United States, GM was the major company using spherical beads as a carrier for the catalytic components, whereas Ford and others used γ-Al_2O_3 washcoated monoliths. Today the monolith technology is exclusively used worldwide in automotive exhaust emission control.

In parallel with all the studies related to catalyst screening, deactivation, and durability, many engineering issues needed to be addressed in the early 1970s. How much back pressure would the presence of a catalytic reactor in the exhaust manifold contribute (increased back pressure translates to a loss in power and fuel economy)? Would the catalyst be able to maintain its physical integrity and shape in the extreme temperature and corrosive environment of the exhaust? How much weight would be added to the automobile, and what would be the effect on fuel economy? Where would the catalytic reactor be located? How would the monolith be enclosed in a suitable housing? Would the heat from the reactions within the monolith affect the heat balance of the vehicle? Another complicating problem was that the exhaust catalyst operation is in a continuously transient (large temperature variations in short time durations) mode, in contrast to normal catalyst operation. These problems were critical because the consumer needed a cost-effective, highly reliable, trouble-free vehicle with readily delivered performance.

6.3.1   The Beaded (or Particulate) Catalyst

An important question that the engine manufacturers had to address was how to house the catalyst in the exhaust. The most traditional way was to use spherical particulate γ-Al_2O_3 particles, anywhere from 1/8 to 1/4 in (0.32 to 0.64 cm) diameter, into which the stabilizers and active catalytic components (i.e., precious metals) would be incorporated. These "beads" would be mounted in a spring-loaded reactor bed downstream, just before the muffler. Since the engine exhaust gas was deficient in oxygen, air was added into the exhaust using an air pump. The rationale was simple: Catalysts had been made on these types of carriers for many years in the petroleum, petrochemical, and chemical industries, and manufacturing facilities to mass produce them were already in place. There were known reactor designs and flow models that would make scale-up easy and reliable. One major concern was the attrition resistance of the γ-Al_2O_3 particles, since they would experience many mechanical stresses during the lifetime of the converter. A typical bead bed reactor design for the early oxidation catalysts is shown in Figure 6.3.

The beads are manufactured with the stabilizers (discussed later) incorporated into the structure. The precious metal salts are impregnated into the bead and, using proprietary methods, fixed in particular locations to ensure adequate performance and durability for 50,000 miles. They are then dried at typically 120 °C and calcined to about 500 °C to their finished state. The finished catalyst usually had about 0.05 wt% of precious metal with a Pt-to-Pd weight ratio of 2.5 to 1. After 1979, the need for NO_x reduction required the introduction of small amounts of Rh into the second-generation catalysts. To

FIGURE 6.3. Bead bed reactor design. Courtesy WR Grace & Co.

control the levels of deactivation and the performance of the bead catalysts, many studies were conducted on varying the location of the active catalysts within the bead structure (Hegedus and Gumbleton 1980). The bead catalyst worked initially and gave excellent efficiency regarding removal of the pollutants of CO, HC, and NO_x. The problem was the durability of the beads themselves as the vibrations of the vehicle eventually ground the beads into a smaller size causing settling of the reactor catalyst bed and bypassing of the reactants resulting in poor performance.

6.3.2   The Washcoated Monolith or Honeycomb Catalyst

An alternative approach for supporting the catalytic components was that of a ceramic honeycomb monolith with parallel, open channels (see Figure 2.3, Chapter 2). In the mid-1960s, Engelhard began investigating the use of monolithic structures for reducing emissions from forklift trucks, mining vehicles, stationary engines, and so on (Cohn 1975). Catalyst preparation studies on these PTX (monolithic catalytic exhaust purifiers) formed the basis for washcoating technology for the automotive applications. The effects of operating temperature and feed impurities on catalyst durability were also determined. Some PTX converters had an operational life of 10,000 hr. This background experience showed that the monolithic support was a viable material for automotive applications. The active catalytic components (e.g., precious metals) were deposited within the γ-Al_2O_3, which was washcoated or deposited onto the walls of the honeycomb channels. A washcoated honeycomb typical of that used in the early catalytic converters is shown in Figure 2.3 in Chapter 2. One major advantage would be low-pressure drop, since the honeycomb structure had a very high open frontal area (about 70%) and parallel channels. Furthermore, given their monolithic structure, they could be oriented in several ways to fit in the exhaust manifold. Also, the monoliths were available in different cell densities or cells per square inch (cpsi). From the experience with forklift trucks, there was a small database from which to design catalytic reactors. They offered potential flexibility, but naturally, the materials and geometries had to be optimized and designed for this new and very demanding application.

The ceramic companies continued to modify the materials and structures to provide sufficient strength and resistance to cracking under thermal shock conditions experienced during rapid accelerations and decelerations. The thermal shock condition was eventually satisfied by mechanical design coupled with the use of a low thermal expansion ceramic material called *cordierite* (synthetic cordierite has a composition approximating $2MgO$, $5SiO_2$, and $2Al_2O_3$). In preparing the catalyst, this desirable property has to be matched by the thermal expansion properties of the washcoat to prevent a mismatch in thermal properties. Monolithic structures were ultimately produced by a novel extrusion technique, which allowed mass production to be cost effective. The first honeycomb catalysts to be used in the auto exhaust had 300 cpsi, with a wall thickness of about 0.012 in (0.3 mm) and an open frontal area of about 63%. These dimensions were finalized, based on mechanical specifications and activity performance requirements, to ensure a high degree of contact between the reactants and the catalyst washcoat (high mass transfer) and the lowest possible lightoff temperature.

Later developments in extrusion technology resulted in a 400-cpsi honeycomb with a wall thickness of 0.006 in (0.15 mm) and an open frontal area of 71%. This increased the geometric surface area, which enhanced the rate of mass transfer, the rate-limiting step under most hot steady-state operations in the vehicle driving cycle.

Catalyst companies began to explore these new structures as catalyst supports. They developed slurries of the catalytic coating that could be deposited onto the walls of the honeycomb producing adherent "washcoats." The washcoat thickness could be kept at a minimum to decrease pore diffusion effects while allowing sufficient thickness for anticipated aging due to deposition of contaminants. The washcoat is about 20 and 60 microns on the walls and corners (fillets), respectively. One method of preparing a washcoated honeycomb is to submerge it in a slightly acidified slurry (slip) containing the γ-Al_2O_3 already impregnated with stabilizers and precious metals. The washcoat bonds chemically and physically to the honeycomb surface, where some of the washcoat fills the large pores of the ceramic. The slurry must have the proper particle size distribution to be compatible with the pores of the ceramic wall. Another method involves first washcoating the honeycomb with the alumina slurry, drying and calcining, and then dipping it into the impregnating solutions. The coated honeycomb is air dried and calcined to about 450 °C to 500 °C, which ensures good adhesion. Typically, the catalyst contains about 0.1% to 0.15% of precious metals. For the oxidation catalysts of the first generation, the weight ratio of Pt to Pd was 2.5 to 1, whereas the second generation contained a weight ratio of 5 Pt to 1 Rh.

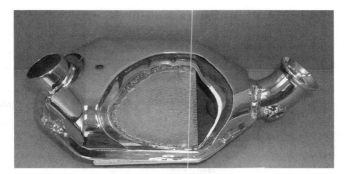

FIGURE 6.4. Monolithic reactor design. Courtesy Engelhard Corporation (now BASF Catalysts).

The honeycomb catalyst is mounted in a steel container with a resilient matting material wrapped around it to ensure vibration resistance and retention (Keith et al. 1969). A positive experience with honeycomb technologies has resulted in increased use of these structures over that of the beads due to size and weight benefits. Today almost all automobiles are equipped with honeycomb-supported catalysts similar to that shown in Figure 6.4.

Although the early honeycombs were ceramic, within the last 10 years, metal substrates have been finding use because they can be made with thinner walls and have open frontal areas of close to 90%, allowing for a lower pressure drop (see Figure 2.3 of Chapter 2). Cell densities greater than 400 cpsi can be used, which permits smaller catalyst volumes when higher cell densities are used. To coat the metallic monoliths, the surface must be heat treated to form an alumina phase on the surface so the washcoat can adhere. The actual depositing of the washcoat is done basically the same as the ceramic counterpart. Because of expense and possible temperature limitations, these catalysts are not preferred. However, they are finding some niche markets because of their low-pressure-drop characteristics, low heat capacity for fast warm-up, and rapid responses to transient temperature operations. They are also finding wide use in stationary applications, which will be discussed later.

By the year 2010, over 40 years of catalyst technology development will have been devoted to the automotive exhaust catalyst. Figure 6.5 shows a typical auto catalyst design.

These technology advances have been driven by the quest for a zero emission vehicle using the spark-ignited engine as the powertrain. Along with the advances in catalyst technology, the automotive engineers were developing new engine platforms and new sensor and control technology. This has resulted in the full integration of the catalyst into the emission control system. The catalyst has become integral in the design strategy for vehicle operation. During this time period, the auto catalyst has progressed through the following development phases:

Three-Way Catalyst (TWC) Design

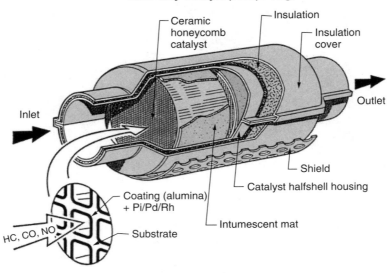

FIGURE 6.5. Schematic of cutaway of typical auto catalyst design. Courtesy Engelhard Corporation (now BASF Catalysts).

Oxidation catalyst
— bead and monolith support
— HC and CO emissions only
— Pt-based catalyst
— stabilized alumina

Three-way catalyst
— HC, CO, and NO_x emissions
— Pt/Rh-based catalyst
— Ce oxygen storage

High-temperature, three-way catalyst
— approaching 950 °C
— stabilized Ce with Zr
— Pt/Rh, Pd/Rh, and Pt/Rh/Pd

All-Palladium three-way catalyst
— layered coating
— stabilized Ce with Zr

Low-emission vehicles
— high-temperature no Ce close-coupled catalyst, approaching 1,050 °C
— with underfloor catalyst

Ultra-low-emission vehicles
— high-temperature no Ce close-coupled catalyst, approaching 1,050 °C
— increased volume underfloor, higher precious metal loading
— optional trap

Engineered catalyst design
— rapid lightoff engine strategy
— fast response sensors
— linear air-to-fuel ratio sensor
— fast-response computer algorithm to closed loop control
— lean start engine
— zoned catalyst
— layered catalyst
— low precious metals
— high cell density
— turbulent flow monolith
— cascade close-coupled catalyst

The implementation of the engineered catalyst design is well underway in the emission control systems is a U.S. vehicle and has progressively advanced in Europe, Japan, and elsewhere. This technology base and experience is now available for implementation worldwide as required.

6.4 FIRST-GENERATION CONVERTERS: OXIDATION CATALYST (1976–1979)

During the early implementation of the Clean Air Act, the catalyst was only required to abate CO and HC. The NO_x standard was relaxed so engine manufacturers used exhaust gas recycle (EGR) to meet the NO_x standards. With EGR, a small portion of the exhaust, rich in high heat capacity containing H_2O, CO_2, and N_2, is recycled into the combustion chamber thereby lowering the combustion flame temperature, which results in less thermal NO_x formation as predicted by the Zeldovich mechanism (Zeldovich 1946). The engine was operated just rich of stoichiometric to reduce the formation of NO_x, and secondary air was pumped into the exhaust gas to provide sufficient O_2 for the catalytic oxidation of CO and HC on the catalyst.

During this period, many catalytic materials were studied and the area of high-temperature stabilization of alumina was explored. It was known that the precious metals, Pt and Pd, were excellent oxidation catalysts; however, the cost and supply of these materials was bothersome. Therefore, many base metal candidates were investigated, such as Cu, Cr, Ni, Mn, and so on. They were less active than the precious metals but substantially cheaper and more readily available.

Table 6.1 shows the relative activities of Pt and Pd versus non-precious metal oxides (base metal oxides) for oxidizing simulated exhaust pollutants at 300 °C (Kummer 1975). From the relative activities, it is clear the precious metals are considerably more active than the base metals. Also, the activity depends on the species to be catalyzed. Palladium is the most active for CO and ethylene oxidation, whereas Pt is equally as active for ethane oxidation. Precious metals would, therefore, be preferred over base metals if not for the expense and limited availability. The base metal oxides could be viable, but

TABLE 6.1. Comparison of relative activities of precious and base metal catalysts for different reactants. Reprinted with permission, copyright © 1975, American Chemical Society (Kummer 1975).

Reactant	1% CO	0.1% C_2H_6	0.1% C_2H_8
Pd	500	100	1
Pt	100	12	1
Co_2O_3	80	0.6	0.05
CuO/Cr_2O_3	40	0.8	0.02
Au	15	0.3	<0.2
MnO_2	4.4	0.04	—
CuO	45	0.6	—
$LaCoO_3$	35	0.03	—
Fe_2O_3	0.4	0.006	—
Cr_2O_3	0.03	0.004	0.008
NiO	0.013	0.0007	0.0008

their lower activity would require larger reactor volumes (lower space velocities). This would be a problem in the engine exhaust underfloor piping where space is at a premium. Studies also showed that the base metal oxides were susceptible to sulfur poisoning (Farrauto and Wedding 1974; Fishel et al. 1974; Taylor 1990). Still interest exists for base metal systems as shown by the Cu-based systems (Kapteijn et al. 1993; Theis and LaBarge 1992); however, to date, no use of base metal oxides as primary catalytic components has successfully been used for vehicle emission control.

Therefore, the first-generation oxidation catalysts were a combination of Pt and Pd and operated in the temperature range of 250 °C to 600 °C, with space velocities varying during vehicle operation from 10,000 to 100,000 1/hr, depending on the engine size and mode of the driving cycle (i.e., idle, cruise, or acceleration). Typical catalyst compositions were Pt and Pd in a 2.5:1 or 5:1 ratio ranging from 0.05 to 0.1 troy oz/car (a troy oz is about 31 g).

6.4.1   Deactivation

The oxidation catalyst was negatively affected by the exhaust impurities of sulfur oxides and tetraethyl lead from the octane booster, both present in the gasoline, and phosphorus and zinc from engine-lubricating oil (Doelp et al. 1975; Acres, et al. 1975). An example of one of these studies (see Figures 6.6 and 6.7) shows the effect Pb has on Pd versus Pt catalysts for the temperature at which 90% of the pollutant is converted (Doelp et al. 1975). The catalyst was formulated to have a constant 0.05 wgt.% of total precious metal.

Clearly, the addition of Pt improved the resistance to Pb poisoning by showing a continuous decrease in the 90% conversion temperature. This study

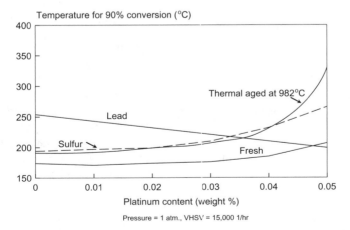

FIGURE 6.6. Effect of lead, sulfur, and thermal aging on carbon monoxide activity for Pt + Pd combined oxidation catalyst. Reprinted with permission, © 1975 American Chemical Society (Doelp et al. 1975).

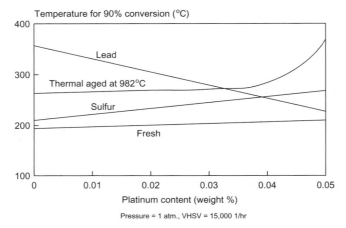

Pressure = 1 atm., VHSV = 15,000 1/hr

FIGURE 6.7. Effect of lead, sulfur, and thermal aging on propylene activity for Pt + Pd combined oxidation catalyst. Reprinted with permission, © 1975 American Chemical Society (Doelp et al. 1975).

also noted an improvement in sulfur resistance for increases in Pd content. When the catalysts were aged in an oxidizing atmosphere at 982 °C, the Pd catalyst retained more activity relative to Pt. Pt catalysts do sinter in oxidizing environments to a much greater degree than Pd (Klimisch et al. 1975), so Pt loses activity relative to Pd after thermal aging at 982 °C, as shown.

As the research was ongoing for improved catalyst compositions, the Pb present as an octane booster continued to deactivate most severely all the catalytic materials. The poisoning of Pt and Pd by the traces of Pb (about 3–4 mg/gallon of Pb were in "unleaded" gasoline) was caused by formation of a low-activity alloy:

$$Pt \text{ or } Pd + Pb \xrightarrow{\text{air}, 900°C} PtPb \text{ or } PdPb \tag{6.8}$$

Electron microscopy studies showed that the Pb deposited primarily in a thin shell (a few microns thick) near the outer edge of the Al_2O_3 carrier and close to the gaseous interface. Studies revealed that the Pt was more tolerant than Pd to Pb poisoning, so preparation processes were developed that permitted the deposition of the Pt slightly below the surface, whereas the Pd had a deeper, subsurface penetration. Additional pore diffusion resistance was then introduced, and although causing a small activity penalty, it provided an improved life for the catalyst.

While the studies on the automotive catalyst were proceeding, other environmental studies were conducted that showed the severe effect of lead present in the environment. There was a growing concern that Pb was a direct poison to humans. These studies contributed to the decision that led the federal government to mandate its removal from gasoline. This proved a benefit for auto emission control. The use of catalysts was now more feasible in meeting the 50,000-mile performance requirements. It is interesting to note that removal of lead from gasoline was still the first step in implementing automobile emission control technologies in the emerging countries such as India and China (MECA a 1998).

However, the operating environment of the catalyst was still hostile in that phosphorus (P) and zinc (Zn), derived from lubricating oil and sulfur (S) from the fuel itself, were present as well as the severe temperature transients and possible temperature exposure of 800–1,000 °C maximum.

The combinations of Pt and Pd dispersed on to high-surface-area γ-Al_2O_3 particles were found to have reasonably good fresh activity. After high-temperature aging (900 °C in air/steam to simulate engine exhaust conditions), the catalyst usually lost some of its activity, as evidenced by increased temperatures for 50% conversion of both CO and HC. Characterization of the partially deactivated catalyst by BET surface area measurements and X-ray diffraction patterns showed that the γ-Al_2O_3 had undergone severe sintering to a lower surface area, more crystalline phase such as α-Al_2O_3 (see Chapter 5: Catalyst Deactivation). The high-area pore structure of the γ-Al_2O_3 effectively collapses and occludes the active catalytic species, making them inaccessible to the reactants (see Chapter 5, Figure 5.5). Naturally, this results in a loss of catalytic performance. Since no other carrier materials had all the desirable properties of γ-Al_2O_3, research was directed toward understanding and minimizing the sintering mechanisms of γ-Al_2O_3 under auto exhaust conditions. It was known that certain contaminants such as Na and K acted as fluxes, accelerating the sintering process of γ-Al_2O_3. Thus, preparations had to exclude these elements. In contrast, small amounts (1% to 3%) of other elements, such as La_2O_3 (Kato et al. 1987; Tijburg et al. 1991), BaO (Machida et al. 1988), and SiO_2, (Beguin et al. 1991) if properly incorporated into the preparation process, had a stabilizing effect on the γ-Al_2O_3 and significantly reduced its sintering rate. Figure 6.8 shows a representative response of the temperature on the BET surface area and gives a comparison of some typical surface areas after high-temperature treatment at 1,200 °C, with and without stabilizers present (Wan and Dettling 1986). Surface areas of 150–175 m^2/g are typical for the aluminas in modern auto catalysts.

Although the precise mechanism for their stabilizing effect is not known, high-resolution surface studies (Tijburg et al. 1991) have indicated that these oxides enter into the surface structure of the γ-Al_2O_3 and greatly diminish the rate of the chemical and physical changes occurring normally during sintering. Other studies have shown that the elements primarily act on free (grain) surfaces, rather than in the bulk volume obtained using first-principles atomistic calculations (Glazoff and Novak 2003).

The development of thermally stable, high-surface-area γ-Al_2O_3 by the incorporation of oxides was a breakthrough in materials technology, and its use uncovered another problem: agglomeration or sintering of the Pt and Pd during high-temperature exposure in the automobile exhaust (Wanke and

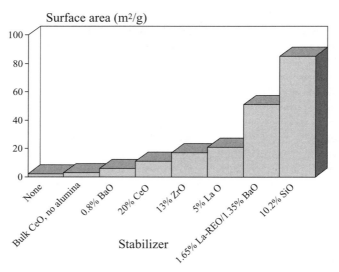

FIGURE 6.8. Thermal stabilization of aluminas for automotive catalyst after 1,200 °C aging (Wan and Dettling 1986).

Flynn 1975). Hydrogen chemisorption and X-ray diffraction (XRD) studies revealed that the Pt and Pd, initially well dispersed on stabilized γ-Al_2O_3, had undergone significant crystallization after high-temperature treatment. The Pd was more stable than the Pt as indicated in Figure 6.9 for CO conversion (Klimisch et al. 1975).

The first-generation oxidation catalysts comprised a 2.5-to-1 weight ratio of Pt to Pd at about 0.05% of total precious metal for beads and at about 0.12% for honeycombs. Stabilizers such as CeO_2, and La_2O_3 were also included in the formulation to minimize sintering of the γ-Al_2O_3 carrier.

6.5   NO$_X$, CO AND HC REDUCTION: THE SECOND GENERATION: THE THREE-WAY CATALYST (1979–1986)

After the successful implementation of catalysts for controlling CO and HC, the reduction of NO_x emissions in the automobile exhaust to less than 1.0 g/mile had to be addressed. NO_x reduction is most effective in the absence of O_2, whereas the abatement of CO and HC requires O_2. The exhaust emanating from the engine can be made sufficiently rich so that a catalyst that can reduce NO_x could be positioned upstream of the air injection system in the exhaust and oxidizing catalyst. With this arrangement, the H_2, CO, and HC could first reduce the NO_x, and whatever amount remaining would be oxidized in the second bed. A primary catalyst for the reduction reaction was Ru; however, on an occasion when the engine exhaust might be oxidizing and the tempera-

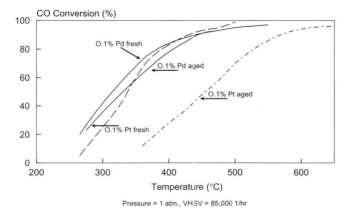

Pressure = 1 atm., VHSV = 85,000 1/hr

FIGURE 6.9. Effect of thermal aging of 70 hr at 900 °C for CO activity on Pt and Pd oxidation catalysts. Reprinted with permission, © 1975, American Chemical Society (Klimisch et al. 1975).

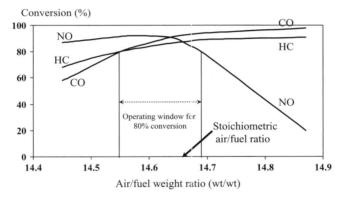

FIGURE 6.10. Simultaneous conversion of HC, CO, and NO$_x$ for TWC as a function of air-to-fuel ratio.

ture exceeded about 700 °C, it was found to volatilize by forming RuO$_2$. This was dropped from further consideration. If Pt or Pd were used instead of Ru, the NO$_x$ was reduced to NH$_3$ and not to N$_2$. The NH$_3$ would then enter the oxidation catalyst and be reconverted to NO$_x$. In previous studies of reducing atmospheres, Rh had also been shown to be an excellent NO$_x$ reduction catalyst (Cohn 1964). It had less NH$_3$ formation than Pt or Pd.

If the engine exhaust could be operated close to the stoichiometric, air-to-fuel ratio $\lambda = 1$, all three pollutants with the right catalyst (Thompson et al. 1979) could be simultaneously converted, and the need for a two-stage reactor with air injection could be eliminated. Figure 6.10 shows such a curve in which the NO$_x$ reduction via reaction with HC and CO occurs readily when the exhaust is rich ($\lambda < 1$), whereas the CO and HC oxidation reactions are prevented by insufficient O_2. As the air-to-fuel ratio approaches the stoichiometric point, there is a narrow window where simultaneous catalytic conversion of all three occurs. On the lean side ($\lambda > 1$), the CO and HC conversions are high, but at the sacrifice of the NO_x conversion since all the reductant is oxidized.

The key to advancing this technology was to control the air-to-fuel ratio of the automobile engine within this narrow window at all times. This was made possible by the development of the O_2 sensor, which was positioned immediately before the catalyst in the exhaust manifold (Wang et al. 1993). The exhaust gas oxygen (EGO) or lambda sensor was composed of an anionic conductive solid electrolyte of stabilized zirconia with electrodes of high-surface-area Pt. Very few solids are like stabilized ZrO_2, which is an anionic 100% oxygen ion conductor. One electrode was located directly in the exhaust stream and sensed the O_2 content, whereas the second was a reference positioned outside of the exhaust in the natural air. The electrode is a catalyst in that it converted the HC and CO at its surface, provided sufficient O_2 was present. For correct air/fuel control, the sensor must at all times equilibrate the exhaust gas. If the exhaust was rich, then the O_2 content at the electrode surface was quickly depleted. For the condition of a lean exhaust, some O_2 remained unreacted, and the electrode sensed its relative high concentration. The voltage (E) generated across the sensor was strongly dependent on the O_2 content and is represented by the Nernst equation. The sensor is an oxygen concentration cell:

$$E = E_o + RT/nF \ \ln(PP_{O2})_{reference}/(PP_{O2})_{exhaust} \qquad (6.9)$$

where

n   = number of electrons transferred
F   = Faraday constant
PP_{O2}= partial pressure of oxygen
E_o = standard state voltage.

The voltage signal generated is fed back to the fuel injection control device (i.e., throttle body injector or multipoint injectors), which adjusts the air-to-fuel ratio. Figure 6.11 shows the response profile for the O_2 sensor. Note that it functions similar to a potentiometric titration curve used in aqueous analytical chemistry.

The total device is a very sophisticated electronic control system to maintain the air-to-fuel ratio within the narrow window, which allows the simultaneous conversion of all three pollutants. This technology, referred to as *three-way-catalysis* (TWC), was first installed in large quantities on vehicles in 1979. Even today, the oxygen sensor is the state of the art in air-to-fuel ratio control in the gasoline internal combustion engine.

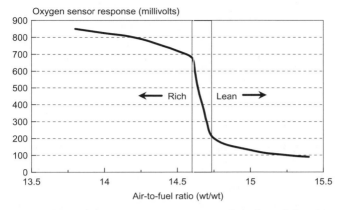

FIGURE 6.11. Oxygen sensor response output as a function of the air-to-fuel ratio.

Modern sensors have been modified to be more poison tolerant to P and Si found in the engine exhaust. Also to improve the operating range of the O$_2$ sensor during driving—particularly in cold start—the heated O$_2$ sensor was developed (Wiedenmann et al. 1984). This is referred to as the *heated exhaust gas oxygen* (HEGO) type sensor. A simple schematic of the vehicle control loop showing the exhaust catalyst and O$_2$ sensor is shown in Figure 6.12.

Because the control system uses "feedback," there is a time lag associated in adjusting the air-to-fuel ratio. This results in a perturbation around the control setpoint. This perturbation is characterized by the amplitude of the A/F ratio and the response frequency (Hz).

A single catalyst technology to convert all three pollutants simultaneously was quickly developed (Thompson et al. 1979). The primary precious metals were Pt and Rh, with the latter being most responsible for reduction of NO$_x$ (although it also contributes to CO oxidation along with the Pt). The oscillatory nature of the air-to-fuel ratio in the exhaust means the catalyst alternatively will see slightly rich and slightly lean conditions.

The effect of this oscillation on an aged Pt/Rh containing TWC catalyst can be measured with an engine dynamometer test stand. The flow rate and temperature were kept constant, and the perturbation amplitude and frequency are varied. At steady state, the curves for CO, NO$_x$, and HC conversion versus A/F ratio are very sharp, and the conversions are high. Increasing the amplitude causes a decrease in the absolute conversions and a broadening of the response curves. Similar results are observed by decreasing the frequency. The best control strategy to adapt to the engine is to have a small amplitude (less than 0.5 A/F) and a frequency greater than 1.5 Hz. Figure 6.13 shows the effect of frequency on the TWC catalyst performance. Note that all three reactants, CO, HC, and NO$_x$, have higher conversions at the lower amplitude (0.5 A/F) and higher frequency (1.5 Hz). Additional studies on the effects of the exhaust

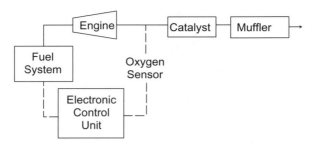

FIGURE 6.12. Major elements of the automotive feedback control system.

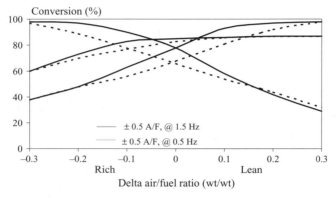

FIGURE 6.13. Increasing the perturbation frequency benefits TWC performance.

gas perturbation can be found in the literature (Heck et al. 1989; Kaneko et al. 1978; Falk and Mooney 1980; Schlatter et al. 1979; Muraki et al. 1985).

Thus, when operating rich, there was a need to provide a small amount of O_2 to consume the unreacted CO and HC. Conversely, when the exhaust goes slightly oxidizing, the excess O_2 needs to be consumed. This was accomplished by the development of the O_2 storage component, which liberates or adsorbs O_2 during the air-to-fuel perturbations (Harrison et al. 1988; Fisher et al. 1993). CeO_2 was found to have the proper redox response and is the most commonly used O_2 storage component in modern three-way catalytic converters. The reactions are indicated below eq. not balanced:

$$\text{Rich condition:} \quad CeO_2 + CO \rightarrow Ce_2O_3 + CO_2 \qquad (6.10)$$

$$\text{Lean condition:} \quad Ce_2O_3 + 1/2\,O_2 \rightarrow CeO_2 \qquad (6.11)$$

Another benefit of CeO_2 is that it is a good steam-reforming catalyst and thus catalyzes the reactions of CO and HC with H_2O in the rich mode. The H_2 formed then reduces a portion of the NO_x to N_2:

$$CO + H_2O \xrightarrow{CeO_2} H_2 + CO_2 \qquad (6.12)$$

$$C_xH_y + 2H_2O \rightarrow (2 + y/2)H_2 + xCO_2 \qquad (6.13)$$

The hydrogen thus formed reduces NO_x via the following:

$$NO_x + xH_2 \rightarrow 1/2 N_2 + xH_2O \qquad (6.14)$$

Other oxide materials with similar oxygen lability, such as NiO/Ni and Fe_2O_3/FeO, have also been used as storage components. Both Figures 6.10 and 6.13 show the response of TWCs fully formulated with Ce. If Ce were not present, the CO and HC conversion on the rich side of stoichiometric would decline dramatically since there would be no O_2 available for the oxidation reactions. This is why a stable Ce-containing compound is required to maintain the activity of the oxygen storage for rich side oxidation reactions. Currently, combinations of oxides of cerium and zirconium are common along with other additional stabilizers. These will be discussed later.

The three-way catalysts of the late 1980s were primarily composed of about 0.1% to 0.15% of precious metals at a Pt-to-Rh ratio of 5 to 1, high concentrations of bulk high-surface-area CeO_2 (10–20%), and the remainder being the γ-Al_2O_3 washcoat. The γ-Al_2O_3 is stabilized with 1% to 2% of La_2O_3 and/or BaO. This composite washcoat is then deposited on a honeycomb with 400 cells per square inch. Typically, the washcoat loading is about 1.5 to 2.0 g/in³ or about 15% of the weight of the finished honeycomb catalyst. The size and shape of the final catalyst configuration varies with each automobile company, but typically, they are about 5 to 6 in (12.7 to 15.2 cm) in diameter and 3 to 6 in (7.6 to 15.2 cm) long.

6.6 VEHICLE TEST PROCEDURES (U.S., EUROPEAN, AND JAPANESE)

The final performance tests for TWCs include the 1975 FTP (Bosch 1996). This procedure was also used for the oxidation catalysts, but the interactions with the driving mode were not as predominate. This was because the vehicle often was lean biased and/or oxygen was injected in the exhaust via an air pump that decoupled the catalyst from the engine operation. However, for TWCs, any fluctuation in the engine operation changes the exhaust gas stoichiometry (air-to-fuel ratio) and dramatically affects catalyst performance.

The actual development of the 1975 FTP using a vehicle on a chassis dynamometer began in the 1950 time period. It evolved through the 1960s and finally, in 1975, was adapted by the EPA as the 1975 FTP. It is basically a driving cycle through Los Angeles, California, and has the following characteristics:

- Cycle Length: 11.115 miles
- Cycle Duration: 1,875 s plus 600-s pause

Bag 1	0–505 s
Bag 2	505–1,370 s
Hot soak	600 s
Bag 3	0–505 s
• Average Speed:	34.1 km/hr (21.2 miles/hr)
• Maximum Speed:	91.2 km/hr (56.7 miles/hr)
• Number of Hills:	23
• Number of Modes:	112

Emissions are measured using a constant volume sampling system. The test begins with a cold-start Phase 1 or Bag 1 (at 20–30 °C) after a minimum 12-hr soak at constant ambient temperature. After 505 s, the vehicle is driving at the speeds indicated in the Phase 2 or Bag 2 hot stabilized portion. The vehicle is then run idle for 600 s, after which the Phase 3 cycle or Bag 3 (hot start) is implemented. This phase is identical to the speeds and accelerations indicated in Phase 1. The actual driving cycle is shown in Figure 6.14.

The driving cycles used for Japan and Europe are different from the 1975 FTP. A comparison of these cycles is shown in the literature (Bosch 1996) and in Figures 6.15 and 6.16. Note that the European test procedure shown in Figure 6.15 was revised to include the "cold-start" or "key-on" portion of the driving cycle. A prior European test did not include emission sampling during the "cold-start" portion. Also, the new Japanese driving cycle will take effect in 2009 for light-duty vehicles (Figure 6.17).

Information on test cycles from other countries is available (ICT 2007).

The EPA conducted a test program to revise the 1975 FTP to reflect the high-speed and acceleration driving habits of vehicle operators in the United States. The new revised Federal Test Procedure (RFTP) contains two new test cycles to include driving encountered in the United States. The following excerpt from the Federal register contains the rationale for these changes:

SUMMARY: This rulemaking revises the tailpipe emission portions of the FTP for light-duty vehicles (LDVs) and light-duty trucks (LDTs). The primary new element of the rulemaking is a Supplemental Federal Test Procedure (SFTP) designed to address shortcomings with the current FTP in the representation of aggressive (high speed and/or high acceleration) driving behavior, rapid speed fluctuations, driving behavior following startup, and use of air conditioning. An element of the rulemaking that also affects the preexisting "conventional" FTP is a new set of requirements designed to more accurately reflect real road forces on the test dynamometer.

The U.S. SFTP portion that simulates high-speed and rapid acceleration is called the US06, whereas the portion that simulates the added engines load from air conditioner operation is the SC03. These new test cycles are shown in Figures 6.18 and 6.19.

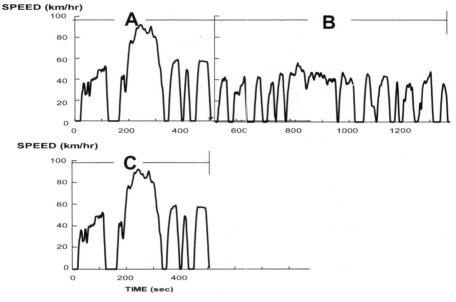

Speed trace for 1975 Federal Test Procedure
    Segment A: cold transient
    Segment B: hot stabilized
    Segment C: hot transient (after 10-min soak)

FIGURE 6.14. 1975 U.S. Federal Test Procedure.

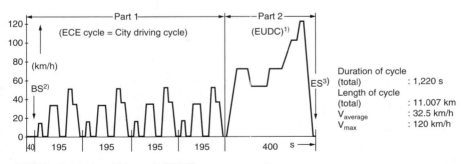

1) EUDC = Extra urban driving cycle (EUDC)
2) Begin of sampling (after 40 s)
3) End of sampling (1,220 s)

Note: 40-s idle eliminated for Stage III (2000) and Stage IV (2005) standards

FIGURE 6.15. European test driving cycles include ECE and EUDC.

1. 11-mode cycle

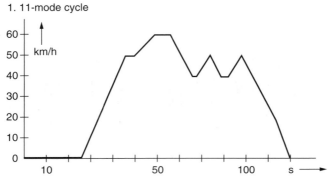

Duration of cycle	: 120 s
Distance per cycle	: 1.021 km
No. of cycles / test	: 4
Total test time	: 505 s
Total distance	: 4.084 km
Average speed	: 30.6 km/h (39.1 km/h) *)
Max. speed	: 60 km/h

*) without idle phases (idle time = 21.7%)

11-mode cold start test: The 11-mode cycle is to be run 4 times, measurements are taken in all 4 cycles. After cold start 25 s idle. The transmission gears to be used are specified for 3- and 4-speed transmissions. For special transmissions, the gear ratios to be used are specified individually; for automatic transmissions only position "Drive." Exhaust emission analysis with CVS-system.

(a)

2. 10 • 15-mode cycle

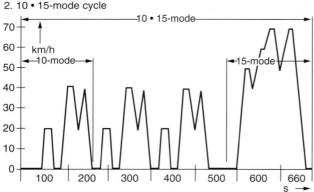

Duration of cycle	: 660 s
Distance per cycle	: 4.16 km
No. of cycles / test	: 1
Average speed	: 22.7 km/h (33.1 km/h) *)
Max. speed	: 70 km/h

*) without idle phases (idle time = 31.7%)

10 • 15-mode hot start test: Preconddtioning 15 min with 60 km/h, followed by 5 min at 60 km/h and one 15-mode cycle. After the preconditioning, the combined 10 • 15-mode (3 cycles 10-mode plus 1 cycle 15-mode) is to be run and measured once.

(b)

FIGURE 6.16. Japanese test driving cycle has two modes: a) 11-mode cycle and b) 10–15-mode cycle.

The transients in each engine chassis dynamometer driving cycle consist of changes in exhaust composition, temperature, speed, and air-to-fuel ratio. Figure 6.19 shows the indicated transients for speed, temperature, and air-to-fuel ratio for a 1996 Lincoln Towncar during the U.S. FTP. The actual vehicle

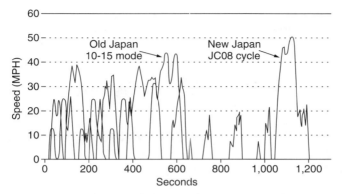

FIGURE 6.17. New Japanese driving cycle JC08 compared with the 10–15 mode.

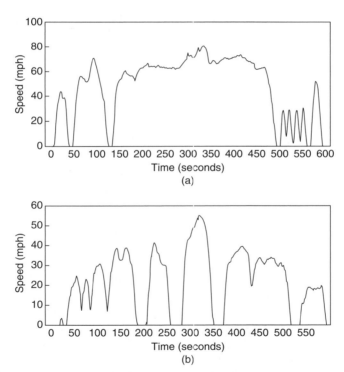

FIGURE 6.18. a) U.S. SFTP test cycle—US06 portion simulates high-speed driving. b) U.S. SFTP test cycle—SC03 portion simulates high-speed driving with air conditioner.

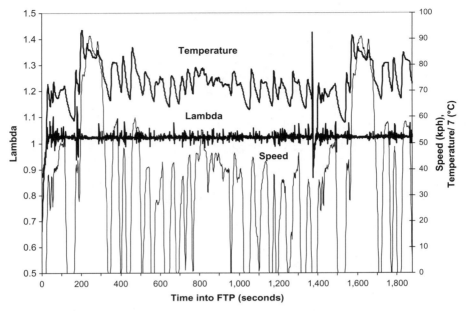

FIGURE 6.19. Transient response of air-to-fuel ratio, exhaust temperature, and speed of 1996 Lincoln Towncar during 1975 FTP.

control strategy is different for each model and manufacturer. This results in different perturbations and different transient responses in the exhaust environment of the TWC catalyst. The effect of these variables has been studied for commercial vehicles with widely different control strategies (Adomaitis and Heck 1988).

6.7 NO$_X$, CO, AND HC REDUCTION: THE THIRD GENERATION (1986–1992)

The latter part of 1980 and the early 1990s required improvement in technology because of the automobile's changing operational strategies: Fuel economy was important; yet operating speeds were higher. This situation resulted in higher exposure temperatures to the TWC catalyst. Higher fuel economy was met by introducing a driving strategy whereby fuel is shut off during deceleration. The catalyst, therefore, is exposed to a highly oxidizing atmosphere that results in deactivation of the Rh function by reaction with the γ-alumina, forming an inactive rhodium–aluminate species.

To simulate these modes of deactivation in the laboratory, engine dynamometer aging cycles were set up to simulate 50,000 to 100,000 miles of performance. These aging cycles consisted of repetitive steps of changing the engine speed/load, air-to-fuel ratio, and exhaust temperature. In some cases,

Engine Aging Cycles

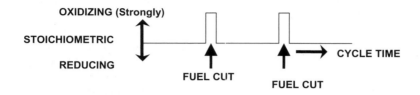

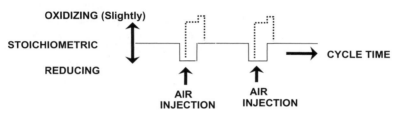

FIGURE 6.20. Description of engine dynamometer aging cycles.

air was injected downstream into the exhaust, whereas in others, the engine was connected to a flywheel and an actual fuel cut was simulated. In addition, some cycles were isothermal, whereas others were exothermal, generating a large temperature rise on the catalyst surface. Figure 6.20 gives a summary of the engine out stoichiometry for the two different types of aging cycles.

For aging cycles simulating a fuel cut, the engine load is adjusted to produce catalyst inlet temperatures ranging from 750 °C to 1,000 °C. This cycle is essentially isothermal with respect to the catalyst surface temperature. This is sometimes referred to as the fuel-cut aging cycle. The European example of fuel-cut aging is the ZDAKW cycle. The aging cycles generating an exotherm on the catalyst surface usually have an inlet temperature from 650 °C to 850 °C, with catalyst bed temperatures ranging from 850 °C to 1,100 °C, depending on the concentration of the CO in the engine exhaust. The bed temperature is measured 1.0 to 1.5 in (2.54 to 3.8 cm) from the start of the catalyzed honeycomb. The GM rapid aging cycle (RAT) is an example of the exothermal cycle. Still other cycles looked at accelerated poisoning of the catalyst by doping the lubricating oil with high concentrations of phosphorous (Kumar et al. 2003; Favre and Zidat 2004).

The EPA has proposed to replace the current on-road aging cycle with a manufacturer-designed durability process, whereby each manufacturer would be required to design a durability process that would match the in-use deterioration of the vehicles they produce. Allowing the manufacturers to develop and validate their own durability programs would reduce the level of emission noncompliance in-use, thus improving the overall ambient air quality. Each automotive manufacturer would have its rapid aging cycle qualified by the EPA, and then this cycle would be used to age the catalyst on an engine dyno for certification performance tests on the vehicle chassis dyno. The EPA issued a notice of proposed rulemaking (NPRM) to propose procedures to be used by manufacturers of light-duty vehicles, light-duty trucks, and heavy-duty vehicles to demonstrate, for the purposes of emission certification, that new motor vehicles will comply with EPA emissions standards throughout their useful lives. The NPRM proposed emissions certification durability procedures known as "CAP 2000" (Compliance Assurance Program) to be used by manufacturers to demonstrate the expected rate of deterioration of the emission levels of their vehicles (EPA 2004; EPA 2006). The proposed straw man test cycle has an exothermal aging step as part of the rapid aging protocol.

Several studies have been conducted on the effect these various aging cycles have on the TWC performance (Carol et al. 1989; Heck et al. 1992; Skowron et al. 1989; Hammerle and Wu 1984; Taylor 1993). One clear observation is the strong effect of aging temperature and exhaust gas oxygen concentration as shown in Figures 6.21 and 6.22 (Heck et al. 1992).

Studies have been conducted to reduce the reaction of Rh with high-surface-area carriers such as stabilized γ-Al_2O_3. At temperatures in excess of 800 °C to 900 °C, in an oxidizing mode, the Rh reacts with the Al_2O_3 forming the inactive aluminate. Fortunately, this reaction is partially reversible, so in the rich mode, some of the Rh is released from the aluminate (reactions not balanced):

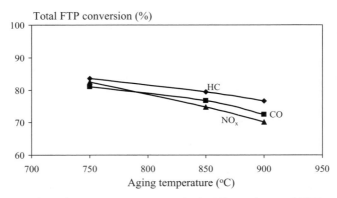

FIGURE 6.21. The aging temperature negatively affects the total FTP performance with fuel-cut aging. Reprinted with permission, © 1992 Society of Automotive Engineers, Inc. (Heck et al. 1992).

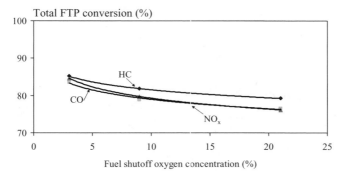

FIGURE 6.22. The fuel shutoff oxygen concentration affects total FTP performance with fuel-cut aging. Reprinted with permission, © 1992 Society of Automotive Engineers, Inc. (Heck et al. 1992).

$$Rh_2O_3 + \gamma\text{-}Al_2O_3 \xrightarrow{\;800°C,\,Lean\;} RhAl_2O_3 \qquad (6.15)$$

$$Rh + \gamma\text{-}Al_2O_3 \xleftarrow{\;Rich\;} RhAl_2O_3 + H_2 \text{ or (CO)} \qquad (6.16)$$

Figure 6.23 illustrates the effect of rich and lean treatment on the performance of a modern TWC catalyst. This catalyst was aged on a fuel-cut engine aging cycle at 850 °C inlet temperature. It was subsequently exposed to rich engine exhaust at 850 °C and evaluated using a sweep test. A sweep test is a convenient test conducted on an engine dynamometer, and the conversion across the catalyst is measured for various air-to-fuel ratios going from a lean to a rich engine operation. Exhaust temperature and flow rates are kept constant during the test sequence, whereas the perturbation amplitude and frequency can be varied. The test catalyst was then exposed to lean engine exhaust, then again evaluated, and then again exposed to rich engine exhaust. The catalyst essentially recovers its activity. This reversible cycling suggests some form of facile complex is being formed with the Rh. The mechanism for Rh catalyst deactivation is still a subject of ongoing research (Wong and Tsang 1992; Wong and McCabe 1989; Hu et al. 1998; Lassi et al. 2004).

Several approaches to minimize these undesirable reactions are currently under investigation. A promising route seems to be to deposit the Rh on a less reactive carrier such as ZrO_2. The catalyst companies consider the exact technology used to be highly confidential.

Another observation with regard to Rh stabilization is its possible interaction with CeO_2, the oxygen storage component (Wan and Dettling 1986). During high-temperature lean excursions, the Rh can react with the rare earth oxide, reducing the activity of both species. Segregating the Rh is suggested as a way to improve tolerance to high-temperature lean excursions. This has resulted in multiple layers of washcoats with the Rh and CeO_2 in different layers. Figure 6.24 shows micrographs of a single- and double-washcoat TWC

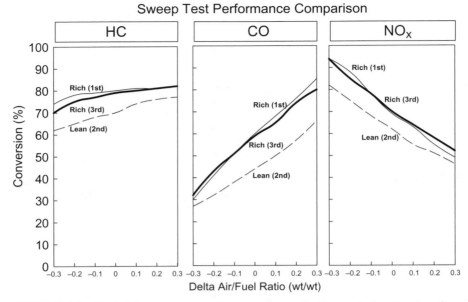

FIGURE 6.23. Fuel-rich engine operation regenerates rhodium activity after deactivation from lean exposure.

catalyst for comparison. Studies have also been conducted on ways to stabilize the Ce with oxides of Zr, Ba, and La, and to prevent interaction with the Rh catalyst (Funabiki et al. 1991; Usmen, et al. 1992; Kubsh et al. 1991).

Additional studies centered on optimizing the ceria location (Ihara et al. 1990). These authors recommended placing the Ce adjacent to the opening of the aluminum pores. They also recommended separating the Ce from the precious metal in a layered catalyst.

Catalyst deactivation and reaction inhibition due to P and S, respectively, are still concerns in modern TWC catalysts (Gandhi and Shelef 1991; Brett et al. 1989). The phosphorous present in the lubricating oil as zinc dialkydithiophosphate (ZDDP) deposits on the catalyst and results in deactivation. The P originates from the lubricating oil from three sources: engine blowby, oil sump vaporization, and malfunctioning PCV (positive crankcase ventilation) valve. When it is engine blowby, it may pass through the combustion chamber and usually deposits as a P_2O_5 film or polymeric glaze on the outer surface of the Al_2O_3 carrier, causing pore blockage and severe masking. This introduces pore diffusion limitations and, in the worst case, masking, which physically blocks reactant molecules from access to the active Pt, Rh, and Pd sites within the pore structure of the Al_2O_3. This can be a serious problem in many converters, but optimization of the pore structure of the Al_2O_3 provides improved life to the catalyst. A small loss in activity results from the decrease in surface area as the smaller pores are occluded (Jobson et al. 1991; Williamson et al. 1985).

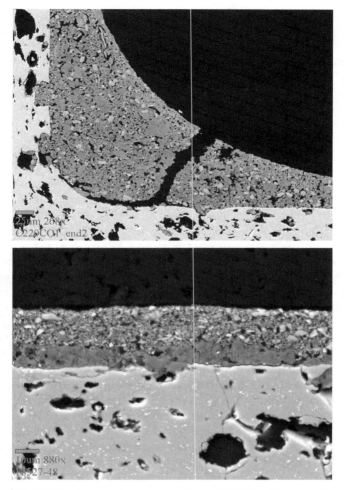

FIGURE 6.24. SEM micrographs: preparation of TWC with a single-coat and a double-coat layer of the catalyst.

If it deposits as a mist as from the PCV valve, then it will deposit only on the surface and mask the surface itself. Since the percentage or amount of engine blowby has declined dramatically with the new low-emission vehicles, masking from oil misting seems to be the most common form of P deactivation. Some P may also interact with the Al$_2$O$_3$ or Ce (Kumar et al. 2003; Rokosz et al. 1991; Xua et al. 2004). Some studies have also looked at the effect of silicon from various lubricants on catalyst performance (Gandhi et al. 1986). Currently, there are several efforts in the industry to develop an effective method to measure the amount of phosphorus "escaping" into the exhaust gas phase that ends up interacting with the catalysts. One of the efforts resulted in the development of a phosphorus emission index (PEI), which measured the mass of phosphorus that escapes from engine oil at 250 °C. Studies have been conducted with oil additives that can modify the impact of phosphorus present in the exhaust gas on catalysts (Guinther and Danner 2007; Guevremont et al. 2007).

Gasoline in the 1990s averaged anywhere from 200 to 500 ppm (0.02–0.05%) and contained up to 1,200 ppm of organo-sulfur compounds that convert to SO_2 and SO_3 during combustion. The SO_2 adsorbs onto the precious metal sites at temperatures below about 300 °C and inhibits the catalytic conversions of CO, NO_x, and HC. At higher temperatures, the SO_2 is converted to SO_3, which either passes through the catalyst bed or can react with the Al_2O_3 forming $Al_2(SO_4)_3$. The latter is a large-volume, low-density material that alters the Al_2O_3 high-area surface leading to catalyst deactivation. In addition, the SO_3 can react with Ce and other rare earths. A study of the effect of sulfur on the TWC catalyst has been conducted (Beck et al. 1991). The actual effect of sulfur on a TWC catalyst performance in an FTP test is given in Figure 6.25 for nominally 60 wppm and 900 wppm of sulfur in the gasoline. The impact of sulfur in the fuel has also been studied in fleet tests, and preliminary findings indicate a significant reduction in both HC and CO emissions is obtainable by reducing the fuel sulfur level from 50 ppm down to 10 ppm (Colucci and Wise 1993). A study with the California Phase 2 gasoline further points out the benefits of lowering the fuel sulfur content to 40 ppm (Lippincott et al. 1994; Calvert et al. 1993). This study looked at reformulated fuels, and the major impact was identified as a reduction in sulfur and fuel vapor pressure. A conclusive studied was conducted that showed the only way to achieve the ultra-low emissions was to lower the S to 30 ppm (MECA b 1998).

Typical profiles of S, P, and Z on deposits in the washcoat at an inlet and outlet section of a vehicle-aged catalyst are shown in Figures 6.26 and 6.27. The concentrations of S, P, and Zn are much greater in the inlet than in the outlet section, indicating the former serves as a filter. The sulfur is uniformly present throughout the washcoat, suggesting an interaction between it and the Al_2O_3. The P and Zn are concentrated near the outer periphery of the wash-

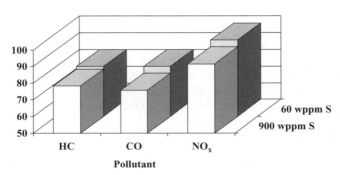

FIGURE 6.25. Sulfur in gasoline negatively affects FTP performance of TWC.

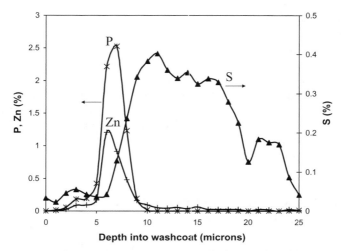

FIGURE 6.26. Concentration profiles of sulfur, phosphorus, and zinc in the washcoat at the inlet section of a vehicle-aged TWC. Reprinted with permission of the Royal Society of Chemistry (Heck and Farrauto 1994).

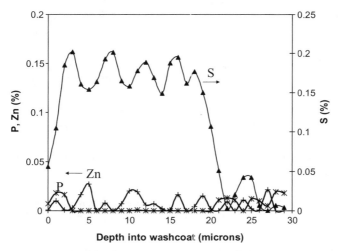

FIGURE 6.27. Concentration profiles of sulfur, phosphorus, and zinc in the washcoat at the outlet section of a vehicle-aged TWC. Reprinted with permission of the Royal Society of Chemistry (Heck and Farrauto 1994).

coat, but only in the inlet section. The drop in poison concentrations at about 20 microns is at the washcoat/monolith interface.

The TWC composition has an interesting effect on the sulfur compounds present in the exhaust. The SO_2 is oxidized in the lean mode, forming SO_3, which reacts with Al_2O_3 or CeO_2 forming $Al_2(SO_4)$ or $Ce(SO_4)_2$. When the exhaust becomes rich, the sulfate compounds are reduced and form the unpleasant-smelling H_2S. Catalyst companies have developed technology to capture the H_2S before it is released to the atmosphere.

It seems that the main mechanism of release of H_2S is through the Ce. Studies have looked at modifications of the Ce through thermal treatment (Lox et al. 1989). This treatment changes the crystallite size and reduces the number of surface hydroxyl groups, thus lowering the retained sulfur oxides. Other studies have looked at scavengers that actually remove the H_2S. Materials such as Ni have been found to be effective (Golunski and Roth 1991; Dettling et al. 1990).

6.8 PALLADIUM TWC CATALYST: THE FOURTH GENERATION (MID-1990s)

Throughout the development of the automotive catalyst, the use of Pd as a replacement for Pt and/or Rh has been desirable because it is considerably less expensive than either. Its effectiveness for reducing hydrocarbon emissions for the more stringent standards has also been recognized (Yamada et al. 1993). One of the first changes in the conventional TWC technology was the substitution of Pd for Pt and the use of the Pd/Rh catalyst for certain commercial automotive applications (Lui and Dettling 1993; Dettling and Lui 1992). In the early 1990s, an effort was made to reduce the cost of TWCs, further so Pd was substituted for Rh, yielding a Pt/Pd catalyst (Lui and Dettling 1993). Performance of these catalysts was satisfactory under restrictive operating conditions. However, when the cost of Rh reached over $5000/tr oz in the late 1980s, more studies with Rh replacement in mind were conducted (Summers et al. 1988; Muraki 1991). This was not an easy task, and many researchers concluded that there was not a direct substitution for Rh among the precious metal groups (Fisher et al. 1992). Palladium in particular was much less active, and thus, if it were to replace Pt and Rh, larger amounts would be needed. However, several changes were occurring in the strategies of the automobile for emission control to meet TLEV and LEV, some of which proved beneficial for substituting Pd for Rh. Basically, the vehicles were being designed to have a very tight air-to-fuel perturbation, resulting in very small amplitudes and a high-frequency response. Fuel quality also was improving by further reductions in the lead content. Concurrently, the catalysts were being placed closer to the manifold, giving faster heatup of the catalyst and higher steady-state operating temperatures. This diminished the adsorption of impurities such as sulfur and phosphorous (Heck et al. 1989).

There were some false starts during this period relative to announcements of the suitability of all Pd catalysts. However, the first real evidence of data giving performance and road durability was published in 1993 (Summers and Williamson 1993). This study showed that the Pd technology gave good performance after fuel-cut aging cycles and was not sensitive to sulfur in the

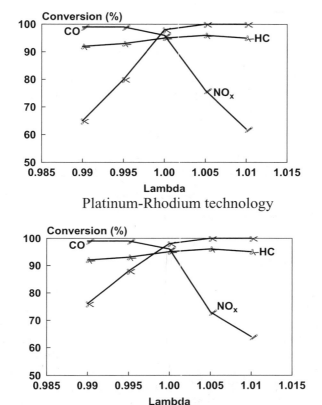

FIGURE 6.28. Performance of a Pd-only TWC is comparable with Pt/Rh TWC on sweep test (Dettling et al. 1994).

exhaust gas. The advances in Pd technology continued at a rapid pace, and the first real commercial installations were in the 1995 model year for Ford (Hepburn et al. 1994). The key success to this new Pd technology is shown in Figure 6.28, which compares the performance of the all Pd technology with the commercial Pt/Rh technology (Dettling et al. 1994). The rich side activities for HC, CO, and NO_x are equivalent for both catalyst technologies. All previous results on Pd catalysts showed lower performance on the rich side and therefore a much narrower operating window for higher performance (Heck et al. 1989). The special combinations of stabilized ceria and washcoat components give this Pd catalyst equivalent performance at both loose and tight control perturbation conditions. The use of the Pd catalyst continued to expand especially for vehicles having catalysts at the higher operating temperatures and possessing tighter air-to-fuel ratio control strategies.

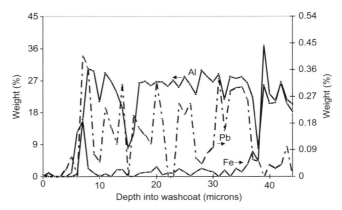

FIGURE 6.29. Pd catalyst performance is affected by residual Pb in gasoline pool (Sung et al. 1998).

One caution continues with a Pd-only TWC catalyst. In geographic locations where Pb continues to be in the gasoline source, Pd-only catalysts are susceptible to Pb poisoning. In one study of modern Pd TWC formulations, catalysts were aged with leaded fuel to determine the effect on activity (Sung et al. 1998b). Lead was found on the aged catalysts and was on the surface of the washcoat coatings and did penetrate within the washcoat; it was more predominant in the inlet section of the catalyzed monolith (see Figure 6.29). The impact of the Pb was mainly on the NO_x performance. Transmission election microscope (TEM) elemental mapping and X-ray photoelectron spectroscopy (XPS) results suggest that Pb preferentially associates with the Pd. Adding Rh to the Pd catalyst improved the resistance to Pb and the catalyst performance especially for NO_x conversion.

At the end of the twentieth century, the shift to a higher price of Pd combined with the short supply from the mine source resulted in a reevaluation of the use of Pd. Pt began to be substituted for Pd particularly in underfloor locations. So it has become very obvious that any future shift in technologies will be governed again by the price/supply issues among the various precious metals: Pd, Pt, and Rh.

6.9 LOW-EMISSION CATALYST TECHNOLOGIES

To accomplish the low emissions formulated by CARB for ULEV and SULEV vehicles, several approaches were investigated in the mid-to-late 1990s. The emphasis of the new regulations was the reduction of HCs in the exhaust. Most hydrocarbon emissions (60–80% of the total emitted) are produced in the cold-start portion of the automobile, i.e., in the first 2 minutes of operation. A typical composition of the hydrocarbons during cold start is as follows (Kumitake et al. 1996):

Hydrocarbon type:	Sampling time (seconds after cold start) Approximate hydrocarbon composition (%)	
	3 s	30 s
Paraffins	20	35
Olefins	45	20
Aromatics, C_6, C_7	20	20
Aromatics, $>C_8$	15	25

For a typical four-cylinder gasoline engine, the engine out emissions in the FTP are shown in Figure 6.30. Also shown on this figure are the LEV and ULEV emission regulations. Note that the emissions control device must be functional in 50 s (for ULEV) to 80 s (for LEV) to meet the low-emissions standards.

The technology race to develop suitable methods to control cold-start HCs included both catalytic and some unique system approaches:

1. Close-coupled catalyst
2. Electrically heated catalyzed metal monolith
3. Hydrocarbon trap
4. Chemically heated catalyst
5. Exhaust gas ignition
6. Preheat burners
7. Cold-start spark retard or post-manifold combustion
8. Variable valve combustion chamber
9. Double-walled exhaust pipe

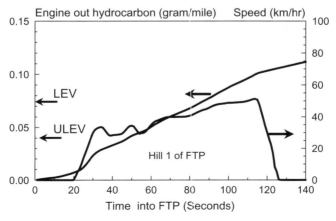

FIGURE 6.30. Cold-start emission control is the key to low-emission vehicles.

All of these approaches contain underfloor catalysts of various compositions. It has become clear that with the development of a high-temperature, close-coupled catalyst, that this is the leading technology for most LEV, ULEV, and SULEV applications. A brief description of each of the various candidate technologies follows along with the details of the close-coupled catalyst development.

6.9.1   Close-Coupled Catalyst

The concept of using a catalyst near the engine manifold or in the vicinity of the vehicle firewall to reduce the heatup time has been published (Ball 1994; Bhasin et al. 1993; Summers et al. 1993). However, these were relatively low-temperature operations, and in any case, the maximum temperature was severely limited by the use of "overfueling" or "acceleration enrichment" to control the temperature (Wards 1993). The practice of "overfueling" or "acceleration enrichment" results in high HC and CO emissions and has come under pressure to be reduced or eliminated. Reducing this protection combined with an increase in the higher speed driving habits of the United States and the existing autobahn driving habits increases the engine manifold discharge temperatures to around 1,050 °C especially for the four- and six-cylinder engines, thus changing the operating envelope of the close-coupled catalyst. In fact, a technology assessment published by CARB in 1994 showed the projected technologies for achieving the low emissions and the close-coupled catalyst was not a viable option. The dominant technology was the electrically heated catalyst. It is interesting that a cost study prepared in 1965 showed that the close-coupled approach was the least expensive solution; however, no suitable technology existed for the 100,000+ mile durability requirements (Eade et al. 1995).

Catalyst manufacturers were once again exploring new carriers capable of retaining high surface areas and metal combinations that resist deactivation due to sintering after high-temperature exposure. A shift in the technology for a close-coupled catalyst occurred when a close-coupled catalyst capable of sustained performance after 1,050 °C aging was developed and shown to give LEV and ULEV performance in combination with an underfloor catalyst (Hu and Heck 1995; Hu et al. 2000). The close-coupled catalyst was designed mainly for HC removal, whereas the underfloor catalyst removed the remaining CO and NO_x. The concept inherent in this technology was to have lower CO oxidation activity for the close-coupled catalyst, thus eliminating severe temperatures when high CO concentrations occur in the rich transient driving cycle. Subsequent studies on a 1.9-l, four-cylinder engine showed that a cascade-designed, close-coupled catalyst could meet ULEV regulations as shown in Figure 6.31 (Heck et al. 1995). This shift in technology was significant enough such that when CARB conducted another survey for projected technologies in 1996 for achieving the low emissions, the close-coupled catalyst dominated the technologies, basically eliminating the electrically heated catalyst (CARB 1996).

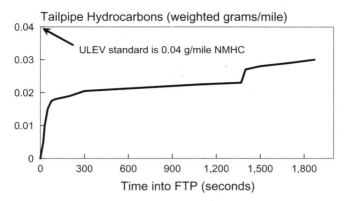

FIGURE 6.31. Close-coupled catalyst solves low-emission requirements. Reprinted with permission, © 1995 Society of Automotive Engineers, Inc. (Heck et al. 1995).

With this benchmark set, other catalyst manufacturers worked on catalyst technologies capable of sustained operation and good HC lightoff after exposure to 1,050 °C (Takada et al. 1996; Waltner et al. 1998; Williamson et al. 1999; Smalling et al. 1996). The characteristic of these close-coupled technologies is that Ce is removed. Ce is an excellent CO oxidation catalyst that stores oxygen, which then can react with CO during the rich transient driving excursions. If the oxygen is stored on the catalyst, then during severe rich excursions (such as fuel enrichment or heavy accelerations) the CO can react at the catalyst surface causing a localized exotherm resulting in very high catalyst surface temperatures. As a rule, every percent of CO oxidized gives a 90 °C rise in temperature. This can cause severe sintering of the catalyst surface reducing the activity.

One study looked at the geometric effect of the close-coupled catalyst on both performance and lightoff. The cross-sectional area and volume of the close-coupled catalyst in the so-called "cascade design" were studied (Heck et al. 1995). It was found that the smaller cross-section, close-coupled catalyst with less volume significantly improved the lightoff characteristics and could be combined with a larger cross-section catalyst with more volume to achieve ULEV-type emissions. This concept is now being practiced as either a cascade design or a small-volume, stand-alone, close-coupled catalyst.

The early lightoff of the close-coupled catalyst can be accomplished by several methods related to the engine control technology during cold start (Hu et al. 2001). One of the initial methods was to control the ignition spark retard that would allow unburned gases to escape the engine combustion chamber and continue to burn in the exhaust manifold, thus providing heat to the catalytic converter (Chan and Zhu 1996). These engine control methods have become more sophisticated as illustrated by the studies on actually controlling the degree of exothermic reactions in the exhaust manifold (Nishizawa et al. 1997; Marsh et al. 2000). In all of these control strategies, it is important to

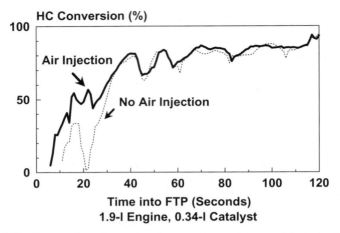

FIGURE 6.32. Oxygen in exhaust is a key component to a cold start of the close-coupled catalyst. Reprinted with permission, © 1995 Society of Automotive Engineers, Inc. (Heck et al. 1995).

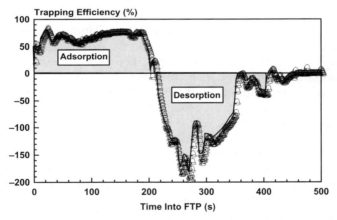

FIGURE 6.33. A hydrocarbon trap stores cold-start unburned hydrocarbons from a vehicle.

have oxygen present in the exhaust gas for early catalyst lightoff as shown in Figure 6.32 (Heck et al. 1995).

6.9.2   Hydrocarbon Traps

Another approach investigated was the hydrocarbon adsorption trap in which the cold-start HCs are adsorbed and retained, on an adsorbent, until the catalyst reaches the lightoff temperature (Figure 6.33). Hydrocarbon trap materials considered to date have been mainly various types of zeolites (silicalite,

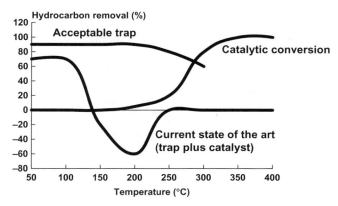

FIGURE 6.34. The lightoff of the catalyst is too late for cleanup of hydrocarbons released from the hydrocarbon trap.

mordenite, Y-type, ZSM-5, and beta zeolite) with some studies on carbon-based material. Studies have been conducted to quantify the hydrocarbon species during the vehicle driving cycle (Takei et al. 1993; Kubo et al. 1993). For an in-line hydrocarbon trap system to work, the hydrocarbons must be eluted from the trap at the exact time the underfloor catalyst reaches a reaction temperature >250 °C as shown in Figure 6.34. These HCs are then desorbed and oxidized in the normal TWC catalyst. No hydrocarbon trap materials have been found capable of retaining HCs at this temperature. Consequently, hydrocarbons pass through the underfloor catalyst unreacted and then out the tailpipe.

However, some unique system designs have been proposed. A crossflow heat exchanger trap system demonstrated a 70% reduction in the non-methane cold-start hydrocarbons during FTP cycle #1 (Hochmuth et al. 1993; Sung and Burk 1997; Burk 1995; Burk et al. 2001; Shore et al. 2001). Another trap design uses a cylinder with a central hole to allow passage of exhaust gas (Patil et al. 1996). This design contains a lightoff catalyst, the trap, and a downstream catalyst. Air is injected in the hole during cold start to divert the majority of exhaust flow to the trap in the cylinder annulus. The small amount bypassed through the hole preheats the downstream catalyst. When the lightoff catalyst is functional and the downstream catalyst up to temperature, the air is turned off and the trap desorbs the HCs. These trap designs are interesting but have not been commercialized (Noda et al. 1997; Buhrmaster et al. 1997; Noda et al. 1998; Patil et al. 1998). Some commercial niche markets for trap systems will be used in combination with close-coupled catalysts or electrically heated catalysts. However, the trap will be designed for only a small portion of the cold start, in the first 10 s or so of the FTP, and not the entire cold start as was the intention of these original studies.

6.9.3   Electrically Heated Catalyst (EHC)

Another approach to overcoming the cold temperatures during startup is to provide heat to the exhaust gas or the catalytic surface using resistive materials and a current/voltage source. Studies began prior to 1990 to develop an electrically heated monolith capable of providing *in situ* heat to the cold exhaust gas. If the electrically heated monolith is also catalyzed (an EHC), then the preheat is directly applied to the catalyst surface. The EHC is placed in front of a small lightoff catalyst, which receives the preheated gas and thus provides an efficient reaction during the cold-start period. Figure 6.35 shows the cold-start performance of an EHC. An underfloor catalyst being much larger in volume supplies the reaction efficiency during the rest of the driving cycle after the cold start.

Electrically heated monolithic catalysts are made two ways (Socha and Thompson 1992; Gottberg et al. 1991; Abe et al. 1996; Kubsh and Brunson 1996; Shimasaki et al. 1997). One approach is to make a metal foil with the foils arranged to form an electrically resistive element. The catalytic washcoat is deposited on the metal. The element is attached to electrical connection points and can be heated quickly to the lightoff temperature of the catalyst (see Figure 6.36). Another approach is to make a monolith from extruded sintered metal and then to deposit the catalytic washcoat. The base material in both cases is ferritic steel with varying amounts of Cr/Al/Fe with additives of rare earths. Substantial advances have been made in reducing the power requirements for EHCs, and studies have shown that the extra battery possibly may be eliminated; the EHC can be powered off the vehicle alternator (Laing 1994). Electrically heated catalysts having low mileage have been shown to achieve ULEV; however, the durability to 100,000 miles is still an open issue

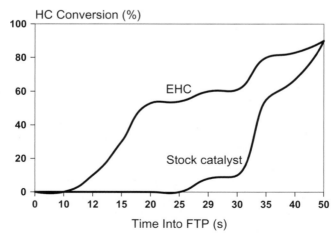

FIGURE 6.35. Electrically heated catalysts achieve rapid lightoff in the FTP test cycle.

FIGURE 6.36. Cutaway of electrically heated catalyst. Courtesy Emitec GMBH.

(Kubsh and Brunson 1996). Actual in-use experience is being gained on larger vehicles, and these studies will give the needed "on the road" experience with EHCs (Hanel et al. 1997), but overall, this approach does not have widespread use.

6.9.4 NonCatalytic Approaches

Three system design approaches have been proposed and studied in combination with catalysts to reduce the cold-start emissions. The preheat burner uses the gasoline fuel in a small burner placed in front of the catalyst. The burner is turned on during cold start, and the heat generated warms up the catalyst so the catalyst is hot when the cold exhaust from the manifold reaches the catalyst (Oser et al. 1994).

The exhaust gas igniter involves placing an ignition source (e.g., glow plug) in between two catalysts. During cold start, the engine is run rich and a small amount of air is injected to make the mixture flammable. This is then ignited and heats the catalyst (Eade et al. 1996). The chemically heated catalyst uses highly reactive species, usually H_2, which is generated in a device onboard, the vehicle. Since this reacts at room temperature over the catalyst, the heat of the reaction warms up the catalyst to react during cold start (Kamada et al. 1996).

These system approaches rapidly heat the catalyst during cold starts resulting in a low-emission operation; however, little is known of the system durability, and they are complex and expensive (Langen et al. 1994). None of these design options are currently being used in the new low-emission vehicles.

6.10   MODERN TWC TECHNOLOGIES FOR THE 2000s

The modern TWC catalysts have taken the lead from the design of the Pd technology and are using a layered catalyst with washcoat architectures to enhance the TWC reactions, and the high-temperature durability of the catalyst. The major components in a modern TWC are as follows:

- Active component—precious metal
- Oxygen storage component (OSC)
- Base metal oxide stabilizers
  - Precious metal
  - Alumina washcoat
  - Oxygen storage component
- Moderator or scavenger for H_2S
- Layered structure
- Segregated washcoat

The Ce is now made as a Ce/Zr/La/X mixture (X being a proprietary component), which further stabilizes the OSC component for the high-temperature operations. Ce is now added to the catalyst in various forms for several reasons:

- Oxygen storage
- Improved precious metal dispersion
- Improved precious metal reduction
- Catalyst for water gas shift reaction, steam reforming, and NO reduction
- Improve lightoff characteristics

More details on the beneficial effects of Zr on Ce began to be publicized in the 1996 time frame. One study showed that 10% to 40% Ce in Zr was beneficial (Culley et al. 1996). Another study looked at the Ce/Zr ratio and the effect on performance after aging (Cuif et al. 1996). For a Pd catalyst, it was found that a 70/30 CeO_2/ZrO_2 solid solution showed superior redox properties compared with pure cerias or ceria with 10% zirconia as shown by OSC capacity and temperature-programmed reduction (TPR) measurements (Cuif et al. 1998). Another study with a Pd and Pt/Rh/Pd catalyst showed that the stabilization of the Ce/ZrO_2 was bimodal and that the complex showed higher surface area maintenance at a 10/90 weight % ratio and at a 90/10 weight % ratio (Yamada et al. 1997). This same study looked at stabilizing the Zr with different components of Al, Ba, Ca, Co, Cr, Cu, Mg, La, and Y. One study showed improved surface area stability by adding 15% SiO_2 or 6% La_2O_3 to a 30% CeO_2/70% ZrO_2 system (Norman 1997). Strong enhancement of the activity of Ce-rich (Ce, Zr)O_2 solid solutions was shown when Pt was added (Cuif et al. 1997). The actual phase or structure of the Ce/Zr mixture has been the subject of some study and controversy (Egami et al. 1997). An unusual destabilization of the oxygen structure occurs at 450 °C, pointing at an ease of the structure to give up oxygen and to generate defects. This intermediate solution approximates a solid solution at macroscopic levels and a physical mixture at the nanometer scale. It seems that the best Ce/Zr ratio is application specific and depends on the automobile calibration, performance requirements, aging conditions, and catalyst formulation.

Also, the precious metals are segregated in the washcoat and are many times prepared associated with a specific compound such as Rh/Ce/Zr and Pt/Al. In these catalyst designs, the deleterious interaction between Rh and Al is reduced. An illustration of the complex relationships between the precious metal catalysts Pd and Rh and the Ce and Zr as well as the location in the bottom coat or top coat was done to show the interaction with fuel sulfur (Rabinowitz et al. 2001). The role of the ceria was complex. NO_x conversion was sharply improved by ceria, especially in combination with rhodium. However, under certain conditions, ceria, due to its ability to store and release sulfur, can be shown to increase the negative impact of sulfur. More details from the study are available in the original article, but this serves to illustrate the complex interactions present in today's scientifically designed auto catalyst. Some studies have been conducted on inserting Pt into the Ce/Zr matrix using sol gel chemistry. The specific compound $Ce_{0.66}Zr_{0.32}Pt_{0.02}O_2$ had some interesting characteristics regarding low temperature (poorer TWC performance) and high temperature (better thermal stability) than a Pt doped sample. The researchers believe some sort of regeneration is occurring after the high-temperature exposure of 950 °C, 20 hr (Wu et al. 2005). So the major components of Ce, Al, and precious metals are stabilized in some sort for the modern TWC.

The effect of sulfur continues to affect the modern catalyst technologies. The effect has been quantified to be an actual selective poisoning effect and not a long-term effect on the catalyst performance (Bjordal et al. 1995). The sulfur affects mostly the lightoff characteristics of the TWC catalyst. As the sulfur is removed from the gasoline, the catalyst gradually regenerates and the performance is regained. The P and Zn in the lubricating oil continue to be an issue. One study looked into regenerating the TWC catalyst using chemical washing to remove the P and Zn deposits (Beck et al. 1995). The study concluded that the P and Zn deposits could be removed using the chemical wash procedure, and once removed, the lightoff performance and conversion of the TWC catalyst improved. So, for the new low-emissions standards, it will be important to reduce the fuel sulfur levels and oil blowby from the SI engine. Efforts should also be made to reduce the ZDDP level of the lubricating oil.

6.11   TOWARD A ZERO-EMISSION STOICHIOMETRIC SPARK-IGNITED VEHICLE

This section could be entitled "putting it all together" or the "complete engineered system." During the later part of the twentieth century, the cooperation among the automobile manufacturers, the monolith suppliers, the exhaust system fabricators, the sensor manufacturers, and the catalyst manufacturers intensified so that a complete engineered system approach was taken to achieve the goal of a "zero-emission stoichiometric spark-ignited (SI) vehicle." With the advent of a durable close-coupled catalyst, many of the engine modifications to provide rapid heatup (e.g., spark retard on a cold start) of the catalyst were now implemented. The ULEV performance requirement for a four-cylinder vehicle, which may range from a hydrocarbon engine out emissions of 1.5 g/mile to 2.0 g/mile, is around a 98% hydrocarbon conversion. Of course, a SULEV vehicle is greater than a 99% hydrocarbon conversion. As an aside, the tailpipe HC emissions from a SULEV vehicle may be less than 5 vppm HC, whereas the background level of ambient HCs is in the same range of 1 to 5 vppm, so the measurement of these low-emission vehicles presents another challenge.

Because of these high-emission reduction efficiencies and, hence, a requirement for more geometric surface area, monolith suppliers began to make higher cell density substrates approaching 1200 cpsi (Kikuchi et al. 1999; Gulati 1999). In addition, the exhaust piping was redesigned to minimize heat loss during the critical cold start with fabrication of the low heat capacity piping (Kishi et al. 1998). Of course, engine control must be better to minimize the perturbation effects on the TWC operation. To accomplish this, a new sensor was developed based on the operating principles of the oxygen sensor but with more sophisticated design and electronics to give a gradual response curve to changes in A/F ratio or oxygen content in the engine exhaust. This universal exhaust gas sensor (UEGO) has the following operating characteristics as compared with the HEGO sensor as shown in Figure 6.37 (Anderson 1993; Bush et al. 1994). With this better control, the operating window for the TWC is narrowed as shown in Figure 6.38. Of course, the narrow operating window gives better overall HC, CO, and NO_x conversion over the TWC.

Finally, demonstration of these ultra-low-emission systems became a reality (Sung et al. 1998; Kishi et al. 1998; Kishi et al. 1999; Nishizawa et al. 1997; Nishizawa et al. 2000). LEV vehicles became common in the late 1990s, and ULEV vehicles were supplied to the California market in 1998. In 1999, a ZLEV (zero-level emission vehicle) vehicle was demonstrated after 100,000 mile aging (Kishi et al. 1999).

The following table represents the main characteristics for the Honda ULEV vehicle (Kishi et al. 1998).

Engine modifications	Exhaust modifications
VTEC L-4 with variable valve timing (VVT)	Low heat capacity manifold
ECU 32-bit microprocessor	UEGO sensor
Air assist fuel injectors	Low heat capacity exhaust pipe
Precise A/F control with self-tuning regulation (STR)	Underfloor catalyst with Pd-only front on 600 cpsi
Electric-controlled EGR valve	Secondary HEGO sensor
Individual cylinder A/F control	
Lean air/fuel cold start	

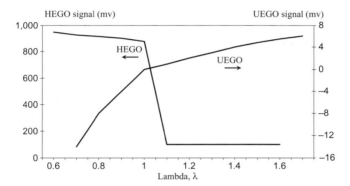

FIGURE 6.37. The response of the UEGO sensor is different than the UEGO sensor.

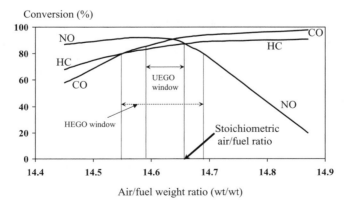

FIGURE 6.38. Example of how UEGO sensor control improves overall TWC performance.

The key features in regard to catalyst performance are the use of an engine-designed, lean, cold-start and fuel management system to supply oxygen for the catalytic oxidation reactions and the reduction of heat loss during cold start. The new sensor for control is a UEGO sensor, which is a sensor that has a different response than the HEGO. So with the UEGO sensor, the actual air/fuel ratio in the exhaust is known directly. The first underfloor catalyst is a 600-cpsi Pd catalyst for a high-temperature operation, and the remaining underfloor catalysts take care of emissions during normal operation. After 100,000 miles on a vehicle, the emissions measured were as follows:

NMOG	CO	NO_x
0.03 g/mile	0.35 g/mile	0.12 g/mile

The Honda so-called ZLEV vehicle is based on the same VTEC platform with additional controls for cold start for hydrocarbons and air/fuel control for NO_x. (Kishi et al. 1999) The characteristics are summarized below:

Engine modifications	Exhaust modifications
VTEC L-4 with VVT	Low heat capacity manifold
ECU 32-bit microprocessor	UEGO sensor
Improved atomization fuel injectors	Low heat capacity exhaust pipe
Precise A/F control STR	Pd close-coupled catalyst on 1200 cpsi
Individual cylinder A/F control	Underfloor catalysts TWC and HC trap
Ir and Pt spark plugs	hybrid catalyst
Lean air/fuel cold start with spark retard	Two secondary HEGO sensors
Electric-controlled EGR valve	
Catalyst condition predicted control	

The engine uses spark retard during a cold start to aid in catalyst heatup and lightoff. Also, the Pd close-coupled catalyst is 1200 cpsi followed by an underfloor catalyst system having a separate TWC and a trap/catalyst hybrid to manage the hydrocarbons during the first 10 s during cold start. Recall that the trap technology considered in the early 1990s to address the entire cold start without a close-coupled catalyst had to function for >100 s into the FTP because the engine technology of that period gave a very slow heatup rate to the exhaust gas. The exhaust heatup rate of these late 1990 engines is faster, thus allowing the catalyst to become active very early in the FTP cycle approaching <30 s.

After 100,000 miles on a vehicle, the emissions measured were as follows:

NMOG	CO	NO_x
<0.004 g/mile	<0.17 g/mile	<0.02 g/mile

Nissan has also demonstrated and now offers for sale the world's first certified partial zero emission vehicles (PZEVs) (Nishizawa et al. 2000). This vehicle not only meets the SULEV tailpipe emissions but also has a zero evaporative emissions system and qualifies as a PZEV. The engine emission control technology consists of close-coupled catalysts followed by a series of trap catalyst combinations to reduce cold-start emissions. The Nissan Sentra CA vehicle has shown 150,000-mile durability and meets the requirements for the on-board diagnostics (OBDs) of the catalysts system. This vehicle is also equipped with a PremAir® catalyst-coated radiator, which removes ambient ozone (Hoke et al. 1996). This catalytic device can be used to offset hydrocarbon tailpipe emissions or evaporative emissions. The PremAir technology will be covered in more detail in Chapter 15. This technology was first commercialized in December 1999 by Volvo who used the concept for a "green" image on the V-70 model, and not for credits to offset tailpipe emissions. Nissan followed the same approach in 2000 by implementing a PremAir catalyst on the radiator of the Sentra CA. This PZEV vehicle exhaust emission control system has been redesigned based on on-road experience and new catalyst and HC trap technology, and the third-generation design layout is shown in Figure 6.39 (Oguma et al. 2003).

Toyota also has published its ULEV and PZEV designs. The following table is a comparison of the catalyst technologies for their four-cylinder engine.

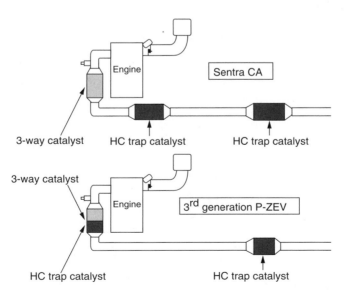

FIGURE 6.39. Progression of Nissan PZEV design layout. Reprinted with permission, © 2003 Society of Automotive Engineers, Inc. (Oguma et al. 2003).

	2002 ULEV	2003 PZEV
Close-coupled catalyst	1.1 liter ceramic 3 mil 600 cpsi	1.1 l ceramic 2 mil 900 cpsi
Underfloor catalyst	0.9 liter ceramic 4 mil 400 cpsi	1.3 l 3 mil 600 cpsi (HC
		adsorbing three-way
Where 1 mil = 0.001 in		catalyst)

In addition to new catalyst technology, the Toyota PZEV engine has variable-valve-timing to open the intake valve early on cold starts, providing oxygen for the catalytic reactions. Downstream of the engine, a compact double-wall exhaust manifold retains the gas heat as it travels from the exhaust port to the three-way catalytic converter, thus providing earlier catalyst light-off. The hydrocarbon absorptive catalyst is prepared with the bottom layer having the hydrocarbon adsorbent material and the top having the three-way catalyst material (Kidokoro et al. 2003).

The excellent benefits of the increased cell density are shown in Figure 6.40. This study looked at the effect of changing the monolith cell density from 100 cpsi to 1200 cpsi and the effect on HC tailpipe emissions. The previous monolith technology was 400 cpsi, and the figures show the benefits of going from 600 cpsi up to 1200 cpsi. Approximately a 30% reduction is possible with 600 cpsi, whereas a 75% reduction is possible with 1200 cpsi. Another study was conducted using the European driving cycle, and this resulted in a similar conclusion as follows:

Cell density (cpsi)	Wall thickness (mils)	Emission reduction (%)
400	6.5	Reference
400	4.3	12
600	3.6	35

These reductions in emissions were measured after the close-coupled catalyst of the different cell geometries.

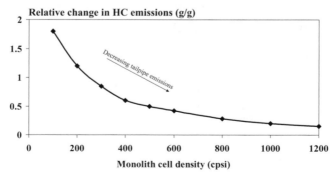

FIGURE 6.40. Increasing cell density required for low-emission vehicles. Courtesy Emitec Corp.

The HC regulations for California are unique in that, for the first time, the ozone-forming potential or reactivity of organic gas emissions will be considered. So the ozone reactivity can be reduced catalytically or with changes in the fuel type or composition. Also, a reactivity adjustment factor is used for fuels other than gasoline for a uniform comparison. Studies are being conducted to understand the source of speciated hydrocarbons from different fuel components and the effect that a catalyst plays in changing the speciated components (Siegl et al. 1992; McCabe et al. 1992). Additional studies have also been conducted on the effect a catalyst has in reducing toxins and mutagens from the passenger car (Cooper and Shore 1989).

The critical effect of fuel properties on the near-zero emissions levels for the advanced technologies has been studied (Takei et al. 2001). Fuels were prepared with <1 ppm S and up to 600 ppm S for tests. The fuel sulfur affects the HC and NO_x emissions most dramatically for the SULEV vehicles. This study also considered the distillation properties and oxygenates on tailpipe emissions. Fifteen fuels with various 50% evaporation and 90% evaporation as well as oxygenates and sulfur contents were prepared. HC emissions increased for both LEV and ULEV with the increase in temperature for 50% evaporation for temperatures >100 °C. NO_x emissions increased with decreasing temperature for 50% evaporation.

6.12 ENGINEERED CATALYST DESIGN

In the early 2000s, it became apparent that the engine manufacturers were finally convinced of the efficacy of the system design approach that married the catalyst technology with the engine technology. Much of this engineering began to occur with some engine manufacturers in the early development of the ULEV and SULEV platforms. Note that the initial designs addressed only four-cylinder engines because they had low engine out emissions and it was possible to then meet the tailpipe emissions. However, as the emission regulations were applied to six- and eight-cylinder engines, the goal became more daunting because of the inherent higher engine out emissions. Of course as in the four-cylinder engines, the first task was to lower the engine out emissions through improved engine design addressing better combustion efficiency to reduce engine out HC with improved EGR strategies to lower the engine out NO_x. The EGR could be accomplished externally to the engine or internally within the engine. Next the cold-start strategies were refined to achieve higher temperatures for catalyst lightoff within 10 s in the FTP, and lean start strategies were implemented to provide the necessary oxygen for the oxidation reactions. Finally the rapid achievement of closed-loop control early in the FTP cycle and more rapid response during vehicle accelerations was accomplished to minimize NO_x and HC breakthrough. So as the engine manufacturers continued to refine the vehicle emission control strategies, they realized that the catalyst technology could be changed. Specifically, the precious metals could be lowered in the catalyst technologies for both close-coupled and underfloor applications. This could be accomplished by reducing the metal loading or using a zone-coating process such that the front part of the monolith had higher loadings and the back part had lower loadings (as well as composition differences). Of course, catalyst performance and durability could not be compromised.

An example of this close integration between engine operation and catalytic emission system was shown in a study of a 6.0-l V8 engine, 2004 model year. The catalytic system consisted of a close-coupled catalyst (900 cpsi, 0.024 ft^3) and an underfloor catalyst (600 cpsi, 0.055 ft^3) on each bank of the engine. A total catalyst volume of 0.157 ft^3 (4.46 l) was used. The cold-start operation was modified with spark timing retard and air-to-fuel ratio biasing. This allowed the close-coupled catalyst to reach 350 °C in 10 s of the FTP and faster closed-loop control to control HC and NO$_x$ emissions in cold start. The steady-state engine operation was also modified with adjustments in timing and air-to-fuel ratio to reduce NO$_x$ breakthrough during high-speed, high-load conditions. The resulting FTP on an aged catalyst system was within the ULEV standards (Anthony et al. 2007). The researchers mention that this work is by no means finished as there are many more engine system parameters that can be changed on new V8s, such as lighter weight exhaust manifolds to heat up the catalyst faster; more sophisticated air-to-fuel control strategies to improve cold-start performance and high-speed, high-load performance; and more advanced fuel injection systems for a lean cold start. Another example of this system integration type study showed that the heat-up strategy could accomplish 400 °C at the inlet to the close-coupled catalyst within 10 seconds (Watanabe et al. 2006). Of course, the engine designer will also be working on reducing engine out emissions through better combustion strategies and EGR. This study points out the system interaction that will be occurring with every vehicle platform design. In taking the system design integration one step further, studies have been done on locations of the HEGO and UEGO sensors using computational fluid dynamics (CFD) predictions (Schuerholz et al. 2004).

The studies that addressed optimizing the catalyst configuration of close-coupled and underfloor catalysts considered many variables, including precious metal type, loading location (layered or zone coating), cell density (cpsi and wall thickness), Ce/Zr package, volume, location, and so on. The tools that had become available in the 2000s included not only experimentation but extensive modeling based on previous databases accumulated over the past 10 years. One such study not only considered these variables but also looked at the interaction of the upstream oxygen sensor and downstream oxygen sensor as well as at the oxygen storage capacity of the catalyst. Directionally this study showed that a small-volume, close-coupled catalyst was sufficient for lightoff and that the underfloor catalyst precious metal could be reduced. Based on the results of a detailed experimental and modeling investigation of a dynamic catalyst operation, an optimized washcoat formulation for use in underfloor applications was developed. Compared with the then-current production-type catalysts, more than 50% reduction of precious metal loading without any loss in catalytic activity was achieved (Vostmeir et al. 2002). So a total system approach was taken using the catalyst composition and catalyst monitoring strategy. As a complement to this type of study, the effect of monolith diameter and cell density was studied for both close-coupled and underfloor catalysts as well as for Pd and Rh loading. Higher cell density in the close-coupled position gave the most effect and could result in precious metal reduction (Ball et al. 2007). So the interaction between monolith cell density and oxygen storage will now have to be weighed in a catalyst system design.

Along with these configuration optimization studies, improvements in the Ce/Zr stability have been achieved in synergy with the lower precious metal formulations. New formulations have shown impressive stability up to 1,100 °C in a wide composition range from Ce-rich to Zr-rich, and doped Zr without Ce. One paper shows that catalytic performance is related to the type of "precious metal–support interaction". For instance, Ce-rich oxides show improved lightoff activity for Pt model catalysts, whereas Zr-rich oxide with 20% Ceria show improved lightoff activity for Pd model catalysts. Zr-rich oxide with 40% Ceria showed the best trade-off between lightoff and TWC conversion for Rh model catalysts. The best TWC conversion was observed for Rh-supported materials containing 60% and 40% Ce, respectively. Rh supported on doped Zr without Ce showed outstanding NO_x and C_3H_6 lightoff (Rohart et al. 2005).

Zone coating of the monolith along its length is becoming a useful way to have a catalyst with high durability and yet low precious metal. For instance, one study looked at Pt/Pd/Rh combinations for LEV 2 emission standards. One of the candidates tested was a zone-coated technology having a close-coupled catalyst of 900 cpsi and 43 in^3 on both banks of the engine exhaust. The close-coupled catalyst was zone-coated Pd/Rh followed by Pt/Rh with a total of 2.07 g of precious metal in the front zones and a total of 0.89 g in the rear zones (a total of 60 g/ft^3). The low PGM underfloor catalyst was 400 cpsi and 41 in^3 with 0.06 g of precious metal (a total of 1.9 g/ft^3) (Andersen et al. 2004).

To layer or not to layer is certainly being pursued as new materials and preparation procedures are being invented. A study looked at layering strategies for different precious metal combinations and the method of addition of the precious metal. Also a new Ce/Zr technology was also investigated and a Zr-rich quaternary solid solution material was used in this work. The concept considered the precious metal addition to the washcoat as the last step and claimed to give more flexibility in the design of the catalyst technologies. The precious metal loadings in this study ranged from 10 to 20 g/ft^3 and included Pt/Pd/Rh catalyst combinations. They also considered the trade-offs between close-coupled and underfloor technologies (Ball et al. 2005; Williamson et al.

2007). Of course the location of the precious metals within the layer to stabilize these materials thermally is still a subject of research. A study looked at Pt and Rh addition within a washcoat. This is another example of the so-called segregated washcoat. The interaction between different carriers for the Pt and Rh were studied with the Ce/Zr oxygen storage material. Also the effect of aging atmosphere (lean vs. rich/lean) was investigated. An optimized formulation of a CeO_2–ZrO_2 and a ZrO_2 material was developed to have excellent durability, improved OSC, and enhanced interaction between precious metals and support materials as well as increased thermal stability (Yoshida et al. 2006).

There have been several publications since 2000 on the use of perovskites in an automotive catalyst. Perovskites are catalytic materials having a densely packed cubic lattice of the formula ABO_3. Metal ions having different valence can replace both A and B ions. Much of this work started with Catalytic Solutions Inc. (CSI), which has patented technology in this area (Golden 2002; Golden 2003). It was originally thought that the use of perovskite materials would give some efficacy and allow less precious metal to be used (Matsuzono et al. 2003). This study looked at an advanced control technology for cold-start and steady-state operations and showed that with a perovskite material, a low precious metal catalyst could be used for LEV 2. The amount of precious metals used in the catalyst, per vehicle, is expected to be 50% less than in conventional systems. It has since been found that conventional TWC formulations (as explained above) can be used when the advanced control systems are used for fast lightoff and tight air-to-fuel ratio control, so the perovskite system does not seem to have any inherent benefits over other more conventional advanced TWC technologies. One perovskite technology that has become commercial is the so-called "intelligent catalyst" (Sato et al. 2003; Taniguichi et al. 2004; Naito et al. 2006). The researchers claim the catalyst function for self-regeneration of Pd, from the less active metal to its active oxide state, is realized through the solid solution and segregation of Pd in a perovskite crystal. The actual configuration is a $LaFePdO_3$ perovskite. It was studied in the washcoat by comparing single- and double-layer washcoats as well as different loading locations for precious metals to maximize the catalyst function. The catalysts were aged on an engine exhaust system at $1,050\,°C$ and maintained good performance and thermal stability. The optimum design of the washcoat was a double layer with a tri-metal (Pt, Rh, and Pd) system. The perovskite was located in the lower layer. Subsequent studies addressed Rh technology. They found that a $CaTiRhO_3$ perovskite has an excellent capacity for the self-regenerative function of Rh. Due to the self-regenerative function of Rh and Pd, the full-scale Rh–Pd-intelligent catalyst containing a new $CaTiRhO_3$ perovskite and a Pd-perovskite ($LaFePdO_3$) exhibited excellent catalytic activity even after aging and this despite having reduced the amount of precious metals sharply. Similar results were found for Pt in the form of $CaTiPtO_3$ (Tanaka et al. 2006). Both of these technologies can be used with lower precious metal concentrations since the catalyst is thermally stable due to the regeneration phenomena during rich-to-lean excursions. The catalyst has been commercial since 2002 and has been applied to over 3 million vehicles, mainly in Japan with some applications in Europe.

The close-coupled catalyst is currently a key component to successful low-emission, long-term performance. It is critical for early lightoff and must sustain activity even after exposure to 1,050 °C. Numerous studies exist in the literature attempting to optimize the close-coupled catalyst. Several studies have looked at the question of whether Ce or an oxygen storage component should be located in the close-coupled catalyst. If Ce is present, then the Ce/Zr complex must have very high-temperature stability, And even if you achieve this stability, the catalyst exposure to the engine oil additives remains a question. So the issues encountered in any close-coupled catalyst design are layering, loading, precious metal type, volume, thermal durability, and poisoning. Some research has shown that the close-coupled catalyst with a lower loading of oxygen storage material (i.e., Ce/Zr) gave better lightoff and lower HC emissions (Watanabe et al. 2006). One study looked at these effects and concluded that it is better to locate the entire catalyst system close coupled (Eckhoff et al. 2004). Of course, in the end system design, there is always a trade-off between close-coupled technology and underfloor technology with the end goal always being to meet the emission target with sustainable durability for any engine size.

In general, the alumina washcoat has a loading of 1.5 to 2.0 g per in^3. Ceria can range up to 10% depending on the manufacturer. Some cerium oxide is bound with zirconia oxide, and some is free. The Ce/Zr-containing complexes are proprietary and usually are quaternary systems. The PGMs vary all over the map and range from 10 to 100 g per ft^3. Many of the close-coupled catalysts are all Pd, some are Pt/Rh, and others are Pt/Pd/Rh. Most close-coupled catalysts have a higher concentration of Pd for high-temperature durability. Some close-coupled catalysts have no ceria in the composition. The ratios usually follow the current PGM price structure. Either Pd or Rh is needed for NO_x conversion.

6.13 LEAN-BURN, SPARK-IGNITED GASOLINE ENGINE

The world is becoming much more conscience of growing environmental problems as well as of the need to conserve energy. This concern should result in greater efficiencies in transportation but without the harmful effects of increased pollution. The gasoline-fueled automobile is a great luxury for the average citizen of the world, but it is a large fossil fuel consumer, and it directly generates enormous amounts of emissions. The modern TWC converter is efficient in the simultaneous reduction of CO, HC, and NO_x, but the engine itself generates large amounts of CO_2 emissions, possibly contributing to global warming or what is called the "greenhouse effect."

A requirement for automotive three-way catalysis is that the air-to-fuel combustion ratio be at the stoichiometric point, which for gasoline engines is about 14.6 on a weight basis. Leaner ratios greater than 14.6 would result in a decrease in fuel consumption and consequently, in less generation of CO_2, but the TWC cannot reduce NO_x in excess air. This can be seen in Figure 6.10. Thus, the challenge is clear: Develop a catalytic system for a lean-burn engine that will reduce all three pollutants. The other emissions are easily removed because catalytic technology already exists that uses Pt and Pd for the abatement of CO and HC from a lean environment, since the first automotive catalytic converters performed this function.

The so-called GDI (gasoline direct injection) engine is a lean-burn, spark-ignited engine with fuel efficiencies 20–30% percent higher than conventional stoichiometric, spark-ignited engines. If a lean NO_x system could be developed, the modern TWC would be replaced by a new catalytic technology, whereas lean-burn engines would operate with a commensurate decrease in CO_2 emissions and fuel consumption.

Catalyst strategies for lean-burn engines are of extreme importance now for controlling diesel emissions and will be thoroughly discussed in Chapter 8.

6.13.1 NO$_x$ Reduction

Since the conventional spark-ignited gasoline operates near the stoichiometric air/fuel ratio, TWC catalyst technology can remove simultaneously HC, CO, and NO_x. However, the so-called lean-burn, gasoline, spark-ignited engines operate on the lean side of stoichiometry, and the excess oxygen removes the reductants CO and HC so no NO_x removal is possible. The rationale for the lean-burn operation is to improve fuel economy since operating an internal combustion engine lean of stoichiometric improves the combustion efficiency and power output. This operating strategy also decreases the emissions of the greenhouse gas CO_2. Since the lean environment inhibits NO_x reduction in the three-way catalyst, a new catalyst would have to be developed.

The reduction of NO_x in lean environments is a technology still currently under investigation, and it is important for both GDI engines as well as diesel engines (see Chapter 8 "Diesel Engine Emissions"). The dominant reaction is as follows:

$$HC + NO_x + O_2 \rightarrow N_2 + CO_2 + H_2O \qquad (6.17)$$

A lean NO_x reduction system must be integrated with the engine so the exhaust stream will have the type and amount of hydrocarbons needed to reduce these oxides at the optimum temperature for the particular hydrocarbon (Iwamoto and Hamada 1991). For example, propane is effective at 500 °C with Cu/ZSM-5 (ZSM-5 is a zeolite structure), but it is ineffective at a lower temperature. In contrast, ethylene reduces NO_x at 160 °C to 200 °C.

In the last ten years of the twentieth century, scientists tried to develop a lean NO_x catalyst (LNC) technology but have continued to fail in this regard because of a combination of the following issues:

- Hydrothermal aging—In the presence of water vapor, the catalytic materials lost activity through a sintering mechanism or lost selectivity through competitive adsorption.
- Sulfur deactivation—Most of the catalytic materials were sensitive to sulfur and lost activity in the presence of even very small amounts of sulfur in the gasoline
- Poor selectivity—The hydrocarbon reductant had to be added to the exhaust stream for the NO_x reduction since none were present from the combustion process under lean engine operation. Only certain species of hydrocarbons would work, and an excess of these particular hydrocarbons had to be added. The amount of hydrocarbons added was well in excess of that needed for the stoichiometric reduction of NO_x (anywhere from 5/1 to 10/1 HC/ NO_x ratios).
- Narrow temperature window—The temperature range of operation was narrow for most technologies for good selectivity. Therefore, a combination of technologies was required for operation over the range of operating temperatures for normal engine operation.

The following is a partial list of the materials investigated for the lean NO_x reaction:

- Cu/ZSM-5
- Pt/ZSM-5
- Fe/ZSM-5
- Co/ZSM-5
- Ir/ZSM-5
- Protonated zeolites, H-ZSM-5, H-Y zeolites
- Noble metals
- Perovskites

Additional materials have also been studied (Fritz and Pitchon 1997).

Different hydrocarbons have also been tried as reductants ranging from CO to low-molecular-weight parafins to partially oxygenated hydrocarbons. The typical response curves for these lean NO_x technologies are shown in Figure 6.41 for a Pt technology and Cu/ZSM5. Some auto manufacturers actually had a lean NO_x catalyst installed on certain GDI engines (Automotive Engineering 1994; Takami et al. 1995). These initial catalysts had in-use durability issues and are no longer being used.

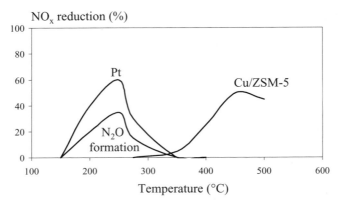

FIGURE 6.41. Performance of lean NO_x catalysts.

So the search for a lean NO_x catalyst remains elusive, and the technology search seems to be at a standstill for technologies with high conversion!

6.13.2 NO_x Traps for Direct-Injected Gasoline Engines

The NO_x trap (called the lean NO_x trap or LNT) seems to be a promising solution for NO_x reduction for direct-injected, gasoline, lean-burn engines and for diesel engines. The TWC catalyst is not effective in reducing NO_x when the engine is operated lean of the stoichiometric air-to-fuel ratio ($\lambda > 1$). An alkaline metal oxide trap adsorbs the NO_x in the lean mode during the lean-burn operation (Miyoshi et al. 1995). During steady-state driving, the engine operates lean for improved fuel economy up to 15–20% relative to continuous stoichiometric operation. The NO must first be converted to NO_2 over the Pt in the three-way catalyst during the lean operation:

$$NO + 1/2O_2 \xrightarrow{\text{Pt}} NO_2 \qquad (6.18)$$

At temperatures above about $500\,^\circ C$, NO_2 is not thermodynamically favored; however, because the trap continuously removes the NO_2 from the gas stream, the equilibrium is shifted toward more NO_2. Two kinds of Pt sites seem to operate: The sites closer to the BaO crystallites are active in barium nitrate formation, whereas the other sites are responsible for NO_2 formation (Mahzoul et al. 1999).

The NO_2 is trapped and stored on an alkaline metal oxide such as BaO or K_2CO_3, which is incorporated within the precious metal-containing washcoat of the three-way catalyst:

$$NO_2 + BaO \rightarrow BaO\text{—}NO_2 \qquad (6.19)$$

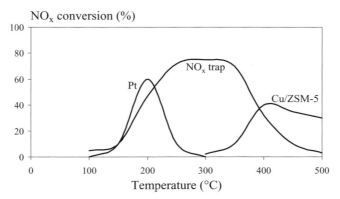

FIGURE 6.42. Performance of the NO_x trap function in comparison with the lean NO_x catalyst.

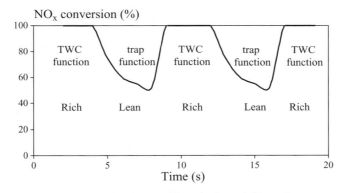

FIGURE 6.43. Performance of NO_x trap in typical partial lean-burn operating cycle.

The performance of just the trapping function of the LNT is shown in Figure 6.42 in comparison with the two lean NO_x catalysts.

Of course the trap function will be finally saturated with the adsorbed NO_x so the trap function will have to be regenerated and the NO_x will have to be reduced. The engine will typically operate in the fuel economy lean mode for up to about 60 s, after which the engine is commanded into a fuel-rich mode ($\lambda < 1$) for less than 1 s, where the adsorbed NO_2 is desorbed and reduced on the Rh in the three-way catalyst:

$$BaO\text{—}NO_2 + 2H_2 \xrightarrow{\ Rh\ } BaO + 1/2\,N_2 + 2H_2O \qquad (6.20)$$

This is the so-called partial lean-burn engine operation. The lean-rich NO_x conversion versus time profile is shown in Figure 6.43.

Note that this LNT operation for a gasoline direct injection (GDI) is totally different than a diesel engine since the GDI control system can easily return the engine to stoichiometric operation with no issues of flame stability and really is not a complicated control strategy. Sulfur oxides derived from the fuel form alkali compounds more stable than the nitrates and are not removed during the rich excursion. Therefore, the LNT progressively becomes less effective for NO_2 adsorption due to poisoning by the SO_x (MECA 1999; Engstrom 1999).

$$BaO + SO_x \rightarrow BaO—SO_x \qquad (6.21)$$

$$BaO—SO_x + H_2 \rightarrow No \ Reaction \qquad (6.22)$$

A more complicated engine control strategy is needed to desulfate the poisoned trap by operating the engine at a high temperature (>650 °C) and rich of the stoichiometric air-to-fuel ratio for a short time to remove the adsorbed sulfur oxides (Guyon et al. 1998; Nakatsuji et al. 1999). But again this is a gasoline engine that is comfortable operating at stoichiometric. The air-to-fuel ratio must be controlled to prevent H_2S from forming at excessively rich conditions. Conditions slightly rich of stoichiometric result in formation of more acceptable SO_2 emissions. Reductions in NO_x up to 90% are possible provided the gasoline has less than 10 wppm of sulfur. Therefore, reduction in sulfur content of the fuel is essential to make such a system practical. The effect of fuel sulfur on NO_x storage catalyst performance was quantified for fuels blended with 8 wppm to 500 wppm of sulfur (Takei et al. 2001). As fuel sulfur increased, conversion efficiency deteriorated. The difference in conversion efficiency at 20,000 miles with 8-ppm and 30-ppm sulfur fuels resulted in 150% more tailpipe emissions for the 30-ppm fuel. This sulfur actually inhibits the reaction and long-term exposure results in regeneration issues. One study looked at reducing the retention SO_x on the catalyst surface by changing the washcoat from γ-alumina to a mixture of γ-alumina and TiO_2 and various washcoat dopants (Matsumoto et al. 2000). They found that a Li doped γ-alumina had the lowest SO_x desorption temperature. The final catalyst formulation contained a combination of 33 mole % TiO_2 and 67 mole % Li doped γ-alumina to maintain the amount of NO_x storage and to minimize the amount of SO_x deposit.

Another problem to be solved is the thermal deactivation of the catalyst due to negative washcoat, precious metal, and ceramic monolith interactions with the alkaline metal oxide trap component after high-temperature excursions.

The GDI engine has not become as popular as predicted in the latter part of the twentieth century; however, the LNT technology is being used in the hybrid gasoline engines. A good example of this technology application published in the open literature is the Honda Insight hybrid vehicle (Honda 2007). The Insight, which is a PZEV vehicle, is designed to operate at a lean air/fuel ratio during light throttle cruise conditions around 22:1. The Insight uses a dual catalyst system with the TWC being a close-coupled catalyst to minimize heat-up, and the underfloor catalyst is an LNT. The LNT comprises a ceramic Al_2O_3 substrate, a platinum (Pt) catalyzing surface, and a titanium-sodium (Ti-Na) NO_x storage surface. For control, the Insight uses two oxygen sensors to monitor the condition of the three-way catalyst as required by OBD-II regulations. However, the detection system cannot directly monitor the condition of the lean NO_x catalyst, so if the monitor detects a code for a deteriorated catalyst, both catalysts have to be replaced. It must be assumed that the same conditions that caused the TWC to deteriorate would also damage the lean NO_x catalyst.

Alternative fuels are another area of active study (Farrauto et al. 1992; Minteer 2006). Fuels such as compressed natural gas (White et al. 1993; Subramanian et al. 1993), liquid petroleum gas, and alcohols (McCabe and Mitchell 1988) are attractive alternatives to gasoline because they are potentially less polluting (although they still produce gaseous emissions). Natural gas-fueled vehicles are discussed in Chapter 8. Engines that burn alternative fuels could be run lean with an oxidation catalyst as the abatement system (although they may need an LNT for NO_x), or they may be able to use a conventional three-way catalyst to meet the new standards for carbon monoxide, hydrocarbon, and nitrogen oxides emissions for stoichiometric engines. Vehicle and catalyst systems have been developed for fuels that contain 85% methanol, and prototype vehicles with catalysts that operate with dual fuels of gasoline and neat methanol and ethanol have been studied (McCabe et al. 1987). Blends of ethanol and gasoline are being used in Brazil and the United States as the transportation fuel. One drawback to the use of alcohol fuels is the potential aldehyde emissions. Studies have shown that these aldehyde emissions can be abated by using small starter catalysts located near the engine (Hochmuth and Mooney 1993). In the United States, it is common to blend in ethanol to the gasoline to make oxygenated fuels. Most of this ethanol is purchased as blends of 10% ethanol and 90% gasoline, known as gasohol or E10, and it is used as an octane enhancer to improve air quality. Flexible fuel vehicles can use E85, a blend of 85% ethanol and 15% gasoline. Since these engines operate at stoichiometric, a TWC is used to meet emission standards.

REFERENCES

Abe, F., Hashimoto, S., and Katsu, M. "An extruded electrically heated catalyst: From design concept through proven reliability," SAE 960340 (1996).

Acres, G., Cooper, B., Shutt, E., and Malerbi, B. "Platinum catalysts for exhaust emission control: The mechanism of catalyst poisoning by lead and phosphorous," pp 54–71, ACS Series 143, *Catalysts for the Control of Automotive Pollutants*. Editor J. McEvoy, American Chemical Society, Washington, D.C. (1975).

Anderson, M. "A feedback A/F control system for low emission vehicles," SAE 930388 (1993).

Andersen, P., Ballinger, T., Bennett, C., and Lafyatis, D. "Development and system application of an ultra-low loaded precious metal catalyst technology on Lev2 vehicles," SAE 2004-01-1271 (2004).

Adomaitis, J., and Heck, R. "Vehicle control strategies effect on catalyst performance," SAE 881597 (1988).

Anthony, J., and Kubsh, J. "The potential for achieving low hydrocarbon and No$_x$ exhaust emissions from large light-duty gasoline vehicles," SAE 2007-01-1261 (2007).

Automotive Engineering, "Mazda lean-burn catalyst," pp 49-51, December 1994.

Ball, D. "A warm-up and underfloor converter parametric study," SAE 932765 (1994).

Ball, D., Nunan, J., Blosser, P., Wilson, J., Mitchell, G., Davis, S., and Zammit, M. "Flexmetal catalyst technologies," SAE 2005-01-1111 (2005).

Ball, D., Zammit, M., and Mitchell, G. "Effects of substrate diameter and cell density on FTP performance," SAE 2007-01-1265 (2007).

Beck, D., Krueger, M., and Monroe, D. "The impact of sulfur on three-way catalyst: Storage and removal," SAE 910844 (1991).

Beck, D., Monroe, D., DiMaggio, C., and Sommers, J. "Axial characterization of lightoff and underfloor catalystic converters vehicle-aged on a 5.7 L corvette," SAE 952416 (1995).

Beguin, B., Garbowski, E., and Primet, M. "Stabilization of alumina toward thermal sintering by silicon addition," *Journal of Catalysis* 127: 595–604 (1991).

Bhasin, M., Nagaki, D., Koradi, P., Sherman, D., and Ankrum, C. "Novel catalyst for treating exhaust gases from internal combustion and stationary source engines," SAE 930254 (1993).

Bjordal, S., Goodfellow, C., Beckwith, P., Bennett, P., Brisley, R., and Wilkins, A. "An evaluation of the long term effects of gasoline sulphur level on three-way catalyst activity," SAE 952421 (1995).

Bosch, G., Robert, H. *Automotive Handbook*. Third Edition, Robert Bentley, Cambridge, MA, 1986.

Brett, P., Neville, A., Preston, W., and Williamson, J. "An investigation into lubricant related poisoning of automotive three-way catalysts and lambda sensors," SAE 890490 (1989).

Buhrmaster, C., Locker, R., and Socha, L. "Passive underbody adsorber complexity and size reductions," SAE 970267 (1997).

Burk, P. "Cold-start hydrocarbon emissions control," pp 47–52, *Automotive Engineering*, October 1995.

Burk, P., Hochmuth, J., Dettling, J., Heck, R., Steger, J., and Tauster, S. "Method and apparatus for treating an engine exhaust gas stream," U.S. Patent 6,171,556 (2001).

Bush, K., Adams, N., Dua, S., and Markyvech, C. "Automatic control of cylinder by cylinder air-fuel mixture using a proportional exhaust gas sensor," SAE 940149 (1994).

State of California, Air Resources Board, Notice of public meeting to consider the status of implementation of the low-emission vehicle program, 1996.

Calvert, J., Heywood, J., Sawyer, R., and Seinfeld, J. "Achieving acceptable air quality: Some reflection on controlling vehicle emissions," *Science* 261: 37–45 (1993).

CARB (California Air Resources Board), *Notice of Public Meeting to Consider the Status of Implementation of Low-Emission Vehicle Program, Nov. 21, 1996.*

Carol, L., Newman, N., and Mann, G. "High temperature deactivation of three-way catalysts," SAE 892040 (1989).

Chan, S., and Zhu, J., "The significance of high value of ignition retard control on the catalysts lightoff," SAE 962077 (1996).

Cohn, G. "Method of removing nitrogen oxides from gases," U.S. Patent 3,118,727, 1964.

Cohn, G. "Catalytic converters for exhaust emission control of commercial equipment powered by internal combustion engines," *Environmental Health Propectives* 10: 159–164 (1975).

Colucci, J., and Wise, J. "Preliminary report auto/oil air quality improvement research program effects of gasoline sulfur content below 50 PPM," Memorandum, 1993.

Cooper, B., and Shore, P. "Catalytic control of mutagenic exhaust emissions from gasoline passenger cars," SAE 890494 (1989).

Cuif, J.-P., Blanchard, G., Touret, O., Marczi, M., and Quemere, E. "New generation of rare earth compounds for automotive catalysis," SAE 961906 (1996).

Cuif, J.-P., Blanchard, G., Touret, O., Seigneurin, A., Marczi, M., and Quemere, E. "(Ce, Zr)O2 solid solutions for three-way catalysts," SAE 970463 (1997).

Cuif, J.-P., Deutsch, S., Touret, O., Marczi, Jen, H.-W., Graham, G., Chun, W., and McCabe, R. "High temperature stability of ceria-zirconia supported Pd model catalysts," SAE 980668 (1998).

Culley, S., McDonnell, T., Ball, D., Kirby, C., and Hawes, S. "The impact of passenger car motor oil phosphorus levels on automotive emissions control systems," SAE 961898 (1996).

Dettling, J., and Lui, Y. "A non-rhodium three-way catalyst for automotive applications," SAE 920094 (1992).

Dettling, J., Hu, Z., Lui, Y., Smaling, R., Wan, C., and Punke, A. "SMART Pd TWC technology to meet stringent standards," CAPoC3, Third International Congress on Catalyst and Automobile Pollution Control, Brussels, Belgium (1994).

Dettling, J., Hwang, H., Pudick, S., and Tauster, S. "Control of H2S emissions from high-tech TWC converters," SAE 900506 (1990).

Doelp, L., Koester, D., and Mitchell, M. "Oxidative automotive emission control catalysts-selected factors affecting catalyst activity," pp 24–31, ACS Series 143, *Catalysts for the Control of Automotive Pollutants*, Editor J. McEvoy, American Chemical Society, Washington, D.C. (1975).

Eade, D., Hurley, R., Ruffer, B., Inman, G., and Bakshi, R. "Fast light-off of underbody catalysts using exhaust gas ignition (EGI)," SAE 952417 (1995).

Eade, D., Hurley, R., Rulter, B., Inman, G., and Bakshi, R. "Exhaust gas igniter," pp 70–73, *Automotive Engineering*, 1996.

Eckhoff, S., Mueller, W., Lindner, D., Leyrer, J., Kreuzer, T., Vent, G., Schoen, C., Schmidt, J., and Franz, J. "Catalyst design for high-performance engines capable to fulfill future legislation," SAE 2004-01-1276 (2004).

Egami, T., Dmowski, W., and Brezny, R. "Characterization of the local structure of CeO2/ZrO2 by pulse neutron scattering," SAE 970461 (1997).

Engstrom, P., Amberntsson, A., Skoglundh, M., Fridell, E., and Smedler, G. "Sulphur dioxide interaction with NO_x storage catalyst," *Applied Catalysis B: Environmental* 22: L241 (1999).

EPA. "40 CFR Part 86, RIN: 2060-AE27, final regulations for revisions to the federal test procedure for emissions from motor vehicles," Environmental Protection Agency, 1996.

EPA. "Assessment and standards division office of transportation and air quality," U.S. Environmental Protection Agency, EPA420-R-07-003, 2007.

EPA. "Emission durability procedures for new light-duty vehicles, light-duty trucks and heavy-duty vehicles," 40 CFR Part 86 [FRL-7638-8] RIN 2060-AK76, 2004.

EPA. "Component durability procedures for new light-duty vehicles, light-duty trucks and heavy-duty vehicles," *Federal Register* 71(10): 2843–2855, 2006.

Falk, C., and Mooney, J. "Three-way conversion catalysts: Effect of closed-loop feedback control and other parameters on catalyst efficiency," SAE 800462 (1980).

Farrauto, R., Heck, R., and Speronello, B. "Environmental catalysts," *Chemical and Engineering News* 70(36): 34–44 (1992).

Farrauto, R., and Wedding, B. "Poisoning by sulfur oxides of some base metal oxide auto exhaust catalysts," *Journal of Catalysis* 33(2): 249–255 (1974).

Favre, C., and Zidat, S. "Emission systems optimization to meet future European legislation," SAE 2004-01-0138 (2004).

Fishel, N., Lee, R., and Wilhelm, F. "Poisoning of vehicle emission catalysts by sulfur compounds," *Environmental Science and Technology* 8(3): 260–268 (1974).

Fisher, G., Theis, J., Casarella, M., and Mahan, S. "The role of ceria in automotive exhaust catalysis and OBD II catalyst monitoring," SAE 931034 (1993).

Fisher, G., Zummit, M., and LaBarge, W. "Rhodium alternatives in emissions catalysts," *Automotive Engineering* 100(7): 37–40 (1992).

Frankel, M. "Deodorization of the exhaust gases in motor vehicles," *Journal of the Society of Chemical Industry* 28: 692 (1909).

Fritz, A., and Pitchon, V. "The current state of research on automotive lean NO_x catalysis," *Applied Catalysis B: Environmental* 13: 1–25 (1997).

Funabiki, M., Yamada, T., and Kayano, K. "Auto exhaust catalysts," *Catalysis Today* 10: 33–43 (1991).

Gandhi, H., Williamson, W., Goss, R., Marcotty, L., and Lewis, D. "Silicon contamination of automotive catalysts," SAE 860564 (1986).

Gandhi, H., and Shelef, M. "Effects of sulfur on noble metal automotive catalysts," *Applied Catalysis* 77(2): 175–186 (1991).

Glazoff, M., and Novak, J. "Nanostructured gamma alumina from amorphous precursors," 18th North American Catalysis Society Meeting, Cancun, Mexico (2003).

Golden, S. "Perovskite-type metal oxide compounds," U.S. Patent 6,352,955 (2002).

Golden, S. "Catalytic converter comprising perovskite-type metal oxide catalyst," U.S. Patent 6,531,425 (2003).

Golunski, S., and Roth S. "Identifying the functions of nickel in the attenuation of H_2S emissions from three-way catalysts," *Catalysis Today* 9: 105–112 (1991).

Gottberg, I., Rydquist, J., Backlund, Wallman, S., Maus, W., Bruck, R., and Swars, H. "New potential exhaust gas aftertreatment technologies for 'clean car' legislation," SAE 910840 (1991).

Greger, L., Berequist, M., Gottberg, I., Wirmark, G., Heck, R., Hoke, J., anderson, D., Rudy, W., and Adomaitis, J. "PremAir® catalyst system," SAE Paper 982728 (1998).

Guevremont, J., Gunither, G., Jao, T., Herlihy, T., White, R., and Howe, J. "Total phosphorus detection and mapping in catalytic converters," SAE 2007-01-4078 (2007).

Guinther, G., and Danner, M. "Development of an engine-based catalytic converter poisoning test to assess the impact of volatile ZDDP decomposition products from passenger car engine oils," SAE 2007-01-4079 (2007).

Gulati, S. "Thin wall ceramic catalyst supports," SAE 1999-01-0269 (1999).

Guyon, M., Blejean, F., Bert, C., and Le Faou, P. "Impact of sulfur on NOx trap catalyst activity—a study of regeneration conditions," SAE 982607 (1998).

Hammerle, R., and Wu, C. "Effect of high temperature on three-way automotive catalyst," SAE 840549 (1984).

Hanel, F., Otto, E., Bruck, R., Disinger, J., and Nagel, T. "Practical experience with the EHC system in BMW alpina B 12," SAE 970263 (1997).

Harrison, B., Diwell, A., and Hallett, C. "Promoting platinum metals by ceria—metal-support interaction in autocatalysts," *Platinum Metals Review* 32(2): 73–83 (1988).

Heck, R., Hochmuth, L., and Dettling, J. "Effect of oxygen concentration on aging of TWC catalysts," SAE 920098 (1992).

Heck, R., Hu, Z., Smaling, R., Amundsen, A., and Bourke, M., "Close coupled catalyst system design and ULEV performance after 1,050°C aging," SAE 952415 (1995).

Heck, R., Patel, K., and Adomaitis, J. "Platinum versus palladium three-way catalysts—effect of closed-loop feedback parameters on catalyst efficiency," SAE 892094 (1989).

Hegedus, L. and Gumbleton, J., *General Motors Research Publication GMR-3211, PCP-121* (1980).

Hepburn, J., Patel, K., Meneghel, M., and Ghandhi, H. "Development of Pd-only three way catalyst," SAE 941058 (1994).

Hightower, J. "Catalytic converters for motor vehicles-general overview", 67th Annual Meeting AIChE, Dec. 2, 1974 (1974).

Hochmuth, J., Burk, P., Tolentino, C., and Mignano, M. "Hydrocarbon traps for controlling cold start emissions," SAE 930739 (1993).

Hochmuth, L., and Mooney, J. "Catalytic control of emissions from M-85 fueled vehicles," SAE 930219 (1993).

Hoke, J., anderson, D., Heck, R., Poles, T., and Steger, J. "New approach for ambient pollution reduction—premair™ catalyst systems," SAE Paper 960800 (1996).

Hoke, J., Heck, R., and Poles, T. "Premair® catalyst system—a new approach to cleaning the air," SAE Paper 1999-01-3677 (1999).

Honda, "Advanced catalyst system," http://www.insightcentral.net/encyclopedia/encatalytic.html(2007).

Hu, Z., Allen, F., Wan, C., Heck, R., Steger, J., Lakis, R., and Lyman, C., "Performance and structure of Pt-Rh three-way catalysts: Mechanism for Pt/Rh synergism," *Journal of Catalysis* 174: 13–21 (1998).

Hu, Z., and Heck, R. "High temperature ultra stable close-coupled catalyst," SAE 950254 (1995).

Hu, Z., Heck, R., and Rabinowitz, H. "Close coupled catalyst," U.S. Patent 6,044,644, 2000.

Hu, Z., Heck, R., and Rabinowitz, H., "Method for using a close coupled catalyst", U.S. Patent 6254842, July 3, 2001 (2001).

Hydrocarbon Processing, "Auto emission control systems", *Hydrocarbon Processing*, pp. 85–88, May 1971 (1971)

ICT, 2007 Global Emission Standards, International Catalyst Technology, http://www.ictcatalyst.com/emission.html (2007).

Ihara, K., Murakami, H., and Ohkubo, K. "Improvement of three-way catalyst performance by optimizing ceria impregnation," SAE 902168 (1990).

Iwamoto, M., and Hamada, H. "Removal of nitrogen monoxide from exhaust gases through novel catalytic processes," *Catalysis Today* 10: 57–71 (1991).

Jobson, E., Hogberg, E., Weber, K., Smedler, G., Lundgren, S., Romare, A., and Wirmark, G. "Spatially resolved effects of deactivation on field-aged automotive catalysts," SAE 910173 (1991).

Kamada, Y., Hayashi, M., Akaki, M., Tsuchikawa, S., and Isomura, A. "Hydrogen added after-burner system," SAE 960346 (1996).

Kaneko, Y., Kobayashi, H., Komagome, R., Hirako, O., and Nakayama, O. "Effect of air-fuel ratio modulation on conversion efficiency of three-way catalysts," SAE 780607 (1978).

Kapteijn, F., Stegenga, S., Dekker, N., Bijsterbosch, J., and Moulijn, J. "Alternatives to noble metal catalysts for automotive exhaust purification," *Catalysis Today* 16: 293–287 (1993).

Kato, A., Yamashita, H., Kawagoshi, H., and Matsuda, S. "Preparation of lanthanum beta alumina with high surface area co-precipitation," *Communications of the American Ceramic Society* 70(7): C159 (1987).

Keith, C., Schreuders, T., and Cunningham, C. "Apparatus for purifying gases of an internal combustion engine," U.S. Patent 3,441,381, 1969.

Kidokoro, T., Hoshi, K., Hiraku, K., Satoya, K., Watanabe, T., Fujiwara, T., and Suzuki, H. "Development of pzev exhaust emission control system," SAE 2003-01-0817 (2003).

Kikuchi, S., Hatcho, S., Okayama, T., Inose, S., and Ikeshima, K. "High cell density and thin wall substrate for higher conversion ratio catalyst," SAE 1999-01-0268 (1999).

Kishi, N., Hashimoto, H., Fujimori, K., Ishii, K., and Komatsuda, T. "Development of the ultra low heat capacity and highly insulating (ULOC) exhaust manifold for ULEV," SAE 980937 (1998a).

Kishi, N., Kikuchi, S., Seki, Y., Kato, A., and Fujimori, K. "development of the high performance L4 ULEV system," SAE 980415 (1998b).

Kishi, N., Kikuchi, S., Suzuki, N., and Hayashi, T. "Technology for reducing exhaust gas emissions in zero level emission vehicles (ZLEV)," SAE 1999-01-0772 (1999).

Klimisch, R., Summers, J., and Schlatter, J. "The chemistry of degradation in automotive emission control catalysts," pp 103–115, ACS Series 143, *Catalysts for the Control of*

Automotive Pollutants. Editor J. McEvoy, American Chemical Society, Washington, D.C. (1975).

Kubo, S., Yamamato, M., Kizaki, Y., Yamazaki, S., Tanaka, T., and Nakanishi, K. "Speciated hydrocarbons emissions from a Si engine during cold start and warm up," SAE 932706 (1993).

Kubsh, J., Rieck, J., and Spencer, N. "Cerium oxide stabilization: Physical property and three-way activity considerations," pp 125–135, in *Catalysis and Automotive Pollution Control II*. Editor A. Crucq, Elsevier Sciences Publishers B.V., Amsterdam, The Netherlands (1991).

Kubsh, J., and Brunson, G. "EHC design options and performance," SAE 960341 (1996).

Kumar, S., Rogalo, J., Deeba, M., Burk, P., and Ferrari, V. "Influence of phosphorous poisoning on TWC catalysts," SAE 2003-01-3735 (2003).

Kumitake, K., Wanatanbe, T., Uchida, K., Yaegashi, T., and Ito, H. "Analysis of exhaust hydrocarbon compositions and ozone forming potential during cold start," SAE 961954 (1996).

Kummer, J. "Oxidation of CO and C2H4 by base metal catalysts for honeycomb supports," pp 178–192, ACS Series 143, *Catalyst for the Control of Automotive Pollutants*. Editor J. McEvoy, American Chemical Society, Washington, D.C. (1975).

Kummer, J. "Catalysts for automobile emission control," *Progress in Energy Combustion Science*, 6: 177–199 (1980).

Laing, P. "Development of an alternator powered electrically-heated catalyst system," SAE 941042 (1994).

Langen, P., Theissen, M., Malloy, J., and Zielinski, R. "Heated catalytic converter," *Automotive Engineering*, pp 31–35 (1994).

Lassi, U., Polvinen, R., Suhonen, S., Kallinen, K., Savimäki, A., Härkönen, M., Valden, M., and Keiski, R. "Effect of ageing atmosphere on the deactivation of Pd/Rh automotive exhaust gas catalysts: Catalytic activity and XPS studies," *Applied Catalysis A: General* 263: 241–248 (2004).

Lippincott, A., Segal, J., and Wang, S. "Advanced-technology vehicle emissions with California phase 2 gasoline," *Automotive Engineering* 59–61, March 1994.

Lox, E., Engler, B., and Koberstein, E. "Development of scavenger-free three-way automotive emission control catalysts with reduced hydrogen sulfide formative," SAE 890795 (1989).

Lui, Y., and Dettling, J. "Evolution of a Pd/Rh TWC catalyst technology," SAE 930249 (1993).

Machida, M., Eguchi, K., and Arai, H. "Preparation and characterization of large surface area BaO•6Al$_2$O$_3$," *Bulletin of the Chemical Society of Japan* 61(10): 3659–3665 (1988).

Mahzoul, H., Brilhac, J., and Gilot, P. "Experimental and mechanistic study of NO$_x$ adsorption over NO$_x$ trap catalyst," *Applied Catalysis B: Environmental* 20: 47–55 (1999).

Marsh, P., Gottberg, I., Thorn, K., Lundgren, M., Acke, F., and Wirmark, G. "SULEV technologies for a five cylinder N/A engine," SAE 2000-01-0894 (2000).

Matsumoto, S., Ikeda, Y., Suzuki, H., Ogai, M., and Miyoshi, N. "NO$_x$ storage-reduction catalyst with improved tolerance against sulfur poisoning," *Applied Catalysis B: Environmental* 25: 115–124 (2000).

Matsuzono, Y., Sakanushi, M., and Kitagawa, H. "Development of a low precious-metal automotive perovskite catalytic system for Lev-II," SAE 2003-01-0814 (2003).

McCabe, R., and Mitchell, P. "Exhaust catalyst development for methanol fueled vehicles," *Applied Catalysis* 44: 73–93 (1988).

McCabe, R., Mitchell, P., Liapri, F., Scruggs, W., and Warburton, R. "Catalyst evaluation on a Detroit diesel allison 6V-92TA methanol-fueled engine," SAE-872138 (1987).

McCabe, R., Siegl, W., Chun, W., and Perry, J. "Speciated hydrocarbon emissions from the combustion of single component fuels, II. catalyst effects," *Air Waste Management Assoc.* 42: 1071–1077 (1992).

MECA, a Manufacturers of Emission Control Association, "The case for banning lead in gasoline," (1998).

MECA, b "The impact of gasoline fuel sulfur on catalytic emission control systems," (1998).

MECA, Manufacturers of Emission Control Association, "The impact of sulfur in diesel fuel on catalyst emission control technology," (1999).

Minteer, S., Editors. *Alcohol Fuels.* Taylor and Frances, Boca Raton, FL (2006).

Miyoshi, M., Matsumoto, S., Katoh, K., Tanaka, T., Harada, J., Takahashi, N., Yokato, K., Sigiura, M., and Kasahara, K. "Development of new concept three-way catalyst for automotive lean burn engines," SAE 950809 (1995).

Muraki, H., "Performance of palladium automotive catalyst", SAE 910842 (1991).

Muraki, H., Shinjoh, H., Sobukawa, H., Yokota, K., and Fujitani, Y. "Behavior of automotive noble metal catalysts in cycled feedstreams," *Journal of I&EC* 24(1): 43–49 (1985).

Muraki, H. "Performance of palladium automotive catalyst," SAE 910842 (1975).

Naito, K., Suzuki, H., Tanaka, H., Taniguichi, M., Uenishi, M., Tan, I., Kajita, N., Takahashi, I., Narita, K., Hirai, A., Kimura, M., Nishihata, Y., and Mizuki, J. "Development of a Rh-intelligent catalyst," SAE 2006-01-0851 (2006).

Nakatsuji, T., Yasaukawa, R., Tabata, K., Ueeda, K., and Niwa, M. "A highly durable catalytic NO$_x$ reduction in the presence of SO$_x$ using periodic two steps, an operation in oxidizing conditions and a relatively short operation in lean conditions," *Applied Catalysis B: Environmental* 21: 121 (1999).

Nishizawa, K., Yamada, T., Ishizuka, Y., and Inoue, T. "Technologies for reducing cold-start emissions of V6 ULEVs," SAE 971022 (1997).

Nishizawa, K., Momoshima, S., Koga, M., Tsuchida, H., and Yamamoto, S. "Development of new technologies targeting zero emissions for gasoline engines," SAE 2000-01-0890 (2000).

Noda, N., Takahashi, A., and Mizuno, A. "In-line hydrocarbon adsorber system for cold-start emissions," SAE 970266 (1997).

Noda, N., Takahashi, A., Shibayaki, Y., and Mizuno, K. "In line hydrocarbon adsorber for cold start emissions—part II," SAE 980423 (1998).

Norman, C. "Zirconium oxide products in automotive systems," SAE 970460 (1997).

Oguma, H., Koga, M., Momoshima, S., Nishizawa, K., and Yamamoto, S. "Development of third generation of gasoline P-ZEV technology," SAE 2003-01-0816 (2003).

Oser, P., Mueller, E., Hartel, G., and Schurfeld, A. "Novel emission technologies with emphasis on catalyst cold start improvements status report on VW-Pierburg burner/catalyst systems," SAE 940474 (1994).

Patil, M., Herth, W., Williams, J., and Nagel, J. "In-line hydrocarbon adsorber system for cold-start emissions," SAE 970266 (1996).

Patil, M., Peng, Y., and Morse, K. "Airless in-line adsorber system for reducing cold start emissions," SAE 980419 (1998).

Rabinowitz, H., Tauster, S., and Heck, R. "The effects of sulfur & ceria on the activity of automotive Pd/Rh catalysts," *Applied Catalysis A: General* 212: 215–222 (2001).

Rokosz, M., Chenb, A., Lowe-Mac, C., Kucherov, A., Benson, D., Paputa Peck, M., and McCabe, R., "Characterization of phosphorus-poisoned automotive exhaust catalysts", *Applied Catalysis B: Environmental* 33 (2001) 205–215 (2001).

Rokosz, M., Chenb, A., Lowe-Mac, C., Kucherov, A., Benson, D., Peck, M., and McCabe, R. "Characterization of phosphorus-poisoned automotive exhaust catalysts," *Applied Catalysis B: Environmental* 33: 205–215 (2001).

Rohart, E., Larcher, O., Ottaviani, E., Pelissard, S., and Allain, M. "High thermostable hybrid Zirconia materials for low loading precious metal catalyst technology," SAE 2005-01-1554 (2005).

Rudy, W., Ober, R., Durilla, M., anderson, D., Hoke, J., and Heck, R. "New approach for ambient pollution reduction—premair™ catalyst systems field experience," 89th Annual Meeting of Air & Waste Management Association, Paper 96-MP4A.08 (1996).

Sato, N., Narita, K., Kimura, M., Taniguichi, M., Kajita, N., Uenishi, M., Tan, I., and Tanaka, H. "Design of a Practical Intelligent Catalyst," SAE 2003-01-0813 (2003).

Sawyer, R. "Reformulated gasoline for automotive emissions reduction," Twenty-Fourth Symposium (International) on Combustion, pp 1423–1432, Sydney, Australia (1992).

Schlatter, L., Sinkevitch, R., and Mitchell, P. "A Laboratory Reactor System for Three-Way Catalyst Evaluation," General Motors, GMR-2911, PC P-87 (1979).

Schuerholz, S., Ellmer, D., and Siemuns, S. "Exhaust gas management," SAE 2004-01-0647 (2004).

Searles, R. "Car exhaust Pollution: The role of precious metal catalysts in its control," *Endeavor: New Series* 13(1): 2–7 (1989).

Shimasaki, Y., Kato, H., Abe, F., Hashimoto, S., and Kaneko, T. "Development of extruded electrically heated catalyst for ULEV standards," SAE 971031 (1997).

Shore, L., Farrauto, R., Deeba, M., Lampert, J., and Heck, R. "Composition for abatement of volatile organic compounds and apparatus and methods using the same," U.S. Patent 6.319,484, 2001.

Siegl, W., McCabe, R., Chun, W., Kaiser, E., Perry, J., Henig, Y., Trinker, F., and anderson, R. "Hydrocarbon emissions from the combustion of single component fuels I. effect of fuel structure," *Air and Waste Management Association* 42: 912–920 (1992).

Skowron, L., Williamson, W., and Summers, L. "Effect of aging and evaluation on the three way catalyst performance," SAE 892093 (1989).

Smalling, R., Sung, S., and Bartlett, R. "Washcoat technology and precious metal loading study targeting the California LEV MDV2 standard," SAE 961904 (1996).

Socha, L., and Thompson, D. "Heated metal converters for low emission vehicles," *Automotive Engineering* 100(7): 21–25 (1992).

Subramanian, S., Kudia, R., and Chattha, M. "Treatment of natural gas vehicle exhaust," SAE 930223 (1993).

Summers, J., and Williamson, W. "Palladium-only catalysts for closed-loop control: a review," ACS Meeting, Denver, CO (1993).

Summers, J., Skowron, J., and Miller, M. "Use of light-off catalyst to meet the California LEV/ULEV standards," SAE 930386 (1993).

Summers, J., Williamson, W., and Henk, M. "Uses of palladium in automotive control catalysts," SAE 880281 (1988).

Sung, S., Ober, R., Casper, R., and Miles, G. "A 45% engine size catalyst system for MDV2 ULEV applications," SAE 982553 (1998a).

Sung, S., Smaling, R., and Brungard, N.M. "Pb poisoning on Pd-only TWC catalysts," *Studies in Surface Science and Catalysts*, 116: 165–174 (1998b).

Sung, S., and Burk, P. "Method and apparatus for treatment of exhaust streams," U.S. Patent 5,603,215, 1997.

Takada, T., Hirayama, H., Itoh, T., and Yaegashi, T. "Study of divided converter catalytic system satisfying quick warmup and high heat resistance," SAE 960797 (1996).

Takami, A., Takemoto, T., Iwakuni, H., Saito, F., and Komatsu, K. "Development of lean burn catalyst," SAE 950746 (1995).

Takei, Y., Hirohiko, H., Okada, M., and Abe, K. "Effect of gasoline components on exhaust hydrocarbon components," SAE 932670 (1993).

Takei, Y., Kinugasa, Y., Okada, M., Tanaka, T., and Fujimoto, Y. "Fuel properties for advanced engines," *Automotive Engineering International*, pp 117–120, July 2001.

Tanaka, H., Taniguchi, M., Uenishi, M., Kajita, N., Tan, I., Nishihata, Y., Mizuki, J., Narita, K., Kimura, M., and Kaneko, K. "Self-regenerating Rh- and Pt-based perovskite catalysts for automotive-emissions control," *Angewandte Chemic International Edition* 45: 5998–6002 (2006).

Taniguichi, M., Uenishi, M., Tan, I., Tanaka, H., Kimura, M., Mizuki, J., and Nishihata, Y. "Thermal properties of the intelligent catalyst," SAE 2004-01-1272 (2004).

Taylor, K. "Catalysts in cars," *Chemical Technology* 551–555 (1990).

Taylor, K. "Nitric oxide catalysis in automotive exhaust systems," *Catalysis Reviews* 35(4): 457–481 (1993).

Theis, T., and LaBarge, B. "An air/fuel algorithm to improve the NOx conversion of copper-based catalysts," SAE 922251 (1992).

Thompson, C., Mooney, J., Keith, C., and Mannion, W. "Polyfunctional catalysts," U.S. Patent 4,157,316, 1979.

Tijburg, I., Geus, J., and Zandbergen, H. "Application of lanthanum to pseudo-boehmite and gamma-Al_2O_3," *Journal of Material Science* 26: 6479–6486 (1991).

Usmen, R., McCabe, R., Haack, L., Graham, G., Hepburn, J., and Watkins, W. "Incorporation of La 3+ into a Rh/γ-Al_2O_3 catalyst," *Journal of Catalysts* 134: 702–712 (1992).

Viala, A. "Health effects of air pollutants emitted by automotive vehicles," *Petroleum Technology* 351: 25–27 (1993).

Votsmeier, M., Bog, T., Lindner, D., Gieshoff, J., Lox, E., and Kreuzer, "A systematic approach towards low precious metal three-way catalyst application," SAE 2002-01-0345 (2002).

Waltner, A., Loose, G., Hirschmann, A., Mubmann, L., Lindner, D., and Muler, W. "Development of close-coupled catalyst systems for European driving conditions," SAE 980663 (1998).

Wan, C., and Detting, J. "High temperature catalyst compositions for internal combustion engine," U.S. Patent 4,624,940, 1986.

Wan, C., and Dettling, J. "Effective rhodium utilization in automotive exhaust catalyst," SAE 860566 (1986).

Wang, T., Soltis, R., Logothetis, E., Cook, J., and Hamburg, D. "Static characteristics of ZrO2 exhaust gas oxygen sensors," SAE 930352 (1993).

Wanke, S., and Flynn, P. "The sintering of supported metal catalysts," *Catalysis Review, Science and Engineering* 12(1): 93–135 (1975).

Wards, "Acceleration enrichment may be large source of pollution," Ward's Engine and Technology Update, 1993.

Watanabe, K., Taga, W., Hirota, T., Tanikawa, K., Nagashima, K., Zhang, G., and Muraki, H. "Advanced emission control system for ULEV2 application," SAE 2006-01-0848 (2006).

White, J., Carroll, J., Brady, M., Burkmyre, W., Liss, W., and Church, M. "Natural gas converter performance and durability," SAE 930222 (1993).

Wiedenmann, H., Raff, L., and Noods, R. "Heated zirconia oxygen sensor for stoichiometric and lean air-fuel ratio," SAE 840141 (1984).

Williamson, W., Dou, D., and Robota, H. "Dual-catalyst underfloor lev/ulev strategies for effective precious metal management," SAE 1999-01-0776 (1999).

Williamson, W., Nunan, J., Frownfelter, D., McClaughry, R., Tripp, G., and Huynh, J., "Palladium/rhodium dual-catalyst LEV 2 and Bin 4 close-coupled emission solutions," SAE 2007-01-1263 (2007).

Williamson, W., Perry, J., Gandhi, H., and Bomback, J. "Effects of oil phosphorous on deactivation of monolithic three-way catalysts," *Applied Catalysis* 15: 277–292 (1985).

Wong, C., and McCabe, R. "Effects of high-temperature oxidation and reduction on the structure activity of Rh/Al_2O_3 and Rh/SiO_2 catalysts," *Journal of Catalysis* 119: 47–64 (1989).

Wong, C., and Tsang, C. "Temperature-programmed reduction studies of Al2O3-supported Pt/Rh catalysts," SAE 922337 (1992).

Wu, X., Fan, J., Ran, R., and Weng, D. "Effect of preparation methods on the structure and redox behavior of platinum-ceria-zirconia catalysts," *Chemical Engineering Journal* 109: 133–139 (2005).

Xua, L., Guoa, G., Uyb, D., O'Neill, A., Weber, W., Rokosz, M., and McCabe, R. "Cerium phosphate in automotive exhaust catalyst poisoning," *Applied Catalysis B: Environmental* 50: 113–125 (2004).

Yamada, T., Kayano, K., and Funabiki, M. "The effectiveness of Pd for converting hydrocarbons in TWC catalysts," SAE 930253 (1993).

Yamada, T., Kobayashi, T., Kayano, K., and Funabiki, M. "Development of Zr containing TWC catalysts," SAE 970466 (1997).

Yao, Y-F. "The oxidation of CO and C_2H_4 over metal oxides," *Journal of Catalysis* 39: 104–115 (1975).

Yoshida, T., Sato, A., Suzuki, H., Tanabe, T., and Takahashi, N. "Development of high performance three-way-catalyst," SAE 2006-01-1061 (2006).

Zeldovich, J. "The oxidation of nitrogen in combustion and explosion," *Acta Physiochimica, U.S.S.R.* 21(4): 577–628 (1946).

CHAPTER 6 QUESTIONS

1. Why is it necessary to control emissions from the spark-ignited engine in the automobile?

2. Why are the precious metals Pt, Pd, and Rh used in automotive catalysts?

3. What is meant by three-way catalysis or TWC?

4. Why is the development of the oxygen sensor HEGO so important for implementation of TWC technology?

5. Why is Ce important in the utilization of the oxygen sensor HEGO and implementation of TWC technology?

6. What technologies were needed to allow the use of Pd as a NO_x reduction catalyst versus Rh?

7. What operating elements are important in the use of a close-coupled catalyst for cold-start performance?

8. How come TWC technologies cannot be used for lean-burn engines such as GDI?

9. What are the major mechanisms for deactivation of TWC technology? Describe how it affects the TWC performance.

10. Design the complete washcoated monolith for low-temperature lightoff during the cold-start process and for hot-start and cruise conditions of the Federal Test Procedure (FTP). Feel free to explore approaches that may be different from what you have learned in class.

11.
    a. Calculate the stoichiometric point ($\lambda = 1$) on a weight–weight basis, for E85 (15% gasoline and 85% ethanol). Assume the volume % of O_2 in air is 21%.

    Assume

$$\text{Gasoline: } C_8H_{18} + 25/2\,O_2 \rightarrow 8CO_2 + 9H_2O$$

$$\text{Ethanol } C_2H_5OH + 3O_2 \rightarrow 2CO_2 + 3H_2O$$

b. What are the consequences for fuel economy for using pure gasoline versus E85?

12. Explain the importance of the oxygen sensor. Why does the signal oscillate in amplitude?

13. What are the catalytic reactions for $\lambda = 1$, $\lambda < 1$, and $\lambda > 1$?

14. What alternative materials might be considered as OSC?

15. What laboratory experiments would you perform to determine whether these alternative materials were acceptable replacements for CeO_2? When preparing the outline of your plan, make sure to think about how you would characterize these materials before and after duty cycle testing.

16. Design a single TWC monolith catalyst for a stoichiometric engine that will successfully pass the cold-start, hot-start, and steady-state modes of the U.S. FTP test.

17. Refer to the TWC U.S. Patent 4157,316 (1979).
    a. Summarize the most important qualitative issues in this patent, and why the choice of catalyst metals. No more than 5 sentences.
    b. Al_2O_3 is stable as the preferred support.
       1) What other oxide (s) can be added and why?
       2) Which other oxides are preferred additions to Al_2O_3?
    c. What is meant by "fixing" the components in the catalyst?
    d. How is fixing performed?
    e. What is the role of the metal oxides that have multiple oxidation states?
    f. Which metal oxide is preferred?
    g. What alternative mode of conversion of all three components can be used in the exhaust instead of the control of the stoichiometric air-to-fuel ratio?
    h. Why would one want to do that?
    i. What are the problems?

18. Refer to claims of patent.
    a. Compare 1b to 2.
    b. Compare 1c to 4.
    c. Compare 1d to 3.
    d. Compare 5 to 6.
    e. Compare 9e to 15.
    f. Compare 1 to 9.

7 Automotive Substrates

7.1 INTRODUCTION TO CERAMIC SUBSTRATES

In view of the high exhaust temperature and large temperature gradients caused by exothermic catalytic reactions and engine malfunction, automakers sought ceramic substrates with large surface area, good thermal shock resistance, and low cost. Alumina beads that had been used as substrates outside of the automotive industry met these requirements. Ceramic monoliths with a honeycomb structure offered another alternative, provided they met the low-cost objective and had a low coefficient of thermal expansion to withstand thermal shock. Three companies—W.R. Grace, American Lava Corp.,[1] and Corning Glass Works[2]—developed new compositions and processes for manufacturing these monoliths, but only Corning succeeded in meeting the cost and technical requirements (Bagley et al. 1973). Corning scientists invented the cordierite ceramic with low coefficient of thermal expansion (Lachman and Lewis 1975) and the extrusion process, which provided the flexibility of honeycomb geometry, substrate contour, and substrate size (Bagley 1974). In addition, these monoliths offered a high geometric surface area approaching 90 in^2/in^3 (35 cm^2/cm^3) and a use temperature approaching 1,200°C. Consequently, the cordierite ceramic substrate has become the world standard and is used in 95% of today's catalytic converters. Although alternative metal materials like FeCrAlloy have become available over the past 20 years, automakers around the world continue to use ceramic substrates to meet the ever demanding emissions and durability requirements due to their cost-effectiveness and decades of successful field performance. The metallic honeycombs will continue to have a niche market as a catalyst support.

Although only pelleted catalysts had been certified in California, ceramic honeycomb substrates with attractive properties had been developed and provided an intriguing alternative to pellets (Bagley et al. 1973). Both types of substrates had to be durable and resistant to attrition and catalyst poisoning. They had to meet performance requirements, including lightoff, high-temperature resistance, efficient heat and mass transport, and low back

[1] Subsidiary of 3M Company.
[2] Now Corning Incorporated.

FIGURE 7.1. Cercor® ceramic heat exchanger. Reprinted from Gulati 1996a; courtesy of Marcel Dekker, New York.

pressure. In addition, they had to meet the space requirements by modifying their shape and size. Although each type of substrate had its advantages and disadvantages, GM and certain foreign automakers chose pellets, whereas Ford, Chrysler, and others decided to use monolithic supports for 1975 MY vehicles.

Figure 7.1 shows the earliest thin-wall ceramic honeycomb structure, invented and manufactured by Corning Glass Works, for use in rotary regenerator cores for gas turbine engines (Hollenbach 1963). This product, trademarked as Cercor®, was shown to Ed Cole of General Motors in early 1970 by Corning's CEO Amory Houghton and President Tom MacAvoy for possible automotive applications.

This interesting structure was formed by wrapping alternate layers of flat and corrugated porous cellulose paper, coated with a suitable glass slurry, until a cylinder of desired diameter and length was obtained. The unfired matrix cylinder was then processed through a firing cycle up to a temperature approaching 1,250 °C to effect sintering and subsequent crystallization. The sintered structure was then cooled to 100 °C at a controlled rate and removed from the furnace. Such a sintered honeycomb structure comprised 400 cells/in^2 (400 cpsi) and offered some attractive properties, namely an open frontal area of 80%, geometric surface area of 125 in^2/in^3 (49 cm^2/cm^3), hydraulic diameter of 0.0256″ (0.065 cm), and a bulk density of 7.9 g/in^3. Furthermore, its coefficient of thermal expansion (CTE) from room temperature to 1,000 °C was low, 20×10^{-7} in/in/°C, due to a special glass-ceramic composition (lithium aluminosilicate, Corning Code 9,455). Mr. Cole was most impressed with the high surface area (per unit volume) of the honeycomb structure and suggested that this would make an ideal substrate for oxidation catalyst. Of course, the other properties were equally beneficial for this application considering high exhaust temperature and appreciable temperature gradients due to catalytic exotherms. Corning was given the challenge to produce such substrates in huge volume and to do so in a cost-effective manner. Unfortunately, the Cercor® process was too slow to be cost-effective. Furthermore, the ceramic composition had to be more refractory than that of Cercor® to withstand temperatures approaching 1,400 °C due to misfiring or other engine malfunctions.

With a massive research and development (R&D) effort, Corning invented both a new ceramic composition (Lachman and Lewis 1975) and a unique forming process (Bagley 1974), which together permitted the manufacture of monolithic cellular ceramic substrates in a cost-effective manner. Under the trademark of Celcor®, these substrates are manufactured by the extrusion process from cordierite ceramic ($2MgO \cdot 2Al_2O_3 \cdot 5SiO_2$) with low CTE and high melting temperature ($\approx 1,450 °C$). One of the unique features of these substrates is their design flexibility, which is discussed in the next section.

7.2 REQUIREMENTS FOR SUBSTRATES

The substrate is an integral part of the catalytic converter system. Its primary function is to bring the active catalyst into maximum effective exposure with the exhaust gases (see Chapter 2: "The Preparation of Catalytic Materials"). In addition, it must withstand a variety of severe operating conditions, namely rapid changes in temperature, gas pulsations from the engine, chassis vibrations, and road shocks. As noted, both pellets of cylindrical and spherical geometry and honeycomb monoliths became available for substrates.

Each type of substrate had its strengths and weaknesses. The pellets were made of porous γ-Al_2O_3 with a density of $0.68 \, g/cm^3$ and Brunauer, Emmett, and Teller (BET) area of 100 to $200 \, m^2/g$. They measured about 1/8″ in diameter and were available in spherical or cylindrical shape with different aspect ratios. They were selected for their crush and abrasion resistance; they also promoted turbulent flow, which improved the contact of reactant gases with noble-metal catalyst deposited predominantly on the outer surface. The latter also improved the rate of pore diffusion mass transfer to the catalyst (Summers and Hegedus 1978). Pellets were also replaceable by refilling the container after, say, 50,000 miles. However, the pellet converter was much heavier and slow to warm up (see Figure 7.2). It also generated high back pressure and had a severe problem of pellet attrition due to rubbing against one another during vehicle use. The use of low-density pellets to improve lightoff performance would aggravate the attrition problem further.

A honeycomb substrate, on the other hand, with γ-Al_2O_3 washcoat, is considerably lighter[3] and can be brought to lightoff temperature rapidly by locating it closer to the engine due to its compact size. The gas flow is laminar with

---

[3] A 4.2-L pellet converter weighed about 10 kg in 1975 compared with 5 kg for the honeycomb converter.

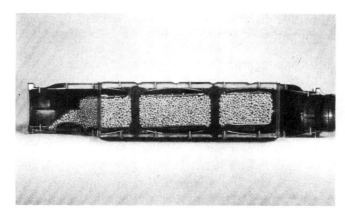

FIGURE 7.2. Pellet converter. Reprinted from Gulati 1996a; courtesy of Marcel Dekker, New York.

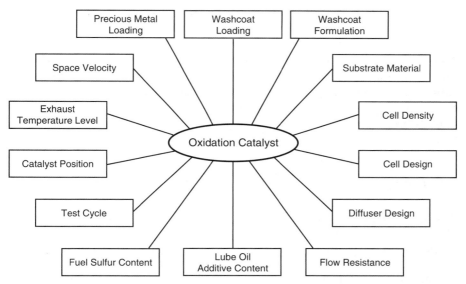

FIGURE 7.3. Parameters affecting the performance of the catalytic converter. Reprinted from Gulati 1996a; courtesy of Marcel Dekker, New York.

relatively large passage ways that result in substantially lower back pressure. Furthermore, the honeycomb substrate can be mounted more robustly, thereby avoiding the attrition problem associated with pellets (Kummer 1980; Harned and Montgomery 1973). As pointed out, GM, American Motors, and certain foreign automakers chose the pellet-type substrate, whereas others went with the honeycomb substrate.

Figure 7.3 shows the key parameters that affect the performance of a catalytic converter. Many of these parameters are influenced by the substrate design, which will be discussed in the next section. As a preface to that, we will list the requirements that an ideal substrate must meet:

1. It must be coatable with a high BET area washcoat.
2. It must have low thermal mass, low heat capacity, and efficient heat transfer to permit gaseous heat to heat up the catalyst-carrying washcoat quickly, notably during lightoff.
3. It must provide high-surface-area-per-unit volume to occupy minimum space while meeting emissions requirements.
4. It must withstand high use temperature.
5. It must have good thermal shock resistance due to severe temperature gradients originating from fuel mismanagement and/or engine malfunction.
6. It must minimize back pressure to conserve engine power for rapid response to transient loads.
7. It must have high strength over the operating temperature range to withstand vibrational loads and road shocks.

The above requirements can be met by optimizing both the geometric and physical properties of the substrate, which in the case of extruded honeycomb substrates, can be independently controlled—a significant design advantage over pellet-type substrates. In view of their design flexibility and other inherent advantages, we will focus on honeycomb substrates in the remainder of this chapter.

7.3   DESIGN/SIZING OF SUBSTRATES

The cell shape and size, which can be designed into the extrusion die, affect the geometric properties and hence the size of honeycomb substrate. Two cell shapes, which proved to be cost-effective in terms of extrusion die cost, were square cell and equilateral triangle cell, as shown in Figure 7.4.

The cell size has a strong bearing on cell density (n), geometric surface area (GSA), open frontal area (OFA), hydraulic diameter (D_h), bulk density (ρ), thermal integrity (TIF), mechanical integrity (MIF), resistance to flow (R_f), bulk heat transfer (H_s), and lightoff (LOF), which in turn affect both the performance and the durability of the catalytic converter. Simple expressions for these geometric properties of square, triangular, and hexagonal, cell substrates are given in the Appendix (Gulati 1988a). In these expressions, TIF is a measure of temperature gradient the substrate can withstand prior to fracture; MIF is a measure of crush strength of substrate in the diagonal direction; R_f is a measure of back pressure; H_s is a measure of steady-state heat transfer; and LOF is a measure of lightoff performance. An ideal substrate must offer high GSA, OFA, TIF, MIF, H_s, and LOF values as well as low D_h, ρ, and R_f

FIGURE 7.4. Honeycomb substrates with square and triangular cells. Reprinted from Gulati 1996a; courtesy of Marcel Dekker, New York.

values. A close examination of the above expressions indicates that certain compromises are necessary in arriving at the optimum substrate.

The initial substrates in 1975, with a square-cell configuration, were designed to have 200 cells/in^2 and a wall thickness of 0.012″ (0.0305 cm) (thus designated by 200/12). They were extruded from cordierite composition (EX-20), which had a fired wall porosity of 35%. Similarly, the triangular cell substrates had a cell density of 236 cells/in^2 and a wall thickness of 0.0115″ (236/11.5). They were extruded from a lower CTE cordierite composition (EX-32) with a fired wall porosity of 40%. The geometric properties of these two substrates as well as current versions of standard substrates are summarized in Table 7.1.

To facilitate comparison of various substrates, we must first express the substrate requirements in terms of their geometric properties. For good light-off performance, the substrate must have a high LOF value. For high conversion efficiency under steady-state or warmed-up conditions, the substrate must have high n, high GSA, and high H_s values. For low back pressure, the substrate must have high OFA, large D_h, and low R_f values. And finally, for high mechanical and thermal durability, the substrate must have high MIF and TIF values. A close examination of Table 7.1 shows the evolution of substrate optimization from 1975 to 1983:

1. The initial substrate, namely 200/12 □, had low values of n, GSA, and LOF, and yet it met the conversion efficiency requirements, which were less stringent; it had large D_h, low R_f, and modest OFA values, which helped minimize the back pressure and conserve engine power; and it had high MIF and modest TIF values, which met the durability requirements.

TABLE 7.1. Geometric properties of honeycomb substrates (fillet radius R = 0).

Designation*	200/12	300/12	400/6.5	236/11.5
Cell shape	square	square	square	triangle
Wall porosity (%)	35	35	35	40
L (cm)	0.18	0.15	0.13	0.25
T (cm)	0.030	0.030	0.016	0.029
D_h (cm)	0.15	0.12	0.11	0.12
GSA (cm^2/cm^3)	18.5	21.5	27.4	22.0
OFA (%)	68.9	62.9	75.7	63.8
ρ (g/cm^3)	0.51	0.61	0.40	0.55
$R_f \times 10^{-2}$	120	216	198	200
H_s	2,885	4,320	5,760	3,675
MIF × 100	3.5	5.4	1.9	3.4
TIF	5.9	4.8	7.7	7.1
LOF (in °C cal^{-1})	373	471	955	550

*Honeycomb geometry is designated by a cell density and wall thickness combination. Thus, 200/12 designates a substrate with 200 cells/in^2 and a 0.012″ wall thickness. Reprinted from Gulati 1996a; courtesy of Marcel Dekker, New York.

2. The subsequent development of 236/11.5 Δ and 300/12 □ substrates resulted in significantly higher LOF, GSA, H_s, and MIF values to meet more stringent conversion and durability requirements, but the lower D_h and higher R_f values had an adverse effect on back pressure.

3. The standard substrate, namely 400/6.5 □, was developed in the early 1980s to meet even more stringent emissions regulations (see Chapter 6); this substrate offered the highest values of LOF, GSA, H_s, OFA, and TIF without any compromise in D_h and R_f values; its MIF value was adequate to meet the durability requirement.

Note that the nomenclature □ refers to the geometric shape of the monolith channel, which in this example is square.

It is clear from the above discussion that honeycomb substrates offer the unique advantage of design flexibility to meet the ever changing performance and durability requirements. Since these requirements can often be conflicting, certain trade-offs in geometric properties may be necessary as illustrated in Table 7.1. New advances in substrates, in terms of both composition and cell geometry, necessitated by more stringent performance and durability requirements for 1995+ vehicles (equivalent to LEV and ULEV standards[4]), are summarized in Table 7.2.

A comparison with Table 7.1 shows that these advanced substrates are designed for close-coupled application where fast lightoff, low back pressure, and compact size are most critical (Gulati and Then 1994; Gulati et al. 1994).

[4] Low-emission vehicle and ultra-low-emission vehicle.

TABLE 7.2. Geometric properties of advanced honeycomb substrates (fillet radius R = 0).

Designation	300/6.7	350/5.5	470/5
Cell shape	triangle	square	square
Wall porosity (%)	35	24	24
L (cm)	0.22	0.14	0.12
T (cm)	0.017	0.014	0.013
D_h (cm)	0.11	0.12	0.10
GSA (cm²/cm³)	27.0	26.4	30.5
OFA (%)	75.4	80.5	79.4
ρ (g/cm³)	0.40	0.37	0.39
$R_f \times 10^{-2}$	183	153	210
H_s	4,680	5,035	6,765
MIF × 100	1.3	1.2	1.3
TIF	10.8	9.7	9.2
LOF (in °C cal⁻¹)	918	888	1,140

Reprinted from Gulati 1996a; courtesy of Marcel Dekker, New York.

Recent developments in ultra-thin-wall substrates and metallic substrates are also discussed in this chapter.

The size of a substrate depends on many factors. Predominant among these are flow rate, lightoff performance, conversion efficiency, space velocity, back pressure, space availability, and thermal durability. Other factors such as washcoat formulation, catalyst loading, inlet gas temperature and fuel management can also have an impact on the size of a substrate.

Considering the substrate alone, both the conversion efficiency and back pressure depend on its size. The former is related to the total surface area (TSA) of the substrate defined by

$$TSA = GSA \times V \tag{7.1}$$

in which V denotes the substrate volume given by

$$V = A \times \ell \tag{7.2}$$

In Eq. (7.2), A and ℓ are cross-sectional area and length of the substrate, respectively. Similarly, the back pressure is related to flow velocity v through the substrate and its length ℓ, both of which are affected by substrate size:

$$v = \frac{V_e}{A(OFA)} \tag{7.3}$$

$$\ell = V/A \tag{7.4}$$

In Eq. (7.3), V_e denotes the volumetric flow rate and the other terms have been defined previously.

Experimental data show that conversion efficiency η depends exponentially on TSA (Day 1995):

$$\eta = 1 - \left(\frac{E_o + E_i \exp(-\text{TSA}/\text{TSA}_o)}{E_o + E_i} \right) \tag{7.5}$$

In Eq. (7.5), E_o denotes unconverted emissions,[5] E_i denotes converted emissions, $(E_o + E_i)$ denotes engine emissions, and TSA_o denotes that value of TSA that helps reduce convertible emissions by 63%. Thus, increasing the TSA to 3 TSA_o would reduce the convertible emissions by 95%. However, it would also increase the substrate volume by 300%. Obviously, additional increases in substrate volume would have little impact on emissions reduction.

The pressure drop Δp, in the laminar flow regime, across the substrate depends linearly on flow velocity and its length but inversely on the square of hydraulic diameter (Day and Socha 1991; Gulati 1993a):

$$\Delta p = C \frac{v\ell}{D_h^2} = \frac{CV_e\ell}{AD_h^2(\text{OFA})} \tag{7.6}$$

Since the substrate volume controls TSA, which in turn affects conversion efficiency, its cross-sectional area A should be maximized and its length ℓ should be minimized to reduce Δp. Of course, such an optimization of substrate shape will depend on the space availability under the chassis.

In practice, both the laboratory data and the field experience have shown that if the substrate volume is approximately equal to engine displacement, it will meet the conversion requirements. Denoting engine displacement by V_{ed} and engine speed by N (revolutions per minute), the space velocity VHSV may be written as follows:

$$\text{VHSV} = \frac{30NV_{ed}}{V} \quad \text{hr}^{-1} \tag{7.7}$$

which, for $V = V_{ed}$, becomes

$$\text{VHSV} = 30N \quad \text{hr}^{-1} \tag{7.8}$$

The residence time τ for catalytic activity is simply the inverse of space velocity, i.e.,

$$\tau = \frac{120}{N} \quad \text{s} \tag{7.9}$$

---

[5] E_o is estimated to range from 5% to 10% of total engine emissions.

For typical engine speeds ranging from 1,500 to 4,000 RPM, the space velocity would range from 45,000 to 120,000 per hr. The corresponding residence time would range from 0.08 to 0.03 s, which seems to be adequate for catalytic reaction. At lower space velocities, the gas temperature is low and requires a longer time for reaction, whereas at higher space velocities, the gas temperature is high and requires less time for reaction.

7.4   PHYSICAL PROPERTIES OF SUBSTRATES

The physical properties of a ceramic substrate, which can be controlled independently of geometric properties, also have a major impact on its performance and durability. These properties include microstructure (crystal orientation, porosity, pore size distribution, and microcracking), CTE, strength (crush strength, isostatic strength, and modulus of rupture), structural modulus (also called E-modulus), and fatigue behavior (represented by dynamic fatigue constant). These properties depend on both the ceramic composition and manufacturing process, which can be controlled to yield optimum values for a given application, e.g., automotive (Gulati and Then 1994a; Gulati 1985; Gulati 1991a); diesel (Gulati 1992b; Gulati and Lambert 1991b), and motorcycle (Gulati and Scott 1993a; Gulati and Scott 1993b).

The microstructure of ceramic honeycombs not only affects physical properties like CTE, strength, and structural modulus, it has a strong bearing on substrate/washcoat interaction, which in turn affects the performance and durability of the catalytic converter (Gulati et al. 1989; Gulati et al. 1991c; Gulati et al. 1991d). The coefficients of thermal expansion, strength, fatigue, and structural modulus of the honeycomb substrate (which also depend on cell orientation and temperature) have a direct impact on its mechanical and thermal durability (Gulati 1985). Finally, since all of the physical properties are also affected by washcoat formulation, loading, and washcoat processing, they must be evaluated before and after the application of washcoat to assess converter durability (Gulati et al. 1989; Gulati 1991c; Gulati 1991d).

7.4.1   Thermal Properties

The key thermal properties include CTE, specific heat c_p, and thermal conductivity K. The CTE values are strongly dictated by the anisotropy and orientation of cordierite crystal (Lachman and Lewis 1975; Lachman et al. 1981) as well as by the degree of microcracking (Ikawa et al. 1987; Buessem et al. 1952). The latter is controlled by composition and firing cycle and can reduce the CTE significantly. As noted, cordierite substrates have an extremely low CTE by virtue of preferred orientation of anisotropic cordierite crystallites afforded by the extrusion process. Figure 7.5 shows the average CTE values along the three axes of cordierite crystallite; the complete CTE curves along A, B, and

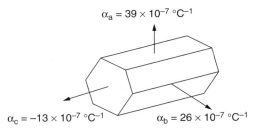

FIGURE 7.5. Average CTE values of cordierite crystallite along its three axes (25–800°C). Reprinted from Gulati 1996a; courtesy of Marcel Dekker, New York.

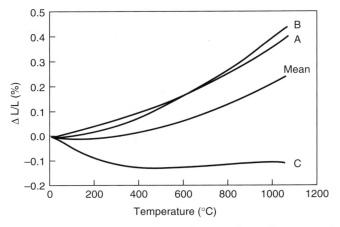

FIGURE 7.6. Thermal expansion curves for orthorhombic cordierite crystal. Reprinted from Gulati 1996a; courtesy of Marcel Dekker, New York.

C axes are shown in Figure 7.6 along with the calculated mean CTE curve, which represents random orientation.

The extrusion process produces preferred orientation in the raw materials that upon firing causes alignment of orthorhombic crystal so that its lowest CTE axis lies in the extrusion direction (also called the axial direction along which the gases flow). Nonaligned crystals generate localized stresses due to expansion anisotropy and lead to microcracking. The combination of preferred orientation and microcracking results in axial CTE values ranging from 1×10^{-7} to $10 \times 10^{-7}/°C$ (over that 25–800°C temperature range) compared with $17 \times 10^{-7}/°C$ for random orientation. This drastic reduction in CTE value makes the cordierite substrate an ideal substrate in that the order of magnitude higher CTE of γ-alumina washcoat[6] can now be managed easily. Tables 7.3a and 7.3b list the CTE values in axial and tangential[7] directions for 400/6.5 □

---

[6] Typical CTE of γ-Al$_2$O$_3$ ≈80 $10^{-7}/°C$ (Gulati et al. 1991a).
[7] Circumferential direction in cross-sectional plane of substrate, ⊥ to axial direxrion.

TABLE 7.3. a) CTE versus temperature data for EX-20, 400/6.5 □ substrate (10^{-7}/°C).

Temp. °C	Axial CTE α_z		Tangential CTE $\alpha\theta$	
	Uncoated	Coated	Uncoated	Coated
400	−1.4	0.5	−0.2	4.2
500	0.8	4.1	2.0	7.5
600	2.6	6.9	3.9	10.3
800	6.1	11.1	7.4	14.3
900	7.3	12.5	8.7	15.6
1,000	8.7	12.5	10.3	15.8

Reprinted from Gulati 1996a; courtesy of Marcel Dekker, New York.

TABLE 7.3. b) CTE versus temperature data for EX-32, 236/11.5 Δ Substrate (10^{-7}/°C).

Temp. °C	Axial CTE α_z		Tangential CTE $\alpha\theta$	
	Uncoated	Coated	Uncoated	Coated
400	−4.7	−6.6	1.5	−1.4
500	−2.4	−3.6	3.9	1.6
600	−0.6	−0.6	5.8	4.5
800	3.5	4.7	9.7	9.8
900	5.3	7.0	11.5	12.0
1,000	7.0	9.1	13.1	13.9

Reprinted from Gulati 1996a; courtesy of Marcel Dekker, New York.

and 236/11.5 Δ substrates[8] with and without the washcoat (Δ refers to triangular-shaped channels). It should be noted that the axial CTE of wash-coated substrates seldom exceeds 12×10^{-7}/°C (which is still below that of an uncoated substrate with random orientation) making it thermal shock resistant. The slightly higher CTE in tangential direction is attributed to random alignment of cordierite crystallites in the cell junction region, which makes the tangential direction less thermal shock resistant than the axial direction (Gulati 1983).

The specific heat and thermal conductivity of extruded cordierite substrates are relatively insensitive to wall porosity and substrate temperature. Their average values are (Gulati 1985):

$c_p = 0.25$ cal/g °C	(400 °C)
K = 0.0005 cal/cm s °C	(in tangential direction)
= 0.0010 cal/cm s °C	(in axial direction)

---

[8] The cordierite composition EX-32 for 236/11.5 Δ is designed to yield lower CTE than EX-20 to compensate for its higher E-modulus (Gulati 1975).

7.4.2 Mechanical Properties

The key mechanical properties include strength, E-modulus, and fatigue constant. The strength is important for withstanding packaging loads, engine vibrations, road shocks, and temperature gradients. Hence, high-strength substrates are more desirable. The E-modulus represents the stiffness or rigidity of the honeycomb structure and controls the magnitude of thermal stresses due to temperature gradients imposed by nonuniform gas velocity and catalytic exotherms. Hence, low E-modulus, which reduces thermal stresses and increases substrate life, is more desirable. The fatigue constant n represents the substrate's resistance to growth of surface or internal cracks when subjected to mechanical or thermal stresses in service. A high n value implies greater resistance to crack growth and hence is more desirable. Much like thermal properties, mechanical properties are also influenced by the washcoat. Moreover, they vary with temperature and cell orientation.

Strength measurements are generally carried out on multiple specimens, prepared carefully from the substrate, to obtain a representative distribution that accounts for the statistical variations associated with specimen preparation and the porous nature of ceramic substrates as well as with their cellular construction. Furthermore, the specimen size should be so chosen as to represent a sufficient number of unit cells, typically 200 to 400 cells across the specimen cross section. Figure 7.7 shows the specimens and load orientation for measuring the tensile and compressive strengths (Gulati 1985) of extruded cordierite ceramic substrates.

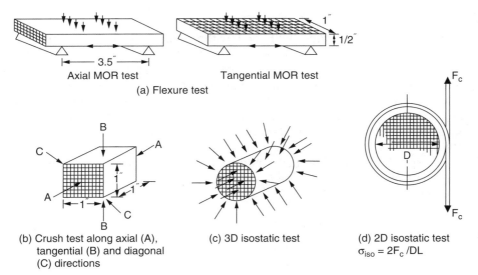

FIGURE 7.7. Cell and load orientation of honeycomb specimens for measuring mechanical strength. Reprinted from Gulati 1996a; courtesy of Marcel Dekker, New York.

TABLE 7.4. a) MOR data for the EX-20, 400/6.5 □ substrate (MPa).

Temp. °C	Axial MOR		Tangential MOR	
	Uncoated	Coated	Uncoated	Coated
25	2.72	3.62	1.38	1.93
200	2.55	3.97	1.24	1.79
400	2.69	4.03	1.31	1.93
600	2.93	4.14	1.38	2.00
800	3.10	4.07	1.52	2.07

Reprinted from Gulati 1996a; courtesy of Marcel Dekker, New York.

TABLE 7.4. b) MOR data for the EX-32, 236/11.5 Δ substrate (MPa).

Temp. °C	Axial MOR		Tangential MOR	
	Uncoated	Coated	Uncoated	Coated
25	3.28	4.76	1.59	2.03
200	2.97	4.65	1.45	2.21
400	3.03	4.65	1.52	2.24
600	3.45	5.00	1.66	2.28
800	3.79	5.28	1.79	2.34

Reprinted from Gulati 1996a; courtesy of Marcel Dekker.

The tensile strength in the axial and tangential directions is measured with the four-point flexure test, Figure 7.7a), by breaking ten specimens in each direction per ASTM specifications (ASTM 1975a). The mean values of modulus of rupture (MOR) at various temperatures are summarized in Tables 7.4a) and 7.4b) for EX-20, 400/6.5 □, and EX-32, 236/11.5 Δ substrates, respectively, before and after the application of a washcoat. These data show, as expected, that the MOR values increase with temperature and washcoat loading. The latter effect is substantial with strength increases of up to 50%. Also, the 15% to 20% higher MOR values of EX-32, 236/11.5 Δ are attributed to higher wall thickness than that of EX-20, 400/6.5 □. It will be shown later that the high MOR values ensure good thermal shock resistance.

The compressive strength, which has a direct relevance to packaging design, is measured in two different ways, namely (1) uniaxial crushing of $1'' \times 1'' \times 1''$ cube (or 2.54-cm cube) specimens in axial (A), tangential (B), and diagonal (C) directions, see Figure 7.7b), and (2) isostatic pressurization of whole substrate, see Figure 7.7c). Although the uniaxial crush test does not provide the absolute compressive strength of whole substrate, it is a simple quality control test for assessing the relative compressive strengths along the A, B, and C axes. Furthermore, it helps design the packaging system for noncircular substrates where the mounting pressure is not uniform. The isostatic test, on the other hand, is more representative of absolute compressive strength and is particularly suited for circular substrates with uniform mounting pressure.

TABLE 7.5. a) Room-temperature compressive strength of the EX-20, 400/6.5 □ substrate (MPa).

	Uncoated	Coated
Crush A	25.3	30.3
Crush B	4.5	5.38
Crush C	0.38	0.52
Isostatic	7.93	11.03

Reprinted from Gulati 1996a; courtesy of Marcel Dekker, New York.

TABLE 7.5. b) Room-temperature compressive strength of the EX-32, 236/11.5 △ substrate (MPa).

	Uncoated	Coated
Crush A	30.1	34.8
Crush B	6.31	7.93
Crush C	2.38	3.28
Isostatic	11.93	18.90

Reprinted from Gulati 1996a; courtesy of Marcel Dekker, New York.

Tables 7.5a) and 7.5b) summarize the mean compressive strength of the two substrates, before and after the application of a washcoat, at room temperature. The orientation of a triangular cell structure defining A, B, and C axes for measuring the crush strength is shown in Figure 7.8. It should be noted in Tables 7.5a) and 5b) that the triangular cell substrate has significantly higher compressive strength, notably along the C-axis and under isostatic loading, than the square cell substrate due to its rigid cell geometry.[9] Although the superior strength of a triangular cell substrate renders it more robust in terms of mechanical durability, its lightoff and steady-state conversion efficiency are inferior to those of the square cell substrate, thus calling for certain trade-offs.

Finally, the typical value of standard deviation for MOR data in Tables 7.4 is ±10% of the mean value and that for compressive strength data in Table 7.5 is ±25% of the mean value. MOR measurements above 800 °C show that the substrate continues to get stronger up to 1,200 °C,[10] after which it behaves like a viscoelastic material (i.e., it exhibits permanent deformation without fracture.) Its strength decreases gradually above 1,200 °C, approaching 40% of its room-temperature value at 1,400 °C, which implies that it can still support a load without failing catastrophically (Gulati and Sweet 1990).

[9] As a consequence of higher cell rigidity the E-modulus of triangular cell substrate is correspondingly higher resulting in a similar strain tolerance as that for square cell substrate.
[10] Strength increases with temperature up to 1,200 °C due to healing of microcracks.

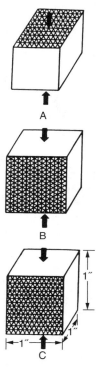

FIGURE 7.8. Crush test specimens along A, B, and C axes of the triangular cell substrate. Reprinted from Gulati 1996a; courtesy of Marcel Dekker, New York.

The fatigue behavior of ceramic substrates is relevant to either predicting a safe allowable stress for ensuring the specified lifetime or estimating the lifetime under a specified stress level. The methodology for obtaining the fatigue constant n has been discussed previously (Helfinstine and Gulati 1985) and involves MOR measurements at operating temperature and relative humidity at five different stress rates each spanning one decade a part. The slope of the axial MOR versus the stress rate plot provides the n value (see Figure 7.9). The n values at 200 °C, when the water vapor in exhaust gas is most active in promoting crack growth due to thermal and mechanical stresses, are shown in Table 7.6.

According to the power law fatigue model, the safe allowable stress σ_s for a converter life of τ_ℓ is given by (Evans 1974).

$$\sigma_s = \text{MOR} \left[\frac{\tau_o}{\tau_\ell(n+1)} \right]^{1/n} \tag{7.10}$$

where τ_o denotes the test duration for measuring MOR, which is typically 40 s. Assuming a converter life of 100,000 miles at an average vehicle speed of 40

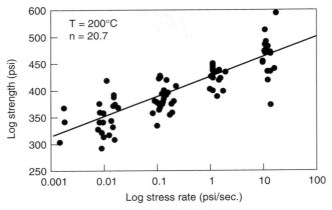

FIGURE 7.9. Axial strength of the EX-20, 400/6.5 substrate as a function of the stress rate at 200 °C (EX-20, 400/6.8). Reprinted from Gulati 1996a; courtesy of Marcel Dekker, New York.

TABLE 7.6. Dynamic fatigue constant n at 200 °C.

Substrate (uncoated)	n	95% confidence interval
EX-20, 400/6.5 □	20.7	18–24
EX-32, 236/11.5 △	36.1	28–50

Reprinted from Gulati 1996a; courtesy of Marcel Dekker, New York.

TABLE 7.7. Safe allowable stress σ_s for 100,000-mile life of a ceramic substrate.

n	σ_s/MOR
15	0.38
20	0.48
25	0.55
30	0.61
35	0.65
40	0.68

Reprinted from Gulati 1996a; courtesy of Marcel Dekker, New York.

miles/hr, we can estimate the safe allowable stress value as a fraction of a substrate's MOR for different values of n; see Table 7.7. Indeed, the higher n value of the EX-32 substrate permits higher allowable stress without a concern for crack propagation. Thus, the safe stress for the EX-32, 236/11.5 △ substrate is 17% higher than that for the EX-20, 400/6.5 □ substrate. As MOR refers to the strength of a substrate specimen with a stressed area A_o, it is necessary to account for the larger stressed area of substrate A_s itself. For a round substrate of diameter d and length l, for example, the area subjected to maximum thermal stresses during oxidation exotherms is approximately 1/3 of the skin area and is given by $(\pi\, d\, l/3)$. To allow for the larger area under stress, we need to multiply the right side of Eq. (7.10) by $[A_o / A_s]^{1/m}$, where m is the slope of Weibull distribution of MOR data. Hence Eq. (7.10) becomes

$$\sigma_s = MOR[\{\tau_o/\tau_\ell(n+1)\}]^{1/n}[A_o/A_s]^{1/m} \qquad (7.11)$$

Assuming a representative value of $m = 12$ for the cordierite ceramic substrate of 4.66″ diameter and 6″ length, the stressed area factor is 0.74, which would reduce the σ_s value by another 26%. In summary, therefore, theoretical considerations require that net tensile stress in the substrate be kept below 37% of its MOR value to ensure 100,000-mile durability (Webb 2004).

The E-modulus of extruded ceramic honeycombs is readily obtained by measuring the resonance frequency of MOR bars over a wide temperature range. Such a measurement is carried out in a high-temperature furnace according to ASTM specifications (ASTM 1975b). The E-modulus data (E_z and E_θ) for the two substrates are summarized as a function of temperature in Tables 7.8a) and 7.8b). In general, the E-moduli increase with temperature

TABLE 7.8. a) E-modulus data for the EX-20, 400/6.5 □ substrate (GPa).

Temp. °C	E_z		E_θ	
	Uncoated	Coated	Uncoated	Coated
25	7.24	9.66	3.66	4.83
200	7.24	11.03	3.66	5.52
400	7.24	11.72	3.66	5.86
600	8.28	12.07	4.14	6.07
800	9.66	12.55	4.84	6.28
1,000	11.03	12.41	5.52	6.21

Reprinted from Gulati 1996a; courtesy of Marcel Dekker, New York.

TABLE 7.8. b) E-modulus data for the EX-32, 236/11.5 △ substrate (GPa).

Temp. °C	E_z		E_θ	
	Uncoated	Coated	Uncoated	Coated
25	10.55	10.62	3.17	3.72
200	10.55	11.24	3.17	3.86
400	11.79	12.55	3.52	4.41
600	12.41	13.03	3.79	4.55
800	12.97	13.24	4.14	4.62
1,000*	14.34	13.45	4.69	4.62

*For temperatures above 1000 °C, the washcoat experiences significant mud cracking due to continued sintering and does not increase the E-moduli any further.
Reprinted from Gulati 1996a; courtesy of Marcel Dekker, New York.

and washcoat loading, which implies higher thermal stresses at higher temperature. However, washcoat formulation and processing and substrate microstructure can modify this trend due to interaction at the substrate/washcoat interface. Also, the triangular cell substrate has a higher E-modulus in the axial direction (E_z) due to higher cell rigidity than the square cell substrate. And finally, the tangential E-modulus (E_θ) is 50% of the axial E-modulus (E_z) for the square cell substrate and only 33% of the axial E-modulus for triangular cell as might be expected from their respective cell geometries.

The foregoing physical properties are key components to ensuring the physical durability of a catalytic converter over its specified lifetime. The next section demonstrates how these properties interact and impact both the mechanical and the thermal durability of the catalytic converter.

7.5   PHYSICAL DURABILITY

In addition to stringent emissions legislation, the automakers are also required to extend the useful life of catalytic converters from 50,000 miles to 100,000 miles. Both the catalytic and the physical durabilities must be guaranteed over 100,000 miles. To meet these requirements simultaneously, it is imperative to use the systems approach wherein each component of the converter assembly is carefully designed, tested, and optimized. The converter designer is therefore challenged with selecting appropriate materials, substrate contour, catalyst volume, washcoat loading, and catalyst formulation, as well as with assembling them into a durable package. In view of the variety of materials, configurations, and microstructures employed in various converter components, the task of designing the total system becomes even more complex and requires cooperative effort by all of the component suppliers to meet the automaker's space, performance, and cost specifications. Such an approach makes best use of each supplier's expertise, while keeping other suppliers' constraints in mind, and it leads to an optimum converter system that meets the total durability requirements.

7.5.1   Packaging Design

Starting with the substrate, its design and size are primarily dictated by performance requirements, which have been discussed earlier. The next step is the washcoat formulation and loading, which must not only provide an adequate BET area for 100,000-mile catalytic durability but also must be compatible with the cordierite substrate in terms of enhancing its physical properties, as discussed in the previous section. The impact of the precious metal catalyst on the physical properties of the substrate is negligible compared with that of the γ-Al_2O_3 washcoat, so that it does not play as critical a role as other components in optimizing physical durability. After washcoating and catalyzing, the substrate must be packaged in a robust housing that ensures its physical durability under severe operating conditions for 100,000 miles.

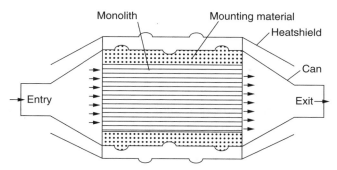

FIGURE 7.10. Schematic of the converter package. Reprinted from Gulati 1996a; courtesy of Marcel Dekker, New York.

Consequently, the packaging design can become the Achilles' heel if not dealt with properly.

The typical converter package, as shown in Figure 7.10, consists of a resilient mat to hold the substrate, end seals to prevent gas leakage (depending on the type of mat), a stainless steel can to house the mat, a substrate, and a heat shield to protect adjacent components, floor pan, and ground vegetation from excessive thermal radiation (Stroom et al. 1990).

A robust converter package provides positive holding pressure on the ceramic substrate, promotes symmetric entry of inlet gas, provides thermal insulation to the substrate thereby heating it rapidly and retaining its exothermic heat for catalytic activity while minimizing the radial temperature gradient, and provides adequate frictional force at the substrate/mat interface to resist vibrational and back pressure loads that would otherwise result in slippage of the substrate inside the can.

These are complex requirements that call for careful packaging design via selection of durable components like the mat. In earlier designs, the mat was made from a stainless steel wire mesh that did not function as a gas seal, did not provide sufficient holding pressure notably at high temperature, did not insulate the substrate against heat loss, and did not provide sufficient frictional force at the substrate/mat interface. Consequently, as emissions regulations became more stringent and converters moved closer to the engine, the need for a thermally insulating and resilient mat with good gas sealing and holding pressure capability became apparent. In the early 1980s, the 3M Company introduced an intumescent ceramic mat, under the trademark Interam™ (Langer and Marlor 1981), containing unexpanded vermiculite that expands upon heating and provides the desired properties for a robust package.

The holding pressure p_m during room-temperature assembly depends exponentially on the mount density of the mat (also known as gap bulk density or GBD) ρ_m as follows (Stroom et al. 1990):

$$p_m = 40,000 \exp(-6.7/\rho_m) \tag{7.12}$$

The above equation estimates p_m in psi if ρ_m is expressed in g/cm³. The mount density or GBD is defined by

$$\rho_m = \frac{W}{g} \tag{7.13}$$

where W denotes basis weight of the mat in g/cm² and g denotes the radial gap between the substrate and the can in cm. As the mat expands upon heating, the holding pressure increases (since it is constrained between the substrate and the can) by 200% to 300% of room-temperature value at temperatures approaching 800 °C, above which the intumescent property of the mat is lost (Gulati 1993a; Gulati 1992b).

The holding pressure must be high enough that the frictional force F_f at the mat/substrate interface exceeds the sum of vibrational (F_v) and back pressure (F_b) forces to prevent relative motion between the substrate and the can (see Figure 7.11):

$$F_f \geq F_v + F_b \tag{7.14}$$

Denoting the friction coefficient between the mat and the substrate by μ, vibrational acceleration by a, catalyst density by ρ_c, catalyst diameter and length by d and ℓ, and back pressure by p_b (see Figure 7.11), Eq. (7.14) yields

$$p_m \geq \left(\frac{d}{4\mu}\right)\rho_c a + \left(\frac{d}{4\mu\ell}\right)p_b \tag{7.15}$$

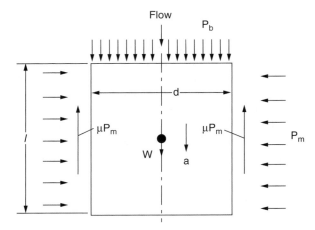

FIGURE 7.11. Schematic of ceramic substrate subjected to inertia, back pressure, and frictional forces. Reprinted from Gulati 1996a; courtesy of Marcel Dekker, New York.

Equation (7.14) helps select the mat and mount density for a given application. For a robust packaging system for underbody application, the nominal mount density should be 0.95 with a range[11] of 0.85 to 1.1 g/cm^3. This value corresponds to a nominal mounting pressure of 35 psi, with a range of 15 to 90 psi, which is adequate to resist both vibrational and back pressure loads (Gulati 1992a). The friction coefficient μ has been measured experimentally and has a value of 0.25. If the mat is not compressed to high enough mount density, it will not act as a good seal. Furthermore, if the temperature and flow velocity of inlet gas are high, direct impingement of the gas onto the edge of the mat could erode it resulting in loss of holding pressure and premature failure of substrate. The loose debris from the eroded mat can lead to plugging of substrates resulting in high back pressure and poor driveability (Stroom et al. 1990). These potential failure modes may be avoided by (1) selecting a mat with a high basis weight, (2) compressing it to a high mount density, (3) maintaining the average mat temperature at <800 °C, and (4) promoting convective cooling of the can via sound heatshield design and optimum converter location under the chassis. Of course, the inlet gas temperature, engine malfunction, and emissions content should also be controlled via proper engine and fuel management to minimize catalytic exotherms and combustion of unburnt fuel within the converter. The high mount density ensures not only the resistance to thermal erosion but also the high holding pressure, which adds to the strength of the substrate and enhances its thermal and mechanical durability. The substrate, as noted, can withstand an isostatic pressure well in excess of holding pressure with a safety factor >5. Finally, a mat with high basis weight enjoys a lower average temperature, thereby preserving its intumescent property critical for total durability, and is more accommodating of dimensional tolerances that would otherwise widen the range of mount density and holding pressure.

The can or clamshells are generally made of ferritic stainless steel, AISI 409, with low CTE to minimize changes in mount density due to expansion of the can at operating temperature. Furthermore, since the can is also subjected to holding pressure, its deformation at operating temperature must be minimized for the same reason. To this end, either the can temperature must be kept below 500 °C by efficient cooling or a better grade of ferritic stainless steel, e.g., AISI 439, should be used (see Figure 7.12).

The latter steel also has excellent resistance to high-temperature corrosion as measured by its weight gain from oxidation (Gulati et al. 1996a). In addition to limiting the can temperature to <500 °C, its flexural rigidity must be high, which can be achieved by designing an adequate number of stiffener ribs protruding inward or outward. The noncircular cans, with oval or racetrack contour, have the lowest rigidity along their minor axis and tend to deform excessively, thereby allowing "blow by" of inlet gas that not only promotes mat erosion but also increases tailpipe emissions. Thus, the stiffener ribs are

[11] The variation of mount density is caused by the tolerance stack-up in converter components.

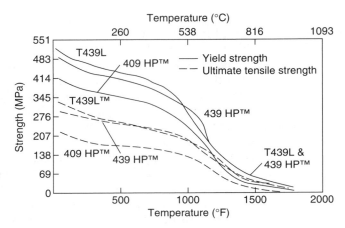

FIGURE 7.12. Yield and ultimate tensile strengths of AISI 409 and 439 stainless steels as a function of temperature. Reprinted from Gulati 1996a; courtesy of Marcel Dekker, New York.

TABLE 7.9. Properties of ferritic and austenitic stainless steels for converter housing.

Property	AISI type 409 (ferritic)	AISI type 316 (austenitic)
Density (g/cm³)	7.84	8.03
E-modulus (GPa)	200.00	193.10
CTE (10^{-7}/°C)	120	180
Thermal conductivity (BTU/ft/hr/°C)	84	55

Reprinted from Gulati 1996a; courtesy of Marcel Dekker, New York.

critical for noncircular cans. In designing inward ribs, care must be taken in controlling the mount density under the rib to limit the localized line pressure to well below the crush strength of the substrate. In some applications, the can temperature may exceed 600 °C, calling for austenitic stainless steel with improved high-temperature corrosion resistance, e.g., AISI 316 steel. However, as shown in Table 7.9, its CTE is 50% higher than that of ferritic steel and can reduce the mat mount density significantly at high temperatures. Care must be taken in selecting the initial mount density to compensate for higher thermal expansion of the austenitic stainless steel can.

As mentioned, the ideal contour of the substrate and its housing is circular since it results in a uniform holding pressure all around the substrate. However, space limitations under the chassis or in the engine compartment may require noncircular contours. Figure 7.13 ranks the four basic contours in order of uniformity of holding pressure, isotropy of clamshell stiffness, minimum can deformation, and radial temperature gradient along the minor axis. Every

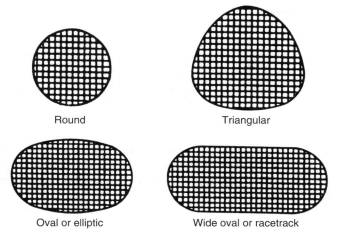

FIGURE 7.13. Ranking of contours of the substrate for optimum durability. Reprinted from Gulati 1996a; courtesy of Marcel Dekker, New York.

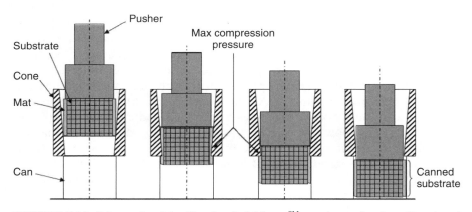

FIGURE 7.14. Schematic of the Corning SoftMount[SM] canning technology. Reprinted from Eisenstock et al. 2002; courtesy of SAE.

effort should be made during the early stages of converter design to select the best contour per Figure 7.13 for optimum durability.

Special canning techniques have been developed for thin-wall and ultra-thin-wall substrates with lower isostatic strength. One of these, SoftMount[SM] canning, has been discussed extensively (Eisenstock et al. 2002) and applied successfully to 600/3, 400/4, and 900/2 substrates. Figure 7.14 shows a schematic of SoftMount[SM] canning, which affords the lowest and most uniform packaging pressure, compared with stuffing and tourniquet techniques, without compromising the long-term mechanical durability of the packaged converter (Gulati et al. 2003).

Advances in support mat technology have also been reported recently. A systematic approach to packaging design using an advanced support mat technology (Fernandes et al. 2006) for converters/filters operating at relatively low temperature was discussed at the 2007 SAE Congress (Nickerson et al. 2007). Their approach resulted in a robust thermo-mechanical design with a 160,000-km life for light-duty vehicles. Their approach was validated by both resistive thermal exposure (Locker et al. 1998; 2000; 2001) and hot vibration tests (Nickerson et al. 2007). The former tests yielded a mat shear strength of up to 51 kPa after 400 thermal cycles at a 280 °C skin temperature of the converter/filter, which is two times higher than the required value. The hot vibration test was carried out at 250 °C and at 735-m/s^2 acceleration. No deterioration of packaging components was observed after 450 cycles, thus attesting to the durable packaging system.

7.5.2 Mechanical Durability

The substrate is subjected to mechanical loads during manufacturing, canning, and in service. It must have sufficient strength to sustain these stresses without the onset of fatigue degradation. In Section 7.4.2, we reviewed the mechanical strength data of ceramic substrates before and after the application of the γ-Al$_2$O$_3$ washcoat. The MOR, crush strength, and isostatic strength were all improved by 30% to 50% after the application of the washcoat. Such an enhancement of strength is critical to mechanical durability. If either the washcoat formulation or the calcining process or the substrate microstructure are not compatible, or if the substrate/washcoat adhesion is too strong, the expansion mismatch can introduce high stresses in the substrate wall and propagate some of the open pores with sharp tips, thereby degrading the mechanical strength. Such a phenomenon manifests itself in the form of large scatter in strength data with standard deviation approaching 30% to 50% of mean strength. Substrates with high variability in mechanical strength exhibit premature cracking either during canning or internal quality control (QC) tests and can lead to early failures in the field.

The canning process applies a biaxial compressive stress on the lateral surface of the substrate via compression of the mat. In the case of a circular substrate, the canning pressure is uniform and well below its biaxial compressive strength p_{2D} (Gulati 1993b). The latter is related to isostatic strength via

$$p_{2D} = p_{3D}\left[1 - \left(\frac{E_\theta}{E_z}\right)\left(\frac{v_{z\theta}}{1 - v_{r\theta}}\right)\right] \tag{7.16}$$

where p_{3D} denotes the isostatic strength given in Table 7.5 and $v_z\theta$ and $v_r\theta$ are Poisson's ratios for a honeycomb structure with values of 0.25 and 0.10, respectively. Substituting E_θ/E_z values from Table 7.8, we find that the biaxial compressive strength ranges from 86% to 91% of isostatic strength for coated substrates. Using the isostatic strength values in Table 7.5, we estimate the

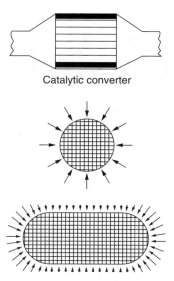

Catalytic converter

FIGURE 7.15. Schematic of pressure distribution during canning of circular and oval substrates. Reprinted from Gulati 1996a; courtesy of Marcel Dekker, New York.

biaxial compressive strength of a 400/6.5 □ substrate to be 1,400 psi and that of a 236/11.5 Δ substrate to be 2,500 psi. Compare these with the maximum holding pressure exerted by the mat during canning, which depends on the mount density per Eq. (7.12). Assuming the maximum mount density of 1.1 g/cm³, the room-temperature holding pressure is estimated to be 90 psi, which at the maximum intumescent temperature of the mat may approach 270 psi. This is only 20% and 11% of the biaxial compressive strength of the 400/6.5 □ and 236/11.5 Δ substrates, respectively. Thus, there is a sufficient safety margin in the compressive strength of substrates to sustain canning loads. In the case of a noncircular substrate, the holding pressure is nonuniform as shown in Figure 7.15. The higher pressure is carried by the semicircular portion (whose biaxial strength is very high as discussed above), whereas the flatter portion experiences much lower pressure due to lower stiffness of the clamshell in that region. When the substrate is not seated or aligned properly in the clamshells, the canning load may lead to localized bending and result in shear crack at the junction of semicircular and flat portions of the contour; as little as 2° to 5° of misalignment can produce shear cracks during canning of 400/6.5 □ substrates (see Figure 7.16).

Chassis vibrations and road shocks are another source of mechanical stresses that the substrate must sustain over its useful lifetime. However, these stresses are damped out to a large extent by the converter package due to the resiliency of the intumescent mat. Thus, the mechanical integrity of the substrate depends heavily on the integrities of the mat and the can. It is therefore imperative that the packaging design be made as robust as possible, taking the

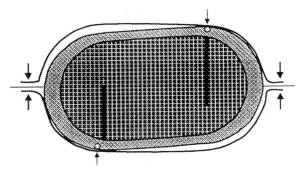

FIGURE 7.16. Shear cracking during clamshell canning of misaligned substrate. Reprinted from Gulati 1999a; courtesy of SAE.

high-temperature limitations of mat and can materials into account. Vibration tests at 800 °C, 80–120-g acceleration, and 100–2,000-Hz frequency sweep have shown no evidence of relative motion or mechanical damage in 400/6.5 □ substrates of racetrack contour when mounted with a holding pressure of 35–90 psi (Maret et al. 1991). Similarly, no service failures due to impact load from stones and other hard objects have been reported when the substrate is properly packaged.

The resonance frequencies of a ceramic substrate in axial and radial directions may be estimated by assuming simple harmonic motion and treating the substrate/mat combination as a spring-mass system. Under these assumptions, the resonance frequency ω of a converter package is given by (Unruh and Till 2004)

$$\omega = (1/2\pi)[K/m]^{0.5} \tag{7.17}$$

where K and m denote mat stiffness and substrate mass, respectively, and are given by

$$K_{\text{axial}} = K_{\text{mat shear}} \qquad \text{for vibrations in axial direction}$$
$$K_{\text{radial}} = K_{\text{mat compression}} \qquad \text{for vibrations in radial direction}$$
$$m = (\pi/4)\,d^2\ell\rho$$

in which d, ℓ, and ρ denote the diameter, length, and mass density of the coated substrate. The units of ω will be cycles/s or Hz. The mat stiffness in shear and compression will depend on a specific value of GBD (gap bulk density) used during canning of coated substrate. Since mat behaves as a nonlinear spring in both shear and compression, load versus mat deformation curves can be measured over a wide range of GBD values to estimate the operational values of K_{axial} and K_{radial}.

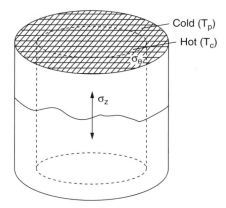

FIGURE 7.17. Development of thermal stresses due to radial temperature gradient. Reprinted from Gulati 1996a; courtesy of Marcel Dekker, New York.

7.5.3   Thermal Durability

One of the key durability requirements of ceramic substrates is to have adequate thermal shock resistance to survive temperature gradients due to non-uniform flow and heat loss to ambient environment. Referring to Figure 7.17, the center region of the substrate experiences higher temperatures than its periphery and induces tensile stresses in the outer region.[12] The magnitude of these stresses depends linearly on CTE, E-modulus, and radial temperature gradient ΔT. These stresses should be kept well below the modulus of rupture of the substrate to minimize premature fracture in the radial and axial directions (see Figure 7.18). It is, therefore, desirable that the substrate exhibit high strength and low E-modulus such that it has high thermal integrity to withstand thermal shock stresses.

Figure 7.19 shows a schematic of finite element mesh for computing thermal stresses in a circular converter with a prescribed temperature field. As a hypothetical example, we assume a uniform temperature of 1,000 °C in the central hot region of 3.6″ (9.1 cm) diameter and a linear radial gradient of 800 °C/in in the external 1″ (2.54 cm) thick region such that the skin temperature of a 5.6″ (14.2 cm) diameter converter is 200 °C. Furthermore, we assume that such a radial temperature profile is constant over the entire 12″ (30.48 cm) length and that the axial temperature gradient is negligible.[13]

The thermal stresses due to the above temperature field are readily computed by using the physical properties similar to those in Section 7.4 (Gulati

---

[12] During cooldown, the center region experiences tensile stresses but these are less detrimental according to the failure modes observed in the field.

[13] The axial temperature gradient depends on oxidation exotherms and ranges from 50 °C to 100 °C over the length of the catalyst.

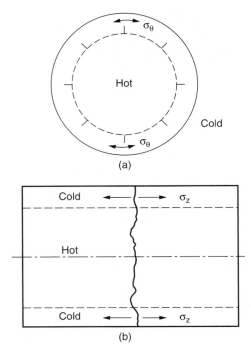

(a)

(b)

FIGURE 7.18. (a) Schematic of radial cracks due to tangential thermal stress; (b) schematic of ring-off cracks due to axial thermal stress. Reprinted from Gulati 1996a; courtesy of Marcel Dekker, New York.

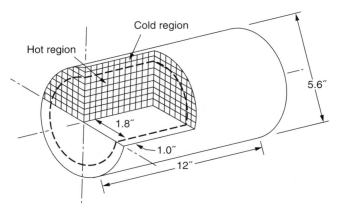

FIGURE 7.19. Finite element model for analyzing thermal stresses. Reprinted from Gulati 1996a; courtesy of Marcel Dekker, New York.

1983). Taking advantage of axial symmetry about the midlength, the axial and tangential stress profiles (σ_z and σ_θ) are shown in Figure 7.20 for this hypothetical example. It should be noted that the axial stress (σ_z) reaches its maximum value on the outer surface at the midlength, whereas the tangential (σ_θ) stress remains relatively constant throughout the length of the converter. Since σ_z is caused by differential elongation of hotter interior relative to the colder exterior, its maximum value depends largely on the aspect ratio (length/diameter) of the converter as indicated in Figure 7.21.

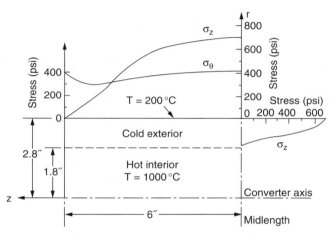

FIGURE 7.20. Thermal stress variation along the converter length. Reprinted from Gulati 1996a; courtesy of Marcel Dekker, New York.

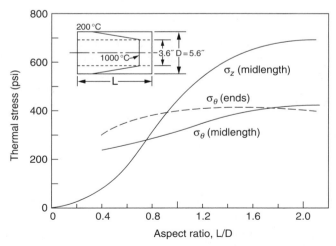

FIGURE 7.21. Effect of aspect ratio on thermal stresses. Reprinted from Gulati 1996a; courtesy of Marcel Dekker, New York.

If σ_z approaches the MOR_z value at skin temperature, a ring-off crack can initiate at the surface as shown in Figure 7.18b). Similarly, if σ_θ approaches the MOR_θ value, a radial or face crack can initiate at inlet or exit faces as shown in Figure 7.18a). To minimize ring-off or facial cracking, the radial temperature gradient should be minimized by good fuel management, proper gas flow, and efficient mat insulation. In particular, ring-off cracking can be eliminated by using a converter with a low aspect ratio. The low aspect ratio is also desirable for minimizing the additional axial stress σ_Z^* due to differential expansion of the substrate and clamshell, namely

$$\sigma_Z^* = \mu p_m (L/D) \tag{7.18}$$

where μ denotes the friction coefficient between the ceramic mat and the monolith and p_m denotes the mounting pressure at skin temperature. With $\mu = 0.25$, σ_Z^* can approach a value of $0.5 p_m$ (for $L/D = 2$) that can be 20% to 30% of MORz value; this should be minimized by limiting the aspect ratio to <1.4.

An alternative and a simple technique for assessing thermal durability is to compute the thermal shock parameter from physical properties. This parameter, defined by Eqs. (7.19) and (7.20), is the ratio of mechanical strain tolerance to thermal strain imposed by radial temperature gradient. The higher this parameter is, the better the thermal shock capability will be.

$$TSP_z = \frac{(MOR_z/E_z)}{\alpha_{cz}(T_c - 25) - \alpha_{pz}(T_p - 25)} \tag{7.19}$$

$$TSP_\theta = \frac{(MOR_\theta/E_\theta)}{\alpha_{c\theta}(T_c - 25) - \alpha_{p\theta}(T_p - 25)} \tag{7.20}$$

In these equations, T_c and T_p denote temperatures at the center and peripheral regions of the substrate, α_{cz} and α_{pz} denote the axial CTE values, and $\alpha_{c\theta}$ and $\alpha_{p\theta}$ denote the tangential CTE values at temperatures T_c and T_p, respectively. We can compute the thermal shock parameter (TSP) values for EX-20, 400/6.5 □ and EX-32, 236/11.5 △ substrates at the steady-state operating conditions defined by $T_c = 825\,°C$ and $T_p = 450\,°C$. Substituting the physical properties at these temperatures from Tables 7.3, 7.4, and 7.8 into Eqs. (7.19) and (7.20), we obtain the TSP values summarized in Table 7.10.

Let us make the following observations:

1. The washcoat reduces the TSP of the uncoated substrate as might be expected from its high CTE.
2. Tangential TSP is generally lower than axial TSP due to higher CTE in that direction.
3. TSP values are very similar for the two uncoated substrates.

TABLE 7.10. TSP values for substrates for $T_c = 825\,°C$ and $T_p = 450\,°C$.

Substrate	TSP_z		TSP_θ	
	Uncoated	Coated	Uncoated	Coated
EX-20, 400/6.5 □	0.70	0.51	0.68	0.40
EX-32, 236/11.5 △	0.65	0.58	0.68	0.66

Reprinted from Gulati 1996a; courtesy of Marcel Dekker, New York.

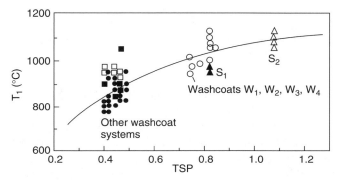

FIGURE 7.22. Correlation between axial TSP and failure temperature in a cyclic thermal shock test. Reprinted from Gulati 1996a; courtesy of Marcel Dekker, New York.

Figure 7.22 plots the failure temperature, measured in a cyclic thermal shock test (Gulati 1991c), for ceramic substrates as a function of their axial TSP values, which were controlled by modifying either the substrate or the washcoat or the substrate/washcoat interaction. There is an excellent correlation between the failure temperature and the TSP value. Most automakers call for a failure temperature in excess of 750 °C although this may depend on the size of the catalyst and the inlet pipe. Thus, a TSPz value of >0.4 is required for the coated substrate. Finally, Figure 7.22 shows that the washcoat may reduce the failure temperature of substrate by 100 °C to 200 °C, which is a trade-off the automakers are well aware.

7.6 ADVANCES IN SUBSTRATES

With stricter emission standards and a low back pressure requirement, substrates have undergone significant developments over the past few years.

7.6.1 Ceramic

The thrust in the ceramics area has centered on thin and ultrathin wall structures with high cell density to minimize thermal mass and maximize surface

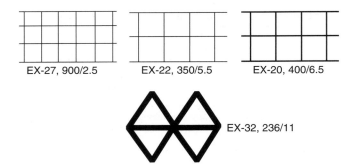

EX-27, 900/2.5   EX-22, 350/5.5   EX-20, 400/6.5

EX-32, 236/11

FIGURE 7.23. Comparison of standard and thin-wall cordierite substrates.

area (Day and Socha 1991; Gulati 1999a; 1999b; 2000; 2001). The thin-wall structures are extruded from cordierite ceramic to provide adequate strength and thermal shock resistance similar to those of standard cordierite substrates (see Figure 7.23). They are available in different cell sizes to achieve faster light-off, higher conversion efficiency and lower back pressure than the standard 400/6.5 □ substrate. The pertinent geometric and physical properties of standard and thin-wall cordierite substrates are summarized in Table 7.11 (Gulati 1991a; 1999a; 1999b; 2000; 2001).

These properties help compare lightoff and steady-state conversion activity through heat capacity, substrate mass and GSA values, and engine performance through flow resistance parameter. It is clear in Table 7.11 that thin-wall substrates enjoy 40% lower heat capacity, 50% lower mass, and 60% higher GSA than standard substrates thereby providing improved lightoff and conversion efficiency. However, high flow resistance parameter implies higher back pressure—a trade-off that goes with improved conversion activity.

The effect of cell shape has also been investigated. Both triangular and hexagonal cells have been extruded for either higher mechanical strength or lower back pressure compared with those of square cell (see Figure 7.24).

In addition, the hex cell may permit a more efficient washcoat distribution with potential improvement in conversion activity. Its higher open frontal area would also help reduce the back pressure. To help compare the relative benefits of different cell shapes, we define lightoff factor, conversion efficiency factor and resistance to flow factor, as follows (Gulati 2001):

$$\text{LOF} = H \times \text{GSA}/M^* \qquad (7.21)$$

$$\text{CEF} = M \times \text{GSA} \qquad (7.22)$$

$$R_f = \lambda f R_e /n D_h^4 \qquad (7.23)$$

where M^*, H, and M denote the thermal mass and heat and mass transfer factors given by

TABLE 7.11. Nominal properties of standard and thin-wall cordierite substrates.

Ceramic Cell density (cell/in^2)	400/6.5	470/5	600/3.5	600/4	600/4.3	900/2.5	1,200/2.5
Substrate diameter (mm)	105.7	105.7	105.7	105.7	105.7	105.7	105.7
Substrate length (mm)	98	88	76	76	76	76	35
Substrate volume (l)	0.86	0.77	0.67	0.67	0.67	0.67	0.31
Material porosity (%)	35	24	35	35	35	35	35
OFA	0.757	0.795	0836	0.814	0.800	0.856	0.834
GSA (m^2/l)	2.74	3.04	3.53	3.48	3.45	4.37	4.98
TSA (m^2)	2.35	2.35	2.35	2.32	2.30	2.91	1.53
Hydraulic diameter (mm)	1.10	1.04	0.95	0.94	0.93	0.78	0.67
Flow resistance (1/cm^2)	3,074 105%	3,274 100%	3,780 100%	3,990	4,122	5,412	7,589
Bulk density (g/l)	395	390	267	303	324	235	269
Heat capacity @200°C (J/K l)	352	348	238	270	289	209	240
Heat capacity @200°C (J/K)	302	269	159	180	193	140	74
Substrate Mass (g)	339	301	178	202	216	156	83

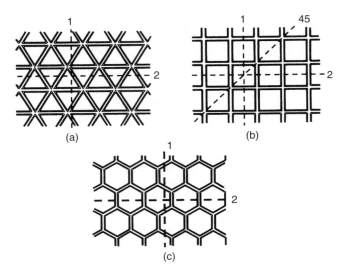

FIGURE 7.24. Triangular, square, and hexagonal cell geometries for ceramic substrates. Reprinted from Gulati 2001; courtesy of SAE.

TABLE 7.12. Nusselt and Sherwood numbers as well as shape and friction factors for different cell shapes.

	Square cell	Hex cell	Triangular cell
Nu & Sh (transient)	3.6	4.0	3.0
Nu & Sh (steady state)	3.0	3.3	2.3
fRe	14.2	15.0	13.3
λ	1.0	1.16	0.77

Reprinted from Kays and London 1955; courtesy of McGraw-Hill, New York.

$$M^* = c_p \rho_c (1 - P)(1 - \text{OFA}) \qquad (7.24)$$

$$H = Nu\text{GSA}/D_h \qquad (7.25)$$

$$M = Sh.\text{GSA}/D_h \qquad (7.26)$$

and λ and fRe denote shape and friction factors for gas flow through a single channel. The term P in Eq. (7.24) denotes the fractional porosity of the cell wall, and ρ_c denotes the density of nonporous cordierite ceramic with a value of $2.51\,\text{g/cm}^3$. Both the Nusselt and the Sherwood numbers in Eqs. (7.25) and (7.26) are nearly identical for a fully developed laminar flow away from the channel entrance. Their values for transient versus steady-state conditions, together with those of shape and friction factors, are summarized in Table 7.12 for the three cell shapes (Kays and London 1955). Table 7.13 compares the thermal mass of standard and thin-wall substrates with a wall prosity of 35%.

TABLE 7.13. Thermal mass M* for various cell sizes and shapes (cal/cm³°C).

Cell size	Cell shape		
	Square cell	Hex cell	Triangle cell
400/6.5	0.099	0.0926	0.1117
400/4.5	0.070	0.0653	0.0795
600/3.5	0.067	0.0624	0.0759
600/4.3	0.0816	0.0759	0.0921
900/2.5	0.059	0.0548	0.0668
1,200/2.0	0.0546	0.0508	0.0619

Reprinted from Gulati 2001; courtesy of SAE.

TABLE 7.14. Heat and mass transfer factors H and M for various cell sizes and shapes for transient versus steady-state conditions (1/cm²).

Cell size	Cell shape		
	Square cell transient/ steady state	Hex cell transient/ steady state	Triangle cell transient/ steady state
400/6.5	5,760/4,800	5,543/4,615	6,235/4,884
400/4.5	5,760/4,800	5,543/4,615	6,235/4,884
600/3.5	8,640/7,200	8,314/6,921	9,353/7,327
600/4.3	8,640/7,200	8,314/6,921	9,353/7,327
900/2.5	12,960/10,800	12,471/10,382	14,030/10,990
1,200/2.0	17,280/14,400	16,628/13,843	18,706/14,653

Reprinted from Gulati 2001; courtesy of SAE.

It shows that the hex cell offers 7% lower thermal mass than the square cell, whereas the triangular cell has a 13% higher thermal mass than the square cell. One would conclude from this that the hex cell would have a better light-off performance than square cell, all other things being equal. However, as noted in Eq. (7.21), the lightoff factor also depends on H and GSA. Similarly, the conversion efficiency factor [see Eq. (7.22)] depends on M and GSA. Both the H and M values for transient (i.e., lightoff) and steady-state conditions for different cell sizes and shapes are summarized in Table 7.14, and the GSA values for same cell sizes and shapes are given in Table 7.15.

The resulting LOF and CEF values, obtained by substituting GSA, H, M, and M* values in Eqs. (7.21) and (7.22) are summarized in Tables 7.16 and 7.17, respectively. Let us note that the square cell, despite higher thermal mass, offers equivalent or slightly better performance than the hexagonal cell due to higher GSA (see Table 7.15) and improved heat and mass transfer factors (see Table 7.14).

On the other hand, the triangular cell seems to have even higher LOF and CEF values than the square cell due to its higher GSA and heat and mass transfer factors. However, these are based on the total wetted area of the

TABLE 7.15. Geometric surface area (GSA) for various cell sizes and shapes (cm²/cm³).

Cell size	Cell shape		
	Square cell	Hex cell	Triangle cell
400/6.5	27.4	25.8	30.6
400/4.5	28.7	26.8	32.2
600/3.5	35.3	33.0	39.7
600/4.3	34.5	32.4	38.7
900/2.5	43.7	40.9	49.3
1,200/2.0	50.8	47.5	57.3

Reprinted from Gulati 2001; courtesy of SAE.

TABLE 7.16. Lightoff factor (LOF) for various cell sizes and shapes (10^6).

Cell size	Cell shape		
	Square cell	Hex cell	Triangle cell
400/6.5	4.19	4.07	4.49
400/4.5	6.21	6.01	6.68
600/3.5	12.00	11.50	12.89
600/4.3	9.69	9.28	10.38
900/2.5	25.59	24.37	27.52
1,200/2.0	41.95	40.26	45.22

Reprinted from Gulati 2001; courtesy of SAE.

TABLE 7.17. Conversion efficiency factor (CEF) for various cell sizes and shapes (10^6).

Cell size	Cell shape		
	Square cell	Hex cell	Triangle cell
400/6.5	0.849	0.767	0.964
400/4.5	0.888	0.799	1.016
600/3.5	1.638	1.476	1.875
600/4.3	1.604	1.446	1.829
900/2.5	3.044	2.741	3.492
1,200/2.0	4.714	4.245	5.413

Reprinted from Gulati 2001; courtesy of SAE.

triangular shape, including its acute corners, whereas in reality, there is little or no gas flow in the acute corner regions. Thus, although the triangular cell may seem to be superior to the square cell, its effective performance is no better. Furthermore, as shown later, the pressure drop for a triangular cell substrate is significantly higher than that for a square cell substrate due to its small hydraulic diameter, which is undesirable.

TABLE 7.18. Hydraulic diameter D_h for various cell sizes and shapes (mm).

Cell size	Cell shape		
	Square cell	Hex cell	Triangle cell
400/6.5	1.105	1.199	0.949
400/4.5	1.156	1.249	1.000
600/3.5	0.948	1.025	0.820
600/4.3	0.928	1.004	0.800
900/2.5	0.783	8.846	0.679
1,200/2.0	0.682	0.737	0.592

Reprinted from Gulati 2001; courtesy of SAE.

TABLE 7.19. Resistance to flow factor at constant flow rate across a substrate of constant cross-sectional area and length with various cell sizes and shapes.

Cell size	Cell shape		
	Square cell	Hex cell	Triangle cell
400/6.5	1,536	1,350	2,034
400/4.5	1,283	1,145	1,652
600/3.5	1,890	1,686	2,430
600/4.3	2,061	1,828	2,684
900/2.5	2,707	2,421	3,450
1,200/2.0	3,525	3,154	4,474

Reprinted from Gulati 2001; courtesy of SAE.

The pressure drop across the substrate, represented by resistance to flow factor R_f, is readily estimated by substituting λ fRe and D_h values from Tables 7.12 and 7.18 into Eq. (7.23). The results of this exercise are summarized in Table 7.19.

It is clear from this table that the triangular cell substrate has the highest pressure drop, nearly 30% higher than the square cell substrate, and the hexagonal cell substrate has the lowest pressure drop, approximately 10% to 12% lower than the square cell substrate, due primarily to its larger hydraulic diameter. This is the only advantage of hexagonal cell over square cell. It follows from these analyses that the square cell offers the best compromise in terms of overall performance and manufacturability. The 10% to 12% lower pressure drop afforded by the hexagonal cell is apparently not enough of an advantage to compensate for poorer lightoff permance, lower conversion efficiency, and more complex manufacturability relative to those for square cell.

The foregoing predictions have been borne out by engine tests by several investigators (Kishi et al. 1999; Ichikawa et al. 1999; Williamson et al. 1999; Johnson 2000). The lower thermal mass of a 600/4.3 substrate helped reduce cold-start emissions by 35%, and its higher surface area helped reduce full-test tail pipe emissions by 15% and NO_x emissions by 10% compared with a 400/6.5 substrate (Ichikawa et al. 1999). Thus, thermal mass dominates during cold start, and surface area dominates during steady state. The 1,200/2 substrate helped reduce the already low emissions from a 600/4.3 substrate by an additional 44% due primarily to its higher surface area (Kishi et al. 1999). Furthermore, relative to a 400/6.5 substrate, the 600/3.5 substrate helped reduce HC emissions by 30%, wheseas 900/2.5 and 1,200/2 substrates helped reduce HC emissions by 43% for a Pd-only close-coupled catalyst (Williamson et al. 1999). In an excellent review paper, it was concluded that gains in overall conversion efficienty afforded by advanced substrates were just as significant, if not more so, than those due to catalyst improvements (Johnson 2000). Table 7.20 quantifies the improvements in relative emissions level as higher cell density substrates are used (Johnson 2000). It is clear from this table that advanced substraes help reduce emissions by an additional 50% relative to the standard 400/6.5 substrate. As much as 35% lower cold-start emissions and 44% lower steady-state emissions have been achieved with advanced ceramic substrates (Johnson 2000).

The triangular cell, as noted, does not experience bending and tensile stresses during canning. Hence, its mechanial durability is superior to that of square or hexagonal cell. This is readily borne out by Table 7.21, which provides an estimate of 2D-isostatic strength of whole substrate with 35% wall porosity and various cell sizes of square and triangular shape (Gulati 2001).

TABLE 7.20. Summary of results for advanced catalytic converter substrates.

Substrate cell geometry	Relative HC emissions	Relative NO_x emissions
400/6.5	100	100
400/4.5	88	94
600/3.5	78	74
600/4.3	65–74	74–93
900/2.5	52–66	59–75
1,200/2.0	41–57	57

Reprinted from Johnson 2000; courtesy of SAE.

TABLE 7.21. Estimate of 2D-isostatic strength of whole substrates with 35% wall porosity and various cell sizes and shapes (MPa).

Cell size	Cell shape	
	Square cell	Triangle cell
400/6.5	3.86	4.17
400/4.5	2.69	2.86
600/3.5	2.55	2.76
600/4.3	3.14	3.78
900/2.5	2.24	2.38
1,200/2.0	2.07	2.24

Reprinted from Gulati 2001; courtesy of SAE.

Hence, the triangular cell substrate is about 8% stronger than the square cell substrate with identical cell density and wall thickness—an advantage the canners have long enjoyed. However, the triangular cell results in a stiffer structure than the square cell with the result that its mechanical strain tolerance given by MOR/E is lower than that of the square cell. Consequently, its thermal durability, represented by a thermal shock parameter, is inferior to that of the square cell as indicated in Table 7.22 for both underbody and close-coupled applications (Gulati 2001). In addition to inferior thermal durability, the pressure drop across a triangular cell substrate, as noted earlier (see Table 7.19), is 30% higher than that across the square cell substrate. Hence, the square cell substrates offer the best compromise and are the preferred choice for the automotive industry.

As the need for ultra-thin-wall substrates grows, to meet more stringent emissions legislation, their mechanical strength may be affected adversely, thereby requiring careful handling by coaters and canners. This can, however, be alleviated by reducing the wall porosity (P), which leads to higher wall strength (σ_w) (as well as substrate strength) as noted in Table 7.23 (Coble and Kingery 1956).

TABLE 7.22. Thermal shock parameter of whole substrates with 35% wall porosity and various cell sizes and shapes.

Cell size and shape	Underbody[*]	Close-coupled[**]
600/3.5 square	1.67	1.23
900/2.5 square	1.61	1.15
236/11.5 triangle	0.97	0.92
300/6.7 triangle	0.85	0.86

[*]The center and skin temperatures of the substrate for underbody application were assumed to be 825 °C and 625 °C, respectively.
[**]The center and skin temperatures of the substrate for close-coupled application were assumed to be 1025 °C and 825 °C, respectively.
Reprinted from Gulati 2001; courtesy of SAE.

TABLE 7.23. Mean wall strength of cordierite ceramic for different wall porosities (tensile).

P	σ_w (MPa)
0.2	44.1
0.25	34.5
0.30	26.9
0.35	21.4
0.40	16.6
0.45	13.1
0.50	10.3

Reprinted from Coble and Kingery 1956; courtesy of Am. Cer. Soc.

Indeed, as little as a 5% reduction in wall porosity can increase the wall strength (and substrate strength) by 25%! The lower porosity is achieved by modifying both the raw materials and the manufacturing process. It should be pointed out, however, that too low a porosity would make it difficult for the required washcoat uptake in one pass and lead to higher processing cost. In case the porosity cannot be reduced, a thorough analysis of compressive stresses in the matrix and skin regions during 2D-isostatic loading, such as during canning, revealed that tangential compressive stresses at the matrix/skin interface differ significantly in the matrix versus skin regions due to the latter's higher stiffness (Gulati et al. 2004). Consequently, the compressive strain in the skin region is more severe for thin-wall and ultra-thin-wall substrates, and it can lead to localized bending of interfacial cells. The analysis pointed out critical parameters, namely skin versus matrix stiffness, cell configuration, cell density, and substrate diameter, which can affect isostatic strength.

Peripheral strengthening offers another approach to enhance the mechanical integrity of webs, adjacent to substrate skin, which can help increase the isostatic strength of ultra-thin-wall substrates (Lakhwani and Hughes 2005). This approach permits decreasing the web thickness gradually toward the center of the substrate, thereby reducing its thermal mass and improving light-off performance coupled with potential savings in precious metal usage. These improvements do not compromise either thermal durability or pressure drop. Recall that the matrix region with higher temperature and thinner webs enjoys very low CTE values, which help reduce thermal stresses in the colder peripheral region, thereby preserving thermal durability. Similarly, the higher open frontal area of the matrix region along with the larger hydraulic diameter ensure low back pressure as with the standard 900/2 substrate. Figure 7.25

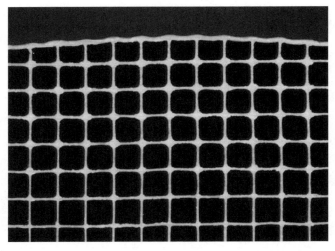

FIGURE 7.25. Peripheral strengthening technology for the 900/2 substrate. Reprinted from Corning Environmental Technology brochure, 3/2006.

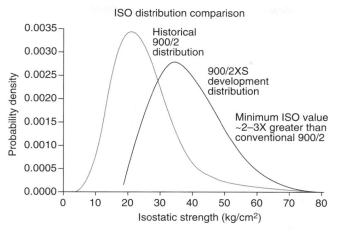

FIGURE 7.26. Comparison of isostatic strength of standard 900/2 and peripherally strengthened 900/2 substrates. Reprinted from Corning Environmental Technology brochure, 3/2006.

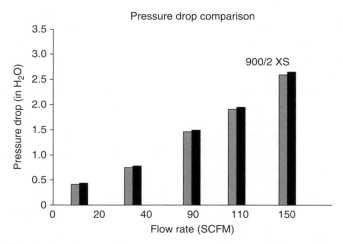

FIGURE 7.27. Comparison of pressure drop versus flow rate data for standard 900/2 and peripherally strenthened 900/2 substrates. Reprinted from Corning Environmental Technology brochure, 3/2006.

illustrates peripheral strengthening for 900/2 XS substrates, wheseas Figure 7.26 shows significant isostatic strength gains compared with those for the standard 900/2 substrate. Figure 7.27 compares back pressure data for standard 900/2 and peripherally strengthened 900/2 substrates at various flow rates indicating insignificant differences. The improved isostatic strength also facilitates system cost savings by enabling the use of conventional mat technologies and canning methods. Moreoever, the higher geometric surface area permits use of substrates in a close-coupled application.

7.6.2 Metallic

The metal foil monoliths of Fe–Cr–Al and Fe–Cr–Al–Ni compositions, with a 400/2 cell structures, for example, offer a larger open frontal area, higher geometric surface area, and bigger hydraulic diameter with potential lightoff, pressure drop, and conversion efficiency advantages relative to a 400/6.5 cordierite substrate (Nonnenmann 1990). However, the field data do not confirm these advantages in a consistent manner due to minimal differences in heat capacity and the hydraulic diameter of ceramic and metal cell structures; there are also certain durability issues with the washcoat/metal adhesion. Furthermore, some early data show that their physical durability can be adversely affected above 800 °C (Maattanen and Lylykangas 1990). They can oxidize and become brittle, and/or they deform permanently, under sustained operating stresses at high temperature (Gulati 1991e).

Figure 7.28 shows three metal monolith designs that have been described (Held et al. 1994; Bruck et al. 1995) as SM (split multiple structure), LS (longitudinal structure), and TS (transversal foil structure).

In the TS structure, flow is restricted within a particular channel, but microcorrugations at 90° to flow direction cause turbulence and so reduce the thickness of the stagnant boundary layer that reactants have to diffuse through. In SM and LS designs, there is the opportunity of flow between channels, which provides the possibility of the whole of the catalytic surface being used even if flow distribution at the catalyst front face is not uniform. However, this may be counterproductive in terms of lightoff due to heat dissipation, although benefits would be expected once the catalyst is at a normal operating temperature. The extent of these effects would depend on the conflicting requirements of low-pressure drop and high geometric surface area. Preliminary data (Bruck et al. 1995) indicated that, at the same cell density, a 14% lower volume of TS catalyst gives similar hydrocarbon and NO_x control as a conventional monolithic one. With equal catalyst volumes, the TS system gives an average of 10% better performance for hydrocarbons and NO_x, whereas at an equal volume, a 300-cells/in^2 TS catalyst gives performance equivalent to a 400-cells/in^2 conventional monolith.

Treating the cell shape as a sine wave, we can compute the geometric properties of metallic substrates with different cell densities and foil thicknesses. Table 7.24 summarizes these properties along with pertinent physical properties similar to those for ceramic substrates (from Table 7.11). A comparison of Tables 7.11 and 7.24 shows that indeed the metallic substrates offer 20% to 30% higher GSA per liter of substrate volume, and 10% to 20% larger OFA than ceramic substrates of identical cell density.

In theory, these properties should help improve conversion efficiency and reduce back pressure. However, other factors negate these advantages. For

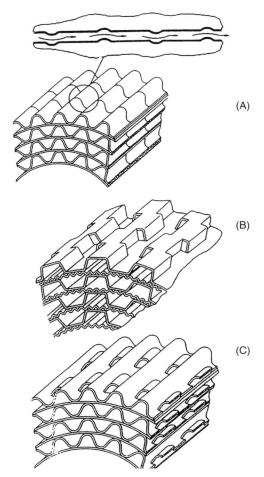

FIGURE 7.28. Structural features of some metal monoliths designed to enhance turbulence. a): Transveral foil structure (TS), in which corrugated layers have microcrorrugations at 90° to the direction of flow. b): In the SM structure, the flow in channels is split into multiple flow paths. c): In the LS structure, interconnecting flow paths are achieved by means of partial countercorrugation of the corrugated layer. Reprinted from Bruck et al. 1995; courtesy of SAE.

example, the mass of metallic substrates is two times greater and their heat capacity, despite lower specific heat value, is 15% to 80% higher than that of ceramic substrates of identical cell density and identical volume. The potential advantage of metallic substrates is the 10% to 15% lower back pressure due to their larger open frontal area, which may figure heavily in certain niche applications like a lightoff converter.

In a recent exhaustive study (Pannone and Mueller 2001), the performance of ceramic and metallic three-way catalysts with 600 cells/in^2 was compared

TABLE 7.24. Nominal properties of standard and thin-wall metallic substrates.

Metal Cell density (cell/in^2)	400/2	500/1.5	500/2	600/1.5	600/2
Substrate diameter (mm)	105.7	105.7	105.7	105.7	105.7
Substrate length (mm)	68	114	114	114	114
Substrate volume (l)	0.60	1.00	1.00	1.00	1.00
OFA	0.890	0.900	0.880	0.890	0.870
GSA (m^2/l)	3.65	4.05	4.00	4.20	4.15
TSA (m^2)	2.18	4.05	4.00	4.20	4.15
Hydraulic diameter (mm)	0.98	0.89	0.88	0.85	0.84
Flow resistance (1/cm^2)	2,646	3,150	3,287	35.03	3,660
Bulk density (g/l)	792	720	864	792	936
Heat capacity @200°C (J/K l)	408	371	445	408	482
Heat capacity @200°C (J/K)	243	371	445	408	482
Substrate mass (g)	473	720	864	792	936

(two 1.2-l catalysts of each). The authors summarized their findings as follows:

1. For the tested catalyst systems, the square-cell ceramic substrate provided superior conversion efficiency relative to equivalent cell density metallic substrates of either equal volume or equivalent surface area. Consequently, the uncoated substrate surface area, a common comparison metric, was not a good predictor of conversion efficiency when comparing substrates of equal cell density but whih different materials and cell shapes.

2. Reducing the specific heat capacity of uncoated substrate material, also a common comparison metric, did not consistently reduce the time to achieve significant oxidation activity. Due to the higher mass of the metallic substrate, the total energy to heat the catalyst system must be compared.

3. When tested on an engine, flow restriction performance was found to be approximately equal between the substrates tested (see Figure 7.29). Consequently, zero-dimensional flow assessments based on open area or hydraulic diameters were not found to be good indicators of actual flow performance. Additionally, when comparing substrates of different materials, cold flow assessments should not be used. It is thought that the cell shape and substrate material play a significant role in flow performance when compared in hot exhaust flow.

More recent advances in metallic substrates relate to partially deformed, interrupted, or laterally offset walls that create flow disturbances and increase mass transfer to the walls. The LS structure, shown in Figure 7.30, consists of secondary corrugation to enhance the mass transfer locally and improve catalytic efficiency (Presti et al. 2006; Dawson and Kramer 2006). In addition, it reduces

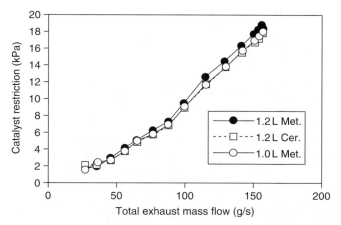

FIGURE 7.29. Catalyst restriction as a function of exhaust gas mass flow. Reprinted from Pannone and Mueller 2001; courtesy of SAE.

FIGURE 7.30. Schematic representation of a single LS channel. Reprinted from Presti et al. 2006, courtesy of SAE.

thermal mass, prevents potential PM (particulate matter) plugging, and makes better use of PGM (platinum group metals). The LS design essentially simulates a disk catalyst and yet is more cost-effective. The PE (perforated foils) structure, shown in Figure 7.31, consists of 8-mm-diameter holes on both flat and corrugated foils that create cavities within the substrate and act as a mixing chamber causing turbulent-like flow distribution. In addition, the cavities allow radial with positive impact on flow distribution and back pressure (Corrado et al. 2005). The PE substrate also offers lower heat capacity.

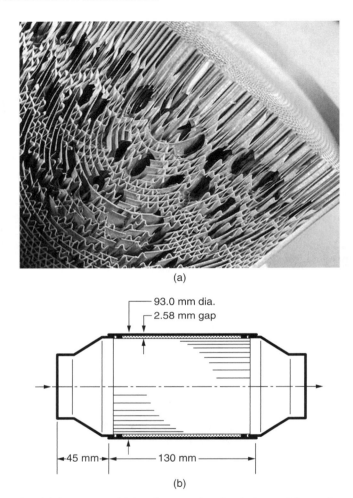

(a)

(b)

FIGURE 7.31. (a) Structure of PE substrate showing holes and formation of cavities. Reprinted from Presti et al. 2006, courtesy of SAE. (b) Schematic of packaging design for preconverter. Reprinted from Gulati and Then 1994a; courtesy of Elsevier, Amsterdam.

7.7 COMMERCIAL APPLICATIONS

Ceramic substrates have performed successfully since their introduction in 1976 in passenger cars. Over a billion units have been installed to date and continue to meet emissions, back pressure, and durability requirements. With new developments in substrate composition, washcoat technology, catalyst formulation, and packaging designs, ceramic substrates are finding new and more severe applications, including in gasoline trucks, motorcycles, and close-coupled converters. The following examples help illustrate their design, performance, and durability.

7.7.1 Underbody Converter

In this first example, we will illustrate the effect of a washcoat microstructure on the thermal integrity of two different 4.66" diameter × 6" (11.84 cm × 15.24 cm) long passenger car substrates. The washcoat microstructure was adjusted through the formulation and processing conditions of the coating slurry. Similarly, the microstructure of a cordierite substrate with a 400/8 square-cell structure was adjusted through compositional and process control. Denote the two different substrate microstructures by S_1 and S_2 and four different washcoat microstructures by W_1, ..., W_4. The compatibility of various substrate/washcoat microstructures was investigated by measuring key physical properties that reflect thermal-shock resistance. For this particular passenger car application, the catalyst is required to pass a thermal cycling test, which imposes a center temperature of 825 °C and a skin temperature of 450 °C. The properties of interest that help assess substrate/washcoat compatibility are MOR and E at 450 °C and differential expansion strain (DES) over the 450 °C to 825 °C range, both in axial and tangential directions. These are summarized in Tables 7.25, 7.26, and 7.27.

TABLE 7.25. Axial and tangential MOR values at 450 °C (MPa).

Washcoat code	MOR_z (450 °C)		MOR_θ (450 °C)	
	S_1	S_2	S_1	S_2
W_1	3.86	3.76	2.07	1.97
W_2	4.45	3.66	2.14	1.97
W_3	4.69	4.00	2.17	1.93
W_4	4.34	3.79	2.24	1.93
Uncoated substrate	3.48	2.93	1.76	1.52

Reprinted from Gulati et al. 1991a; courtesy of SAE.

TABLE 7.26. Axial and tangential E-moduli at 450 °C (GPa).

Washcoat Code	E_z		E_θ	
	S_1	S_2	S_1	S_2
W_1	10.34	9.66	5.03	4.90
W_2	10.21	10.34	5.03	5.03
W_3	9.66	10.00	4.90	5.03
W_4	10.00	9.66	4.97	4.90
Uncoated substrate	7.79	6.34	3.72	3.03

Reprinted from Gulati et al. 1991a; courtesy of SAE.

TABLE 7.27. Axial and tangential DES values over the 450–825 °C range (10⁻⁶ cm/cm).

Washcoat code	DES_z		DES_θ	
	S_1	S_2	S_1	S_2
W_1	520	435	610	515
W_2	565	440	610	520
W_3	555	435	625	525
W_4	565	430	610	500
Uncoated Substrate	500	390	585	420

Reprinted from Gulati et al. 1991a; courtesy of SAE.

TABLE 7.28. Axial and tangential TSP values over the 450–825 °C range.

Washcoat code	TSP_z (825–450 °C)		TSP_θ (825–450 °C)	
	S_1	S_2	S_1	S_2
W_1	0.74	0.89	0.67	0.79
W_2	0.79	0.82	0.70	0.75
W_3	0.86	0.92	0.71	0.74
W_4	0.77	0.90	0.74	0.79
Uncoated substrate	0.88	1.16	0.76	1.00

Reprinted from Gulati et al. 1991a; courtesy of SAE.

Table 7.27 demonstrates good compatibility between either of the two substrates and the four washcoats. The increase in DES is marginal relative to that of the substrate. The impact of these properties on thermal shock parameter is summarized in Table 7.28. The TSP values are reduced by only 5% to 10% for S_1 and by 10% to 20% for S_2.

Several catalysts with above washcoats were thermally cycled at successively higher temperatures until failure occurred. An acoustic technique was employed to detect invisible fractures. The failure temperature (T_f) obtained in this manner is plotted versus the TSP value in Figure 7.22 for each of these coated monoliths. Also included in these data are uncoated substrates, S1 and S2, indicating the superior thermal shock resistance of S_2 relative to S_1. Furthermore, the excellent compatibility between the substrates (S_1 and S_2) and the various washcoats (W_1, …, W_4) results in as good a thermal-shock resistance of coated catalysts as that of bare substrates.

7.7.2   Heavy-Duty Gasoline Truck Converter

As the second example, let us review the impact of the high-temperature washcoat formulation on the durability of an EX-32 cordierite substrate with a 236/11.5 Δ cell structure. The overall dimensions of this oval substrate, designed for heavy-duty gasoline trucks, are 3.38″ × 5.00″ × 3.15″ long (8.59 cm × 12.7 cm × 8 cm). This particular cordierite composition differs in porosity and mean pore size from those of EX-20 cordierite properties, which influence density, E-modulus, and tensile strength of the wall material. The microstructural differences in EX-32, 236/11.5 Δ cell structure provide another opportunity to study substrate/washcoat interaction. This is best done by comparing MOR, E, and α values in axial and tangential directions before and after coating [see Tables 7.4b), 7.8b) and 7.3b)].

The MOR data for the EX-32 Δ cell monolith are summarized in Table 7.4b) as a function of temperature. Note the beneficial effect of coating on both the axial and the tangential MOR values. Compared with the substrate, they are 30% to 40% higher. This improvement in strength is most likely caused by the reduction in wall porosity and stress concentration at the pores, filleting of cell corners by the coating, and reduced microstresses due to the large mean pore size. It is clear that such a beneficial effect on the substrate's strength translates into improved durability of the catalyzed monolith. The data for elastic moduli are shown in Table 7.8b). Note that the coating has a minimal effect on the rigidity of the monolith. Again, this is highly desirable from a durability point of view. The combination of high strength and low modulus increases the strain tolerance of the cellular structure and makes it more thermal shock resistant. A plausible explanation for the minimal effect of coating on a substrate's rigidity is the pore size distribution in the wall that affects the particle size and spatial distribution of the alumina washcoat (intersubstrate vs. intrasubstrate distribution).

The average thermal expansion data in Table 7.3b) show that the coating increases axial thermal expansion by 30% but leaves tangential thermal expansion unaltered. This behavior is most likely related to distribution of the washcoat at the corners of the triangular cell. The minimal effect of coating on tangential thermal expansion, however, is good news from a thermal durability point of view, particularly since the substrate expansion is higher in this direction.

The net effect of the above properties on mechanical and thermal durabilities is reflected by strain tolerance (ST) and TSP values over the operating temperature range of the vehicle. The thermocouple data during dynamometer testing at high engine load yielded a center temperature of 1,025 °C and a skin temperature of 950 °C. The ST and TSP values over this temperature range, both in axial and tangential directions, are summarized in Table 7.29. It is clear from this table that the substrate and washcoat are compatible in that the substrate durability is either preserved or enhanced. Indeed, this particular catalyst exhibited no failures during 2000 hr of accelerated durability testing.

7.7.3   Close-Coupled Converter

In our last example, we present the design and performance data for a ceramic preconverter system which helps meet tighter emission standards—notably

TABLE 7.29. Strain tolerance and thermal shock parameter data for EX-32, 236/11.5 Δ substrate and catalyst during high load operation of a heavy-duty truck engine (T_c = 1025 °C, T_p = 950 °C).

	Axial direction		Tang. direction	
	ST (10^{-6} cm/cm)	TSP	ST (10^{-6} cm/cm)	TSP
Uncoated	300	1.73	410	1.93
Coated	400	1.81	490	1.85

Reprinted from Gulati et al. 1988b; courtesy of SAE.

TABLE 7.30. Key properties of lightoff substrates.

Property	EX-22, 350/5.5 □	EX-22, 340/6.3 Δ
Light-off parameter	5,075	4,790
Back pressure parameter	1.19	0.85
Biaxial compressive strength, MPa	5.69	10.10
High temp. axial modulus of rupture @900 °C, MPa	3.52	4.48

Reprinted from Gulati and Then 1994a; courtesy of Elsevier, Amsterdam.

those corresponding to TLEV[14] and LEV standards. The key requirements of compactness, high surface area, low thermal mass, high exposure temperature, efficient heat transfer, high-temperature strength, prolonged catalyst activity and robust packaging design are best met by thin wall substrates of EX-22 composition.

The pertinent properties of EX-22 substrates with two different cell structures, optimized for lightoff performance, are summarized in Table 7.30. The superior strength of 340/6.3 Δ over that of 350/5.5 □ permits not only a robust packaging design, it contributes to long-term durability. Following washcoat and catalyst application, circular substrates, 3.36″ diameter × 3.15″ (8.53 cm × 8 cm) long, were wrapped with 3100 g/m² Interam™ 100 mat (with high temperature integral seals) and assembled in 409 stainless steel cans, using tourniquet technique, to a mount density of 1.2 g/cm³. The end cones were then welded on to obtain a robust converter package for high temperature testing (see Figure 7.31b).

The mechanical durability of an EX-22, 350/5.5 □ ceramic preconverter was assessed in the high-temperature vibration test, using an exhaust gas generator and electromagnetic vibration table, under the following conditions:

Exhaust gas temperature	1,030 °C
Acceleration	45 g's
Frequency	100 Hz

[14] Transition low-emission vehicle.

TABLE 7.31. Back pressure data for two different preconverters during chassis dynamometer test @ 50 miles/hr (3.66″ diameter × 3.54″ long preconverter with inlet gas temp. = 700 °C).

Cell structure	Back pressure across preconverter (in H_2O)	Total back pressure (in H_2O)
350/5.7 □ ceramic	12.8	26.8
340/6.3 △ ceramic	14.2	28.4

Reprinted from Gulati and Then 1994a; courtesy of Elsevier, Amsterdam.

The test was run for a total of 100 hr. The preconverter was cycled back to room temperature and inspected at 5-hr intervals with no failure detected in the mounting system. The ceramic substrate maintained its original position, the seals rigidized and remained attached to the mat, and the mat was not eroded. In view of the low CTE of an EX-22 substrate and low-temperature gradients in a close-coupled location, thermal durability did not pose a problem.

The back pressure was measured for two different preconverter assemblies (all catalyzed) in a chassis dynamometer test (4-l, six-cylinder. engine with PFI).[15] Both preconverters had identical outside dimensions and catalyst loading but different cell structures. The back pressure, measured with the aid of H_2O manometers, is summarized in Table 7.31. The differences in back pressure across the two preconverters are attributed to different back pressure parameters and to frictional drag of the catalyzed walls to gas flow. It should also be noted that the back pressure across the preconverters is about 50% of the total back pressure from the exhaust manifold to the tailpipe.

The catalytic efficiency was measured on a 1994 vehicle powered by a 4-l, six-cylinder engine with port fuel injection. This vehicle's engine-out emissions were 2.3 g/mile NMHC, 16.9 g/mile CO, and 5.9 g/mile NO_x. Three different exhaust configurations were employed during emissions testing using the FTP cycle (see Figure 7.32). The main converter consisted of two ceramic substrates with a total volume of 3.2 ls. One of these substrates was catalyzed with a Pt/Rh catalyst and the other with a Pd-only catalyst. Figure 7.33 compares the tailpipe HC emissions with and without the ceramic preconverter (350 □), i.e., for configurations 1 and 3. It shows that the preconverter contributes to emissions' reduction during the 40- to 120-s interval. The FTP emissions are reduced significantly due to oxidation of HC and generation of exothermic heat for faster lightoff of the main converter. This benefit is attributed to higher surface area, lower thermal mass, and close proximity of ceramic preconverter to the exhaust manifold, which reduce the time for lightoff temperature to 40 s after cold start. It is also clear in Figure 7.33 that after 120 s, the HC emissions for both configurations are identical, which implies that the main converter is fully effective.

[15] Port fuel injection.

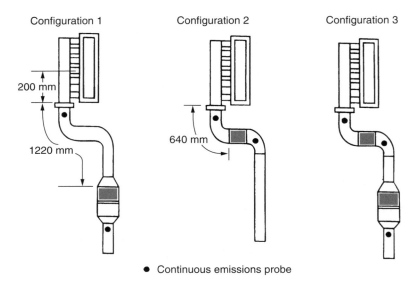

● Continuous emissions probe

FIGURE 7.32. Exhaust configurations during FTP test: (1) main converter only, (2) preconverter only, (3) main converter plus preconverter. Reprinted from Gulati 1996a; courtesy of Marcel Dekker, New York.

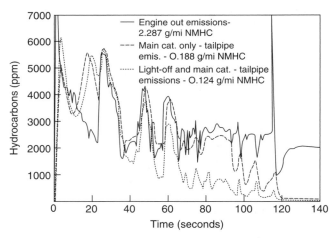

FIGURE 7.33. Continuous HC emissions during cold start from engine and tailpipe with main catalyst-only versus main catalyst-plus preconverter. Reprinted from Gulati 1996a; courtesy of Marcel Dekker, New York.

7.8   SUMMARY

Ceramic substrates offer the advantages of a-high-surface-area-to-volume ratio, large open frontal area, low thermal mass, low heat capacity, low thermal expansion, high oxidation resistance, high strength, and high use temperature—properties that facilitate quick lightoff, high conversion efficiency, low back pressure, good thermal shock resistance, and excellent mechanical durability. Furthermore, the geometric and physical properties of extruded ceramic substrates can be tailored independently to optimize both their performance and their durability. Their microstructure can also be tailored via ceramic composition and manufacturing process so as to be compatible with different washcoat systems for optimum catalytic and physical durability.

The above attributes have made the ceramic substrates ideal for automotive application. Over a billion of these supports have been manufactured since 1975 and successfully implemented in passenger cars, minivans, sport utility vehicles, jeeps, and gasoline trucks worldwide. New advances in ceramic compositions and high-temperature catalysts have lead to lower back pressure, improved performance during lightoff and steady state, and 120,000-mile durability for close-coupled and manifold-mounted applications. The ceramic converter technology is growing rapidly and finding new applications, e.g., slationary engines, marine engines, lawn and garden equipment, and so on, which must now comply with emissions legislation.

This chapter has emphasized the importance of a systems approach during the design phase of a catalytic converter for maximum utilization of chassis or engine space, optimum interaction between substrate and washcoat, and long-term robustness of the total converter package. The systems approach calls for continuous dialogue and prompt feedback among the automaker, substrate manufacturer, catalyst company, mat and seal manufacturers, and canners. In this manner, the automaker's requirements can be best met by taking component suppliers' limitations (e.g., tolerances) into account and arriving at design trade-offs acceptable to all parties. Furthermore, the systems approach provides a rational basis for assessing the probability and warranty cost of converter failure in the field.

The effect of cell shape on performance and durability has also been examined. The key advantage of a hex cell lies in 10% lower back pressure, whereas that of a triangular cell lies in 8% higher isostatic strength relative to those of the square cell. From an overall performance and manufacturability point of view, however, the square cell offers the best compromise and is now considered the industry standard.

The Fe–Cr–Al metallic substrates offer higher GSA and OFA with 10% to 15% lower back pressure than ceramic substrates. Consequently, they are favored in certain niche applications like lightoff catalysts. New cell configurations such as LS and PE have helped promote turbulent flow, thereby improving conversion efficiency and reducing back pressure. Despite their high GSA and thermal conductivity, however, their heat capacity is 15% to 30% higher due to two times greater thermal mass than that of the ceramic substrate. As a result, their lightoff performance and FTP efficiency are similar to those of ceramic substrates with little or no advantage. Their higher cost and complex coatability have limited the use of metallic substrates to primarily close-coupled applications. However, for non-automotive applications such as stationary source applications, the metallic monolith is many times the material of choice such as in ozone abatement (see Chapter 10) and CO abatement from gas turbines (see Chapter 13), where the constraints of a high-temperature operation and the resulting thermal shock are not a consideration and pressure drop and space dominate the design criteria. Also, metallic monoliths are used in a considerable amount for stationary applications in selective catalytic reduction (SCR) NO_x (see Chapter 12) abatement and aircraft ozone abatement (see Chapter 10).

Three different converter applications were discussed to illustrate not only the variable operating conditions, but also the effectiveness of systems approach in optimizing converter design for passenger car (for both close-coupled and underbody locations) and heavy-duty gasoline trucks. These examples should prove valuable in designing other converter systems where automaker's requirements and component supplier's limitations are even more challenging!

REFERENCES

Annual Book of ASTM Standards: Part 17, Designation C158, Philadelphia, PA (1975a).

Annual Book of ASTM Standards: Part 17, Designation C623, Philadelphia, PA (1975b).

Bagley, R., Doman, R., Duke, D., and McNally, R. "Multicellular ceramics as catalyst supports for controlling automotive emissions," SAE 730274 (1973).

Bagley, R. U.S. Patent 3,790,654 (1974).

Bruck, R., et al. "Flow improved efficiency by new cell structures in metallic substrates," SAE 950788 (1995).

Buessem, W., Thielke, N., and Sarakaukas, R. "Thermal expansion hysteresis of aluminum titanate," *Ceramic Age*, 60: 5 (1952).

Corrado, L., et al. "Backpressure optimized close coupled pe catalyst—first application on a maserati powertrain," SAE 2005-01-1105 (2005).

Coble, R., and Kingery, W. "Effect of porosity on physical properties of sintered alumina," *Journal of the American Ceramic Society*, 19: 11 (1956).

Dawson, E. K., and Kramer, J. "Faster is better: the effect of internal turbulence on DOC efficiency," SAE 2006-01-1525 (2006).

Day, J., "Analysis of catalyst durability data from the standpoint of substrate surface area," Ninth International Pacific Conference on Automotive Engineering, Yokohama, Japan (1995).

Day, J., and Socha, L. "The design of automotive catalyst supports for improved pressure drop and conversion efficiency," SAE 910371 (1991).

Eisenstock, G., et al. "Evaluation of SoftMount[SM] technology for use in packaging ultrathin wall ceramic substrates," SAE 2002-01-1097 (2002).

Evans, A. "Slow crack growth in brittle materials under dynamic loading conditions," *International Journal of Fracture*, 10(2) (1974).

Fernandes, S., et al. "Advanced support mat systems for diesel emission control devices: Addressing solutions to cold holding and erosion issues," SAE 2006-01-3507 (2006).

Gulati, S., and Lambert, D. "Fatigue-free performance of wall-flow diesel particulate filter," ENVICERAM'91 Saarbrüken, Germany (1991b).

Gulati, S., and Reddy, K. "Effect of contour, size and cell structure on compressive strength of porous cordierite ceramic substrates," SAE 930165 (1993c).

Gulati, S., and Scott, P. "Design and performance of a ceramic monolithic converter system for motorcycles," Small Engine Technical Conference Proceeding, Pisa, Italy (1993a).

Gulati, S., and Scott, P. "Design optimization for a ceramic converter system for motorcycles," Two-Wheeler Conference Proceedings, Graz Austria (1993b).

Gulati, S., and Sweet, R. "Strength and deformation behavior of cordierite substrates from 70 °F to 2550 °F," SAE 900268 (1990).

Gulati, S., and Then, P. "Design and performance of a ceramic preconverter system," CAPoC-3 Proceedings, Amsterdam, The Netherlands (1994a).

Gulati, S., Geisinger, K. L., Roddy, K., and Thompson, D. F. "High temperature creep behavior of ceramic vs. metallic substrate," SAE 910374 (1991c).

Gulati, S., Sherwood, D., and Corn, S. "Robust packaging system for diesel/natural gas oxidation catalysts," SAE 960471 (1996b).

Gulati, S. "Thermal shock resistance of oval monolithic heavy duty truck converters," SAE 880101 (1988b).

Gulati, S. "Cell design for ceramic monoliths for catalytic converter application," SAE 881685 (1988a).

Gulati, S. "Design and durability of standard and advanced ceramic substrates," SAE 2001-01-0011 (2001).

Gulati, S. "Design considerations for advanced ceramic catalyst supports," SAE 2000-01-0493 (2000).

Gulati, S. "Design considerations for diesel flow-through converters," SAE 920145 (1992b).

Gulati, S. "Durability and performance of thin wall ceramic substrates," SAE 990011 (1999a).

Gulati, S. "Effects of cell geometry on thermal shock resistance of catalytic monoliths," SAE 750171 (1975).

Gulati, S. "Long term reliability of ceramic honeycombs for automotive emissions control," SAE 850130 (1985).

Gulati, S. "New developments in catalytic converter durability," CAPoC-2 Proceedings, Amsterdam, The Netherlands (1991a).

Gulati, S. "Performance parameters for advanced ceramic catalyst supports," SAE 1999-01-3631 (1999b).

Gulati, S. "Thermal stresses in ceramic wall-flow diesel filters," SAE 830079 (1983).

Gulati, S. "Ceramic catalyst supports for gasoline fuel", *Structured Catalysts and Reactors*, in editors. Cybulski, A., and Moulijn, J., Marcel Dekker, New York (1996a).

Gulati, S., Cooper, B., Hawker, P., Douglas, J., and Winterborn, D. "Optimization of substrate/washcoat interaction for improved catalyst durability," SAE 910372 (1991c).

Gulati, S., Lambert, D., Hoffman, M. and Tuteja, A. "Thermal durability of ceramic wall-flow diesel filter for light duty vehicles," SAE 920143 (1992a).

Gulati, S., Socha, L., Then, P., and Stroom, P. "Design considerations for a ceramic pre-converter system," SAE 940744 (1994b).

Gulati, S., Summers, J., Linden, D., and Mitchell, K. "Impact of washcoat formulation on properties and performance of cordierite ceramic converters," SAE 912370 (1991d).

Gulati, S., Summers, J., Linden, D., and White, J. "Improvements in converter durability and activity via catalyst formulation," SAE 890796 (1989).

Gulati, S., Ten Eyck, J., and Lebold, A. "New developments in packaging of ceramic honeycomb catalysts," SAE 922252 (1992c).

Gulati, S., Ten Eyck, J., and Lebold, A. "Durable packaging design for cordierite ceramic catalysts for motorcycle application," SAE 930161 (1993d).

Gulati, S. T., et al. "Performance and durability of advanced ceramic catalyst supports," SAE 2003-26-0015 (2003).

Gulati, S. T. Xu, W., Widjaja, S., Yorio, J. A., and Treacy, D. "Isostatic strength of extruded cordierite ceramic substrates," SAE 2004-01-1135 (2004).

Harned, J., and Montgomery, D. "Comparison of catalyst substrates for catalytic converter systems," SAE 730561 (1973).

Held, W., et al. "Improved cell design for increased catalytic conversion efficiency," SAE 940932 (1994).

Helfinstine, J., and Gulati, S. "High temperature fatigue in ceramic honeycomb catalyst supports," SAE 852100 (1985).

Hollenbach, R. U.S. Patent 3,112,184 (1963).

Ichikawa, S., et al. "Development of low light-off three way catalyst," SAE 1999-01-0307 (1999).

Ikawa, H., Ushimaru, Y., Urabe, K., and Udagawa, S. *Science of Ceramics* 14: (1987).

Johnson, T. "Gasoline vehicle emissions," SAE 2000-01-0855 (2000).

Kays, W., and London, A. *Compact Heat Exchangers*. McGraw-Hill, New York (1955).

Kishi, N., et al. "Technology for reducing exhaust gas emissions in zero-level emission vehicles," SAE 1999-01-0772 (1999).

Kummer, J. *Progress in Energy and Combustion Science*, Vol. 6; Pergamon Press, Oxford, England (1980).

Lachman, I. and Lewis, R. U.S. Patent 3,885,977 (1975).

Lachman, I., Bagley, R., and Lewis, R. "Thermal expansion of extruded cordierite ceramics," *Ceramic Bulletin*, 60, 2 (1981).

Lakhwani, S. G., and Hughes, K. W. "Evaluation of a stronger ultra thin wall corning substrate for improved performance," SAE 2005-01-1109 (2005).

Langer, R., and Marlor, A. U.S. Patent 4,305,992 (1981).

Locker, R. J., et al. "Demonstration of high temperature durability for oval ceramic converter," SAE 980042 (1998).

Locker, R. J., et al. "Low temperature catalytic converter durability," SAE 2000-01-0220 (2000).

Locker, R. J., et al. "Thin film pressure sensor technology applied to catalytic converter technology," SAE 2001-01-0223 (2001).

Maattanen, M., and Lylykangas, R. "Mechanical strength of a metallic catalytic converter made of precoated foil," SAE 900505 (1990).

Maret, D., Gulati, S., Lambert, D., and Zink, U. "System durability of a ceramic race-track converter," SAE 912371 (1991).

Nickerson, S. T., et al. "Advanced mounting system for light duty diesel filter," SAE 2007-01-0471 (2007).

Nonnenmann, M. "New high performance gas flow equalizing metal supports for automotive exhaust catalysis," SAE 900270 (1990).

Pannone, G. M., and Mueller, J. D. "A comparison of conversion efficiency and flow restriction performance of ceramic and metallic catalyst substrates," SAE 2001-01-0926 (2001).

Presti, M., et al. "Turbulent flow metal substrates: A way to address cold start CO emissions and to optimize catalyst loading," SAE 2006-01-1523 (2006).

Stroom, P., Merry, R., and Gulati, S. "Systems approach to packaging design for automotive catalytic converter," SAE 900500 (1990).

Summers, J., and Hegedus, L. "Effects of platinum and palladium impregnation on the performance and durability of automobile exhaust oxidizing catalysts," *Journal of Catalysts*; 51, (185) (1978).

Unruh, J. F., and Till, P. D. "Catalytic converter design from mat material coupon fragility data," SAE 2004-01-1760 (2004).

Webb, J. E. "Reliability and design strength limit calculations on diesel particulate filters," DEER Conference (2004).

Williamson, W. B., et al. "Dual catalyst underfloor LEV/ULEV strategies for effective precious metal management," SAE 1999-01-0776 (1999).

APPENDIX

A. Geometric Properties of Square-Cell Substrates

Referring to Figure A1, the square cell is defined by cell spacing L, wall thickness t, and fillet radius R^*. ρ_c denotes the density of cordierite ceramic (41.15 g/in^3), P the fractional porosity of a cell wall, and C_p the specific heat of a cell

---

*Note that R is normally not specified since it varies with die wear; however, we include its effect on geometric properties for the sake of completeness.

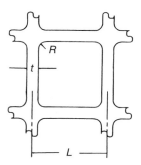

FIGURE A1. Geometric parameters of the square cell. Reprinted from Gulati 1996a; courtesy of Marcel Dekker, New York.

wall $(0.25 \, \text{cal/g}°\text{C})$. The foregoing geometric properties can readily be expressed in terms of L, t, and R [Gulati 1988a]:

$$n = 1/L^2 \quad \text{cells/in}^2 \tag{A1}$$

$$\text{GSA} = 4n[(L-t)-(4-\pi)R/2] \quad \text{in}^2/\text{in}^3 \tag{A2}$$

$$\text{OFA} = n\left[(L-t)^2 - (4-\pi)R^2\right] \tag{A3}$$

$$D_h = 4(\text{OFA}/\text{GSA}) \quad \text{in} \tag{A4}$$

$$\rho = \rho_c(1-P)(1-\text{OFA}) \quad \text{g/in}^3 \tag{A5}$$

$$\text{TIF} = \frac{L}{t}\left(\frac{L-t-2R}{L-t}\right) \tag{A6}$$

$$\text{MIF} = \frac{t^2}{L(L-t-3R)} \tag{A7}$$

$$R_f = 1.775(\text{GSA})^2/(\text{OFA})^3 \quad 1/\text{in}^2 \tag{A8}$$

$$H_s = 0.9(\text{GSA})^2/\text{OFA} \quad 1/\text{in}^2 \tag{A9}$$

$$\text{LOF} = \frac{(\text{GSA})^2}{4\rho_c c_p(1-P)\{\text{OFA}(1-\text{OFA})\}} \tag{A10}$$

$$= \frac{(\text{GSA})^2}{4\rho c_p(\text{OFA})} \tag{A11}$$

B.   Geometric Properties of Triangular-Cell Substrates

Figure A2 defines the parameters L, t, and R for a triangular cell. Its geometric properties are given by

$$n = \frac{4/\sqrt{3}}{L^2} \tag{A12}$$

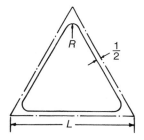

FIGURE A2. Geometric parameters of the triangular cell. Reprinted from Gulati 1996a; Courtesy of Marcel Dekker, New York.

$$\text{GSA} = 4\frac{\sqrt{3}}{L^2}\left[(L-\sqrt{3}t)+\left(\frac{2\pi}{3}-2\sqrt{3}\right)R\right] \qquad (A13)$$

$$\text{OFA} = \frac{1}{L^2}\left[(L-\sqrt{3}t)^2-4\left(3-\frac{\pi}{\sqrt{3}}\right)R^2\right] \qquad (A14)$$

$$D_h = 4(\text{OFA}/\text{GSA}) \qquad (A15)$$

$$\rho = \rho_c(1-P)(1-\text{OFA}) \qquad (A16)$$

$$\text{TIF} = 0.82\frac{L}{t}\left[\frac{L-\sqrt{3}t-2\sqrt{3}R}{L-\sqrt{3}t}\right] \qquad (A17)$$

$$\text{MIF} = \frac{2t^2}{L\left(L-\sqrt{3}t-2\sqrt{3}R\right)} \qquad (A18)$$

$$R_f = 1.66(\text{GSA})^2/(\text{OFA})^2 \qquad (A19)$$

$$H_s = 0.75(\text{GSA})^2/\text{OFA} \qquad (A20)$$

$$\text{LOF} = \frac{(\text{GSA})^2}{4c_p\rho(\text{OFA})} \qquad (A21)$$

C. Geometric Properties of Hex-Cell Substrates

Figure A3 defines the parameters L and t for a hexagonal cell. Its geometric properties are given by

$$n = 0.384/L^2 \qquad (A22)$$

$$\text{GSA} = 6n(L-0.577t) \qquad (A23)$$

$$\text{OFA} = (L-0.577t)^2/L^2 \qquad (A24)$$

$$D_h = 4(\text{OFA}/\text{GSA}) \qquad (A25)$$

$$\rho = \rho_c(1-P)(1-\text{OFA}) \qquad (A26)$$

$$R_f = 1.879(GSA^2)/(\text{OFA})^3 \qquad (A27)$$

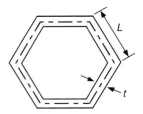

FIGURE A3. Geometric properties of the hexagonal cell. Reprinted from Gulati 2001; courtesy of SAE.

$$H_s = 0.98 GSA^2/\text{OFA} \qquad (A28)$$

$$\text{LOF} = \frac{GSA^2}{4\rho C_p(\text{OFA})} \qquad (A29)$$

REFERENCES

Gulati, S. AIAM Seminar on "Catalytic converters: fresh steps," Bangalore, India (1995).

Gulati, S. "Cell design for ceramic monoliths for catalytic converter application," SAE 881685 (1988).

CHAPTER 7 QUESTIONS

1. Why is cordierite ceramic an ideal material for automotive substrate?
2. Calculate geometric surface area, open frontal area, and bulk density of a cordierite substrate having a 400/8 square-cell structure and 35% wall porosity; assume the density of solid cordierite is 2.51 g/cm^3.
3. Consider a 2.2-l gasoline engine. Assume it requires a substrate having a volume of 60% of engine volume. Assume also that the available space in the exhaust system can accommodate a substrate of 4.16-in diameter. Find the required length of the substrate.
4. Assume that the substrate in Q3 has a square-cell structure of 400/8. Using the answers of Q2 for such a substrate, find the total geometric surface area of the substrate having a diameter of 4.16 in and a length of 5.93 in.
5. A customer is considering using a substrate with either a 600/4 square-cell structure or a 750/3 square-cell structure. The exhaust system can accommodate a 6-in-diameter substrate. If the total GSA of either substrate is kept constant for equivalent conversion efficiency, what will be the relative pressure drop of a 750/3 substrate compared with that of a 600/4 substrate for identical gas flow velocity v?

6. Another customer is considering using an intumescent mat for packaging a coated substrate that has a 2D-isostatic strength of 600 psi. Determine the gap bulk density and compressed thickness of the following mats being considered so as not to exceed a holding pressure of 100 psi:

   a) 3100 mat (3100 g/m²)

   b) 4700 mat (4700 g/m²)

   c) 6200 mat (6200 g/m²)

7. Determine the thermal shock parameter of the following substrate, before and after coating, if its center temperature is 1025 °C and peripheral temperature is 775 °C; use the following properties for bare and coated substrate:

   a) Bare substrates: MOR = 550 psi   and   E = 1.2 × 10⁶ psi   at   775C
      CTE = 8 × 10⁻⁷/°C at 1025C = 6 × 10⁻⁷/°C at 775C

   b) Coated substrate: MOR = 700 psi   and   E = 2.6 × 10⁶ psi   at   775C
      CTE = 13 × 10⁻⁷/°C at 1025C = 9 × 10⁻⁷/°C at 775C

8. Provide a brief comparison of ceramic versus metallic substrates by noting their advantages and disadvantages.

8 Diesel Engine Emissions

8.1 INTRODUCTION

There is no question that driving behind a diesel bus or truck, whether on a highway or in a city, often is an unpleasant experience because of the emissions of black smoke or soot. However, this is only the visible emissions and is usually the characteristic of a poorly maintained engine. The real health issue is the small invisible particulate coming from the diesel engine exhaust. Emissions from a diesel engine are composed of three phases: solids, liquids, and gases. The combined solids and liquids are called *particulates*, or *total particulate matter* (PM or TPM), and are composed of dry carbon (soot), inorganic oxides primarily as sulfates, and organic liquids. When diesel fuel (Song et al. 2000) is burned, a portion of the sulphur is oxidized to sulphate that, upon reaction with the moisture in the exhaust, becomes H_2SO_4. Because of the standard analytical procedure for collection of PM, the sulfates are included in the PM. The organic liquids are a combination of unburned diesel fuel and lubricating oils, called the *soluble organic fraction* (SOF) or the *volatile organic fraction* (VOF), which form discrete aerosols and/or are adsorbed within the dry carbon particles (Zelenka et al. 1990). Gaseous hydrocarbons, carbon monoxide, nitrogen oxides, and sulfur dioxide are the constituents that make up the third phase. Figure 8.1 gives a breakdown of diesel particulates from what would be considered a wet exhaust of a truck engine typical of the early 1990s, i.e., high liquid or SOF content.

The SOF shown is 35% lube and 20% fuel for a total of 55%. Other engines may generate a dry exhaust in which the SOF is lower, with the balance being primarily dry carbon. Diesel emissions are clearly more complex than those from a gasoline engine, and hence, their catalytic treatment is more complicated and requires new technology. The other component of concern in the diesel exhaust is NO_x, which is a source of ozone, through a known photochemical reaction, and this is the major component in SMOG (i.e., a blend of smoke and fog).

The popularity of diesel engines is derived primarily from their fuel efficiency and long life relative to the gasoline spark-ignited engine. They have high compression ratios giving rise to their enhanced efficiency relative to

Catalytic Air Pollution Control: Commercial Technology, by Ronald M. Heck and Robert J. Farrauto, with Suresh T. Gulati.
Copyright © 2009 John Wiley & Sons, Inc.

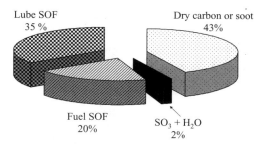

FIGURE 8.1. Diesel particulate emission breakdown. Pie chart showing components of diesel particulates.

gasoline engines. They operate very lean of stoichiometric, with air-to-fuel ratios greater than about 22 compared with <15 for gasoline engines. They have good fuel economy, producing less CO_2. It is not uncommon for a diesel engine to have a life of 1 million miles or about 5–10 times that of the gasoline engine. Diesel fuel as a liquid is injected into the cylinder when the piston has fully compressed the air (at what is referred to as top dead center) resulting in combustion providing the energy for the expansion work stroke. The term "cetane number" (typically 50 or 60) is used to describe the ability of the fuel to combust when injected into the hot compressed air. Straight-chain paraffinic hydrocarbons (C_{14}–C_{20}) have high cetane numbers, and for this reason, diesel fuels have larger concentrations of these molecules. Its lean nature results in a cooler combustion with less gaseous NO_x, CO, and HC emissions than its gasoline counterpart. The design of the combustion process and the fact that fuel is injected as a liquid, results in high particulate emission levels.

8.1.1   The Process of Developing Emission Standards

In the United States, emissions for heavy-duty truck applications are measured over a standardized Federal Test Procedure (FTP), a trace of which is shown in Figure 8.2. Diesel engine emission controls began in the mid-1980s when both the Environmental Protection Agency (EPA) and California Air Resources Board (CARB) began to consider emission standards for the original equipment on road vehicles. This included both NO_x and PM. The Clean Air Amendment of 1990 required that by 1994 particulate emissions be reduced to 0.1 g/bhp-hr for trucks and no greater than 0.07 g/bhp-hr for buses. The units of g/bhp-h reflect the fact that trucks do work, whereas g/mile or g/km are the units used for passenger cars that move people. The first round of standards for heavy-duty truck (HDT) in the early 1990s was achieved using exhaust gas recycle (EGR) and diesel oxidation catalysts (DOCs) (Neitz 1991). Since in HDTs the engine is certified on a test stand, the units of measurement for particulates are grams per brake-horsepower generated in an hour, or g/bhp-hr. For example, in 1986, particulate emissions from a typical diesel truck

| New York Non-freeway 297 s | Los Angeles Non-freeway 300 s | Los Angeles Freeway 305 s | New York Freeway 297 s |

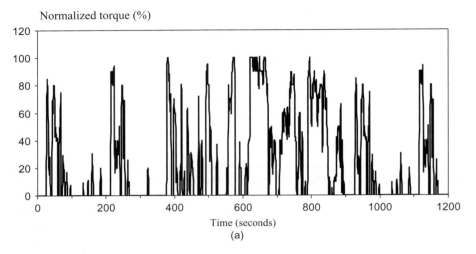

(a)

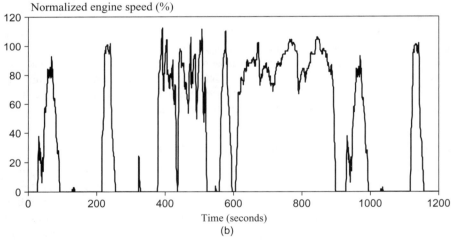

(b)

FIGURE 8.2. U.S. heavy-duty transient cycle for diesel trucks.

engine were 0.6 g/bhp-hr. In the early 1990s, manufacturers had redesigned the engine to reduce the particulates to well below the standards. The HDT engine manufacturers realized that the engine-out emissions of both PM and NO_x from the diesel engine could be controlled within the standards by optimizing fuel delivery, air intake systems, the combustion process, and EGR; the DOCs were in general removed from the emission control design. As the engine manufacturers improved the combustion process to reduce the engine-out PM, it was also found that the PM size distribution was moved to much smaller particles (Baumgard and Johnson 1996).

This approach was effective in meeting the standards up to 2006; however, the new standards proposed for the 2007 phase into the 2010 standards will require catalytic devices combined with a system approach to engine control to meet regulations. In addition, emission standards are being proposed for existing on-road diesel vehicles (so-called retrofit programs) of all classes as well as off-road diesels in all applications. The same trend has occurred in Europe and Japan. Of course, Europe has a more concentrated effort on diesel passenger cars because of the fuel strategy favoring diesel fuel and, therefore, has a much larger population of diesel-powered vehicles versus the United States.

This FTP simulates U.S. truck-driving conditions. Passenger car requirements are measured using the chassis dynamometer procedures described in Chapter 6. The driving cyles for heavy duty vehicles are different in Europe and Japan (DieselNet 2007).

8.2 WORLDWIDE DIESEL EMISSION STANDARDS

Advanced diesel emission control using catalysts is being addressed worldwide since the early 1990. In the United States, Europe, and Japan, essentially all of the trucks and buses operate with diesel-fueled engines. In Europe, the favorable price of diesel-fuel relative to gasoline has resulted in a large number of diesel-fueled passenger cars. In the United States, there has been a revitalization of the diesel passenger cars because of the high cost of gasoline (approaching $3 to $4 per gallon in early 2008) due to enhanced fuel economy. Predictions indicate prices will continue to increase with time as cost of oil increases. Trucks are used to ship goods, buses are used to transport groups of people (usually within urban areas), and passenger cars are used for individual or small groups in both urban and rural locations. Consequently, the allowable emission standards differ for each application. Furthermore, the terrain in the United States, Japan, and Europe differ sufficiently that the test conditions must reflect local driving habits. Each vehicle must meet specific emission standards as measured in standardized driving cycles that reflect the duty cycle anticipated for the particular engine.

Within the United States, a significant reduction in particulates for heavy-duty trucks came in 1994—from 0.25 to 0.10 g/bhp-h, as measured during the U.S. Heavy Duty (HD) FTP Transient Cycle. This test reflects a continuous measure of emissions during various speeds and loads in different U.S. cities, as shown in Figure 8.2. Each task results in a different exhaust temperature and flow condition. Thus, the catalyst must convert the particulates over a broad range of conditions. The U.S. standards for HD trucks for HC, NO_x, and PM remained unchanged until 1998, after which a decrease in NO_x from 5 to 4 g/bhp-hr was required. NO_x removal remains the most significant challenge

TABLE 8.1. Emission regulations for U.S. trucks.

	1994	1998	2003	2007	2009	2010
HC	1.3	1.3	0.5	0.5	0.14	0.14
NO$_x$	5	4	2	2	0.2	0.2
PM	0.1	0.1	0.1	0.01	0.01	0.01

TABLE 8.2. Emissions standards (2009) for Japanese passenger cars.

	Small Cars <1,250 kg 2005	Small Cars <1,250 kg 2009	Medium Car >1,250 kg 2005	Medium Cars >1,250 kg 2009
HC	0.024	0.024	0.024	0.024
CO	0.63	0.63	0.63	0.63
NO$_x$	0.14	0.08	0.15	0.15
PM	0.013	0.005	0.014	0.005

TABLE 8.3. Japanese heavy-duty truck standards for various years (g/kW-hr).

	2005	2009
HC	0.17	0.17
CO	2.22	2.22
NO$_x$	2.0	0.7
PM	0.027	0.01

for which a catalyst solution is currently under intense investigation and will be discussed in detail later in this chapter. The catalyst must decompose or reduce NO$_x$ in diesel exhausts and other lean environments. Currently, the engine control strategy of EGR is being used to reduce the engine-out NO$_x$ emissions. The emissions regulations for HD trucks out to the year 2010 in the units of g/bhp-hr are summarized in Table 8.1.

The years listed show the key changes in emission regulations for HC, NO$_x$, and PM. PM reductions change dramatically in the year 2007, and this will require new technologies. The PM emission standard took effect for the 2007 model year. The NO$_x$ and HC standards are being phased in for diesel engines beyond 2007 into 2010.

The emission standards for Japanese passenger cars through 2009 based on two weight classes are shown in Table 8.2.

The units for Japanese passenger car and truck regulations are in g/km. The HD truck regulations in Japan are shown in Table 8.3.

In Europe, the New Emission Driving Cycle (NEDC) test for passenger cars must meet standards reflecting both urban (ECE) conditions and high-speed extra urban driving (EUDC). These individual cycles are shown as Figure 8.3a) and b). The conbined cycle NEDC is shown in Fig 8.3c.

The NEDC test combines the emissions from both the ECE and the EUDC modes. The catalyst inlet temperatures are cool (200–250 °C) for the

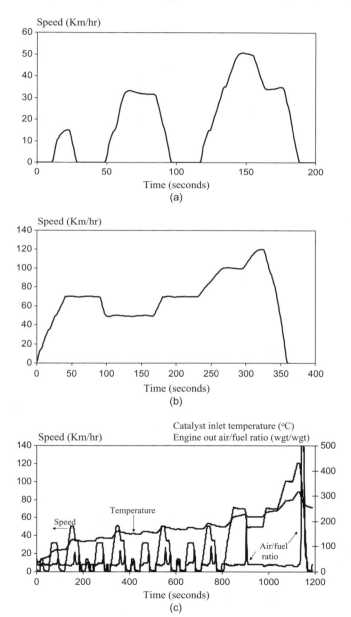

FIGURE 8.3. European cycles: a) EDC portion: urban cycle; b) EUDC: extra urban driving cycle. b) NEDC cycle showing temperature.

TABLE 8.4 Emission standards to 2014 for European passenger cars.

	2000	2005	2009	2014
CO	0.64	0.5	0.5	0.5
$HC + NO_x$	0.56	0.25	0.23	0.17
NO_x	0.50	0.25	0.18	0.08
PM	0.05	0.025	0.005	0.005

TABLE 8.5 Emission standards to 2010 for European trucks.

	1994	1996	2000	2005	2008	2010
HC	1.1	1.1	0.66	0.46	09.46	0.46
CO	8	7	5	3.5	2	2
NO_x	4.5	4	2.1	1.5	1.5	1.5
PM	0.36	0.15	0.1	0.02	0.02	0.02

TABLE 8.6 Next proposal for European emission standards.

	A	B	C	D
PM	0.01	0.02	0.015	0.015
NO_x	0.4	0.2	1.0	0.5

ECE mode but increase to between 300 °C and 550 °C depending on the engine for the EUDC mode. Therefore, the catalyst must, therefore, function over a broad temperature range. For European heavy-duty truck applications, a 13-mode steady-state test with various loads/torques and speeds is used. This test puts emphasis on the high-temperature performance of the catalyst, reflecting high load conditions. At such high temperatures, minimum generation of sulphate by catalytic oxidation of gaseous SO_2 (sulfur originates from its presence in the fuel) becomes the most demanding requirement. The European emission regulations for passenger cars are shown in Table 8.4. The European regulations for HD trucks are shown in Table 8.5.

The emission units are in g/kW-h. The table of information reflects the years where there are changes in the emission regulations. As of the writing of this edition, Europe is considering future regulations for heavy-duty truck and bus engines. The four proposals for PM and NO_x limits are summarized in Table 8.6 in g/kW-h.

The proposal also specifies limits for other pollutants, limits for gas engines, and an ammonia emissions limit of 10 ppm for selective catalytic reduction (SCR) systems. Proposal A has emission regulations comparable with the U.S. 2010 standards.

More details on emission standards on a worldwide basis are available (ICT 2007; Johnson 2008).

8.3   NO$_x$-PARTICULATE TRADE-OFF

The control of particulate emissions and NO$_x$ represent significant challenges to the diesel engine manufacturer because they are coupled inversely. When the engine operates cooler, it produces less NO$_x$ but more particulate. At higher temperatures, combustion is more complete generating less particulate but more NO$_x$. This is referred to as the NO$_x$-particulate trade-off; when one is high, the other is low.

A common engine management strategy to control NO$_x$ is EGR. Part of the exhaust, containing O$_2$, N$_2$, H$_2$O, and CO$_2$, is recycled back to the combustion chamber. The presence of added mass of N$_2$, H$_2$O, and CO$_2$ with its high heat capacity reduces the average combustion temperature limiting NO$_x$ production since combustion flame temperature is decreased. In addition, existing emission control strategies use cooled EGR where the temperature of the EGR gas is reduced through a heat exchanger, further reducing the combustion flame temperature. This process leads to greater particulate and especially to dry soot formation and possibly lower fuel economy.

So a strategy to meet the U.S. 2007 standards is to operate with cooled EGR to reduce the NO$_x$ and to use a diesel particulate filter to remove the PM. However, this strategy alone will not be sufficient to meet the NO$_x$ regulation in 2010. For this reason, there has been an enormous amount of research and development with the goal of finding catalyst technology that when located downstream from the combustion chamber will reduce both NO$_x$ and particulates.

8.4   ANALYTICAL PROCEDURES FOR PARTICULATES

The analysis of the diesel exhaust is much more complicated than that of the IC engine due to the three phases of pollutants present (Cuthbertson and Shore 1988). The FTP defines "particulates" as that which is collected on a filter at 52 °C to condense SOF. The particulates are collected in a dilution tunnel, which cools the exhaust to the required temperature. Because of the hygroscopic nature of the sulfates, the filter must be further conditioned in a controlled atmosphere to equilibrate the water content. The unburned fuel and lubricating oil are extracted from the filter with methylene chloride, producing the SOF. It is then injected into a GC, where a capillary column separates the fuel from the lube fraction. In this manner, the effectiveness of the catalyst toward converting each component of the SOF can be assessed. Some laboratories volatilize the unburned fuel and lube directly from the filter into the GC. This gives a VOF. The results are essentially equivalent to the SOF method.

Another portion of the conditioned filter is then extracted with water to dissolve the sulfate. The aqueous solution is then injected into an ion chromatograph for sulfate analysis. The dry carbon is obtained by difference or, in some cases, can be subjected to thermal analysis for a burn-off after the extraction steps to complete the material balance.

8.5 PARTICULATE REMOVAL

8.5.1 Diesel Oxidation Catalyst Technology

The basic technology for DOCs was developed in the late 1980s, early 1990s to bring U.S. trucks into compliance with standards for 1994 (Farrauto et al. 1992). The manufacturers of trucks sold in the United States preferred a flow through a monolithic honeycomb catalyst structurally similar to that used in the gasoline engine. Given the composition of the diesel exhaust and the particulate reductions required, new and different catalyst formulations would be necessary. The general idea was for the catalyst to oxidize the SOF without oxidizing the SO_2 to sulphate (or SO_3). For such a system to function, the dry soot had to be further reduced to eliminate plugging of the channels of the monolith. New engine design, such as high-pressure injection of the diesel fuel, reduced the droplet size and decreased dry soot formation, which resulted in an increase in the SOF that could then be oxidized by the catalyst.

The cool operating condition of the engine requires that the catalyst first adsorb and retain the SOF at low temperatures (i.e., idle), followed by its combustion as the exhaust temperature reaches lightoff (i.e., 200–250 °C), according to the general reaction (8.1):

$$SOF + O_2 \rightarrow CO_2 + H_2O \tag{8.1}$$

Catalyst design required an organo-philic surface to enhance SOF adsorption at low temperatures. The surface area and pore size also had to be optimized to enhance condensation and to maintain a high adsorption capacity for the low-temperature idle conditions. The SOF would then volatilize and catalytically oxidize as the exhaust temperature heated up (Voss et al. 1994; Farrauto et al. 1995).

An extremely important catalytic property was low activity toward the oxidation of the gaseous SO_2 to SO_3, the latter that quickly forms sulfate particulates after reaction with H_2O.

$$SO_2 + O_2 \rightarrow SO_3 \tag{8.2}$$

This was challenging because catalysts with a high activity for oxidizing the SOF would also enhance SO_2 conversion. Thus, a highly selective catalyst was required.

Catalysts containing precious metal such as Pt and Pd were considered the most likely candidates because they do have good low-temperature activity for hydrocarbon conversions. However, they are also active for the SO_2 oxidation reaction. In one catalyst, it was found that the addition of V_2O_5 suppressed the activity of the Pt for the SO_2 conversion reaction without diminishing the rate of the SOF oxidation reaction (Beckmann et al. 1992; Domesle et al. 1992; Wyatt et al. 1993).

Another approach used Pd as the catalytic metal because its intrinsic SO_2 activity is generally lower than that of Pt but still has reasonable SOF activity (Horiuchi et al. 1991). Yet another approach was to eliminate the use of precious metals and to rely on base metal oxides to oxidize the SOF catalytically (Farrauto et al. 1993, 1995) without oxidizing the SO_2.

Although catalysts with reduced activity for the $SO_2 \rightarrow SO_3$ reaction were found, catalytic generation of sulphate was still a major contributor to particulates at temperatures in excess of about 400 °C. Since some SO_3 was inevitable and Al_2O_3 washcoated catalysts deactivate due to formation of $Al_2(SO_4)_3$, alternative carrier materials had to be used. Titanium oxide (Beckmann et al. 1992; Domesle et al. 1992), silicon dioxide (Ball and Stack 1990), and zirconium oxide (Horiuchi et al. 1991) are sufficiently inactive toward reaction with SO_3 and, thus, could be used as carriers for the Pt and/or Pd metals. They could also be prepared with the proper pore size to adsorb the SOF at low temperatures.

The U.S. Heavy Duty FTP test simulates the driving cycle of a diesel truck, including idle/low-load to high-speed/high-load conditions. The composition and temperature of the exhaust depend on the type of engine. This, in turn, influences the function of the catalyst. For example a 10.3-l engine, typical of the early 1990s, had a particulate composition of 30% SOF, whereas the SOF represents 40% of the particulates of a 5.9-l engine. The temperature histograms generated during the U.S. FTP test show that the exhaust from the 10.3-l engine operates up to 475 °C, whereas the smaller 5.9-l engine never exceeds 300 °C. The performance of the catalyst would have to be much different for treating the exhausts from these two engines. To illustrate the importance of matching the catalyst performance with the exhaust characteristics, Figure 8.4 compares two platinum-containing catalysts installed in the exhausts of the 10.3- and 5.9-l engines.

A completely alternative approach was taken by Farrauto et al. 1993, 1995 when they discovered that a specially activated CeO_2 when physically blended with γ-Al_2O_3 (50:50 wt. %) produced a highly active catalyst for the SOF conversion. The mechanism proposed by the authors was that the SOF was adsorbed into the pore structure of the catalyst during cold (<200 °C) operation. When the engine exhaust exceeded 200 °C, the SOF volatized and catalytically converted on the CeO_2 component to CO_2 and H_2O. Greater detail is available (Farrauto and Voss 1996). Traces of CO and odor-bearing hydrocarbon compounds were produced so a small amount of Pt (0.5 g/ft³) was added to complete the oxidation of the CO and any residual hydrocarbon. Surpris-

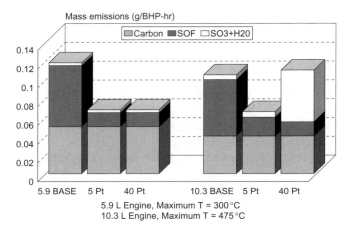

FIGURE 8.4. Particulate emissions exiting from two different truck engines and two different catalysts.

ingly, when the Pt was deposited on the CeO_2 directly, it further decreased the oxidation of SO_2 (Farrauto et al. 1993).

An added benefit of the catalytic treatment for reduction of SOF is the conversion of the small quantities of unburned gaseous HC and CO. Although in the United States the truck engine-out gaseous emissions are well within the standards, their removal further cleans the exhaust. Furthermore, unburned organics are often odor-bearing aldehydes, benzo-pyrenes, ketones, butadienes, and so on. Hence, their removal improves the quality of the exhaust. This is especially important for European diesel passenger car applications where standards require some reduction of the gaseous emissions by the catalyst. For passenger cars, the standards are expressed in g/km.

The European Cycle A, shown in Figure 8.3, embraces the cool urban driving portion, called ECE, where the catalyst inlet temperature typically is below 200 °C, reflecting slow-moving traffic. During these cool conditions, odors caused by partially oxidized hydrocarbons are generated. Odor reduction can be achieved by increasing the concentration of Pt in the catalyst as well as by incorporating hydrocarbon-trapping materials in the formulation (see 8.6.2). In the second portion of the test, reflecting high-speed highway driving and called EUDC, the maximum inlet temperature to the catalyst typically rises above 350 °C. Even with the increased Pt content, the sulfate emissions did not contribute significantly to the total particulate allowing standards to be met.

8.5.1.1   Diesel Oxidation Catalysts for Trucks—1990s Timeframe. The first catalysts used on U.S. trucks were monolithic honeycomb structures of cordierite with 200–400 cells per square inch (cpsi). The catalyzed washcoat is deposited onto the walls of each channel at a loading of about $2 g/in^3$ and is typically no more than about 40–80 microns in the thickest locations (corners or fillets) on the monolith. The first-generation Engelhard (now BASF Catalysts) catalyst was composed of about 50% catalytically active CeO_2 in combination with an equal amount of γ-Al_2O_3 with small amounts of Pt (0.5–$2\,g/ft^3$) for truck and bus applications in the United States and Europe (Farrauto et al. 1993; 1995). A joint venture between Nippon Shokubai and Degussa (now Umicore) promoted the use of a combination of Pd ($40\,g/ft^3$) supported on zirconia with various promoter oxides such as the rare earths for trucks (Horiuchi et al. 1991). Recent studies have shown the key mechanism for Ce in PM oxidation. Using an environmental transmission electron microscope (ETEM), *in situ* observations of CeO_2-catalyzed soot oxidation at the nanoscale were obtained. The results show that the catalytic oxidation reaction involved processes, which were confined to the soot–CeO_2 interface region, and the catalytic reaction resulted in motion of soot agglomerates toward the catalyst surface, which acted to reestablish the soot–CeO_2 interface in the course of the oxidation process (Simonsen et al. 2008).

The DOC catalyst volume as a general rule equals the displacement volume of the engine. Therefore, a 6-l medium-duty truck engine has about 6 liters(l) of catalyst. The catalyst is contained in a steel can with a mounting material made of a ceramic wrapped around its outside diameter to ensure mechanical integrity and resistance to vibration (Clerc et al. 1993). Space velocities vary between about 20,000 and 250,000 hr^{-1}, depending on the duty cycle of the vehicle. Catalyst diameters are 7–10 in (17.78–25.40 cm), with lengths of 5–7 in (12.7–17.18 cm) for trucks.

Studies by the Manufacturers of Emission Controls Association (MECA) show the performance of commercially available technologies for DOC application for trucks. The importance of fuel sulfur was also included in these studies. Although this study did not reveal the composition or the suppliers of the DOCs evaluated, the performance can be viewed in reference to the new emission regulations (MECA 1999b). Following are the conclusions from these studies:

- The test program demonstrated that advanced exhaust emission control technology can be used to meet the program targets of a 0.03-g/bhp-hr PM emission level combined with a 1.5 NO_x + HC emission level for standard No. 2 diesel fuel (368 ppm) and a 0.01-g/bhp-hr PM emission level combined with a 1.5 NO_x + HC emission level for lower sulfur No. 2 diesel fuel (54 ppm)

- Commercially available diesel oxidation catalysts significantly reduced hydrocarbon, carbon monoxide, and particulate emissions from a diesel engine over the heavy-duty engine FTP and off-cycle test points. Optimized catalyst systems can achieve emission reductions in excess of 35% for PM and 70% for HC and CO.

- Diesel oxidation catalysts enable a current heavy-duty engine using standard No. 2 diesel fuel (368 ppm) to meet a particulate emission level of 0.05 g/bhp-hr.

- Diesel oxidation catalysts enable a current heavy-duty engine using 54 ppm of sulfur No. 2 diesel fuel (368 ppm) to meet a particulate emission level of 0.046 g/bhp-hr.
- Diesel oxidation catalysts reduce toxic compounds in diesel exhaust, such as polyaromatic hydrocarbons, by an average of approximately 55% using standard No. 2 diesel fuel (368 ppm S).
- Reducing the sulfur content in the fuel from 54 ppm to 0 improved the performance of the catalyst.
- Fuel sulfur restricts catalyst technology due to the conversion of sulfur dioxide to sulfate at high exhaust temperatures. Gas phase and SOF emission reductions are directly related to the catalyst activity. Highly active gas phase catalysts will make a significant amount of sulfate over the FTP. Although catalyst formulations that can be designed to provide significant HC, CO, and PM control, using low sulfur fuel increases the opportunity to use formulations that can achieve even greater HC, CO, and PM control.

In 1999, most truck manufactures in the United States engineered the DOC catalyst off the exhaust system, with the exception of some fleets in California. A key result from this study was the importance of sulfur level in the diesel fuel and the successful application of catalytic technology in general for controlling emissions from diesel engines. The data from studies such as the MECA study were used to justify the new sulfur standards of diesel fuel at 15 ppm of sulfur (MECA 1999a).

8.5.1.2 Diesel Oxidation Catalysts for Passenger Cars. The initial application for DOCs was in Europe for the many passenger cars in operation. There the emission requirements required removal of PM as well as of gas phase hydrocarbons and carbon monoxide. The effectiveness of increasing amounts of Pt for converting HC and CO is clear from Figure 8.5.

The catalyst designated 2 Pt contains $2 g/ft^3$ of Pt supported on high-surface-area SiO_2. The 20 Pt catalyst contains $20 g/ft^3$ of Pt on the same carrier and shows the largest reduction in HC and CO emissions. The engine-out particulates of 0.15 g/km are composed of about 40% SOF and of the balance dry carbon and traces of sulphate. A conversion of 100% of the SOF results in a 40% conversion of the particulates provided no additional sulphate is formed. The particulate emissions exiting from the 20 Pt catalyst are actually higher than the base emissions with no catalyst, due to formation of sulphate. This illustrates the issue of selectivity. The approach initially taken was to use high concentrations of Pt or Pd with added promoters, designated as I, to inhibit sulphate formation. The effectiveness of this approach is shown in the final data set in Figure 8.5, in which 20 Pt + I is shown to control particulate make (i.e., sulphate make, relative to 20 Pt) while reducing HC and CO. Inhibitors such as vanadium (Beckmann et al. 1992; Domesle et al. 1992) and rhodium

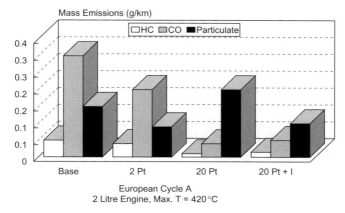

FIGURE 8.5. Passenger car particulates and gas phase emissions using Pt-containing catalysts for European Cycle A. Catalyst 20 Pt + I (Inhibitor) generates less particulate (sulfate) than 20 Pt without inhibitor.

(Fukano et al. 1993) have been shown to be effective in controlling sulphate make at high temperatures. The literature reports that the early Degussa (now Umicore) catalyst was composed of combinations of Pt (20–60 g/ft^3), 3–10% V_2O_3, and balance TiO_2 (Beckmann et al. 1992; Domesle et al. 1992), Johnson-Matthey was believed to be using a variation of Pt and vanadia for selected applications (Wyatt et al. 1993).

Sulfate make was not a major problem during the ECE or cooler urban driving cycle, but the CO and hydrocarbon emissions were. In particular, the odors generated from the partially oxidized hydrocarbons were unacceptable especially in an urban environment. To address this issue, Yavuz and his associates in the 1990s (Yavuz et al. 2001) added specially treated zeolites to the catalyst that acted as traps adsorbing the hydrocarbons at low temperatures. As the exhaust temperature increased, the hydrocarbons desorbed and were oxidized on the Pt function of the catalyst. One typical catalyst was composed of Pt on CeO_2 admixed with a metal-exchanged zeolite such as beta, mordenite, or ZSM-5 all deposited on a monolith. The CeO_2 functioned to oxidize the SOF, whereas the Pt oxidized the CO and hydrocarbons released from the zeolite. A form of this technology is still in use today, but improvements are always being made to improve the low-temperature performance (or lightoff) of the catalyst for CO and hydrocarbon oxidation, using a zeolite function for HC trapping (Standt and Konig 1995).

Typically the diesel oxidation catalyst is 3 in by 5 in (7.6 cm by 12.7 cm) or 4.66 in by 3–6 in long (11.8 cm by 7.6–15.2 cm). They are housed in metallic exhaust systems similar to the gasoline catalyst. The washcoat loadings are similar to three-way catalysis (TWC) and in general tend to have all Pt at about 100 g/ft^3. Recently Pd is being added up to around 20% of Pt as the sulfur level in the fuel is reduced to 15 ppm (Autotech Daily, April 28, 2005). This movement to low sulfur fuel will continue to provide more opportunities for better catalyst compositions since the effect of sulfur on performance will be decreased. DOCs can be designed to remove up to 80% SOF, up to 25% dry carbon, around 60% gaseous HCs and CO, and 50% to 70% of polynucleararomatics (PAHs) and toxic hydrocarbons in the exhaust.

8.5.1.3   Catalyst Deactivation of DOC.   A diesel oxidation catalyst is positioned downstream from the turbocharger in the exhaust manifold and experiences much cooler temperatures (maximum <650 °C) than the gasoline catalyst. Because many of the thermally resistant materials developed for the gasoline catalytic converter have been incorporated into the diesel oxidation catalysts, deactivation due to thermal stresses is not a major problem. Diesel engines do burn larger quantities of oil than their gasoline counterparts, so the catalyst must be more resistant to the oil and its additives. As stated, the unburned oils and their additives deposit within the catalyst structure under the cooler modes of operation. Unlike the organic portion of the oil, the additives remain after the oil is catalytically oxidized. Zinc, phosphorous, sulfur, and calcium oxide accumulate on or within the catalyst (Voss et al. 1994a, 1994b). The microprobe traces shown in Figures 8.6a) and b) were scanned across a 75-micron thick washcoat fillet. Note the scale is in units of 3 microns.

Sulfur accumulates throughout the entire depth of the washcoat, whereas the P-, Zn-, and Ca-containing compounds are concentrated in the outer 15–20 microns. The washcoat interface with the monolith surface is at about 75 microns, as is evident by the drop in S profile to zero. Their accumulation represents the most significant cause of deactivation for gaseous and particulate activity. Knowledge of the poison locations and their interactions with the active catalytic sites allows the catalyst manufacturer to design poison-tolerant catalysts by locating the active catalytic species in protected depths within the washcoat. The washcoat must also be designed to be chemically inert to these types of poisons. Furthermore, the washcoat pore structure must be designed to tolerate large amounts of these metal oxides so as to prevent pore blockage and the subsequent pore diffusion limitations that result. It is not unusual for field-evaluated catalysts to contain 2–3% of these metal oxides and yet continue to function, with little loss in performance, out to the 185,000-mile requirements for medium- to heavy-duty trucks (19,500 to 33,000 pounds). Truly, this is another significant accomplishment in the development of durable catalysts for mobile applications.

A diagram of a typical diesel oxidation catalyst, as it appears in the exhaust manifold, is shown in Figure 8.7.

8.5.1.4   Commercial Diesel Oxidation Catalysts—2010 Timeframe.   The DOC will continue to be used as a primary catalytic device to remove the SOF from the engine in the PM emission control designs for 2007 and 2010, and in addition, the DOC will take on more roles in the integrated emission control system. Of course, the DOC will also be used in the off-road and retrofit applications in some cases as the primary PM control device. The current

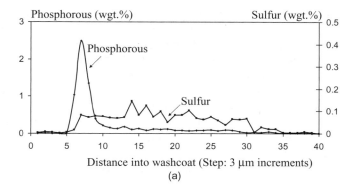

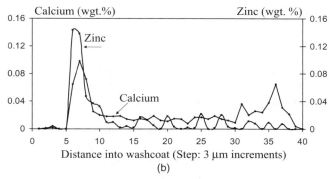

FIGURE 8.6. Electron microprobe scans of the washcoat of an aged diesel oxidation catalyst: a) P and S as well as b) Zn and Ca.

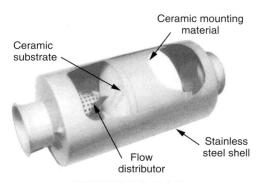

FIGURE 8.7. DOC.

generation of DOCs for 2007 and proposed for 2010 will play a critical role in the regeneration strategy of the diesel particulate filter (DPF) by oxidizing the NO to NO_2, thus lowering the lightoff temperature of the PM on the DPF, and by combusting the unburned fuel from the diesel engine or injected into the exhaust to provide higher temperature gas for regeneration of the DPF or lean NO_x trap. This will be covered in more detail under system designs (see 8.9) for PM removal and combined PM and NO_x removal. The DOC also removes the gaseous hydrocarbons that inhibit the performance of the selective catalytic reduction (SCR) of NO_x with urea and adjusts the NO/NO_2 ratio to the SCR catalyst. For lean NO_x traps (LNT), the DOC catalyst is used to burn HCs and to raise the temperature for desulfation of the LNT catalyst. These technologies will be discussed later in the chapter.

8.5.2 TPM Removal Using DPFs

8.5.2.1 DPFs. In the mid-1980s, many engine manufacturers considered using a device to filter or trap the PM physically on the walls of a wall-flow ceramic honeycomb made of cordierite (Figure 8.8). Envision a flow-through honeycomb where channel #1 is open at its entrance but closed at its exit. The adjacent channel #2 is closed at its entrance but open at its exit. The particulate-bearing gas stream would enter channel #1 flow through the wall and the gas stream exited channel #2. The dry carbon particulates (dry soot), having larger particle size than the pore size of the monolith wall, were trapped on the wall of channel #1. This is known as the DPF.

Since this device had limited capacity before pressure drop became excessive, it is necessary to regenerate the DPF periodically by combustion of the retained PM. The soot required at least 500 °C for ignition in the absence of a catalyst, but the engine exhaust does not frequently or reliably reach these temperatures. Commercial ceramic wall-flow filter devices are made of cordierite, silicon carbide, or alumina titanate (Figure 8.9). The silicon carbide and aluminum titanate have higher temperature capabilities than cordierite but

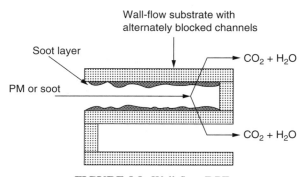

FIGURE 8.8. Wall-flow DPF.

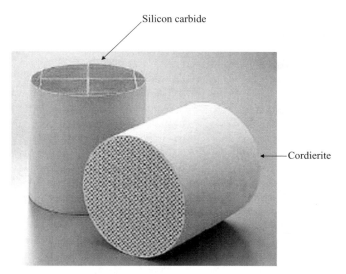

FIGURE 8.9. Cordierite wall-flow filter—Silicon carbide wall-flow filter (Courtesy NGK).

are more expensive and thus are only used when temperatures above about 1,300 °C are expected during regeneration of the soot.

Other approaches with similar mechanisms of particulate trapping are the sintered metal filter and the ceramic Nextel™ cartridge filter (3M 1999). Flow-through metal monoliths constructed with an integral metal mesh are also being used to capture particulates (Brück et al. 2001). These metal monoliths are not wall-flow devices and rely on impingement on a mesh screen material for capture of the PM. All the wall-flow devices provide high-efficiency trapping of the PM (>95%), and all require a regeneration step to combust the collected PM, reduce the back pressure, and restart the PM trapping cycle.

A typical distribution of the PM leaving a diesel engine and the trapping efficiency are shown in Figure 8.10 (Kittelson 2004). Shown in this figure is the size distribution and weight distribution of PM. The weight distribution covers the broad spectrum from the nuclei mode to the accumulation mode and then to the coarse mode. Most particles in the size distribution are in the nuclei mode. Note that the figure mentions that the DPF is efficient over the entire particle size range. In fact it is so efficient that measurement of the particle size leaving the DPF is difficult due to the ability of the measurement equipment.

So the DPF is effective in removing the PM and the issue now becomes the design of a system for *in situ* regeneration of the DPF.

8.5.2.2 Catalyzed DPFs. An effective way to lower the combustion temperature of the collected PM on the DPF is to place a catalyzed washcoat on

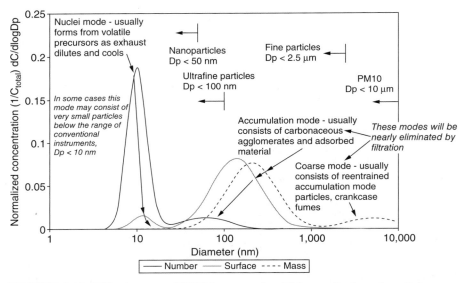

FIGURE 8.10. Effectiveness of DPF for removing PM over the broad particle spectrum (Kittelson 2004).

the filter wall so that the PM deposits in close proximity to the catalyst. As one can imagine, the catalyst technology will resemble the materials used on the DOC since they also catalyze the combustion of the fraction of the PM known as SOF. The reactions on a catalyzed DPF are as follows:

$$SOF, Carbon + O_2 \rightarrow CO_2 + H_2O \tag{8.3}$$

$$HC, CO + O_2 \rightarrow CO_2 + H_2O \tag{8.4}$$

$$SO_2 + O_2 \rightarrow SO_3 \tag{8.5}$$

$$NO + O_2 \rightarrow NO_2 \tag{8.6}$$

$$NO_2 + C \rightarrow CO_2 + H_2O \tag{8.7}$$

A MECA 1999 study summarized the status of diesel particulate traps and found the following for a catalyzed DPF (CDPF) (MECA 1999b):

- Commercially available diesel particulate filters can be used to reduce hydrocarbon, carbon monoxide, and particulate emissions from diesel engines significantly over the heavy-duty engine FTP and off-cycle test points. Optimized filter systems can achieve emission reductions in excess of 70% for PM, 80% for HC, and 60% for CO regardless of fuel sulfur levels. Using low sulfur fuel increases the reductions to in excess of 87%, 95%, and 93% for PM, HC, and CO, respectively.

- Diesel particulate filters enable a current heavy-duty engine using standard road diesel to meet a particulate emission level of less than 0.03 g/bhp-hr using regular sulfur fuel (368 ppm) and of less than 0.01 g/bhp-hr using low sulfur fuel (54 ppm).
- Diesel particulate filters reduce toxic compounds in diesel exhaust, such as polyaromatic hydrocarbons, in excess of 80% whether tested on 368 ppm or 54 ppm of sulfur fuel.
- PM emission levels of 0.005 g/bhp-hr were achieved using a diesel particulate filter with zero sulfur fuel.

Diesel particulate filters used in combination with EGR can achieve NO_x + HC emissions of less than 2.5 g/bhp-hr while maintaining low PM emissions (as low as 0.01 g/bhp-hr) on low sulfur fuel. This MECA study further confirmed the necessity to lower the diesel sulfur level to 15 ppm (ultra-low sulfur diesel or ULSD) to enable high-efficiency removal of the PM.

On-road studies have been conducted with catalyzed DPFs with a driving cycle that provides temperatures in the exhaust of sufficient magnitude such that passive regeneration will occur. In other words, the temperatures in the driving cycle are high enough for the combustion reaction to occur on the catalyst surface. These catalysts accumulated 350,000 operating miles from the years 2000 to 2003 with no issues with back-pressure nor change in engine-out emissions or catalyst performance. The CDPF gave 99% PM reduction. One cleanout of the DPF occurred in 2002 (Kimura et al. 2004).

It is important to have a uniform coating on the CDPF so that the coating does not cause issues with back pressure or PM trapping efficiency. Some studies in the literature address the coating methods for wall-flow filters to optimize the coating with regard to increase in back pressure and alteration of the wall porosity (pore volume and distribution) of the DPF. Both of these can be detrimental to the DPF performance if the coating is done improperly. The goal is to have the catalytic components and washcoat deposited homogeneously within the pore system of the substrate and not deposited on the surface of the substrate (Pfeifer et al. 2005; Solvat et al. 2000). One study looked at changing the DPF physical properties of cell density, wall thickness, wall porosity, and mean pore size (Tao et al. 2003). The new higher porosity cordierite DPFs possess a unique combination of high filtration efficiency combined with a low pressure drop when coated with a catalyst. By controlling the microstructure in combination with the development of new catalyst coating technology, the pressure drop after catalyzing has been minimized. However, if the pores are too large in the higher porosity DPF, the filtration efficiency and mechanical strength are compromised.

The catalyzed DPF has lower washcoat loadings than a DOC or TWC. Washcoat loadings range from 0.06 g/in³ to 1.6 g/in³ with a Pt loading range of 20 g/ft³ to 100 g/ft³ (Huang et al. 2006). With the advent of ULSD, Pd (up to 30% of PGM) is also being used in the CDPF. Some DPFs are zone coated.

The inlet zone of the catalyzed soot filter may have platinum group metals from 5 to $180 \, g/ft^3$, whereas the outlet zone may have loadings from 2 to $90 \, g/ft^3$. The inlet coating in the upstream zone may have a washcoat loading of 0.1 to $2.0 \, g/in^3$, and the outlet coating may be 0.1 to $2.0 \, g/in^3$. The temperature for combustion of the PM approaches $350 \, °C$ with the new catalyst technologies (Alive et al. 2006). Another reason for zone coating is the observation that in a regeneration the highest temperatures occur toward the rear of the CDPF. So the Pt in the rear is severely sintered and not as effective as the front material. Also, at the higher temperatures, less Pt can be used (Punke et al. 2006).

8.5.2.3 *Continuous Regenerable Trap.*

Another approach is to generate NO_2 upstream of an uncatalyzed DPF for combusting the PM. This so-called continuous regenerable trap or CRT™ is based on phenomena first published in 1989 (Cooper and Thoss 1989) where NO_2 was shown to oxidize the dry carbon soot held within the trap at temperatures below the similar oxidation with the O_2 in the air.

$$2NO_2 + C \to CO_2 + 2NO \tag{8.8}$$

$$NO_2 + C \to CO + NO \tag{8.9}$$

The predominate nitrogen species exiting the diesel engine combustion chamber is NO, so it was necessary to oxidize it to NO_2 using a Pt, Pd on Al_2O_3 flow-through ceramic monolith upstream from a wall-flow trap as illustrated in Figure 8.11.

$$NO + 1/2 O_2 \to NO_2 \tag{8.10}$$

The NO_2 is produced using an oxidation catalyst upstream from the trap. Exhaust gas temperatures of $250 \, °C$ to $450 \, °C$ are required for best performance (Zelenka et al. 2001). Of course, the concentration of NO required to form the NO_2 must be high enough to oxidize the trapped soot. Additionally, CO and HC were also oxidized in this same catalyst; however, any SO_2 was oxidized by the precious metals to SO_3. Because of the complex effects of fuel sulfur in this technology (deactivation of NO oxidation catalyst and formation of SO_3), this technology is limited to low sulfur fuels (<15 ppm), which is becoming more and more available throughout the world (Hawkins et al. 1997, 1998). NO_2–C reacts at $250–280 \, °C$, whereas O_2–C reacts at $550–600 \, °C$. Engine-out NO_x–PM ratios should be >10 on a mass basis (Verkiel et al. 2001).

On-road studies have been conducted with CRTs with a driving cycle that provides temperatures in the exhaust of sufficient magnitude such that passive regeneration will occur. In other words, the temperatures in the driving cycle are high enough for the combustion reaction to occur on the CRT. These catalysts accumulated 350,000 operating miles from the years 2000 to 2003 with

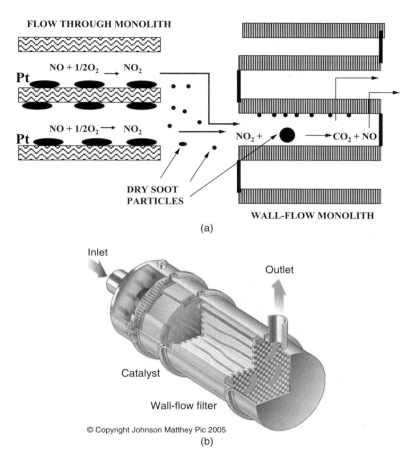

FIGURE 8.11. Conceptual design of a continuous regenerable trap. Courtesy of Johnson Matthey.

no issues with back pressure nor change in engine-out emissions or catalyst performance. The CRT gave 99% PM reduction. One cleanout of the DPF occurred in 2002 (Kimura et al. 2004).

Continued developments for CRTs were conducted in which the Pt catalyst is deposited directly on the trap (Beutel and Punke 1999). The current technology has an improved NO oxidation catalyst upstream (oxicat), which is more sulfur tolerant for conversion to NO_2 and a catalyzed soot filter (CCRT™) (Allansson et al. 2004). This system is again recommended for the low sulfur diesel fuel (ULSD).

8.5.2.4 Fuel-Bound (Borne) Catalysts. An alternative approach to a CDPF where the catalyst is actually paced on the filter is to add certain metallo-organic compounds (fuel-bound or borne catalysts—FBC) containing organo-

TABLE 8.7 Fuel additives lower regeneration temperature (Jelles et al. 1999).

Additive	Concentration (ppm)	Filter	Temperature, °C
None	—	EX-80	537–557
None	—	Pt EX-80	417–427
Cerium	100	EX-80	432
Platinum/cerium	0.5–5	Pt EX-80	327
Platinum/copper	0.5–5	Pt EX-80	347
Platinum/iron	0.5–22	Pt EX-80	357

metallic napthenates of either Cu, Fe, Pt, or Ce to the diesel fuel. During the combustion process, the catalysts are incorporated into the soot matrix and then collected on the filter. This approach improves the solid catalyst–soot contact with the O_2 and reduces the temperature for dry soot combustion (Bloom et al. 1997; McKinnon 1994; Summers et al. 1996). Peugeot in Europe is probably the most advanced in adapting this technology to passenger cars. These systems have a separate dosing apparatus that adds the FBC to the diesel fuel before it goes to the fuel distribution system of the diesel engine. Studies have shown their effectiveness (Richards et al. 1999; Stanmore et al. 1999). Table 8.7 shows the effect on regeneration temperature.

The EX-80 is a wall-flow DPF, and the Pt EX-80 is a catalyzed DPF.

These inorganics must collect on the DPF and not be emitted into the atmosphere. Note that Cu was removed as a commercial product due to volatization off the trap into the diesel exhaust. Because they collect on the trap and do not burn off during the DPF regeneration, the FBCs collect as a residual and gradually increase the pressure drop or engine back pressure of the DPF. For this reason, the trap must be cleaned or serviced to restore the original pressure drop characteristics of the DPF.

The earlier version of the FBC was operated for 80,000 km before cleaning. The newer version with Ce/Fe gives 150,000 km before cleaning. With an Octaquare DPF SiC design and a Ce/Fe FBC, the DPF system can achieve 350,000 km before cleaning. The newer additive requires less dosing concentration (10 ppm vs. 25 ppm) and gives a faster burnoff time (400 s vs. 1600 s) (Campenon et al. 2004; Seguelong et al. 2002).

FBCs can be used with a DOC followed by a catalysts DPF or uncatalyzed DPF (Richards et al. 2001). The exact combination of fuel-borne catalysts and catalyzed or uncatalyzed DPF is dependent on the operating driving cycle of the vehicle.

Another approach that has been used on retrofit technology in the United States is the Clean Diesel approach where a Pt/Ce compound is added to the filling station fuel supply tanks. There is no on-board dosing apparatus with this approach. This is being used in the retrofit market (Peter-Hoblyn and Valentine 1998).

The broad commercialization of FBC still is an open question. In Europe, for passenger cars, there are over a million installations using SiC DPFs with the FBC system installed. But even in Europe, some automobile manufacturers use FBCs on some model vehicles and use CDPFs on other models (Broge 2004). The U.S. market is not as imminent for either passenger cars or original equipment. This approach continues to look attractive for the U.S. retrofit market.

8.5.2.5 Plasma Approaches. The concept of using nonthermal plasma for *in situ* removal of PM in the tailpipe from a diesel engine with no filters present has been studied (Thomas et al. 2000). In the nonthermal plasma, electrons have a kinetic energy higher than the energy corresponding to the random motion of the background gas molecules. The idea is to transfer selectively the input electrical energy to the electrons, which would generate free radicals through collisions and promote the desired chemical changes in the exhaust gas. These reactions can be accomplished at a fraction of the energy that is required in the thermal plasma system. An example of nonthermal plasma is the gas filling a fluorescent tube. Its temperature is only about 40 °C, but the temperature of free electrons in the system exceeds 10,000 °C.

A flow-through device or reactor has to be designed for contacting the engine exhaust gas with the nonthermal plasma device. In this contacting reactor, which is usually a packed bed, the collisions of the free radicals are promoted with the PM and then are decomposed to CO_2 and H_2O. One such device is shown in Figure 8.12 (DieselNet 2007).

It has been shown that >90% reduction in PM is accomplished and essentially that complete removal of the PAHs is accomplished. Any utilization of this approach will require the design of a compact reactor device and the power source for operating the device. The durability of such a device still

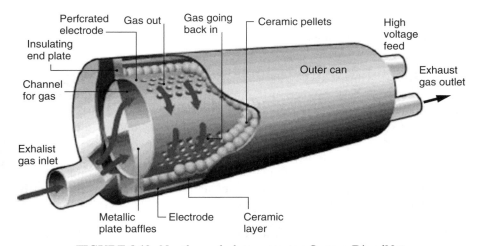

FIGURE 8.12. Nonthermal plasma reactor. Source: DieselNet.

needs to be proven in the end-use application. Currently, there are no commercial applications.

8.5.2.6 DPF Regeneration. Since the most critical step in the satisfactory performance of the DPF is the regeneration procedure where the collected PM is combusted to CO_2 and H_2O, several solutions have been proposed to facilitate the DPF regeneration. These methods involve changing the operation of the diesel engine (so-called active regeneration), modifying the trap composition (so-called passive regeneration), or using external systems to accomplish the regeneration (Konstandopoulos et al. 2000).

Active regeneration methods:
Engine
   exhaust gas recirculation
   postinjection of fuel (in cylinder)
   exhaust pipe injection of fuel
   decrease of boost pressure
   intercooler bypass
   injection timing retard
External method
   fuel burners (full and partial flow)
   electric heating (upstream or embedded in filter)
   microwave heating
   injection of reactive species in exhaust (e.g., H_2O_2)
   generation of reactive species (e.g., nonthermal plasma)
   electrochemical filter reactor

Passive regeneration methods:
   fuel-borne catalyst
   catalytic filter coatings
   in situ reactive species generation (e.g., NO_2)

Many of these external system approaches were also tried in achieving satisfactory cold-start performance in the automobile and were essentially abandoned in favor of catalytic approaches. However, in the diesel emission control world, there are many applications, including the original equipment, off highway and retrofit. In some of these instances, external methods or devices (e.g., fuel burners) may find some use (Syed 2004).

The concept of using a trapping device for removing PM from a diesel engine has been practiced for over 20 years in the mining industry, particularly in Canada. Here the DPF is used during the daily mining operation and then removed for off-line regeneration using some type of hot air device using either combustion air or electrical heaters. Once regenerated, the DPF is reinstalled. The issue with widespread utilization of the DPFs has been a suitable regeneration system design over a broad range of exhaust operating temperatures. And, of course, the DPF material must be durable to withstand the exotherms that occur in the carbon burnoff during the regeneration cycle. Recall that heavy-duty truck applications must be effective for 320,000 miles of operation to meet U.S. regulations. So in the design of a system for onboard regeneration, it is important to understand the engine operating map for the applicable duty cycle because this defines the areas where passive regeneration is possible (using either CDPF, CRT, or FBC) and where active regeneration is required using various engine operating strategies. These engine strategies are aimed at raising the engine exhaust temperature through various approaches. In general, since the exhaust flow is high, it is advantageous to lower the flow to thus reduce the heating requirements and then either to burn fuels in the combustion process or post the combustion chamber using a DOC catalyst or various combinations of the two. Recall that the regeneration process results in a fuel penalty and must be designed optimally (Johnson 2004, 2008; Solvat et al. 2000; Anderson et al. 2004). One study has shown that regenerations are more efficient when done at higher temperatures. This gives a faster reduction in DPF back pressure and a shorter duration for the regeneration event (Pfeifer et al. 2003). In addition, it seems that fresh soot deposits on the trap are easier to regenerate than accumulated or aged soot (Yezerets et al. 2003; Fang and Lance 2004; Nakatani et al. 2002).

A significant increase in exhaust temperature can be achieved through reducing the EGR rate, reducing the boost pressure, retarding the injection timing, and post injection of hydrocarbons. The location of the PM removal system is also important since the thermal inertia of the exhaust line must be taken into account for regeneration. Of course, the optimum strategy for minimizing the fuel penalty depends on engine conditions. The actual fuel penalty may be less than 2% under steady-state conditions (Bouchez and Denenthon 2000). Currently, the regeneration strategy is monitored using the DPF pressure drop as shown in Figure 8.13.

Note that the pressure measurement also becomes the prime diagnostic for monitoring the condition of the DPF and will be used in the control algorithm for triggering regenerations and determining their effectiveness. Also, the pressure measurement will be used to monitor the integrity of the DPF itself (Boretto et al. 2004; Toorisaka et al. 2004).

The presence of Pt does seem to reduce the temperature for the initiation of combustion of the soot, and it does convert NO to NO_2 and will oxidize any CO or HC resulting from the regeneration; for these reason, it will likely be included in the DPF.

8.6   NOₓ REDUCTION TECHNOLOGIES

One of the most important technologies that will impact the design of diesel engine exhaust treatment systems for the 2010 and beyond emission regula-

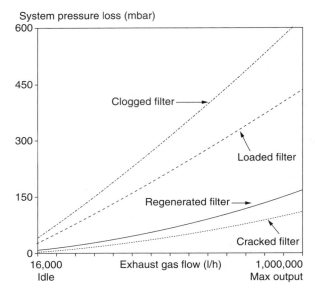

FIGURE 8.13. Pressure drop as a diagnostic for trap regeneration. Reprinted with permission, © 2000 Society of Automation Engineers, Inc. (Solvat et al. 2000).

tions is the method for NO_x reduction. The lean operation of the diesel engine gives rise to high fuel combustion efficiency that, in turn, decreases CO_2 emissions that contribute to the "greenhouse" effect. This advantage, however, may be offset by the inability of existing catalysts to reduce NO_x to N_2 in the high O_2 content environment. The catalytic reduction of NO_x from lean-burn diesel engines has proven to be an even greater challenge relative to the stoichiometric-operated gasoline engine (see Chapter 6). The modern TWC catalyst cannot reduce NO_x in the presence of excess O_2. Consequently, there is a strong driving force to develop a four-way catalytic system capable of reducing NO_x to N_2 and of oxidizing particulates, HC and CO to CO_2 and H_2O, in the presence of excess O_2 (Liu et al. 1996). To date, several methods have been investigated

- NO_x-selective reduction with onboard fuel using a lean NO_x catalyst (LNC)
- NO_x-selective reduction with ammonia-based material such as urea [selective catalytic reduction (SCR)]
- NO_x trapping followed by regeneration of NO_x trap using a lean NO_x trap (LNT)

Note that the catalytic decomposition of NO_x to N_2 and O_2 was extensively studied in the early and the mid-1990s, but catalysts were poisoned by H_2O

and irreversible adsorption of O$_2$ on the catalyst surface rendering them unacceptable (Amiridis et al. 1996).

8.6.1 NO$_x$ Reduction with Onboard Fuel

Since the ideal solution of catalytic NO$_x$ decomposition was not successful, researchers began investigating selective NO$_x$ reduction [(lean NO$_x$ catalyst or LNC)] using the onboard diesel fuel or a derivative. This research, which began in the 1990s, was to find a catalyst capable of reducing NO$_x$ with injection of diesel fuel into the exhaust as a reducing agent (Rice et al. 1996). The major challenge was to find a selective catalyst for the reaction (8.11) and to minimize the nonselective reaction (8.12):

$$HC + NO_x + O_2 \rightarrow CO_2 + H_2O + N_2 \tag{8.11}$$

$$HC + O_2 \rightarrow CO_2 + H_2O \tag{8.12}$$

After almost 20 years of intensive research, only two candidate materials emerged, neither of which are sufficiently active or selective: (1) Pt supported on an Al$_2$O$_3$ or a zeolite and (2) Cu/ZSM-5 (Figure 8.14). The ZSM-5 is a pentasil zeolite with a Si/Al ratio of about 20 and pore size openings of about 5.5 Å. Literally thousands of materials were investigated but none were found (Amiridis et al. 1996; Misono 1998; Petersson et al. 2005; Krantz and Senkan 2004; Burch et al. 2002). Pt is active in a very narrow "temperature window" of 180 °C and 275 °C, whereas the Cu catalyst is active above about 350 °C. Furthermore, the Pt catalyzes the reduction of NO$_x$ to N$_2$O, which is a powerful greenhouse gas. It does have the advantage of being generally insensitive to the presence of SO$_2$ and possesses good thermal stability, but it will oxidize SO$_2$ to SO$_3$. Cu/ZSM-5 catalyzes NO$_x$ reduction with certain hydrocarbons, but it is poisoned by SO$_2$ (Feeley et al. 1995) and lacks hydrothermal stability above 600 °C.

Therefore, the LNC system will not be sufficient to meet emission standards for 2010. Some benefit for NO$_x$ reduction is being realized by the activity of Pt present in diesel passenger car oxidation catalysts because it does convert about 10–15% NO$_x$ without added hydrocarbon. The engine is tuned to permit sufficient slippage of HC exiting the combustion chamber to bring about the reduction. Maximum possible conversions for a Pt-based catalyst with diesel fuel injection in the exhaust are placed between 25% in the U.S. FTP test (Jochheim et al. 1996) and 45% with sulfur fuels <1 ppm (MECA 1999a).

It also needs to be mentioned that a catalyst was commercialized having Ru as the active species and that this gave satisfactory laboratory results but failed in the commercial application. The Ru volatilized off the catalyst support severely deactivating the NO$_x$ performance. This product was removed from the marketplace. Work has also been ongoing recently using a silver-based technology using alcohol materials for the selective reduction. This gives high

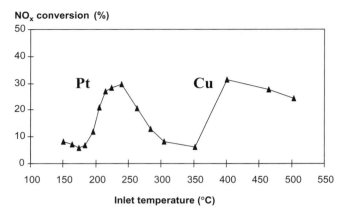

FIGURE 8.14. Performance of lean NO_x catalyst (MECA 1999).

performance but requires another fuel onboard as an alcohol (Thomas et al. 2005).

There have also been some studies using fuel reformer technology to generate H_2 and CO to react over LNCs (Abu-Jrai et al. 2007; Pieterse and Booneveld 2007). This would require operation of the diesel engine to generate H_2 or some type of onboard reformer.

The hydrocarbon LNC still requires a major catalyst breakthrough because of its relatively low NO_x conversion, poor HC selectivity, and poor durability. Most diesel manufacturers and catalyst companies have turned toward other possible solutions.

8.6.2   Reducing NO_x with Urea

The reduction of NO_x using NH_3 is successfully practiced in stationary applications (See Chapter 12). Given the lack of success with the hydrocarbon lean NO_x catalysts, engine manufacturers and catalyst companies are considering the use of SCR with NH_3 as the reductant for NO_x for heavy-duty trucks (Miller et al. 2000). Urea is the source for the NH_3 and is convenient for onboard use as a liquid carrier for ammonia (MECA 1999b). It hydrolyzes in the exhaust system according to the reaction (8.13) at about 200 °C:

$$CO(NH_2)_2 + H_2O \rightarrow 2NH_3 + CO_2 \tag{8.13}$$

The NH_3 in the presence of a suitable catalyst acts as a selective reductant for the NO_x giving conversion between 80% and 90%:

$$NH_3 + NO_x + O_2 \rightarrow N_2 + H_2O \tag{8.14}$$

For heavy-duty diesel applications, the catalyst has been studied extensively in Europe with numerous on-highway studies using the V_2O_5 catalyst sup-

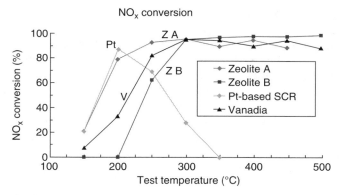

FIGURE 8.15. Comparison of V and zeolite for NO_x reduction with NH_3 as a function of temperature (Walker 2005).

ported on TiO_2 usually with WO_3 added (Amon and Keefe 2001). It was discovered that the SCR catalyst is affected by the unburned diesel fuel in the diesel exhaust, so a DOC-type catalyst is usually placed upstream to remove the unburned hydrocarbons (Gieshoff et al. 2000).

However, some concerns have been raised recently (particularly in the United States) regarding the volatility of vanadium at the higher temperatures in the diesel exhaust particularly during regeneration of the DPF. To address this concern, new metal-exchanged zeolite catalyst technology was developed and most likely will be the commercial catalyst installed in the United States for heavy-duty truck applications (Walker 2005).

The performance of the vanadium catalyst and the zeolite catalyst are shown in Figure 8.15. The performance is comparable. Work is still ongoing addressing the maximum-temperature stability of the zeolite technology. As it now stands, the metal-exchanged zeolite SCR catalyst performance declines above 750 °C (Hammerle 2004).

The reduction of NO_x with ammonia is desired over the undesirable oxidation of ammonia to either N_2 or NO.

$$4NH_3 + 4NO + O_2 \rightarrow 4N_2 + 6H_2O \tag{8.15}$$

$$4NH_3 + 3O_2 \rightarrow N_2 + 6H_2O \tag{8.16}$$

$$4NH_3 + 5O_2 \rightarrow 4NO + 6H_2O \tag{8.17}$$

The desired reaction (8.15) is dominant in the rising portion of each of the curves, whereas the undesirable reactions (8.16 and 8.17) occur on the descending curves for zeolite and V_2O_5 catalysts. These systems are capable of achieving 80% to 90% NO_x reductions. SCR technology can be applied to any sulfur-containing fuel if the correct catalyst is used. Metal-exchanged zeolites and V_2O_5/TiO_2-based catalysts are generally sulfur insensitive. This is especially the case now that sulfur levels in diesel fuel are being reduced. In some parts of the truck driving cycle, the temperatures are still too low for operation. A DOC catalyst will oxidize the NO in the exhaust to NO_2, and this compound is more reactive and extends the operating temperature window for the SCR catalyst to lower operating temperatures (Walker et al. 2004). A NO_2/NO ratio around 1.0 is best, whereas the higher ratios still give high NO_x conversion but shift the performance to higher temperatures. Studies continue regarding the use of a zeolite-based or vanadia-based SCR catalyst. The impact of sulfur and desulfation was studied on zeolite technologies. The impact of sulfur was more significant on Cu than on Fe/zeolite SCR catalysts for NO_x activity with NH_3, and the detrimental effect of sulfur was mainly noted below 300 °C (Cheng et al. 2008). One study considered emerging markets where the sulfur level can be well over hundreds of ppm. This study concluded that vanadia based would be preferred since the Cu- and Fe-based zeolite technologies would lose activity at the higher sulfur concentrations (Girard et al. 2008). The development of new materials for the Cu- and Fe-based zeolite continue to extend their maximum-use temperatures. Fe/zeolite formulations are known to exhibit superior hydrothermal stability over Cu/zeolite formulations. However, current Fe/zeolite formulations are not active for NO_x conversion in the desired 200–350 °C temperature regime under conditions having low NO_2/NO_x ratios. Cu/zeolite formulations have demonstrated never-to-exceed temperatures up to 775 °C. Laboratory flow reactor studies on hydrothermal aging have indicated new Cu zeolite formulations with hydrothermal stability up to 950 °C while maintaining stable, low-temperature NO_x activity (Cavataio et al. 2008).

Issues of unreacted NH_3 breaking through must still be solved. It is proposed to use a catalyst downstream of the SCR to decompose the unreacted NH_3 before it exits into the atmosphere. This catalyst must not produce NO_2 or N_2O (Hirata et al. 2005). Current catalysts have a selectivity between 35% and 55% and lightoff temperatures of 250 °C to 300 °C (Hunnekes et al. 2006). One study incorporated the NH_3 decomposition function in the SCR catalyst using a titania-based material. The catalyst had good NO_x decomposition although over a narrower temperature operating window and >90% NH_3 removal at >200 °C (Hamada et al. 2006). Other studies using a Pt-based NH_3 decomposition catalyst show that a wider window NH_3/NO_x ratio can be used for satisfactory ammonia slip. However, the Pt catalyst must still be optimized to reduce formation on N_2O (Hunnekes et al. 2006). The catalyst technologies under consideration are proprietary, and thus, no compositions can be disclosed at this time.

The urea supply and infrastructure as well as refilling of the urea on the vehicle have been the subject of many studies. In addition, the issues associated with the low-temperature properties of urea are being addressed to ensure proper delivery in cold climates (Grezler 2004; Jackson et al. 2001). In the United States, the EPA has issued a proposed guidance document for SCR systems. The vehicle compliance criteria are divided into five different categories. Manufacturers must satisfy all five categories. The categories are as follows:

(1) driver warning system, (2) driver inducement, (3) identification of incorrect reducing agent, (4) tampers resistant design, and (5) durable design (EPA 2007). Various interlock strategies have been proposed to assure that the urea tank is filled and that it is being supplied to the SCR system. Questions regarding urea stability and decomposition as well as the low-temperature properties have been studied (Kowatari et al. 2006). NO_x sensors are now commercially available for on-board diagnostics (OBD) operation, and NH_3 sensors are being developed (NGK 2005; Lambert 2005). Some studies on urea usage indicate that for a light-duty vehicle, it is 2% relative to fuel, whereas for a heavy duty, it is 1% relative to fuel (Joubert et al. 2004). Some reports indicate that $MgCl_2$ can adsorb appreciable amounts of NH_3, and thus, the reductant can be stored as a solid on board the vehicle (Johnson 2008).

8.6.3 SCR Deactivation

Most of the on-road studies have been with the V_2O_5-based catalyst system. Aged over 250,000 miles of operation, the following post analysis of the catalyst was performed:

- X-ray fluorescence showed the presence of the original catalyst components of TiO_2, V_2O_5, and WO_3.
- There was no change in Al_2O_3, SiO_2, or CaO signals.
- Increasing amounts of P were detected and adsorbed by the catalyst.
- There was more P on the upstream face of the catalyst compared with essentially none on the downstream face.
- The P continued to accumulate on the upstream face throughout the on-road study

From this study, it is apparent that the P from the lubricating oil can cause a decline in catalyst performance and that the oil usage must be monitored and controlled in diesel vehicles (Amon and Keefe 2001). It is anticipated that the zeolite SCR catalyst will behave in a similar fashion with exposure to the ash and phosphorus from the engine lubricating oil. The other issue with the SCR catalyst is thermal durability. The SCR catalyst is exposed to higher temperatures than the normal engine exhaust temperatures during the active regeneration of the DPF. This is true whether the SCR is located upstream or downstream of the DPF since the heat for regeneration comes from the burning of fuel across the DOC, which is upstream of the SCR in all applications.

8.6.4 NOₓ Traps

A modified TWC catalyst with a special alkaline trap material seems to be a promising solution for NO_x reduction for lean-burn diesel. This LNT or NO_x adsorber catalyst (NAC) uses an alkaline metal oxide trap that adsorbs NO_x

during the lean mode of operation (Miyoshi et al. 1995). During steady-state driving, the engine operates lean for improved fuel economy. The exhaust is rich in NO (high-temperature thermodynamically stable species), and thus, it must first be converted to NO_2 over the Pt-containing catalyst during lean operation.

$$NO + 1/2 O_2 \rightarrow NO_2 \tag{8.18}$$

At temperatures above about 500 °C, NO_2 is not thermodynamically favored; however, because the trap continuously removes the NO_2 from the gas stream, the equilibrium is shifted toward more NO_2. Eventually, at high enough temperatures, the formation is too slow for the reaction to NO_2 to occur and the NO_x conversion or trapping declines rapidly.

The NO_2 is trapped and stored on an alkaline metal oxide such as Ba, Na, K, etc. (see Figure 8.16), which is incorporated within the precious metal-containing washcoat of the three-way catalyst (Hepburn et al. 1996; Pfeifer and Kreuger 2007).

$$NO_2 + BaO \rightarrow BaO\text{—}NO_2 \tag{8.19}$$

The engine will typically operate in the fuel economy lean mode for up to about 60 s; after which, hydrocarbon is either injected into the exhaust creating a fuel-rich condition for less than 1 s or the engine can be commanded into a rich mode generating some H_2. The NO_2 is desorbed and reduced (8.20) on the Rh component of the catalyst.

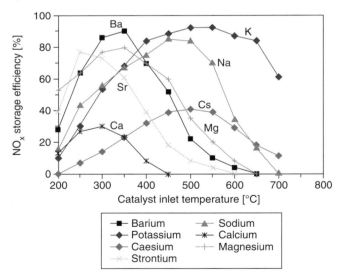

FIGURE 8.16. Various alkaline trap materials for LNT combinations to extend the temperature operating window. Reprinted with permission © 1996 Socity of Automation Engerineer, Inc. (Hepburn et al. 1996).

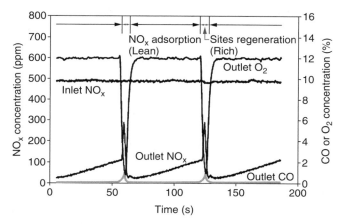

FIGURE 8.17. Typical cycle for lean NO_x (or more precisely NO_2) storage and reduction (Walker 2005).

$$BaO\!-\!NO_2 + HC \rightarrow BaO + N_2 + H_2O + CO_2 \qquad (8.20)$$

Since the diesel engine operates lean, the engine manufactures will need to make adjustments in fuel strategies to generate a rich condition to reduce the adsorbed NO_x. The lean-rich NO_x conversion versus the time profile is shown in Figure 8.17.

NO_x is adsorbed during lean operation and reduced when excess hydrocarbon is injected (Walker 2005).

Sulfur oxides derived from the fuel form alkali compounds (8.21) that are much more stable than the nitrates and are not removed sufficiently (8.22) during the HC injection mode to create a rich atmosphere. Therefore, the trap progressively becomes poisoned by the SO_x (MECA 1999a; Engstrom et al. 1999).

$$BaO + SO_x \rightarrow BaO\!-\!SO_x \qquad (8.21)$$

$$BaO\!-\!SO_x + H_2 \rightarrow \text{No Reaction } (<600\,°C) \qquad (8.22)$$

Complicated engine control strategies are being developed to desulfate the poisoned trap by operating the engine at higher temperatures ($>650\,°C$) and rich of the stoichiometric air-to-fuel ratio for a short time to remove the adsorbed sulfur oxides (Guyon et al. 1998; Johnson 2000). The air-to-fuel ratio must be controlled to minimize the formation of H_2S due to excessive rich conditions. This technology has the capability of removing up to 90% of the NO_x in the exhaust, but in reality, only 60–70% NO_x reductions are achieved (Johnson 2008). Having lower sulfur fuels available will favor the higher conversion levels and reduce the frequency of the desulfation step. Some studies in the literature are looking at sulfur traps that would last around 50,000 miles and minimize the number of desulfations and be disposable (Fang et al. 2003).

Several studies have investigated the use of H_2 as a reductant for the sulfates and have promoted sulfur desorption. The idea is to promote the following reaction.

$$BaSO_4 + 4H_2 \rightarrow BaO + H_2S + 3H_2O \qquad (8.23)$$

This reaction if promoted will make the desulfation step easier to accomplish (shorter times at rich operation, lower temperatures, and more complete desulfation). One study looked at catalyst composition variations to produce H_2 on the catalyst surface via the steam-reforming reaction on the catalyst surface (Hachisuka et al. 2000). Another study used the operation of the diesel engine to generate the H_2. The intake air was throttled, the injection timing was retarded, and the injection duration was increased to create a rich exhaust having H_2 (West et al. 2003). Other studies are looking at H_2-generating devices, including a catalytic reformer, a plasma reformer, and a burner reformer (Betta et al. 2004; Crane and Khadiya 2004; Johannes et al. 2007). These studies using H_2 generation have progressed to various stages, and some of the noted advantages are lower desulfation temperatures and more complete desulfation such that the LNT is returned to complete initial activity. Of course, there is an inherent fuel penalty since diesel fuel is used to generate the hydrogen. One commercial system using a reformer is being offered for commercial applications (Hu 2006). Besides the desulfation issues with loss in NO_x trapping capacity, there is another problem from thermal deactivation of the catalyst due to negative interactions between the alkaline components and the washcoat and monolith after high-temperature excursions. New materials are being researched to resolve this issue.

8.6.5 Plasma

Technologies such as nonthermal plasma (MECA 1999) have been studied for direct NO_x reduction. It has been found that the NO is converted to NO_2 in the plasma, which is the NO_x species suitable for excellent trap adherence and nitrate formation. Since this can be readily accomplished over a Pt catalyst, this technology offers no apparent advantage and is more complex than the catalytic approaches (Aardahl et al. 2002; Vogtlin et al. 1998).

8.7 2007 COMMERCIAL SYSTEM DESIGNS (PM REMOVAL ONLY)

8.7.1 Passenger Cars

The major application for diesel passenger cars is in Europe, and the dominant in-use technology is a system having an FBC with a SiC trap material (Campe-

Filtre à particules (FAP)
particulate filter

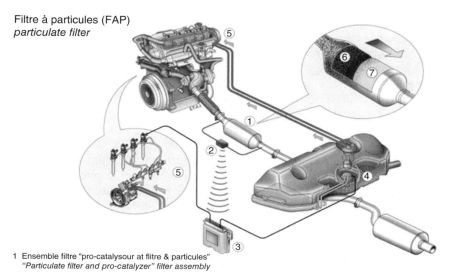

1 Ensemble filtre "pro-catalysour at flitre & particules"
   "Particulate filter and pro-catalyzer" filter assembly

2 Capteurs de pression at do temperature
   Temperature and pressure sensors

3 Calculatour moteur
   Engine ECU

4 Injection de product additif dans le gazole du neservoir
   principal si necessairo
   *Injection of an additive into the fuel in the main tank if
   necessary*

5 Information specifique a la tete d'injectour lorsqu'll
   est necessaire do reallser de la post-combustion
   *Specific information sent to the injector head when post-
   combustion is needed*

6 Pre-catalyseur
   Pre-catalyser

7 Filtre a particules (F.A.P.)
   Particulate filtor

FIGURE 8.18. Schematic of a commercial FBC system (Macaudiere et al. 2004).

non et al. 2004). Figure 8.18 show a schematic of the system design for an FBC on a passenger car in Europe (Macaudiere et al. 2004)

The system control strategy is as follows:

- Amount of soot in the DPF is continuously monitored by measuring the pressure drop across the filter.
- Pressure drop is used as an indicator to start or stop the regeneration process.
- Regeneration process is initiated by increasing the in-cylinder gas temperature by multiple fuel injections to maintain the torque at the same level and thus render the process transparent to the driver.
- Process is accompanied by air-pressure boost.
- Oxidation catalyst upstream of the filter begins to oxidize the hydrocarbons leaving the engine and the amount of hydrocarbon in the exhaust gas increased by multiple injections to create this catalytic postcombustion.

- Temperature rise generated by these two actions, complemented by the catalytic effect of the Ceria-based FBC, ensures the rapid and complete regeneration of the filter.

Even though this technology has been installed in over 850,000 vehicles in Europe, it is not in serial production for all applications (Seguelong and Joubest 2004). For instance, VW uses iron-based additive to perform filter regeneration at 500°C, and VW's later offerings will not have an additive (Broge 2004).

As the system designs for the passenger cars progress, it is anticipated that catalyzed traps (CDPF) will be used more in passenger car application either alone or with FBC technology. The FBC technology uses about 0.5 gallons of additive per 150,000 miles of operation (Christoffel 2007).

8.7.2 Light- and Heavy-Duty Trucks

The implementation for emission control on the original equipment (original equipment manufacturers or OEM programs) for both light- and heavy-duty trucks is moving forward in both Europe and the United States. In addition, many programs address emission controls for off-highway diesel engines and retrofit applications. The retrofit programs are many times connected with the individual state implementation plan (so called SIP) for ambient ozone reduction (Kubsh 2006). In any case, there is substantial momentum in the United States for control of diesel engine emissions through these three programs.

The OEM program for 2007 in general will be using a DPF technology to reduce the PM with an advanced EGR strategy using engine controls and heat exchangers to cool the EGR as indicated in Figure 8.19.

Usually there will be a DOC catalyst upstream of either the CDPF or the CCRT™. The engine operating points where passive regeneration occurs are

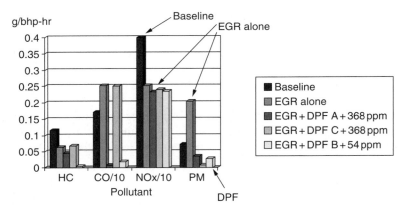

FIGURE 8.19. Emission results indicate control strategy for 2007 with notorious fuel S levels (MECA 1999).

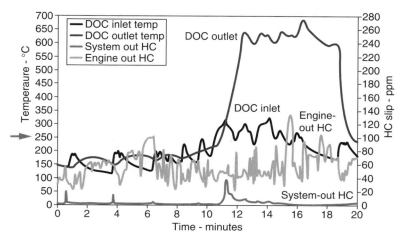

FIGURE 8.20. DOC used to combust fuel catalytically to regenerate DPF (Anderson et al. 2004).

usually determined a priori by measuring an engine operating map of torque and speed and by plotting the isotherms of the temperature. The active regeneration is determined by back pressure readings. At a given pressure drop, the engine control strategy for an active regeneration takes control and fuel is usually combusted across the DOC giving a higher temperature to the filter, which regenerates the trap removing the PM and lowering the back pressure. Figure 8.20 depicts the exotherm across the DOC, which will provide heat to the DPF.

The heat generated from the fuel combusted in the DOC is then transmitted to the DPF, and the deposited soot is then ignited and burned causing an exotherm in the DPF as shown by an increase in the DPF outlet temperature in Figure 8.21.

Light-duty trucks will be using similar technology but a different control strategy since the test is run on a chassis dynamometer to qualify for emission regulations and the cold-start portion of the test will be emphasized as well as steady state operation.

Ash from the engine lubricating oil will accumulate with time on the filter causing a residual buildup in pressure drop and the filter must be cleaned to remove this ash. The engine manufacturers have set up standard cleaning maintenance procedures for cleaning traps (Cummins 2007; MECA 2005).

8.8 2010 COMMERCIAL SYSTEM APPROACHES UNDER DEVELOPMENT (PM AND NO$_x$ REMOVAL)

Because of the lack of progress with the LNC and some durability issues with the LNT regarding desulfation, the leading contender for heavy-duty truck

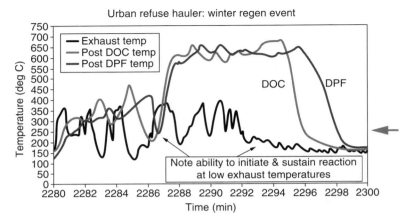

FIGURE 8.21. Exotherm generated across DPF as result of active regeneration using DOC (Anderson et al. 2005).

applications is SCR with urea as the NO_x reduction technology for the 2010 timeframe. This will probably hold true for medium-duty trucks also. Light-duty applications are still considering LNT and SCR technologies. Passenger cars are not well defined, and some manufacturers are going with LNT technology with the option of using SCR later for more stringent regulations. It seems that the LNT technology is good to around 70% NO_x removal, whereas SCR is needed for higher performance (i.e., >70% removal) (Johnson 2007). The EPA has issued guidelines for SCR systems for diesel involving an approach for the certification of light-duty and heavy-duty diesel vehicles and heavy-duty diesel engines (EPA 2007). This covers guidelines for allowable maintenance and adjustable parameters.

8.8.1 Heavy-Duty Trucks

With the sustained studies conducted in Europe over the past ten years, there are many on-road background studies with a DPF in combination with an SCR system using urea. The system design for 2010 envisions a CDPF followed by an SCR system using urea as the NO_x reductant (Amon and Keefe 2001).

The system shown in Figure 8.22 is a design suggested by Volvo that uses the AdBlue system for urea. The oxidation catalyst upstream could be a standard DOC or the NO oxidation catalyst proposed for the CRT™ system designs. Diesel fuel injection into the DOC is not shown in this figure. In any case, gas phase hydrocarbons, SOFs, and NO will be oxidized providing a better exhaust gas to the DPF and the SCR. In addition, the upstream oxidation catalyst will be used for the active regeneration of the DPF. The DPF will most likely be catalyzed to facilitate regeneration. There is also mention of having a catalyst downstream of the SCR to remove any slip ammonia (unre-

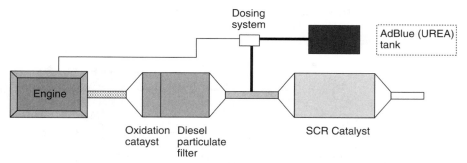

FIGURE 8.22. Representative 2007-type design for heavy-duty trucks (Greszler 2004).

acted NH_3) that passes through the SCR catalyst (Halstrom 2004). The urea requirements are around 1% relative to fuel usage, and the urea tank will be around 5 gallons in size (Greszler 2004).

Engine manufacturers are continually working on control of the NO_x leaving the engine. One engine manufacturer claims that the heavy-duty 2010 engines will not require SCR for NO_x aftertreatment, instead using an integrated system comprising the high-pressure common-rail fuel system, cooled exhaust gas recirculation, advanced electronic controls, and a particulate filter (Costlow 2007).

8.8.2 Light-to-Medium-Duty Trucks

Since the test conditions for light-to-medium-duty trucks (LDT and MDT) are based on a chassis dyno test cycle, the approaches to achieve NO_x control seem to be very similar for all these applications and different than the HDT because of the transient cold-start component of the FTP emission test cycle.

A strategy of one engine manufacturer is to use SCR combining urea and a catalytic converter to reduce NO_x emissions significantly with a DPF. The manufacturer claims that SCR enables the extension of its power range while maintaining excellent fuel economy, maintenance intervals, and overall low cost of ownership. Replenishing the urea supply should be simple, and OEMs will typically provide tanks that hold enough urea to match conventional maintenance periods. Operators should be able to replenish urea at oil change intervals (Costlow 2007).

In considering the DPF plus SCR system, one manufacturer proposes to move the SCR upstream of the DPF to handle these cold-start issues (Figure 8.23).

Because the DOC and the SCR are close to the engine exhaust, these catalytic devices may see higher operating temperatures. The operating temperatures will be even higher when the DOC is being used to provide an active

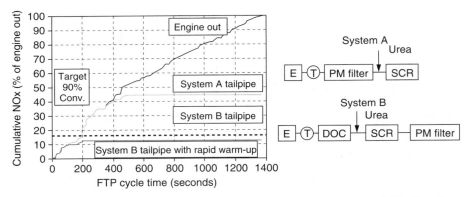

FIGURE 8.23. Passenger cars require the cold-start performance of SCR (Lambert 2006).

regeneration for the soot burnoff from the DPF. If the SCR is downstream of the DPF, then the SCR will be exposed to the DPF outlet temperature during regeneration. The maximum temperature the SCR catalyst is exposed to in LDT or MDT will depend on the regeneration operating strategy. The new base metal-exchanged zeolite catalysts can tolerate occasional exposure to 800–850 °C and still maintain performance. This is usually specified as the "not-to-exceed" temperature range. The maximum performance declines somewhat after 750 °C. Many other automakers have announced the use of SCR systems for SUVs and LDTs, but the configuration of the system in regard to location of SCR has not been disclosed.

The BlueTec-2 uses much the same hardware as a BlueTec-1 (see below) having a DOC and a DPF followed by an SCR. The urea tank will hold 5 to 8 gallons of the so-called AdBlue urea water mixture and will last for at least an oil change interval of roughly 15,000 miles (AEI 2007). The urea SCR system consumes about 1 quart of AdBlue per 600 miles traveled. AdBlue is the registered trademark for AUS32 (Aqueous Urea Solution 32.5%). AUS32 is a 32.5% solution of high-purity urea in demineralized water.

Another approach being proposed is to use LNT technology. The manufacturer proposes using a DOC (which is close coupled to the engine) followed by the LNT and DPF to meet the 2010 emissions standards (Brezonick 2007; Stang 2005). The LNT is regenerated every few minutes at 315 °C to 425 °C for about 3 to 5 s. Also the LNT is desulfated about every four operating hours or approximately after two tanks of fuel. The regenerating cycle is determined by mass flow sensors located in the exhaust. Active regeneration is accomplished by the engine's fuel injection system.

Daimler has combined these components with an SCR catalyst without urea injection. This is the so-called BlueTec-1. During the rich phase of operation when the LNT is being regenerated, NH_3 is generated. The SCR catalyst then uses the NH_3 to reduce the NO_x to N_2 and H_2O. Apparently when the catalyst is new, the NH_3 is production is high and the reduction reaction is effective. However, as the LNT ages due to thermal cycling, the amount of NH_3 produced is reduced. The system gives a high enough performance over the life of the vehicle to meet Tier 2 Bin8 but not good enough for those states adopting California standards (AEI 2007).

An advanced approach by Honda that is somewhat similar in concept is being considered by using a new "LNC" (actually a combination of an LNT and SCR function and not to be confused with hydrocarbon LNC), where NH_3 is generated at the surface of the catalyst and used in the SCR reaction. The emission control system consists of a DOC placed close coupled to the engine followed by a DPF and then the new "LNC" system. It is in the underfloor position to minimize thermal deterioration. An airflow sensor is placed in the exhaust, and a UEGO sensor is placed upstream and downstream of the DOC to monitor the air-to-fuel ratio. The new "LNC" system consists of a double-layer construction in which the upper layer contains a transition metal ion-exchanged zeolite that stores the NH_3, whereas the lower layer contains Pt on Ce, which acts as a NO_x trap. In the LNT function, the NO is converted to NO_2 over Pt in the bottom layer and stored on the Ce in the lean operating mode. In the rich operating mode, the stored NO_x is released and converted to NH_3 and stored in the upper layer of the zeolite. Upon returning lean, the stored NH_3 reacts with the NO_x in the exhaust gas to N_2 as in the normal SCR reaction. The H_2 required for the NH_3 formation in the rich mode is supplied by a water gas shift reaction between CO and H_2O. So the special feature of this new "LNC" is the way it reduces NO_x through an NH_3 SCR in which NO_x adsorbed in the lean mixture condition is converted to NH_3 in the rich mixture condition and is reduced in the following lean mixture condition (Morita et al. 2007).

Nissan is taking a different approach with an LNT that incorporates a HC-trap layer in the NO_x-trap catalyst. The HC-trap layer serves to trap the HC, which is oxidized to generate hydrogen (H_2) and carbon monoxide (CO), which in turn react with the NO_x gases trapped by the NO_x-trap layer to produce nitrogen (N_2) and carbon dioxide (CO_2) gases, in addition to water vapor (H_2O), as end products. The chemical reactions effectively reduce HC and NO_x, resulting in cleaner tailpipe emissions. With this new HC–NO_x trap catalyst technology, Nissan believes it will be able to achieve cleaner diesel emissions in future vehicles that will meet the stringent SULEV-standards set by the state of California (Nissan 2007).

Catalyst companies are always attempting to make the catalytic system as compact as practical and still have good performance and durability. One approach would be to place the LNT function onto the DPF support. The Toyota approach places the LNT catalytic components within the walls of the DPF and is called a diesel particulate NO_x reduction (DPNR) system (Watanabe 2005). It is being used on light-duty trucks and passenger cars. The systems design is shown in Figure 8.24.

An air/fuel ratio sensor, exhaust temperature sensor, and pressure differential sensor regulate catalyst regeneration. To create the rich conditions

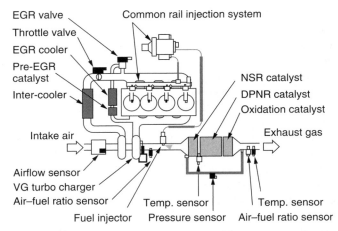

FIGURE 8.24. System design of NO_x storage within the diesel particulate filter (Watanabe 2005).

required for NO_x reduction, Toyota uses a fifth fuel injector, known as the Exhaust Port Injector (EPI), placing it in the exhaust manifold. The injection of fuel creates the requisite rich conditions and is used to perform the catalyst desuphation or sulfur discharge. When the sulfur accumulated in the DPNR catalyst reaches a certain level, the EPI will increase the catalyst's bed temperature to release the sulfur stored on the catalyst (Green Car Congress 2007; Shoji et al. 2004).

8.8.3 Passenger Cars

The diesel passenger car technology for NO_x and PM control will draw heavily from the technology used for LDT and MDT (covered in 8.7.2) since the cold-start portion of the emissions test verification is important. It is interesting to note that some auto manufacturers will have LNTs installed with the control system for SCR in place to allow implementation for SCR when the regulations become more stringent for NO_x in 2010. The idea is to be able to switch easily to SCR as required. An example of this emission control strategy is the announcements by Mercedes-Benz at the North American International Auto Show 2006 in Detroit. They unveiled two new diesel car models: the E 320 and a full-size concept car diesel SUV named Vision GL 320. The E 320 uses an LNT system comprising a close-coupled DOC, followed by the LNT, a diesel particulate filter, and an SCR catalyst. The aftertreatment system reduces NO_x by up to 80%. The Vision GL 320, on the other hand, uses a urea–SCR system for NO_x aftertreatment. The emission system includes a close-coupled DOC/DPF unit, followed by a urea solution injection nozzle and the SCR catalyst in the underfloor position (Mercedes-Benz 2006).

A study by VW looks at LNT versus SCR and concludes that both will have applications, with LNT and SCR being comparable in performance at low vehicle weights; the SCR is favored at higher vehicle weights for low-emission performance (Dorenkamp 2006). Some studies suggest that the LNT is limited by the size of the vehicle, its duty cycle, and the level of engine power and brake effective pressure (BMEP) required for moving the vehicle. The maximum vehicle weight threshold for LNT applications is around 4000 LBS (AEI 2007).

8.9 RETROFIT AND OFF-HIGHWAY

There is an enormous amount of activity ongoing for retrofit applications in many states and large cities mainly for the city bus and school bus applications as well as for large fleet situations. Funding for these retrofits is coming from the EPA, CARB, and local governments. CARB has recently proposed changes to the in-use, off-road diesel vehicles that require implementation of both PM and NO_x emissions controls. The proposed regulations would apply to self-propelled, diesel-fueled vehicles that cannot be registered and licensed to drive on-road. Examples include loaders, crawler tractors, skid steers, backhoes, forklifts, and airport ground support equipment. The regulation would not apply to stationary equipment or to portable equipment such as generators (CARB In-Use Off-Road 2007). In addition, CARB is in the process of adopting regulations for in-use, on-road vehicles (CARB In-Use On-Road 2007). Regulations are already in place for vehicles used in solid waste collection, utility fleets, transit fleets, and heavy-duty drayage trucks. As given in CARB title 13 CCR sections 2021, 2022, and 2023 (EPA 2006).

Usually the initial retrofit for catalytic devices involves a DOC because it is a universal solution (i.e., requires no engine controls or regeneration) albeit it produces lower PM conversion efficiency. For vehicle platforms where the engine operating indicates that the operating temperature cycle is high enough for passive regeneration, then catalyzed traps (CDPFs or CCRT™) will be used. Also, some installations contain NO_x reduction approaches along with PM reduction. Many NO_x reduction systems use SCR with urea. Before any technologies can be implemented, they must qualify and be placed on the EPA's and CARB's verification lists (EPA Retrofit 2007; CARB Retrofit 2007). Similar lists also exist for non-road applications (EPA Non-Road 2007; CARB Non-Road 2007). Similar programs exist in Europe as approved by the European VERT (Verminderung der Emissionen von Realmaschinen im Tunnelbau) program. These lists are not interchangeable.

One issue that has originated with the retrofit technologies is the increase in NO_2 emissions in the tailpipe due to the catalytic devices. Since NO_2 is a reactive gas in the production of ground-level ozone, CARB is striving to minimize it in the local environment. All NO_x leaving the tailpipe eventually converts to NO_2 due to the thermodynamic equilibrium reaction of the low temperatures present in the atmosphere. CARB recently placed limits on the amount of NO_2 by approved retrofit systems. Beginning in 2007, the relative increase in NO_2 leaving the tailpipe is limited to 30% and in 2009 to 20%. Verified systems that do not meet these criteria will be removed from the verification list.

Several studies have been conducted on off-road vehicles for PM and NO_x control, but unfortunately, little information is published. One example for PM control was a study by the South Coast Air Quality Management District (SCAQMD). The project focused on the installation of 21 PM filters onto 15 engines used on 12 heavy-duty construction vehicles (some vehicles use 2 engines—and certain engines required 2 filters). Engelhard supplied 12 filters, whereas JM provided the other 9. The DPFs were on six bulldozers and six scrapers. Although basic particulate trap technology (the self-regeneration process) was validated for use on heavy-duty diesel construction equipment, significant challenges remain regarding installation and mounting of the very large particulate filters on these types of equipment. The heavy filters, combined with severe vehicle vibration that is typical of large off-road construction equipment, likely led to mechanical issues with filter canning and mounting on many of the installations (SCAQMD 2005). This effort is moving forward with many studies being conducted at various locations throughout the United States (EPA/NESCAUM 2003).

8.10   NATURAL GAS ENGINES

Natural gas engines fall under two categories: lean burn and stoichiometric (Dunn 2003; Eaves 2006). The lean-burn emission control options are similar to the diesel engine to remove CO, HC, and NO_x, whereas the stoichiometric engine can use conventional auto catalyst TWC technology to remove CO, HC, and NO_x. Usually the PM is so low in both engine designs that a DPF-type technology may not be required. Because of the form of the natural gas, there are no issues with vaporization or droplet burning, so in both engine designs, the NO_x and PM emissions levels are very low.

Lean-burn, natural gas-fueled (compressed natural gas or CNG) vehicles are currently the dominant technology in the United States and are used for service operations such as delivery trucks, buses, and so on. Its popularity is derived primarily from its clean burning characteristics relative to uncontrolled diesels with lower emissions of NO_x and particulates allowing emission standards, especially for urban bus applications (Frailey et al. 2000; White 1991; Samsa 1991; Kubsh 2003) to be met. These vehicles still need catalysts to control particulate emissions, primarily derived from lubricating oil, and specific gaseous non-methane hydrocarbon emissions, called ROGs (reactive organic gases), which participate with NO_x and sunlight in the generation of smog (White et al. 1993). Catalysts for lean-burn engines are typically Pd and/or Pt supported on $CeO_2 + \gamma\text{-}Al_2O_3$ deposited on a ceramic monolith. The diameter can be as large as 10 in (25.4 cm) with a length of 7 in (17.8 cm) for an engine with a 10-l displacement.

Methane is the major hydrocarbon in natural gas. It is nonreactive and does not participate in photochemical smog-generating reactions, and so in the United States, it is unregulated. If it were to be required (CH_4 is a greenhouse gas), a major breakthrough in catalyst technology would be needed (Lampert et al. 1997; McCormick et al. 1996) to pass the U.S. FTP test due to inhibition by sulfur oxides. For the European bus market, a Pd-based catalyst is successful for abating methane emissions because the R-49, 13-mode test favors high-temperature modes where the deactivation by fuel and oil sulfur have a minimum effect on the performance of the catalyst.

In addition to the hydrocarbon emissions, the lean-burn CNG buses must meet the 2010 NO_x emissions. This means using either an LNT or an SCR approach since the engine is operated in a lean combustion mode (Kubsh 2003).

Recent combustion designs for natural gas, heavy-duty applications use stoichiometric engine designs (Dunn 2003; Boyce 2006). These systems can use standard TWC technology that is used for the spark-ignited gasoline engine for HC, CO, and NO_x control. This stoichiometric engine equipped with TWC technology gives the lowest emissions relative to alternative natural gas engine designs. A recent study showed that a stoichiometric natural gas engine with TWC emission controls will be highly competitive with diesel engine life-cycle costs when considering comparable vehicles that meet 2010 emission requirements (Schubert and Fable et al. 2006).

REFERENCES

3M. 3M Diesel Filter Cartridges Technical Bulletin, Automotive Division, 3M Center (1999).

Aardahl, C., Birnbaum, J., Rappe, K., Tran, D., Park, P., and Singh, G. "Plasma-Activated Lean NOx Catalysis for Heavy-Duty Diesel Emissions Control," 2002 Diesel Engine Emissions Reduction (DEER) Conference Presentations (2002).

Abu-Jrai, A., Tsolakisa, A., and Megaritisb, A., et al. "The influence of H_2 and CO on diesel engine combustion characteristics, exhaust gas emissions, and after treatment selective catalytic NOx reduction," *International Journal of Hydrogen Energy* (2007).

AEI. "Diesels' brave new world," pp 48–50, November 2007 (2007).

AEI. "Creating the Bin-5 diesel," *Automotive Engineering International* 38–40 (2007).

Alive, K., Baudoux, A., Golden, S., and Iretskaya, S. "Platinum group metal-free catalysts for reducing the ignition temperature of particulates on a diesel particulate filter," U.S. Patent Application 20060120936 (2006).

Allansson, R., Walker, A., Goersmann, C., Lavenius, M., Phillips, P., and Uusimaki, A. "The development and in-field performance of highly durable particulate control systems," SAE 2004-01-0072 (2004).

Amiridis, M., Zhang, T., and Farrauto, R. "Selective catalytic reduction of NO$_x$ by hydrocarbons," *Applied Catalysis B: Environmental* 10: 203 (1996).

Amon, B., and Keefe, G. "On-road demonstration of NO$_x$ emission control for heavy-duty trucks using SINOX^{TM} urea SCR technology—long-term experience and measurement results," SAE 2001-01-1931 (2001).

Anderson, M., and Angelo, T. "Donaldson active regeneration PM system," 2005 Diesel Engine Emissions Reduction (DEER) Conference Presentations (2005).

Anderson, M., Angelo, T., Hou, J., Protas, M., Steinbreuck, E., Wagner, W., Way, P., and Zhang, W. "Development of an active regeneration diesel particulate filter system Donaldson Company Inc.," 2004 Diesel Engine Emissions Reduction (DEER) Conference Presentations (2004).

Ball, D., and Stack, R. "Catalyst considerations for diesel converters," SAE 902110 (1990).

Baumgard, K., and Johnson, J. "The effect of fuel and engine design on diesel exhaust particle size distributions," SAE 960131 (1996).

Beckmann, R., Engeler, W., Mueller, E., Engler, B., Leyrer, L., Lox, E., and Ostgathe, K. "A new generation of diesel oxidation catalysts," SAE 922330 (1992).

Betta, R., Sheridan, D., and Cizeron, J. "Fuel processor enabled NO$_x$ adsorber after-Treatment system for diesel engine emissions control," 2004 Diesel Engine Emissions Reduction (DEER) Conference Presentations (2004).

Beutel, T., and Punke, A. "Status of diesel technologies," International Automobile Exhibit, Frankfurt, Germany (1999).

Bloom, R., Brunner, N., and Schroeer, S. "Fiber wound diesel particulate filter durability experience with metal based additives," SAE 970180 (1997).

Boretto, G., Imarisio, R., Rellecati, P., Barucchi, E., and Sanguedolce, A. "Serial application of catalyzed diesel particulate filter on common rail DI diesel engines for passenger cars," FISITA 2004 World Automotive Congress, Barcelona, Spain (2004).

Bouchez, M., and Denenthon, J. "Strategies for the control of particulate trap regeneration," SAE 2000-01-0472 (2000).

Boyce, B. "Every alternative," Cummins Westport Inc., www.cumminswestport.com (2006).

Brezonick, M. "An adsorbing chapter in emissions," Diesel Progress North American Edition (2007).

Broge, J. "The diesel is coming, the diesel is coming," Automotive Engineering International, 33–37 (2004).

Brück, R., Hirth, P., Reizig, M., Treiber, P., and Breuer, J. "Metal supported flow-through particulate trap; a non-blocking solution," SAE 2001-01-1950 (2001).

Burch, R., Breen, J., and Meunier, F. "A review of the selective reduction of NO$_x$ with hydrocarbons under lean-burn conditions with non-zeolitic oxide and platinum group metal catalysts," Applied Catalysis B: Environmental 39: 283–303 (2002).

Campenon, T., Wouters, P., Blanchard, G., and Seguelong, T. "Improvement and simplification of DPF system using a ceria-based, fuel-borne catalyst for diesel particulate filter regeneration in serial applications," SAE 2004-01-0071 (2004).

CARB In-Use Off-Road. "Technical support document proposed regulation for in-use off-road diesel vehicles," http://wwwarbcagov/regact/2007/ordiesl07/TSDpdf, (2007).

CARB In-Use On-Road. "Draft proposed regulation to reduce emissions RAFT PRO-POSED of diesel particulate matter. And other pollutants from in-use on-road heavy-duty diesel-fueled vehicles," http://www.arb.ca.gov/msprog/onrdiesel/documents/draft_reg_for_in-use_on-road_HDD_vehicles_8_07.pdf, (2007).

CARB Non-Road. "California Air Resources Board (CARB) verified nonroad engine retrofit technologies," http://www.arb.ca.gov/diesel/verdev/vt/vt.htm, (2007).

CARB NO_2. "Public hearing notice and related material," http://www.arb.ca.gov/regact/verpro06/verpro06.htm, (2007).

CARB Retrofit. "Currently verified technologies," http://www.arb.ca.gov/diesel/verdev/vt/cvt.htm, (2007).

Cavataio, G., Jen, H., Warner, J., Girard, J., Kim, J., and Lambert, C. "Enhanced durability of a Cu/Zeolite based SCR catalyst," SAE 2008-01-1025 (2008).

Cheng, Y., Montreuil, C., Cavataio, G., and Lambert, C. "Sulfur tolerance and DeSOX studies on diesel SCR catalysts," SAE 2008-01-1023 (2008).

Christoffel, J. "Green Diesel Platform develops off-the-shelf exhaust aftertreatment solution," *Autromototive Engineering International* p. 28 (2007).

Ciambelli, P., Parma, V., Russo, R., and Vaccaro, S. "The effect of NO on Cu, V, K, Cl catalyzed soot combustion," *Applied Catalysis B: Environmental* 22: L5-10 (1999).

Clerc, J., Miller, R., McDonald, A., and Schlamadinger, H. "A diesel engine/catalyst system for pick-up and medium-duty trucks," SAE 932982 (1993).

Cooper, B., and Thoss, J. "Role of NO in diesel particulate emission control," SAE 890404 (1989).

Costlow, T. "Cummins engines meet 2010 emissions regs," Automotive Engineering International Online, Technical Newsletter, (2007).

Crane, S., and Khadiya, N. "A fast start-up onboard diesel fuel reformer for NO_x trap regeneration and desulfation," 2004 Diesel Engine Emissions Reduction (DEER) Conference Presentations (2004).

Cummins. "Exhaust aftertreatment advanced emission controls," Training Manual, http://www.cumminsfiltration.com/files/aftertreatment_training.pdf, (2007).

Cuthbertson, R., and Shore, P. "Direct capillary gas chromatography of filter-borne particulate emissions from diesel engines," *Journal of Chromatographic Science* 26: 106–112 (1988).

DieselNet, http://www.dieselnet.com/, (2007).

Domesle, R., Engler, B., Koberstein, E., and Voelker, H. "Catalyst for the purification of exhaust gases of diesel engines and method of use," U.S. Patent 5,157,007 (1992).

Dorenkamp, R. "LNT or urea SCR technology: Which is the right technology for Tier 2 Bin 5 passenger vehicles?" 12th Diesel Engine-Efficiency and Emissions Research (DEER) Conference, Detroit, MI, (2006).

Dunn, M. "State of the art and future developments in natural gas engine technologies," 2003 Diesel Engine Emissions Reduction (DEER) Conference Presentations (2003).

Eaves, M. "Natural gas products for 2007," California Natural Gas Vehicle Coalition, http://64.143.64.21/mobile/cff/Events/20060912NatGasProducts.pdf, (2006).

Engstrom, P., Amberntsson, A., Skoglundh, M., Fridell, E., and Smedler, G. "Sulfur dioxide interaction with NO_x storage catalysts," *Applied Catalysis B: Environmental* 22: L241 (1999).

EPA. U.S. Environmental Protection Agency, Heavy-Duty Highway Diesel Program, http://www.epa.gov/otaq/highway-diesel/regs.htm, (2006).

EPA. "Certification procedure for light-duty and heavy-duty diesel vehicles and heavy-duty diesel engines using selective catalyst reduction (SCR) technologies," U.S. Environmental Protection Agency, http://www.regulations.gov/fdmspublic/component/main, (2007).

EPA Non-Road. U.S. Environmental Protection Agency, Diesel Retrofit Technology Verification, Verified Nonroad Engine Technologies List, http://www.epa.gov/otaq/retrofit/nonroad-list.htm, (2007).

EPA Retrofit. Diesel Retrofit Technology Verification, Verified Technologies List, http://www.epa.gov/otaq/retrofit/verif-list.htm, (2007).

EPA. "Certification procedure for light-duty and heavy-duty diesel vehicles and heavy-duty diesel engines using selective catalyst reduction (SCR) technologies," CISD-07-07(LDV/LDT/MDPV/HDV/HDE), (2007).

EPA/NESCAUM. EPA/NESCAUM Diesel Retrofit Workshop, New York City (2003).

Fang, H., and Lance, M. "Influence of soot surface changes on DPF regeneration," SAE 2004-01-3043 (2004).

Fang, H., Wang, J., Yu, R., Wan, C., and Howden, K. "Sulfur management of NO_x adsorber technology for diesel light-duty vehicle and truck applications," SAE 2003-01-3245 (2003).

Farrauto, R., Adomaitis, J., Tiethof, J., and Mooney, J. "Reducing truck emissions: A status report," *Automotive Engineering* 100: 19–23 (1992).

Farrauto, R., Heck, R., and Speronello, B. "Environmental catalysts," *Chemical and Engineering News* 70(36): 34–44 (1992).

Farrauto, R., and Voss, K. "Monolithic diesel oxidation catalysts," *Applied Catalysis B: Environmental* 10: 29–51 (1996).

Farrauto, R., Voss, K., and Heck, R. "A base metal oxide catalyst for reduction of diesel particulates," SAE 932720 (1993).

Farrauto, R., Voss, K., and Heck, R. "Diesel oxidation catalyst," U.S. Patent 5,462,907 (1995).

Feeley, J., Deeba, M., and Farrauto, R. "Abatement of NO_x from diesel engines: Status and technical challenges," SAE 950747 (1995).

Frailey, M., Norton, P., Clark, N., and Lyons, D. "An evaluation of natural gas versus diesel in medium-duty buses," SAE 2000-01-2822 (2000).

Fukano, I., Sugawara, K., Sasaki, K., Honjou, T., and Hatano, S. "A diesel oxidation catalyst for exhaust emissions reduction," SAE Truck and Bus Symposium, Detroit, MI, (1993).

Gieshoff, J., Schäfer-Sindlinger, A., Spurk, P., van den Tillaart, J., and Garr, G. "Improved SCR systems for heavy-duty applications," SAE 2000-01-0189 (2000).

Girard, J., Montreuil, C., Kim, J., Cavataio, G., and Lambert, C. "Technocal advantages of vanadium SCR systems for diesel NO_x control in emerging markets," SAE 2008-01-1029 (2008).

Green Car Congress. "Toyota enhancing its diesel particulate NO_x reduction system for smaller light-duty vehicles," http://www.greencarcongress.com/, (2007).

Greszler, A. "Clean diesel development for heavy duty vehicles," CAPCOA, (2004).

Guyon, M., Blejean, F., Bert, C., and Le Faou Ph. "NO_x traps," SAE 982607 (1998).

Hachisuka, I., Hirata, H., Ikeda, Y., and Matsumoto, S. "Deactivation mechanism of NO_x storage-reduction catalyst and improvement of its performance," SAE 2000-01-1196 (2000).

Halstrom, K. "Catalyst based diesel emission control technology," AVEC Conference, (2004).

Hamada, I., Kato, Y., Imada, N., Fujisawa, M., Yamada, A., and Mukai, T. "A preliminary evaluation of unregulated emissions during low temperature operation of a small diesel engine with a multi-function SCR catalyst," SAE 2006-01-0641 (2006).

Hammer, T., and Broer, S. "Plasma enhanced selective catalytic reduction of NO_x for diesel cars," SAE 982428 (1998).

Hammerle, R. "Urea SCR and DPF system for diesel sport utility vehicle meeting tier 2 bin 5," Diesel Engine Emission Reduction Conference, DEER 2004, (2004).

Hawkins, P., Myers, W., Huthwohl, G., Vogel, H., Bates, B., Magnussson, L., and Bronnenberg, P. "Experiments with a new particular trap technology in Europe," SAE 970182 (1997).

Hawkins, P., Huthwohl, G., Henns, J., Koch, W., Luders, H., Lueng, B., and Stommel, P. "Effective continuous regeneration of diesel particulate filter in non-regulated emissions and particle size distributions," SAE 980189 (1998).

Hepburn, J., Thanasiu, E., Dobson, D., and Watkins, W. "Experimental and modeling investigations of NO_x trap performance," SAE 962051 (1996).

Hirata, K., Masaki, N., Ueno, H., and Akagawa, H. "Development of urea-SCR system for heavy-duty commercial vehicles," SAE 2005-01-1860 (2005).

Horiuchi, M., Ikeda, Y., and Sato, K. "Exhaust gas purification catalyst," U.S. Patent 5,000,929 (1991).

Hu, H., "Advanced NO_x aftertreatment system for commercial vehicles," SAE 2006-01-3552 (2006).

Huang, Y., Dang, Z., and Barllan, A. "Catalyzed diesel particulate matter filter with improved thermal stability," U.S. Patent 7,138,358 (2006).

Hunnekes, E., van, der Heijden, P., and Patchett, J. "Ammonia oxidation catalysts for mobile SCR systems," SAE 2006-01-0640 (2006).

ICT. "2007 Global Emission Standards," International Catalyst Technology, http://www.ictcatalyst.com/emission.html, (2007).

Jackson, M., Venkatesh, S., and Fable, S. "Supplying urea for the on-road vehicle market," 2001 Diesel Engine Emissions Reduction (DEER) Conference Presentations (2001).

Jelles, S., Makkee, M., and Moulijin, J. "Diesel particulate control-application of an activated particulate trap in combination with fuel additives at an ultra low dose rate," SAE 1999-01-0113 (1999).

Jochheim, J., Hesse, D., Duesterdiek, T., Engeler, W., Neyer, D., Warren, J., Wilkins, J., and Twigg, M. "Results from testing 4-way catalysts for diesel exhaust treatment," SAE 962042 (1996).

Johannes, E., Li, X., and Towgood, P. "Performance of a non-catalytic syngas generator for LNT and DPF regeneration," SAE 2007-01-1238 (2007).

Johnson, T. "Diesel emission control—last 12 months in review," SAE 2000-01-2817 (2000).

Johnson, T. "Diesel emission control technology-2003 in review," SAE 2004-01-0070 (2004).

Johnson, T. "Diesel emission control in review," SAE 2007-01-0233 (2007).

Johnson, T. "Diesel engine emissions and their control," *Precious Metal Reviews* 52(1): 23–37, (2008).

Joubert, E., Seguelong, T., and Weinstein, N. "Review of SCR technologies for diesel emission control: European experience and worldwide perspectives," 2004 Diesel Engine Emissions Reduction (DEER) Conference Presentations, (2004).

Kimura, K., Alleman, T., Hallstrom, K., and Chatterjee, S. "Long-term durability of passive "diesel particulate filters on heavy-duty vehicles," SAE 2004-01-0079 (2004).

Kittelson, D. "Ultrafine and nanoparticle emissions: An ongoing challenge," CAPCOA 2004, (2004).

Konno, M., Chikahisa, T., Murayama, T., and Iwamoto, M. "Catalytic reduction of NO_x in actual diesel engine exhaust," SAE 920091 (1992).

Konstandopoulos, A., Kostoglou, M., Skaperdas, E., Papaioannou, E., Zarvalis, D., and Kladopoulou, E. "Fundamental studies of diesel particulate filters: Transient loading, regeneration and aging," SAE 2000-01-1016 (2000).

Kowatari, T., Hamada, Y., Amou, K., Hamada, I., Funabashi, H., Nakagome, K., and Takakura, T. "A study of a new aftertreatment system (1): A new dosing device for enhancing low temperature performance of urea-SCR," SAE 2006-01-0642 (2006).

Krantz, K., and Senkan, S. "Systematic evaluation of monometallic catalytic materials for lean-burn NOx reduction using combinatorial methods," *Catalysis Today* 98: 413–421 (2004).

Kubsh, J. "Advanced emission controls for natural gas vehicles," DOE Natural Gas Vehicle Technology Forum, Manufacturers of Emission Controls Association, www.meca.org) (2003).

Kubsh, J. "Retrofit emission controls for off-road diesel engines," ARB Off-Road Fleet Workshops, www.meca.org, www.dieselretrofit.org, (2006).

Lambert, C. "Urea SCR and DPF system for diesel sport utility vehicle meeting tier 2 bin 5," 2005 Diesel Engine Emissions Reduction (DEER) Conference Presentations (2005).

Lambert, C. "Combined SCR/DPF system for tier 2 LDT," Ninth CLEERS Workshop, (2006).

Lampert, J., Kazi, S., and Farrauto, R. "Palladium catalyst performance for methane emissions abatement from lean burn natural gas engines," *Applied Catalysis B: Environmental* 14(2,3): 211 (1997).

Li, Y., and Armor, J. "Catalytic reduction of NO_x using methane in the presence of oxygen," *Applied Catalysis B: Environmental* 1: L31–40 (1992).

Liu, Y., Dettling, J., Weldlich, O., Krohn, R., Neyer, D., Engler, W., Kahman, G., and Dore, P. "Smart catalyst technology for 4-way conversion of diesel exhausts," SAE 962048 (1996).

Macaudiere, P., Woulters, P., Kunstmann, O., Thompson, J., and York, C. "Improvement and simplification of diesel particulate filter system using a ceria-based fuel-borne catalyst in serial applications," 2004 Diesel Engine Emissions Reduction (DEER) Conference Presentations (2004).

McCormick, R., Newlin, A., Mowery, D., and Grabowski, M. "Abating methane from natural gas fueled lean burn engines," SAE 961976 (1996).

McKinnon, D., Tadrous, T., Shepard, D., and Pavlich, D. "Results of North American field trials using filters with a copper additive for regeneration," SAE 940455 (1994).

MECA Manufacturers of Emission Controls Association. "The impact of sulfur in diesel fuel on catalyst emission control technology," www.meca.org, (1999a).

MECA Manufacturers of Emission Controls Association. "Demonstration of advanced emission control technologies enabling diesel powered heavy-duty engines to achieve low emission levels," www.meca.org, (1999b).

MECA. "Diesel particulate filter maintenance: Current practices and experience," www.meca.org, (2005).

Mercedes-Benz, "Mercedes-Benz at the North American International Auto Show 2006: GL-Class and E 320 BLUETEC in the spotlight," http://wwwsg.daimlerchrysler.com/SD7DEV/GMS/TEMPLATES/GMS_PRESS_RELEASE/0,2941,0-1-75909-1-1-text-1-0-0-0-0-0-0-0-0-0,00.html, (2006).

Miller, W., Klein, J., Mueller, R., Doelling, W., and Zuerbig, J. "Urea selective catalytic reduction," *Automotive Engineering International*, pp 125–128 (2000).

Misono, M. "The state of lean-NO_x technology with hydrocarbons," *CATTECH* 4: 183 (1998).

Miyoshi, M., Matsumoto, S., Katoh, K., Tanaka, T., Harada, J., Takahashi, N., Yokato, K., Sigiura, M., and Kasahara, K. "A new approach to NO_x reduction in lean burn engines," SAE 950809 (1995).

Morita, T., Suzuki, N., Wada, K., and Ohno, H. "Study on low NO_x emission control using newly developed lean NO_x catalyst for diesel engines," SAE 2007-01-0239 (2007).

Nakatani, K., Hirota, S., Takeshima, S., Itoh, K., Tanaka, T., and Dohmae, K. "Simultaneous PM and NO_x reduction system for diesel engines," SAE 2002-01-0957 (2002).

Nakatsuji, T., Yasaukawa, R., Tabata, K., Ueeda, K., and Niwa, M. "Regenerating sulfated NO_x traps," *Applied Catalysis B: Environmental* 21: 121 (1999).

Neitz, A. "Development potential for improvement of exhaust quality of commercial vehicle diesel engines. In worldwide engine emission standards and how to meet them," Proceedings of the International Seminar of the International Mechanical Engineering Society, London, England, (1991).

NGK. "Technical Trend, Usage and Performance of Smart My Season," CES-041416, NGK Internation Ltd. (2005).

NGK, "OBD Source: NGK Insulators presentation to EPA, CARB & SwRI," (2006).

Nissan, http://www.nissan-global.com/EN/NEWS/2007/_STORY/070806-01-e.html, (2007).

Peter-Hoblyn, J., and Valentine, J. "Method for reducing emissions from a diesel engine," U.S. Patent 6,023,928 (1998).

Petersson, M., Holma, T., Andersson, B., Jobson, E., and Palmqvist, A. "Lean hydrocarbon selective catalytic reduction over dual pore system zeolite mixtures," *Journal of Catalysis* 235(20050): 114–127 (2005).

Pfeifer, A., Krueger, M., and Tomazic, D. "U.S. 2007-Which Way to Go?-Possible Technical Solutions," SAE 2003-01-0770 (2003).

Pfeifer, M., Votsmeir, M., Spurk, P., Kogel, M., Lox, E., and Knoth, J. "The second generation of catalyzed diesel particulate filter systems for passenger cars—particulate filters with integrated oxidation catalyst function," SAE 2005-01-1756 (2005).

Pfeifer, M., and Kreuger, M. "U.S. 2007-Which Way to Go?-Possible Technical Solutions," SAE 2003-01-0770 (2003).

Pieterse, J., and Booneveld, S. "Catalytic reduction of NO_x with $H_2/CO/CH_4$ over PdMOR catalysts," *Applied Catalysis B: Environmental* 73: 327–335 (2007).

Porter, B., Doyle, D., Faulkner, S., Lambert, P., Needham, P., Andersson, S., Fredholm, S., and Frestad, A. "Engine and catalyst strategies for 1994," SAE 910604 (1991).

Punke, A., Grubert, G., Li, Y., Dettling, J., and Neubauer, T. "Catalyzed soot filters in close coupled position for passenger vehicles," SAE 2006-01-1091 (2006).

Querini, C., Uullsa, M., Requejo, F., Soria, J., Sedran, U., and Miro, E. "Catalytic combustion of diesel soot particles," *APCAT B*: 15: 5–19 (1998).

Rice, G., Deeba, M., and Feeley, J. "NO_x abatement for diesel engines: Reductant effects; Engine vs. reactor tests," SAE 962043 (1996).

Richards, P., Terry, B., Vincent, M., and Cook, S. "Assessing the performance of diesel particular filters with fuel additives for enhanced regeneration," SAE 1999-01-0112 (1999).

Richards, P., Terry, B., Vincent, M., and Chadderton, J. "Combining fuel borne catalyst, catalytic washcoat and diesel particulate filter," SAE 2001-01-0902 (2001).

Samsa, M. "Potential for compressed natural gas vehicles in centrally fueled automobile, truck and bus fleet applications," GRI Report-91/0183, (1991).

SCAQMD. "Demonstration of particulate trap technologies on existing off-road heavy-duty construction equipment," (2005).

Schubert, R., and Fable, S. "Comparative costs of 2010 heavy-duty diesel and natural gas technologies—final report," Report to the California Natural Gas Vehicle Partnership, South Coast Air Quality Management District and Southern California Gas Company, 2005, http://www.tiaxllc.com/reports/HDDV_NGVCostComparison Finalr3.pdf, (2006).

Seguelong, T., and Joubert, E. "Diesel particulate filters market introduction in Europe: Review and status," 2004 Diesel Engine Emissions Reduction (DEER) Conference Presentations (2004).

Seguelong, T., Rigaudeau, C., Blanchard, G., Salvat, O., Colignon, C., and Griard, C. "Passenger car series application of a new diesel particulate filter system using a new ceria-based, fuel-borne catalyst: From the engine test bench to European vehicle certification," SAE 2002-01-2781 (2002).

Shirk, R., Bloom, R., Kitahara, Y., and Shinzawa, M. "Fiber wound electrically regenerable diesel particulate filter cartridge for small diesel engines," SAE 950153 (1995).

Shoji, A., Kamoshita, S., Watanabe, T., Tanaka, T., and Yabe, M. "Development of a simultaneous reduction system of NO_x and particulate matter for light-duty truck," SAE 2004-01-0579 (2004).

Simonsen, S., Dahl, S., Johnson, E., and Helveg, S. "Ceria-catalyzed soot oxidation studied by environmental transmission electron microscopy," *Journal of Catalysis* 255: 1–5 (2008).

Solvat, O., Marez, P., and Belot, G. "Passenger car serial application of a particulate filter system on a common-rail, direct-injection diesel engine," SAE 2000-01-0473 (2000).

Song, C., Hsu, C., and Mochida, I. *Chemistry of Diesel Fuel*, Applied Energy Technology Series, Taylor and Francis, Oxford, England, (2000).

Standt, U., and Konig, A. "Performance of zeolite—based diesel catalysts," SAE 950749 (1995).

Stang, J. "Cummins work toward successful introduction of light-duty clean diesel engines in the U.S.," 2005 Diesel Engine Emissions Reduction (DEER) Conference, Chicago, (2005).

Stanmore, B., Prilhac, J., and Gilot, P. "The ignition and combustion of cerium doped diesel soot," SAE 1999-01-0115 (1999).

Summers, J., Van Houtte, S., and Psaras, D. "Simultaneous control of particulate and NO_x emissions from diesel engines," *Applied Catalysis B: Environmental* 10(1–3): 139 (1996).

Syed, M. "Diesel exhaust emission control for the Asian market," AVECC 2004, http://www.meca.org/page.ww?name=Asian+Vehicle+Emission+Control+Conference+2004&section=Resources, (2004).

Tao, T., Cutler, W., Voss, K., and Wei, Q. "New catalyzed cordierite diesel particulate filters for heavy duty engine applications," SAE 203-01-3166 (2003).

Thomas, J., Lewis, S., Bunting, B., Storey, J., Graves, R., and Park, P. "Hydrocarbon selective catalytic reduction using a silver-alumina catalyst with light alcohols and other reductants," SAE 2005-01-1082 (2005).

Thomas, S., Martin, A., Raybone, D., Shawcross, J., Ng, K., and Whitehead, J. "Non thermal plasma aftertreatment of particulates-theoretical limits and impact on reactor design," SAE 2000-01-1926 (2000).

Toorisaka, H., Narita, H., Minamikawa, J., Muramatsu, T., Kominami, T., and Sone, T. "DPR developed for extremely low PM emissions in production commercial vehicles," SAE 2004-01-0824 (2004).

Verkiel, M., Verbeek, R., and van Aken, M. "DAF Euro-4 heavy-duty diesel engine with TNO EGR system and CRT particulates filter," SAE 2001-01-1947 (2001).

Vogtlin, G., Merritt, B., Hsiao, M., Wallman, P., and Penetrante, B. "Plasma-assisted catalytic reduction system," U.S. Patent 5,711,147 (1998).

Voss, K., Rice Gary, Y. B., Hirt, C., and Farrauto, R. "Performance characteristics of a novel diesel oxidation catalyst," SAE 940239 (1994a).

Voss, K., Lampert, J., Farrauto, R., Rice, G., Punke, A., and Krone, R. "Catalytic oxidation of diesel particulates with base metal oxides," Third International Congress on Catalysis and Automotive Pollution Control, Brussels, Belgium, (1994b).

Walker, A. "Diesel emission control: past, present and future," 19th NACSM, Philadelphia, PA (2005).

Walker, A., Blakeman, P., Ilkenhans, T., Magnusson, B., McDonald, A., Kleijwegt, P., Stunnenberg, F., and Sanchez, M. "The development and in-field demonstration of highly durable SCR catalyst system," SAE 2004-01-1289 (2004).

Watanabe, S. "An improvement of diesel PM and NO_x reduction system," 2005 Diesel Engine Emissions Reduction (DEER) Conference Presentations," Chicago IL (2005).

West, B., Thomas, J., Kaas, M., Storey, J., and Lewis, S. "NO_x adsorber regeneration phenomena in heavy-duty applications," 2003 Diesel Engine Emissions Reduction (DEER) Conference Presentations (2003).

White, J. "Low emission catalysts for natural gas engines," GRI Report-91/0214, SwRI-3178-2.2, (1991).

White, J., Carroll, J., Brady, M., Burkmyre, W., Liss, W., and Church, M. "Natural gas converter performance and durability," SAE 930222 (1993).

Wyatt, M., Manning, W., Roth, S., D'Aniello, M., Andersson, E., and Fredholm, S. "The design of flow-through diesel oxidation catalysts," SAE 930130 (1993).

Yavuz, B., Voss, K., Deeba, M., Adomaitis, J., and Farrauto, R. "Zeolite containing oxidation catalyst and method of use," U.S. Patent 6, 248, (2001).

Yezerets, A., Currier, N., and Eadler, H. "Experimental determination of the kinetics of diesel soot oxidation by OD2—modelling consequences," SAE 2003-01-0833 (2003).

Zelenka, P., Kriegler, W., Herzog, P., and Cartellieri, W. P. "Ways toward the clean heavy-duty diesel," SAE 900602 (1990).

Zelenka, P., Egert, M., and Cartellierl, W. "Meeting future standards with diesel SUVs," *Automotive Engineering International* pp 146–150 (2001).

CHAPTER 8 QUESTIONS

1. What are the issues with the emissions from a diesel engine?
2. What are the functions of a diesel oxidation catalyst (DOC)?
3. Why does a DPF have to be regenerated? List all possible methods to achieve this regeneration.
4. List the methods that have been tried for NO_x reduction in the lean diesel exhaust?
5. Explain how an LNT works. Design an LNT.
6. What are the system design issues for implementation of the LNT technology?
7. Describe a catalytic system for NO_x and PM control from a diesel exhaust. Draw the system and auxiliary equipment needed for implementation.
8. A technology describes the combustion of soot from a diesel soot filter called the continuous regenerable trap (CRT). It operates on the prin-

ciple that NO_2 can directly oxidize the soot (C) forming CO_2 and NO at temperatures as low as $250\,°C$. The NO_2 is produced by the oxidation of NO using a Pt catalyst.

a. Is it best to put the Pt on a separate monolith upstream of the filter or directly on the wall-flow soot filter? What are the advantages and disadvantages to both locations of the Pt?

b. How would you simulate this in the lab?

c. How would you manage the NO generated in the trap?

9.

a) Describe how the NO_x trap works in a diesel engine?

b) What are the anticipated deactivation modes?

c) What are the pollutants that could be liberated when the engine operates too rich (lambda much less than 1).

10.

a) What happens when the soot builds on the filter?

b) How is the soot regenerated?

c) Why doesn't the catalyst decrease the ignition temperature?

d) Why add Pt?

e) What deactivation modes are expected for filter and catalyst?

11. Please refer to the article by Farrauto and Voss, "Monolith diesel oxidation catalysts," *Applied Catalysis B: Environmental*, 10: 29–51 (1996). Note that emissions for diesel trucks are measured in units related to work expressed as grams per brake horse power—hour (g/bhp-h). Total particulate matter (TPM) is composed of solids (dry soot and small amounts of sulfates) and liquids [diesel fuel + oil = soluble organic fraction (SOF)]:

a) What can you conclude from Table 1 of *Applied Catalysis B: Environmental*, 10: 29–51 (1996) regarding the emissions and regulations?

b) What was the role of the diesel oxidation catalyst (DOC)?

12.

a) Describe the laboratory test the authors developed for screening candidate catalysts (see Table 2 of *Applied Catalysis B: Environmental*, 10: 29–51 (1996).

b) Explain the data generated and how it was used to evaluate the effectiveness of the catalyst candidates.

13.

a) Why do regulations vary with vehicle and country for trucks and passenger cars?

b) Give some names of the different tests.

14. What was the main conclusion for the best catalyst in Table 2 of *Applied Catalysis B: Environmental*, 10: 29–51 (1996).

    a) What were the catalytic results when a cerium solution was impregnated into Al_2O_3 and calcined? This would make the surface rich in CeO_2.

15.

    a) Explain the importance of Figure 5 of *Applied Catalysis B: Environmental*, 10: 29–51 (1996). What does Figure 8 in this article tell us about poisoning?

    b) Why is $Pt/(CeO_2 + Al_2O_3)$ considered a dual-function catalyst?

16. How was the truck catalyst modified to meet passenger car standards in Europe? Refer to Figures 9 and 10 of *Applied Catalysis B: Environmental*, 10: 29–51 (1996).

17. Explain the mechanism of catalysis using the CeO_2 Al_2O_3 catalyst referring to Figure 11 in *Applied Catalysis B: Environmental*, 10: 29–51 (1996).

18.

    a) Describe the NO_x-particulate trade-off for diesel emissions?

    b) What catalysts and operational procedures are to be used to regenerate soot filters?

    c) Why can't a TWC be used to abate NO_x from a diesel engine?

    d) What unique performance do V_2O_5 and urea (NH_3) have that make them useful for NO_x reduction in diesel exhausts?

9 Diesel Catalyst Supports and Particulate Filters

9.1 INTRODUCTION

Since the invention by Rudolf Diesel in 1893, the application of the diesel engine has become widespread across the world. The popularity of the diesel engine is a result of its attractive characteristics such as fuel economy, engine durability, low maintenance requirements, and large indifference to fuel specification. Fuel efficiency for the diesel engine is 30% to 50% higher than that for the gasoline engine with comparable power output. In other words, CO_2 emissions will be 30% to 50% lower for the diesel engine for the same amount of power output. Since CO_2 is claimed to be one of the greenhouse gases that may contribute to global warming, the transition from a gasoline-powered engine to a diesel-powered engine is a logical choice for maintaining mobility via transportation. Transport applications of the diesel engine are found in passenger cars, trucks, construction equipment, and ships in addition to stationary power sources. Many electricity and hydraulic power plants are equipped with diesel engines. Unfortunately, however, diesel engines encounter incomplete combustion that results in emission of undesirable pollutants.

During operation, the diesel fuel is injected into the cylinder where it atomizes into small droplets that vaporize and mix with intake air under pressure and combust to deliver power. In general, the fuel distribution is nonuniform and leads to unwanted emissions. Carbonaceous soot is formed in the center of fuel spray where the air/fuel ratio is low. Nonoptimized mixing of fuel and air creates small pockets of excess fuel where the carbonaceous soot particles, comprising both a solid and a soluble organic fraction (SOF), are formed (Yanowitz et al. 2000; Abdel-Rahman 1998; Heywood 1988). Associated with carbonaceous soot, adsorbed hydrocarbons and small amounts of sulfates, nitrates, metals, trace elements, water, and unidentified compounds make up the so-called particulate matter (PM). The adsorbed hydrocarbons, sulfates, and water act like "glue" causing multiple particles to agglomerate, thereby shifting the particle size and mass distribution upward (Mark and Morey 1999). Together with particulate emissions, CO, hydrocarbons (HCs), and NO_x are also emitted as gaseous pollutants.

Catalytic Air Pollution Control: Commercial Technology, by Ronald M. Heck and Robert J. Farrauto, with Suresh T. Gulati.
Copyright © 2009 John Wiley & Sons, Inc.

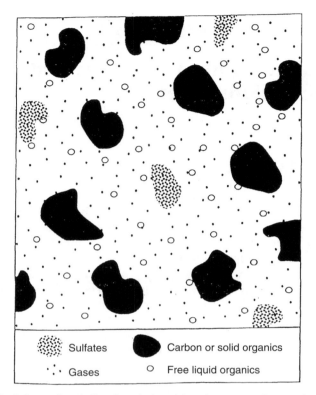

Sulfates

Gases

Carbon or solid organics

O   Free liquid organics

FIGURE 9.1. Schematic of diesel emissions showing approximate sizes. Reprinted from Gulati 1996b; courtesy of Marcel Dekker, New York.

Thus, diesel exhaust consists of gaseous, liquid, and solid emissions. Gaseous emissions, whose approximate size is shown in Figure 9.1, comprise N_2, CO_2, CO, H_2, NO/NO_2, SO_2/SO_3, $HC(C_2-C_{15})$, oxygenates, and organic nitrogen and sulfur compounds. Liquid emissions include H_2O, H_2SO_4, $HC(C_{15}-C_{40})$, oxygenates, and polyaromatics. Solid emissions are made up of dry soot, metals, inorganic oxides, sulfates, and solid hydrocarbons. The physical and chemical processes responsible for soot formation are illustrated in Figure 9.2, which also shows the soot size and production time. Similarly, the formation of diesel particulate, with soot carbon at the center surrounded by other organic and inorganic combustion products, is illustrated in Figure 9.3. The relative contribution of the above emissions to total emissions in Europe, where diesels are most popular, has been estimated as follows (Lox et al. 1990): CO at 2.2%, HC at 6.5%, NO_x at 23%, and PM at 11.4%. As noted, particulates originate from soot, sulfates, and hydrocarbons (from both fuel and oil). Two methods are commonly employed for reducing the particulate matter from diesel engines: diesel oxidation catalysts (DOCs) and diesel particulate filters (DPFs)

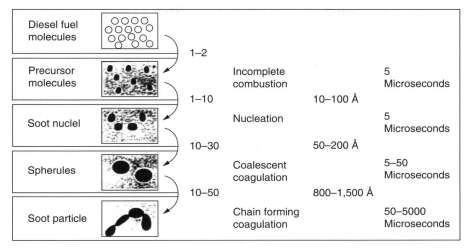

FIGURE 9.2. Physical and chemical processes leading to soot formation. Reprinted from Gulati 1996b; courtesy of Marcel Dekker, New York.

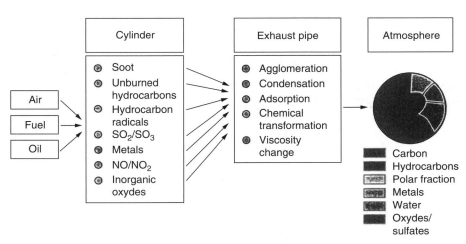

FIGURE 9.3. Physical and chemical processes responsible for diesel particulate formation. Reprinted from Gulati 1996b; courtesy of Marcel Dekker, New York.

or traps. The supports used in these applications are completely different. The DOC oxidizes SOF, whereas the DPF traps these particulates, through a wall-flow filter, ceramic fiber filter, or ceramic foam.

With advances in engine design, low sulfur fuel, sensor technology, and ceramic compositions, the DOCs and DPFs are becoming standard equipment for control of particulate emissions from passenger cars, buses, and trucks.

9.2   HEALTH EFFECTS OF DIESEL PARTICULATE EMISSIONS

Particulate matter from diesel engines emitted directly into the air is one of the origins of air pollution. Together with biomass combustion, fuel combustion contributes to excessive soot particles at the lower troposphere level (Cooke and Wilson 1996). In urban areas, where the exposure to diesel exhaust may be especially high, diesel engines become a major source of PM (Faiz et al. 1996).

The presence of soot has serious consequences for human health. In general, particles inhaled by humans are segregated by size during deposition within the respiratory system. Larger particles deposit in the upper respiratory tract, whereas smaller inhalable particulates travel deeper into the lungs and are retained for longer periods of time. When present in greater numbers, smaller particles have a greater total surface area than larger particles of equivalent mass. Hence, the toxic material carried by small particles is more likely to interact with cells in the lungs than that carried by larger particles (Health Effect Inst. 2002; Farleigh and Kaplan 2000).

Diesel PM, smaller than $10\,\mu m$, PM_{10}, not only penetrates deeper and remains longer in the lungs than larger particles, but also it contains large quantities of organic materials that may have significant long-term health effects. Both linear- and branched-chain hydrocarbons with 14 to 35 carbon atoms, together with polynuclear aromatic hydrocarbons (PAHs), alkylated benzenes, nitro-PAHs, and a variety of polar, oxygenated PAH derivatives are common, particulate-bound compounds. Diesel emissions legislation has forced vehicle manufacturers to comply and reduce the levels of diesel emissions over time as shown in Table 9.1.

9.3   DIESEL OXIDATION CATALYST SUPPORTS

Recent advances in diesel technology have lowered exhaust particulates significantly such that DOCs provided the required incremental particle removal for MY 1994[+] vehicles. The diesel oxidation catalysts use flow-through cordierite substrates with a large frontal area (see Figure 9.4) (Gulati 1992b). Depending on the type of engine and its exhaust, they oxidize 30–80% of the gaseous HC and 40–90% of the CO present. They do not alter NO_x emissions. DOCs have been used in more than 60,000 diesel forklift trucks and mining vehicles since 1967 for HC and CO emissions (Farrauto et al. 1992).

DOCs have little effect on dry soot (carbon), but engine tests show that they typically remove 30–50% of the total particulate load. This is achieved by oxidizing 50–80% of the SOFs present. DOCs are less effective with "dry" engines in which particulates have a very low SOF content (Farrauto et al. 1992).

Diesel oxidation catalysts are different from those used for gasoline (see Chapter 8 "Diesel Engine Emissions"); however, they do use a monolithic

TABLE 9.1. **U.S. Diesel engine emission standards (g/kWh): Past, present, and future.**

Year	THC	CO	NO$_x$	Particulate matter (trucks)	Particulate matter (urban buses)
1974–1978	21.5	53.6	—	—	—
1979–1984	2.0	33.5	13.4[a]	—	—
1985[b]–1987	1.8	20.8	10.7	—	—
1988–1989	1.8	20.8	10.7	0.8	0.8
1990	1.8	20.8	8.1	0.8	0.8
1991–1992	1.8	20.8	6.7	0.3	0.3
1993	1.8	20.8	6.7	0.13	0.13
1994–1995	1.8	20.8	6.7	0.13	0.09
1996–1997	1.8	20.8	6.7	0.13	0.07
1998–2003	1.8	20.8	5.4	0.13	0.07
2004	—	20.8	3.4[a]	0.13	0.07
2007	—	20.8	1.48[a]	0.014	0.014
2010	—	20.8	0.27	0.014	0.014

[a]THC + NO$_x$ (THC = total hydrocarbons).
[b]Test cycle changed from steady state to transient operation.

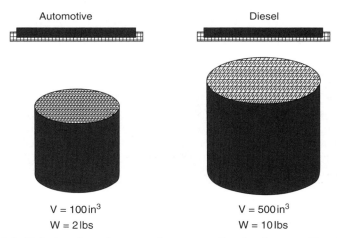

FIGURE 9.4. Relative size of automotive versus diesel substrates. Reprinted from Gulati 1996b; courtesy of Marcel Dekker, New York.

honeycomb support. Gases flow through the honeycomb with minimum pressure drop and react with the catalyst on the walls of the channels.

Although both metal and ceramic supports have been used, ceramic substrates offer stronger catalyst adhesion, less sensitivity to corrosion, and lower cost. The use of ceramics in automotive converters gives additional confidence to their performance.

Catalytic sites promote the reaction between HC gases, including those that would condense as SOFs downstream, and oxygen to form carbon dioxide and water. The sites also oxidize liquid SOFs, whether they are droplets that contact the catalyst or SOF gases that adsorb on them. SOFs adsorbed or condensed on the porous carrier are volatilized and then oxidized at the catalytic sites.

DOCs must function in a demanding environment. Although diesel exhaust temperatures are well below those in auto exhaust (200 °C to 600 °C vs. 300 °C to 1,000 °C), diesel catalysts must contend with solids, liquids, and gases (not just gases) and with deposits of non-combustible additives from lubricating oil. The latter contain zinc, phosphorous, antimony, calcium, magnesium, and other contaminants that can shorten catalyst life below that mandated. Contamination also can come from sulfur dioxide in the exhaust.

Catalyst life might be extended by the development of low-ash lubricating oils and by modifying carrier properties such as surface chemistry, pore structure, and surface area, to create contamination-resistant catalysts. Another avenue is periodic regeneration to remove contaminants.

The physical durability of DOCs depends heavily on mechanical and thermal properties of catalyzed substrates and their operating conditions (see Table 9.2a). Since the diesel exhaust is substantially colder than that from a gasoline engine, and since the conversion temperature for diesel emissions is considerably lower, the thermal stresses associated with radial and axial tem-

TABLE 9.2. a) Operating conditions for automotive versus diesel catalysts.

	Automotive	Diesel
Temperature range	300–1,100 °C	100–550 °C
Temperature gradient	100–300 °C	100–200 °C
Relative humidity	<100%	100%
Space velocity	30,000–100,000 hr^{-1}	60,000–150,000 hr^{-1}
Vibration acceleration	28 g's	10–20 g's

TABLE 9.2. b) LFA substrate sizes available in different cell geometries.

Size (inch)	200/12	300/8	400/7	300/5	400/4
Diameter × length					
7.5 × 7	•	•	•		
9.0 × 6	•	•	•	•	
9.5 × 6	•	•	•		
10.5 × 6	•	•	•	•	•
11.25 × 6		•	•		
12.0 × 6		•	•		
12.5 × 8				•	

perature gradients in a large frontal area (LFA) converter are well below the threshold strength, thereby eliminating the thermal fatigue potential and ensuring crack-free operation over the required 435,000 vehicle miles. With thermal durability under control, the mechanical durability takes on a major focus to ensure total durability. To this end, it is necessary to ascertain the high mechanical strength of the coated monolith, build in a resilient packaging system, and ensure positive and moderately high mounting pressure to guard against vibrational and impact loads. Much like automotive catalysts, the DOCs can continue to function catalytically even in the fractured state as long as there is a sufficient mounting pressure to keep the cracks shut and adequate catalytic activity to oxidize organic particulates over the required 435,000 vehicle miles; i.e., the packaging design is just as important as catalyst formulation to meet the 435,000 vehicle mile durability.

The standard cell geometry for LFA DOC support is 300/8. Additional cell geometries for LFA substrates include 200/12, 400/7, 400/4, and 300/5. The latter substrate, with thinner walls, offers the additional advantages of 15% lower pressure drop and 5% higher geometric surface area than the standard 300/8 DOC support. Table 9.2b) provides the standard sizes available for LFA DOC supports.

The circular contour of the DOC is ideal from a packaging point of view because it experiences a uniform mounting pressure and an axisymmetric temperature distribution, both of which are beneficial to long-term durability. However, since the DOC is both larger and heavier and may experience different vibrational loads than the automotive catalyst, its mounting design requires special considerations. Moreover, since its operating temperature is lower than the intumescent temperature of a ceramic mat, the converter assembly may have to be preheated to remove the organic binder and to ensure adequate mounting pressure before installation (Stroom et al. 1990).

Finally, since packaging plays a key role in preserving the mechanical integrity of the diesel converter, both the mat thickness and its mounting density must be carefully tailored to provide substantial mounting pressure on the monolith to meet the 435,000-mile durability. Guided by the successful packaging designs for European automotive converters and North American heavy-duty gasoline truck converters, both subjected to harsher driving conditions, a 6,200-g/m^2 mat with a mount density of 1 g/cm^3 would be a good starting point. Such a design would result in a nominal mounting pressure of 50 psi, which is ten times the minimum required value. It would also enhance the initial tangential strength of the DOC and help contain any partial fragments over the required 435,000 vehicle miles should the converter experience any cracking. Additional mounting support, if necessary, may be provided in the axial direction by welding two end rings to the can at the entry and exit faces. The loss of active catalytic area in the peripheral cells, however, should be weighed against the improved mounting design afforded by the additional axial constraint.

9.4   DESIGN/SIZING OF A DIESEL PARTICULATE FILTER

The ceramic wall-flow filter is an innovative extension of extruded honeycomb catalyst support described in Chapter 7. The filter is a critical component in the continuous regenerable trap (CRT™) in which NO is oxidized upstream with a Pt-based catalyst and the product NO_2 reacts and combusts the soot trapped on the walls of the filter. This technology is thoroughly discussed in Chapter 8 ("Diesel Engine Emissions").

This filter concept, shown in Figure 9.5, involves having the alternate cell openings on one end of the unit plugged in checkerboard fashion. The opposite end or face is plugged in a similar manner, but one cell displaced, allowing no direct path through the unit from one end to the other as indicated in Figure 9.6. The exhaust gas entering the upstream end is therefore forced through the porous wall separating the channels and exits through the opposite end by way of an adjacent channel. In this way, the walls of the honeycomb are the filter medium (Howitt and Montierth 1981; Howitt et al. 1983). They can be made sufficiently porous to allow exhaust gas to pass through without excessive

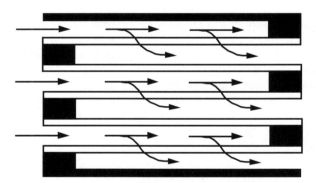

FIGURE 9.5. Wall-flow filter concept with alternate plugged cells. Reprinted from Gulati 1996b; courtesy of Marcel Dekker, New York.

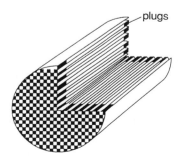

FIGURE 9.6. Schematic of a diesel filter with a checkerboard plug pattern. Reprinted from Gulati 1996b; courtesy of Marcel Dekker, New York.

pressure drop. This wall-flow concept offers a large amount of filter surface area in a reasonably compact volume together with high filtration efficiency. Periodically, the soot that is collected is oxidized to CO_2—a process known as regeneration—which renders the filter clean. The fact that the filter is constructed of special ceramic materials results in its capability of withstanding high temperatures while being chemically inert. One of these materials is a porous cordierite ceramic with magnesia/alumina/silica composition ($2MgO_2 \cdot 2Al_2O_3 \cdot 5SiO_2$). A key property of this composition is a low coefficient of thermal expansion. The material used to plug the cell openings in the faces is similar in nature to the body in composition and thermal characteristics. It is a high-temperature foaming cement that during firing seals to the cell walls and is impervious to gas flow. More advanced materials capable of withstanding even higher temperature are discussed in Section 9.7. The walls contain a series of interconnected pores of a volume and size sufficient to enable the exhaust gas to flow completely through but restrain most of the particles.

The functional characteristics of wall-flow filters can be varied and managed. Its collection efficiency can be controlled to a large degree by the properties of the walls that form the channels. These characteristics include total pore volume, pore size distribution, and the thickness of the wall itself. The flow through the wall can be made more restrictive by adjusting the porosity of the wall. A smaller pore volume creates a highly efficient filter but at the same time restricts the flow and produces high back pressure. Conversely, with porosity adjusted in the opposite direction, low back pressure is achieved, but at the expense of reduction in collection efficiency.

9.4.1 Functionality Requirements

The four basic requirements that the DPF must meet are as follows (Wade et al. 1981):

1. Adequate filtration efficiency to satisfy particulate emissions legislation
2. Low-pressure drop to minimize fuel penalty and conserve engine power
3. High thermal shock resistance to ensure filter integrity during regeneration
4. High surface area per unit volume for compact packaging

Although a high filtration efficiency would make the filter more effective, it must not be accomplished at the expense of high back pressure or low thermal integrity. Indeed, the microstructure and plugging pattern of a ceramic filter can be tailored to obtain filtration efficiencies ranging from 50% to 95% per engine manufacturers' specification. Furthermore, recent advances in ceramic composition have led to filters with high filtration efficiency, acceptable back pressure, and excellent thermal integrity (Murtagh et al. 1994; Cutler and Merkel 2000; Merkel et al. 2001b; Miwa 2001; Ohno et al. 2000).

As the particulate matter is trapped in the filter walls, it begins to build up on the surface of open cells forming a soot layer that also acts as a filter. With increasing thickness of the soot layer, the hydraulic diameter of the channel decreases resulting in higher back pressure. Obviously, the initial and final channel size must be controlled via filter design and soot accumulation level to limit the back pressure to an acceptable value. Again, this can be accomplished by designing the microstructure, the cell geometry, the plugging pattern, and the size of ceramic filter, which in turn are dictated by engine size, flow rate, and engine out emissions.

The dominant component of trapped particulates is soot carbon that is formed during combustion of a fuel-rich mixture in the absence of adequate oxygen. Although some soot may be oxidized to CO_2 during the latter part of a power stroke, a major portion does not get oxidized due to the slow process (Amann et al. 1980). The other major component of particulate matter consists of heavy unburned hydrocarbons. Since the chemical energy of soot carbon and heavy hydrocarbons is high, once they are ignited during regeneration, they release a great deal of heat that, if not dissipated continuously, can result in high-temperature gradients within the filter (Weaver 1984). The thermal stresses associated with such gradients must be kept below the fatigue threshold value of the filter material to ensure thermal integrity over its lifetime (Gulati and Helfinstine 1985; Gulati and Sherwood 1991). This is best accomplished by using a ceramic composition with ultra-low thermal expansion and a modestly high fatigue threshold value (Murtagh et al. 1994). Other approaches to improving thermal integrity include the use of fuel additives and/or catalysts to effect regeneration at lower temperatures (Weaver 1984). Alternatively, more frequent regenerations can also reduce the temperature gradients and enhance thermal integrity but at the expense of fuel penalty if a burner is used for regeneration.

The honeycomb configuration of ceramic filters offers high surface area per unit volume thereby permitting a compact filter size (Gulati 1996a). The absolute filtration surface area depends on cell size, filter volume, and plugging pattern, all of which are design parameters whose optimization, as will be shown shortly, calls for trade-offs in pressure drop, filtration efficiency, mechanical durability, thermal integrity, and space availability.

9.4.2 Composition and Microstructure

The filter composition that has performed successfully over the past two decades is cordierite ceramic with the chemical formula of $2MgO \cdot 2Al_2O_3 \cdot 5SiO_2$. Its unique advantages include low thermal expansion, ideal for thermal shock resistance, and tailorable microstructure to meet filtration and pressure drop requirements. The extrusion technology for producing automotive catalyst

TABLE 9.3. Properties of extruded cordierite diesel filters with 100/17 cell structure and checkerboard plugging pattern.

Filter designation	Wall porosity	Mean pore size (μm)	Open frontal area	Specific filtration area (in^2/in^3)
EX-47	50%	13.4	34.4%	16.6
EX-54	50%	24.4	34.4%	16.6
EX-66	50%	34.1	34.4%	16.6
DHC-221	48%	13.0	34.4%	16.6
C-356E	50%	20.0	34.4%	16.6
DHC-141E	48%	35.0	34.4%	16.6

supports also helps manufacture diesel filters. Consequently, the unit cell design can be achieved via die design, whereas the porosity and microstructure are best controlled by composition and process modifications.

The most common cell density employed for diesel filters is 100 cells/in^2 with a 0.017″ (0.043 cm) thick cell wall. This choice offers the best compromise in terms of filtration area and back pressure. Although a 200 cells/in^2 structure offers 41% larger filtration area and has been used for diesel filters, it results in a higher pressure drop. Similarly, thicker cell walls (0.025″ or 0.064 cm thick) offer 50% higher strength, but they too result in a higher pressure drop. Another parameter that affects pressure drop is mean pore size, which can range from 12 μm to 35 μm. Although the pressure drop decreases with increasing pore size, so does filtration efficiency. Hence, a compromise is necessary in tailoring the pore size, wall thickness, and cell density. The wall porosity also affects pressure drop and mechanical strength. Both pressure drop and mechanical strength decrease as the wall porosity increases, thus calling for a compromise in selecting the wall porosity. Most filter compositions and manufacturing processes are designed to yield a wall porosity of 48% to 50%. Table 9.3 summarizes the geometric and microstructural properties of six different candidates in the early development of diesel filters (Howitt and Montierth 1981; Murtagh et al. 1994; Kitagawa et al. 1990; 1992). Filters with low mean pore size, namely EX-47 and DHC-221, were designed to offer high filtration efficiency (>90%); those with intermediate mean pore size, namely EX-54 and C-356E, were designed for medium filtration efficiency (80–90%); and those with large mean pore size, namely EX-66 and DHC-141E, were designed for low filtration efficiency (60–75%).

9.4.3   Cell Configuration and Plugging Pattern

Figure 9.6 shows the wall-flow filter with square cell configuration and checkerboard plugging pattern. The open frontal area (OFA) and specific filtration area (SFA) for such a filter are defined in terms of cell spacing L and wall thickness t:

$$OFA = 0.5\left(\frac{L-t}{L}\right)^2 \tag{9.1}$$

$$SFA = \frac{2(L-t)}{L^2} \tag{9.2}$$

Since the cell density N for a square-cell structure is given by

$$N = \frac{1}{L^2} \tag{9.3}$$

it follows from Eq. (9.2) that the specific filtration area is directly proportional to cell density. On the other hand, as the cell density increases, the hydraulic diameter defined by

$$D_h = L - t \tag{9.4}$$

decreases. Hence, a portion of the total pressure drop due to gas flow through the open channels of the filter, which depends inversely on the square of hydraulic diameter, increases. Thus, care must be exercised in selecting the appropriate cell density (Gulati 1996a). Other factors that play a key role in designing the filter are its mechanical integrity and filtration capacity. The former is defined by the mechanical integrity factor (MIF), which for a given wall porosity depends on cell geometry via

$$MIF = \frac{t^2}{L(L-t)} \tag{9.5}$$

The filtration capacity is the total amount of soot that can be collected prior to safe regeneration. It is directly related to the total filtration area (TFA) defined by the product of a specific filtration area and filter volume; i.e.,

$$TFA = \frac{2(L-t)}{L^2}V_f \tag{9.6}$$

where the filter volume V_f is given by

$$V_f = \frac{\pi}{4}d^2\ell \tag{9.7}$$

in which d and l denote filter diameter and length, respectively.

As noted, most filter compositions enjoy 50% wall porosity to limit the pressure drop to acceptable levels. The mean pore size, which also has a bearing on pressure drop due to gas flow through the wall, is primarily dictated by filtration efficiency requirement. As the emissions legislation becomes more stringent, filtration efficiencies ≥90% become desirable calling for a mean pore diameter of 12 μm to 14 μm. With the microstructure fixed in this manner, the two common cell configurations for diesel filters that have been manufactured are 100/17 and 200/12. It may be verified that they have identical open frontal area and mechanical integrity factor. However, the specific filtration area of 200/12 is 41.5% greater than that of a 100/17 configuration, which implies a lower[1] filter volume for the former that may be desirable to meet space constraints. On the other hand, the hydraulic diameter of 200/12 is 30% smaller than that of a 100/17 configuration, which implies a higher pressure drop for the former that may not be acceptable. Furthermore, the 200/12 configuration may also experience fouling due to ash buildup after several regenerations. The model for total pressure drop will be discussed in a later section; however, for a comparison of two different cell configurations, we need to write the expression for pressure drop, under fully developed laminar flow conditions, due to gas flow through open channels, ΔP_{ch}, namely

$$\Delta P_{ch} = \frac{CV_{ch}\ell}{D_h^2} \tag{9.8}$$

where C is a constant and V_{ch} denotes gas velocity through the channel, which is given by

$$V_{ch} = Q/A_o \tag{9.9}$$

Here Q is the flow rate through the filter and A_o is the open cross-sectional area given by

$$A_o = \frac{\pi}{4} d^2 \times \text{OFA} \tag{9.10}$$

In view of an identical open frontal area, filters with 100/17 and 200/12 cell configurations will have an identical open cross-sectional area and gas velocity through their respective channels under constant flow rate conditions. Thus, the pressure drop ΔP_{ch} will now be proportional to ℓ/D_h^2 according to Eq. (9.8). We define this ratio as a "back pressure index" or BPI:

$$\text{BPI} = \ell/D_h^2 \tag{9.11}$$

Since the specific filtration area of a 200/12 configuration is 41.5% greater, the filter length with such a configuration can be 58.5% smaller than that of the filter with a 100/17 cell configuration for an identical total filtration area. In

---

[1] For constant total filtration area.

TABLE 9.4. Properties and functional parameters of diesel filters with two different cell configurations and constant total filtration area.

Property and functional parameter		100/17 cell	200/12 cell
L	(in)	0.100	0.071
T	(in)	0.017	0.012
N	(cells/in^2)	100	200
OFA		0.345	0.345
MIF		0.035	0.035
D_h	(in)	0.083	0.059
SFA	(in^2/in^3)	16.6	23.5
BPI	(in^{-1})	1.0	1.17
TFA	(in^2)	X	X
l	(in)	l	0.585l

this manner, Eq. (9.11) helps estimate the back pressure penalty[2] due to the smaller hydraulic diameter of a 200/12 cell configuration. The results of this exercise are summarized in Table 9.4, which compares the properties and functional characteristics of filters with two different cell configurations. It shows that despite the compact volume of a 200/12 filter, it will experience 17% higher back pressure than the 100/17 filter. Such a back pressure penalty, as will be shown later, may well exceed 17% as the soot membrane begins to build up on the surfaces of open channel walls. In addition, the pressure drop through porous walls can also be significant. It is clear from Table 9.4 that filter design often calls for trade-offs in functional parameters that, in turn, require prioritization of durability and functional requirements on the part of the filter designer.

9.4.4   Filter Size and Contour

Both mechanical and thermal durability requirements favor a circular contour for the filter since it lends itself to robust packaging and at the same time experiences less severe temperature gradients during regeneration. Furthermore, circular filters are easier to manufacture and to control tolerances, making them more cost-effective than noncircular contours. Indeed, the latter have also been manufactured for special applications where space constraint is the dominating factor.

Filter size is generally dictated by engine capacity and is normally equal to engine volume. This "rule of thumb" for designing filter size has worked well in both mobile and stationary applications in that it helps control soot collection and regeneration without impairing filter durability and imposing high

---

[2]The pressure drop due to channel flow is a significant fraction of total pressure drop through the filter.

back pressure penalty. We will illustrate these benefits with a realistic example.

Consider a 10-l, 230-hp, diesel engine for a medium-to-heavy-duty truck for an urban area. We design the total filter volume to be 10 l with a microstructure commensurate with 90% filtration efficiency. Based on prior experience, we limit the soot loading to 10 g/l of filter volume to ensure safe regeneration at 2-hr intervals. Then,

$$\text{total soot collected} = \frac{10 \times 10}{2} = 50 \, \text{g/hr}$$

$$\text{rate of soot emitted by engine} = \frac{50}{0.9} = 55.5 \, \text{g/hr}$$

which in standard units works out to 0.24 g/bhp-hr. This is a good representation of soot output of new modern-day diesel engines. Let us note that the filter will help reduce the soot emissions from 0.24 to 0.024 g/bhp-hr due to its 90% collection efficiency.

We will develop the pressure drop model in the next section and estimate the back pressure due to the above loading.

9.4.5 Pressure Drop Model

The pressure drop model is based on the following assumptions (Brown 1955):

1. Incompressible gas
2. Laminar flow
3. Constant density and viscosity at a given temperature
4. Cylindrical pores in filter walls
5. No crossflow between pores

Referring to Figure 9.7, the total pressure drop across the filter is made up of five components, namely:

$$\Delta P_{\text{total}} = \Delta P_{\text{en}} + \Delta P_{\text{ch}} + \Delta P_{\text{w}} + \Delta P_{\text{s}} + \Delta P_{\text{ex}} \tag{9.12}$$

The entrance and exit losses, ΔP_{en} and ΔP_{ex}, are relatively small compared with other losses. Hence, they will be neglected.

The remaining three losses can be estimated from the generic equation for a circular pipe:

$$\Delta P = \frac{32 \mu v l}{g d^2} \tag{9.13}$$

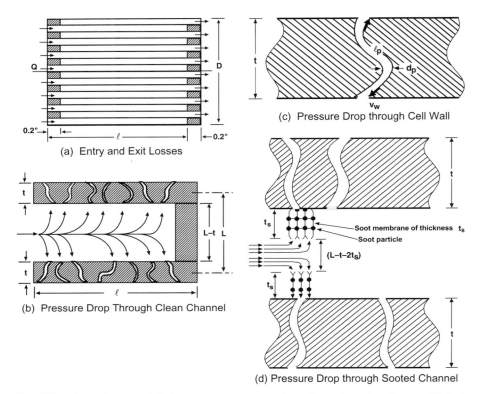

(a) Entry and Exit Losses

(b) Pressure Drop Through Clean Channel

(c) Pressure Drop through Cell Wall

(d) Pressure Drop through Sooted Channel

FIGURE 9.7. Flow model for pressure drop calculations. Reprinted from Gulati 1996b; courtesy of Marcel Dekker, New York.

where

μ	= gas viscosity	(lb/s ft)
v	= gas velocity through pipe	(ft/s)
l	= effective length of pipe	(ft)
d	= effective diameter of pipe	(ft)
g	= gravitational acceleration	(ft/s^2)

We will apply the above equation to estimate each component of the pressure drop through a 10-l filter, 10.5″ diameter × 7″ long (26.67 cm × 17.78 cm), with a 100/17 cell configuration. To this end, we assume

engine size	= 10 l = 0.353 ft^3
engine speed	= 1,500 RPM
gas temperature	= 325 °C

Then

$$Q = \text{flow rate} = 0.353 \times 750 \quad = 265\,\text{ft}^3/\text{min at } 325\,^\circ\text{C}$$
$$= 531\,\text{SCFM}$$
$$\mu \quad = 2 \times 10^{-5}\,\text{lb/s ft}$$

9.4.5.1   Estimate of ΔP_{ch}. For the checkerboard plug pattern,[3]

$$A_{open} = \frac{\pi}{4}\left(\frac{10}{12}\right)^2\left(\frac{0.083}{0.100}\right)^2 \times 0.5 = 0.188\,\text{ft}^2$$

$$v_{ch} = \frac{Q}{A_{open}} = \frac{531}{0.188 \times 60} = 47.2\,\text{ft/s}$$

$$d_{ch} = (L - t) = 0.0069\,\text{ft}$$

$$\Delta P_{ch} = \frac{32\mu v_{ch}\ell}{g d_{ch}^2}$$

$$= 10.83\,\text{lb/ft}^2$$
$$= 0.153''\text{Hg} \tag{9.14}$$

9.4.5.2   Estimate of ΔPw. The effective length of pores in the filter wall depends on their tortuosity and mean pore diameter that for the filter will be taken as 12.5 μm. The effective pore length is approximately 3 t, with t being the wall thickness (Carman 1956). The gas velocity through the pores is readily obtained by the continuity equation, namely:

$$V_{ch}(L - t)^2 = v_w \times 4P(L - t)\ell$$

where P denotes the fractional porosity of the filter walls, which will be taken as 0.5. Substituting $L = 0.1''$, $t = 0.017''$, $P = 0.5$, and $l = 6.6''$ in the above equation, we obtain

$$v_w = 0.0063\,v_{ch} = 0.297\,\text{ft/s}$$

Substituting $l_p = 3\,t = 0.00425\,\text{ft}$, $d_p = 12.5\,\mu\text{m} = 0.000041\,\text{ft}$, and $v_w = 0.297\,\text{ft/s}$ in Eq. (9.13), we obtain

$$\Delta Pw = 14.91\,\text{lb/ft}^2 = 0.211''\text{Hg} \tag{9.15}$$

Thus, the pressure drop through the wall is 38% higher than that through the channel. The above estimate of ΔP_w is based on clean and open pores. As these

---

[3] We assume a diameter of 10″ (25.4 cm) for the checkerboard region due to the fully plugged peripheral region, 0.25″ (0.635 cm) wide.

pores accumulate soot, their mean diameter will decrease, the flow velocity will increase, and ΔP_w will go up. To reestimate ΔP_w, we can still use Eq. (9.13) once we know the amount of soot trapped in the pores.

9.4.5.3 *Estimate of* ΔPs.

The pressure drop through the soot membrane is negligible due to both its open structure and its small thickness. However, as the membrane thickness increases with continuous deposition of soot, the hydraulic diameter of the sooted channel decreases and the gas velocity increases, thereby contributing to ΔP_{ch}. To estimate the incremental pressure drop due to soot membrane we must first study the kinetics of soot deposition.

Recall that the maximum allowable soot accumulation for safe regeneration is typically 10 g per liter of filter volume. For a filter volume of 10 l, the total soot collected prior to regeneration is 100 g over a 2-hr filtration cycle. With a filtration efficiency of 90%, the soot output of a 230-hp engine is given by

$$\text{soot output} = \frac{100}{2 \times 2300.9} = 0.242\,\text{g/bhp-hr}$$

$$\text{soot accumulation rate} = \frac{50}{60} = 0.833\,\text{g/min}$$

$$\text{active filter volume} = \frac{\pi}{4} \times 100 \times 6.6 = 518\,\text{in}^3$$

$$\text{total filtration area} = \text{SFA} \times V_f = 8605\,\text{in}^2$$

The soot density has been reported in the literature and is approximately 0.056 g/cm³ or 0.917 g/in³ (Wade et al. 1981). Using this value we can estimate the rate at which soot volume, hence, the soot membrane thickness, builds up.

$$\text{rate of soot volume collected per filter} = \frac{0.833}{0.917} = 0.909\,\text{in}^3/\text{min}$$

$$\text{rate of increase in soot membrane thickness} = \frac{0.909}{8605} = 0.00011\,\text{in/min}$$

$$\text{total thickness of soot membrane after 2 hr} = 0.013\,\text{in}$$

$$d_h = 0.083 - 0.025 = 0.058'' = 0.00483\,\text{ft}$$

$$A_{\text{open}} = 0.188 \times \left(\frac{0.058}{0.083}\right)^2 = 0.0918\,\text{ft}^2$$

$$v_{ch} = \frac{Q}{A_{\text{open}}} = \frac{531}{0.0918 \times 60} = 96.4\,\text{ft/s}$$

TABLE 9.5. Comparison of pressure drop for two filters with identical total filtration area but different cell configurations (in Hg).

	10.5″ diameter × 7″ long filter (100/17 cell)	10.5″ Diameter × 5″ long filter (200/12 cell)
ΔP_{ch}	0.153	0.211
ΔP_w	0.211	0.302
ΔP_s	0.639	2.180
Total ΔP	1.003	2.693

Substituting into Eq. (9.13), we obtain

$$\Delta P_s = \frac{32 \times 2 \times 10^{-5} \times 96.4 \times 0.55}{32.2 \times (0.00483)^2}$$
$$= 45.2 \, lb/ft^2$$
$$= 0.639″Hg \qquad (9.16)$$

Thus, the pressure drop through a sooted channel (with 10 g/l of soot loading) is three times as large as that through the wall and over four times as large as that through the clean channel.

The above computations were also carried out for a 10.5″ diameter × 5″ long filter with a 200/12 cell configuration (and identical total filtration area as the 10.5″ diameter × 7″ long filter with a 100/17 cell configuration). Table 9.5 compares the individual pressure drop components for the two filters. It is clear from this table that the largest contribution comes from flow through the sooted channel. Furthermore, the small hydraulic diameter of a 200/12 cell results in a nearly three times higher pressure drop than that for the 100/17 cell, which explains the popularity of the 100/17 cell configuration for filter applications.

The foregoing pressure drop model is only an approximation that helps quantify the effect of flow rate, open frontal area, and hydraulic diameter. It also provides the relative contributions of open and sooted pores in the wall as well as those of open and sooted channels to the total pressure drop. A more refined model is needed that must correlate with the experimental data.

9.5 REGENERATION TECHNIQUES

In addition to high filtration efficiency, the pressure drop across the filter must be controlled to minimize the fuel economy penalty. Early tests by Ford Motor Company showed that for a pressure drop of 1″ Hg, the typical diesel-powered vehicle traveling at 40 mph would incur a 1% loss in fuel economy (Wade et al. 1981). To prevent a further increase in pressure drop, the filter must be regenerated periodically by oxidizing the soot carbon. To facilitate oxidation of soot carbon, the exhaust gas entering the filter must be above 540 °C and contain sufficient oxygen to initiate the reaction. Since the exhaust temperature under normal driving conditions is <540 °C, this requirement can be relaxed downward by either catalyzing the filter or using fuel additives. We provide a brief review of common regeneration techniques.

9.5.1   Throttling

The exhaust temperature of a diesel engine can be increased by throttling the air flow supplied to the engine. By reducing the air flow, the overall air/fuel ratio is decreased, which increases the average combustion temperature and, hence, the exhaust temperature. The effect of intake throttling on the exhaust temperature of a 5.7-l engine is shown in Figure 9.8. The exhaust temperature for the standard engine seldom exceeds 300 °C (up to speeds approaching 55 mph), which is too low to incinerate particulate matter. Throttling not only reduces the heat loss to unused excess air, but also lowers the manifold pressure thereby increasing the pumping loss. This increases the fuel input by the driver to maintain power and results in hotter exhaust temperature (Ludecke and Dimick 1983). Figure 9.8, which is based on laboratory tests, shows two different particulate regeneration regions: one with a catalyst assist and one without. Let us note that regeneration cannot be achieved below

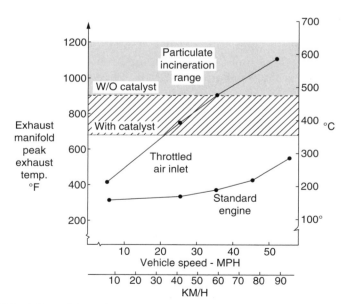

FIGURE 9.8.   Effect of throttling on an exhaust temperature of a 5.7-l engine. Reprinted from Gulati 1996b; courtesy of Marcel Dekker, New York.

37 mph without the catalyst and below 21 mph with the catalyst. However, vehicle tests showed that regeneration by throttling was not achieved even in a catalyzed filter due, primarily, to limited oxygen availability. Throttling for a diesel engine is limited to an exhaust oxygen concentration of 2% to 5%. Under heavy loads and adequate oxygen availability, throttling can lead to a high enough temperature to effect regeneration. Thus, throttling regeneration cannot be applied at steady-state vehicle speeds and light loads. Furthermore, throttling can have an adverse effect on engine out emissions and fuel consumption. It increases HC, CO, and smoke emissions while decreasing the NO_x emissions (Wade et al. 1981; Rao et al. 1985). Fuel consumption increases due to richer air/fuel ratio, higher heat losses, and negative pumping work.

In the absence of sufficient oxygen, over-throttling is employed sometimes to raise the exhaust temperature to a regeneration level. However, with insufficient oxygen, soot combustion does not occur, and when the engine returns to an unthrottled condition defined by lower engine speed and higher oxygen content, the overheated filter can experience "run-away" regeneration with a peak temperature approaching 1,400 °C (due to insufficient exhaust flow), which can lead to thermal cracking and melting (Wade et al. 1981). Failures of this nature can be prevented by reducing particulate loading and increasing the flow rate.

9.5.2 Burner Regeneration

A diesel-fueled burner placed in the exhaust system, in front of the filter, can affect its regeneration at nearly all engine speeds and load conditions (Wade et al. 1981; 1983; Ludecke and Dimick 1983; Rao et al. 1985). However, such a system, shown schematically in Figure 9.9a), is not only complex in terms of sophisticated electronic controls needed to ensure filter reliability, it requires high fuel consumption and a large air pump to heat up the entire exhaust gas to 540 °C. Second, a modulated, high-pressure burner fuel flow system is required to maintain the filter inlet temperature at a safe level. Third, air must be circulated continuously through the burner nozzle to minimize fouling by particulate deposits.

Isolating the filter from engine exhaust during regeneration using a bypass system, as shown in Figure 9.9b), overcomes many of the above issues. First, the regeneration process is independent of the engine operating conditions; second, the electronic control system to effect regeneration is considerably simplified; and third, the energy required to heat up the entry face of the filter to 540 °C is reduced by nearly an order of magnitude due to much reduced exhaust flow as shown in Figure 9.10. In the bypass mode, little or none of the exhaust flow is allowed to flow through the filter. Thus, the air pump is considerably smaller and the outlet pressure requirement is considerably lower than those for the in-line burner, particularly when the filter is located after the muffler. The regeneration process is initiated when the back pressure across the filter reaches a specified level, e.g., 3″ Hg. Alternatively or simulta-

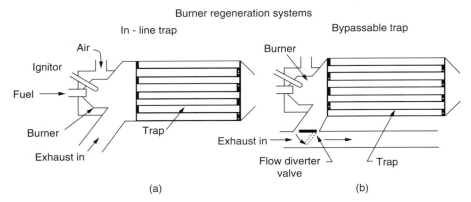

FIGURE 9.9. Schematic of the diesel-fueled burner regeneration system: a) in-line burner and b) burner with bypass system. Reprinted from Gulati 1996b; courtesy of Marcel Dekker, New York.

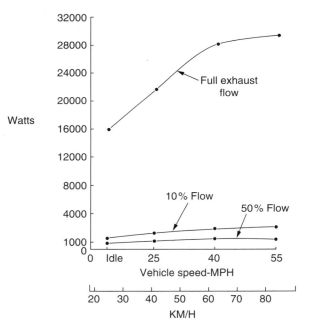

FIGURE 9.10. Energy required to raise the exhaust temperature of a 5.7-*l* diesel engine to 540 °C. Reprinted from Gulati 1996b; courtesy of Marcel Dekker, New York.

neously, the exit temperature is monitored by a thermocouple. A sudden and significant increase in exit temperature is a good indicator of the onset of regeneration.

When the entry face of a filter approaches 650 °C, soot oxidation begins with the combustion front propagating slowly toward the exit face. Conse-

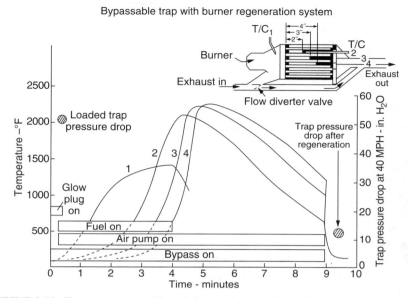

FIGURE 9.11. Temperature profiles of the combustion front during burner regeneration of the cordierite filter with a bypass system. Reprinted from Gulati 1996b; courtesy of Marcel Dekker, New York.

quently, the burner may be shut off halfway through the regeneration cycle and the increased oxygen concentration may be used to accelerate the remainder of the soot oxidation process. Care must be taken to limit the midbed temperature to <1,400 °C and temperature gradients to 35 °C/cm to prevent melting or cracking of the filter. Figure 9.11 shows temperature profiles of the combustion front during burner regeneration with a peak value of 1,260 °C. As the combustion front moves toward the exit face, the peak temperature increases progressively due to preheating of the combustion air in the preceding region where regeneration had been completed. The interval between regenerations is dictated by not only the fuel economy penalty but also the back pressure and critical soot mass for sustained self-regeneration.

9.5.3   Electrical Regeneration

Since the bypass system has demonstrated good feasibility, the complex burner with its electronic controls may be replaced by a simple resistance heater powered by the alternator on the vehicle (see Figure 9.12). A small air pump (3 to 5 CFM) is used to transfer heat from the heater to the filter and to provide sufficient oxygen for soot oxidation. Typical heaters are fabricated from two nichrome resistance elements contained in MgO powder insulation with a 0.260″ diameter stainless steel sheath. The energy requirement for

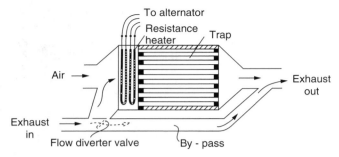

FIGURE 9.12. Schematic of a cordierite filter with a bypass and electrical regeneration system. Reprinted from Gulati 1996b; courtesy of Marcel Dekker, New York.

regeneration is typically 3 KW and readily achieved from a 90-amp alternator at a speed of 6,400 rpm.

The heater is turned on when the back pressure across the filter reaches 5″ Hg. Once the heater temperature reaches 760 °C, the air pump is turned on to transfer heat and oxygen to the entry face of the filter. The combustion front moves slowly toward the exit face much like burner regeneration (see Figure 9.11). At the end of regeneration, which may last 8 to 10 minutes, the air pump is shut off and the flow diverter valve is returned to its original position to allow all of the engine exhaust gas to pass through the filter. The completion of the regeneration cycle is best indicated by the back pressure value approaching that of a clean filter. In contrast to burner regeneration, where the air pump is on from the beginning, electrical regeneration does not trigger the air pump until the heating element reaches a temperature of 760 °C. Consequently, both the peak temperature and the temperature gradients are lower (100 °C to 200 °C lower) in the case of electrical regeneration, which alleviates cracking and melting issues (Wade et al. 1981; Ludecke and Dimick 1983; Rao et al. 1985; Wade et al. 1983).

9.5.4 Catalytic Regeneration

Catalytic additives in diesel fuel have also proven effective in reducing the regeneration temperature sufficiently to effect self-regeneration under normal driving conditions (Wade et al. 1981; 1983; Ludecke and Dimick 1983; Rao et al. 1985). Fuel additives consisting of organometallic compounds of Cu, Ni, Ce, Mo, Mn, Zn, Ca, and Ba have been evaluated with respect to their effectiveness in lowering regeneration temperature while remaining chemically stable in diesel fuel. Screening tests show that octoate-based compounds of Cu, Cu and Ni, as well as Cu and Ce are effective. At elevated temperatures, these compounds decompose into oxides of metals that are ideal for promoting soot oxidation. Additional detail and references are available in Chapter 8: "Diesel Engine Emissions".

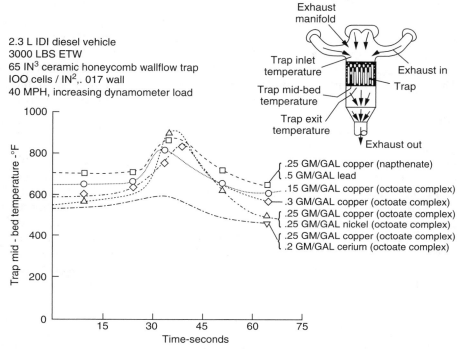

FIGURE 9.13. Peak midbed temperature versus time during regeneration with different fuel additives. Reprinted from Gulati 1996b; courtesy of Marcel Dekker, New York.

To evaluate the effectiveness of these additives, steady-state tests were conducted on a chassis dynamometer at an equivalent vehicle speed of 40 mph. Under normal load conditions, the exhaust temperature was only 200 °C. After loading the filter with soot to a back pressure of 6″ Hg, the chassis dynamometer was operated at successively higher loads at the constant vehicle speed of 40 mph to increase the filter inlet temperature in 30 °C increments. Different additives in Phillips 2D diesel fuel helped complete regeneration of a sooted filter at different temperatures. For example, 0.25 g/gallon of Cu and 0.2 g/gallon of Ce reduced the regeneration temperature to 280 °C, which is nearly 230 °C lower than that without the additive. The peak midbed temperature seldom exceeded 480 °C with other combinations of fuel additives as shown in Figure 9.13. The temperature gradients were also very low with little or no concern over filter durability. Thus, regeneration with fuel additives is a viable option.

9.6 PHYSICAL PROPERTIES AND DURABILITY

The physical properties of ceramic diesel filters, which can be controlled independently of geometric properties, have a major impact on their functionality and durability. These properties include microstructure (porosity, pore size distribution, and microcracking), coefficient of thermal expansion (CTE), strength (crush strength, isostatic strength, and modulus of rupture), structural modulus (also called E-modulus), fatigue behavior (represented by dynamic fatigue constant), thermal conductivity (k), specific heat (c_p), and density (ρ). These properties depend on both the ceramic composition and the manufacturing process, which can be controlled to yield optimum values for a given application.

The microstructure of diesel filters not only affects physical properties like CTE, strength, and structural modulus, but also it has a strong bearing on the filter/catalyst interaction that, in turn, affects the functionality and the durability of a catalyzed filter. The coefficient of thermal expansion, strength, fatigue, and structural modulus of the diesel filter, which also depend on cell orientation and temperature, have a direct impact on its mechanical and thermal durability (Gulati 1983; 1986; Vergeer et al. 1985; Gulati and Lambert 1991; Gulati et al. 1992a). Finally, since all of the physical properties are affected by washcoat formulation, washcoat loading, and washcoat processing, they must be evaluated before and after the application of the washcoat to assess filter durability.

9.6.1 Physical Properties

The initial filter compositions, shown in Table 9.6, were designed to offer different microstructures to meet the different filtration efficiency and back pressure targets set by engine manufacturers (Howitt and Montierth 1981). However, they were not optimized with respect to thermal durability, which became a critical requirement to survive regeneration stresses. A secondary process such as acid leaching helped reduce the coefficient of thermal expansion, thereby improving thermal durability significantly, but it had an adverse effect on mechanical durability and filter cost. A more advanced filter composition, EX-80, with excellent functionality was developed in 1992. This material is a stable cordierite composition with low CTE and has demonstrated improved long-term durability over a wide range of operating conditions. Moreover, it offers high filtration efficiency and low pressure drop. The low CTE reduces thermal stresses, thereby permitting numerous regeneration cycles without impairing filter's durability. This composition is now one of the industry standards for diesel exhaust aftertreatment (Murtagh et al. 1994). For the sake of completeness, however, we will summarize the properties of early filter compositions together with those of EX-80 composition. The specimen size, orientation, and test technique for measuring the physical properties are identical to those for ceramic catalyst supports, which are described in Chapter 7.

Table 9.6 compares the nominal physical properties of four different filter compositions that differ in their mean pore diameter. The strength and modulus of elasticity data are those measured at room temperature. The axial coeffi-

TABLE 9.6. Physical properties data for cordierite ceramic diesel filters.

		EX-47 100/17	EX-54 100/17	EX-66 100/25	EX-80 100/17
Wall porosity	(%)	50	46	45	48
Mean pore size	(μm)	12	25	35	13.4
Weight density	(g/cm^3)	0.39	0.45	0.54	0.42
Crush A strength	(psi)	1,130	900	1,300	1,595
Crush B strength	(psi)	380	300	430	325
3D-isostrength	(psi)	585	460	660	500
2D-isostrength	(psi)	500	390	560	425
Axial MOR	(psi)	365	320	418	410
Tang. MOR	(psi)	142	131	176	184
Axial MOE	(10^6 psi)	0.81	0.83	1.06	0.75
Axial CTE 25–800°C	(10^{-7}/°C)	8.8	8.7	10.5	3.3

TABLE 9.7. MOR data for cordierite ceramic diesel filters with a 100/17 cell structure.

Temp. °C	Axial MOR (psi)				Tangential MOR (psi)			
	EX-47	EX-54	EX-66	EX-80	EX-47	EX-54	EX-66	EX-80
25	365	305	300	408	145	115	120	162
200	370	260	270	385	150	120	110	156
400	372	260	240	363	160	120	100	156
600	390	265	260	380	170	125	110	166

cient of thermal expansion is the average value over the 25–800°C temperature range. It is clear from Table 9.6 that the EX-80 filter offers an optimum combination of properties, namely small mean pore size, high strength, low modulus of elasticity (MOE), and low CTE, which together provide excellent functionality compared with that of the other three filter compositions.

The axial and tangential modulus of rupture (MOR) values at high temperature that impact the filter's thermal durability are summarized in Table 9.7. Again, the EX-80 filter excels in this property relative to other filter compositions. Similarly, the axial[4] E-modulus data in Table 9.8 show that the EX-80 filter has similar stiffness as EX-47 and EX-54 filters notably at or below 600°C (representative of the filter's peripheral temperature during regeneration).

Finally, the axial CTE data for the four filter compositions are summarized in Table 9.9. These data also demonstrate that the EX-80 filter has the lowest expansion coefficient over the entire temperature range, which implies low thermal stresses during regeneration and excellent thermal durability for this filter.

[4]The tangential E-modulus is approximately 50% of the axial E-modulus and, hence, is not included in Table 9.8.

TABLE 9.8. Axial E-modulus data for cordierite ceramic diesel filters (10^6 psi).

Temp. (°C)	EX-47 (100/17)	EX-54 (100/17)	EX-66 (100/25)	EX-80 (100/17)
100	0.63	0.61	1.02	0.63
200	0.63	0.61	1.03	0.64
300	0.63	0.62	1.04	0.64
400	0.64	0.62	1.05	0.65
500	0.65	0.62	1.06	0.66
600	0.66	0.65	1.08	0.68
700	0.68	0.68	1.11	0.71
800	0.70	0.70	1.16	0.75
900	0.75	0.77	1.31	0.83
1,000	0.83	0.91	1.43	0.92

TABLE 9.9. Axial CTE data for cordierite ceramic diesel filters (10^{-7} in/in/°C).

Temp. (°C)	EX-47 (100/17)	EX-54 (100/17)	EX-66 (100/25)	EX-80 (100/17)
100	−4.1	−9.4	−7.7	−1.33
200	−1.6	−5.1	−4.4	−9.6
300	0.8	−1.4	−1.5	−6.5
400	2.8	1.2	0.6	−4.0
500	4.4	3.2	2.4	−1.9
600	6.0	5.0	4.3	−0.2
700	7.8	6.8	6.3	1.2
800	9.4	8.5	8.4	2.9
900	10.9	9.7	10.5	4.3
1,000	12.6	10.5	12.6	5.4

9.6.2   Thermal Durability

Thermal durability refers to the filter's ability to withstand both axial and radial temperature gradients during regeneration. These gradients depend on soot distribution, soot loading, O_2 availability, and flow rate and give rise to thermal stresses that must be kept below the fatigue threshold of the filter material to prevent cracking. A detailed analysis of thermal stresses requires temperature distribution, which is readily measured with the aid of 0.5-mm-diameter, Type K, chromel-alumel thermocouples during the regeneration cycle (Gulati 1983; 1996a). Several examples illustrating this technique are given in Section 9.8. To assess the relative thermal durability of different filter candidates, we will compute the thermal shock parameter using physical properties data and Eq. (9.17):

$$\text{TSP} = \frac{(\text{MOR}/\text{E})\,T_p}{\alpha_c(T_c - 25) - \alpha_p(T_p - 25)} \tag{9.17}$$

TABLE 9.10. Axial thermal shock parameter for cordierite ceramic diesel filters (for $T_p = 400\,°C$).

Temp. (°C)	EX-47 (100/17)	EX-54 (100/17)	EX-66 (100/25)	EX-80 (100/17)
600	2.42	1.32	1.49	4.20
700	1.38	0.77	0.83	2.40
800	0.93	0.52	0.53	1.50
900	0.68	0.40	0.37	1.10
1,000	0.52	0.33	0.28	0.80

In the above equation, T_c and T_p denote the temperature of center and peripheral regions of the filter during regeneration and α_c and α_p denote the corresponding CTE values. In view of plugging around the peripheral region, there is no gas flow in that region and the temperature T_p is typically 400 °C. The center temperature, on the other hand, is higher depending on soot loading, O_2 content, and flow distribution. We will assume T_c to range from 600 °C (low soot loading) to 1,000 °C (high soot loading) and will compute thermal shock parameter (TSP) values for each of these T_c values while keeping $T_p = 400\,°C$. The results of this exercise are summarized in Table 9.10. Let us note that the TSP values for EX-80 filter are nearly 50% to 250% higher than those for other filters due, primarily, to its two- to three-fold lower CTE values (see Table 9.9). The higher TSP value signifies improved thermal shock resistance and extended thermal durability. Alternatively, it permits higher regeneration stresses without impairing the filter's durability.

The power law fatigue model (Wiederhorn 1974; Ritter 1978) helps estimate the safe allowable regeneration stress for a specified filter life. Denoting the filter's short-term modulus of rupture by S_2, the safe allowable stress S_1 is given by

$$S_1 = S_2[t_2/t_1]^{1/n} \qquad (9.18)$$

where t_1 denotes the specified filter life, t_2 denotes the equivalent static time for measuring short-term modulus of rupture, and n denotes the dynamic fatigue constant of filter composition. The latter is obtained by measuring MOR as a function of the stress rate at temperature T_p. Table 9.11 summarizes n values for the four filters at $T_p = 200\,°C$ and 400 °C (Murtagh et al. 1994). Both the mean value and the 95% confidence interval are listed for n. For a conservative estimate of S_1, the lowest value of n should be used in Eq. (9.18). The equivalent static time t_1 is defined as the actual test duration for measuring MOR divided by $(n + 1)$. Since the typical test duration is 30 s. and the lowest value of n is approximately 29 (with the exception of the EX-47 filter), $t_1 \cong \dfrac{30}{30} \cong 1\,\text{s}$. Filter life is generally specified in terms of the number of regeneration cycles over the vehicle's lifetime. We will assume a filter life of 120,000 miles with a regeneration interval of 200 miles and a regeneration

TABLE 9.11. Dynamic fatigue constant "n" for cordierite ceramic diesel filters in the axial direction.

Temp. (°C)	EX-47 (100/17)	EX-54 (100/17)	EX-66 (100/17)	EX-80 (100/17)
200	21	34	27	—
400	24	44	42	53
Combined data	23	38	38	—
95% confidence interval	17–30	27–62	28–60	32–92

TABLE 9.12. Safe allowable stress in the axial direction for long-term durability.

Filter composition	S_2 (psi)	n	t_2 (sec.)	S_1 (psi)
EX-47	372	17	2	183
EX-54	260	27	1	166
EX-66	240	28	1	156
EX-80	363	32	1	249

duration of 10 minutes. This translates to $t_1 = 6,000$ minutes $= 360,000$ s. Substituting these values in Eq. (9.18), we arrive at the safe allowable stress in the axial direction as shown in Table 9.12. Again, it is clear from this table that the EX-80 filter can sustain higher regeneration stresses than the other filters. This superiority derives from its higher fatigue constant and MOR values that, in turn, are related to its optimized microstructure.

9.6.3 Mechanical Durability

The mechanical durability of the ceramic filter depends not only on its tensile and compressive strengths but also on its packaging design (Gulati 1996a). In addition to mechanical stresses due to handling and processing, the filter package must be capable of withstanding in-service stresses induced by gas pulsation, chassis vibration, and road shocks. The design of a robust packaging system for catalyst supports as discussed in Chapter 7 is equally applicable to the filter. Tables 9.6 and 9.7 demonstrate more than adequate strength for tourniquet canning, which is recommended for long-term mechanical durability. In addition, preheat-treatment of an intumescent mat also promotes mechanical durability (Gulati 1996b).

9.7 ADVANCES IN DIESEL FILTERS

Both the stringent diesel emission legislation in Japan, North America, and the United States (introduced in 2007), and the popularity of diesel passenger cars in Europe have led to new advances in diesel filter technology. With new legislation in the offing, one of the automotive manufacturers in Europe (PSA) decided to introduce a noncordierite DPF in MY 2001 diesel passenger cars (Miwa et al. 2001; Eastwood 2000). This created a great opportunity for new filter materials (Cutler and Merkel 2000; Merkel et al. 2001a, 2001b; Miwa 2001; Ohno et al. 2000), new filter designs (Hickman 2000, Miwa 2001), and improved detection techniques for soot deposits through increased pressure drop (Johnson 2001). The motivation for developing new materials stemmed from the need for higher thermal conductivity, higher melting temperature, and higher heat capacity than those of cordierite ceramic to facilitate regeneration under uncontrolled conditions (Cutler and Merkel 2000).

Uncontrolled regeneration is most often described as an unplanned regeneration in which the combustion of a large amount of accumulated soot occurs under conditions in which the exhaust gas has a low flow rate but a high oxygen content, resulting in temperatures that far exceed those of a normal controlled regeneration. For example, operation of the diesel engine at high loads and speeds could produce exhaust temperatures that are sufficiently high to initiate combustion in a filter that is heavily loaded with soot. If the engine were to continue running at these conditions throughout combustion, the low oxygen content of the exhaust gas would result in a slow burn, whereas the higher flow rate of the exhaust would serve to transfer heat effectively away from the filter. Thus, only moderately high regeneration temperatures would be achieved. However, if the engine load were to be dramatically reduced soon after combustion was initiated, such as might occur under near-idling conditions, then the exhaust flow rate would decrease and the oxygen content of the exhaust gas would increase. The increased oxygen would accelerate soot combustion, whereas the lower exhaust flow rate would be less effective in removing heat from the system to cool the filter. Consequently, excessively high temperatures could be achieved within the filter during this uncontrolled regeneration, potentially causing cracking or melting of the filter.

Similarly, the motivation for new designs stemmed from the need for reducing thermal stresses during uncontrolled regeneration by either limiting the peak regeneration temperature to 1,000 °C via higher heat capacity (Hickman 2000) or by incorporating stress-relief slits in the filter albeit at the risk of impairing mechanical integrity (Miwa 2001). In this section, we will compare new materials like improved cordierite RC, SiC, and AT with the standard cordierite. This will be followed by new filter designs and how such designs impact their functionality, including pressure drop.

9.7.1   Improved Cordierite "RC 200/19" Filter

Because there is no known compositional modification that can be made to a cordierite-based ceramic to increase its refractoriness without also increasing its CTE and compromising its thermal shock resistance, survival of a cordierite filter must rely on modifications in filter design that reduce the maximum temperature the filter will experience during an uncontrolled regeneration.

The temperature increase experienced by the filter during regeneration is inversely proportional to the heat capacity of the filter per unit volume for a given exhaust gas flow rate and mass of soot burned per unit volume. The volumetric heat capacity of a filter is equal to the product of bulk density of the filter and specific heat of ceramic comprising the filter. Thus, the temperature increase during regeneration can be reduced simply by increasing the mass per unit volume of the cordierite filter (Hickman 2000). An increase in filter mass per unit volume may be achieved by increasing the filter cell density (cells per unit area) or wall thickness, or by decreasing the percent porosity in the filter walls. However, changes in cell geometry or porosity will also have an effect on pressure drop across the filter. An increase in cell density acts to decrease pressure drop by virtue of the increase in geometric surface area, whereas an increase in wall thickness increases pressure drop due to the increased path length through the wall. Increases in wall thickness are generally limited by the permeability of ceramic comprising the wall. A decrease in porosity may increase pressure drop unless the effect can be offset by simultaneous modification in pore size or pore connectivity.

Development of a cordierite ceramic that exhibits a reduced soot-loaded pressure drop for a given filter geometry, cell density, and wall thickness requires modification of the pore microstructure of the ceramic. This may be achieved, for example, by a change in raw materials, forming parameters, or firing conditions (such as furnace atmosphere, heating rates, peak temperature, and hold time at peak temperature.) The best candidate resulting from such modification was designated "RC Filter" with a cell structure of 200/19 (Merkel et al. 2001a). Table 9.13a) compares its properties with those of the EX-80, 100/17, cordierite filter. Figure 9.14 compares their pore microstructures. It is clear from these data that RC 200/19 filter offers 25% higher filtration area, 84% higher wall permeability, 52% higher weight density, and 52% higher heat capacity. The latter helps reduce the peak temperature of the RC 200/19 filter during uncontrolled regeneration, thereby compensating for its slightly higher CTE value relative to that of the EX-80, 100/17 filter. Furthermore, both the larger filtration area and the higher wall permeability of the RC 200/19 filter should result in a more uniform soot distribution and lower temperature gradient, thereby preserving or improving its thermal shock resistance as will be shown later.

Table 9.13b) provides the available sizes for cordierite filters (CO and RC) with different cell geometries, weight densities, and heat capacities.

9.7.2   AT and SiC Filters

Aluminum titanate (AT) is a novel ceramic oxide composite with high heat capacity, high melting temperature, and a unique microstructure that results in low CTE as well as in low elastic modulus (E-mod). These attributes permit

TABLE 9.13. a) Properties of Corning cordierite diesel particulate filters.

Property	EX-80 (100/17)	RC (200/19)
SFA (in^2/in^3)	16.6	20.7
Wall porosity (%)	48	45
Median pore size (micro;m)	13	13
Wall permeability $(10^{-12}\,m^2)$	0.61	1.12
Weight density (g/cm^3) including plugs	0.46	0.70
Axial MOR (psi)	410	677
Axial E-modulus $(10^6\,psi)$	0.63	1.32
Heat capacity per unit volume of filter at 500 °C (Joules/cm³ °C)	0.54	0.82
Axial CTE from 25–800 °C $(10^{-7}\,°C^{-1})$	3.3	6.0

TABLE 9.13. b) Cordierite filter sizes available in different geometries.

Filter Size	CO 100/17	CO 200/12	RC 200/19
Diameter × length			
5.66 × 6	•	•	•
7.5 × 8	•	•	
7.5 × 12	•	•	
9.0 × 12	•	•	•
9.5 × 12	•	•	•
10.5 × 12	•	•	•
11.25 × 12	•	•	•
11.25 × 14	•	•	
12.0 × 15	•	•	
15.0 × 15 (assembled)	•	•	
18.0 × 15 (assembled)	•	•	
20.0 × 15 (assembled)	•	•	

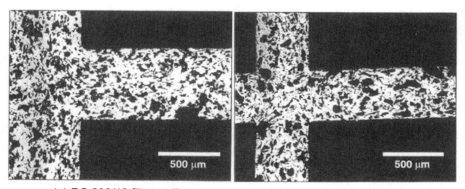

(a) RC 200/19 filter wall   (b) EX-80, 100/17 filter wall

FIGURE 9.14. Scanning electron micrograph of polished section of a) RC 200/19 filter wall and b) EX-80, 100/17 filter wall showing improved pore connectivity of the RC 200/19 filter. Reprinted from Merkel et al. 2001a; courtesy of SAE.

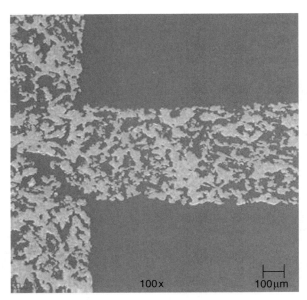

FIGURE 9.15. SEM image of a polished section of an AT filter wall showing a well-connected microstructure. Reprinted from Ogunwumi et al. 2005, courtesy of SAE.

monolithic construction of the filter and high regeneration temperature without compromising its thermal shock resistance (Ogunwumi et al. 2005). The AT filter is ideally suited for light-duty applications where it has proven successful in engine durability tests (Heibel et al. 2005).

The desired porosity and pore size distribution for reducing back pressure, while maintaining high filtration efficiency, are achieved by a judicious choice of raw materials and pore formers. Figure 9.15 is the SEM image of a polished section of an extruded aluminum titanate filter showing a well-connected microstructure that contributes to low back pressure while providing adequate strength (modulus of rupture or MOR).

Table 9.14 compares physical properties of cordierite (EX-80), SiC, and AT filters. It is interesting to note that AT's thermal shock index, defined by Eq. (9.19), is an order of magnitude higher than that of SiC due, primarily, to its low E-mod and CTE afforded by its unique microstructure. Alternatively, since thermal stresses during regeneration are proportional to the product of E-mod and CTE, the AT filter will experience substantially lower stresses and preserve its thermal durability over the vehicle's lifetime.

$$\text{Thermal Shock Index} = MOR/[E \times CTE] \qquad (9.19)$$

The well-connected pore structure of the AT filter results in lower pressure drop during filtration as demonstrated in Figure 9.16. A comparison of pressure drop data shows equivalent functionality in the clean state for both SiC

TABLE 9.14. Intrinsic and physical properties of various DPF materials.

Property	Cordierite (200/12)	SiC (200/18)	AT (300/13)
Intrinsic material properties			
Melting point (°C)	~1,460	~2,400	~1,600
Density (g/cm^3)	2.51	3.24	3.40
Specific heat @ 500°C (J/g°C)	1.11	1.12	1.06
DPF material properties			
CTE (22° to 1,000°C) (10^{-7}/°C)	7	45	9
Wall porosity (%)	50	43	50
Mean pore size (μm)	12	9	16
Permeability (10^{-12} m^2)	0.61	1.24	
Axial E-modulus (10^6 psi)	0.68	4.83	0.21
Axial MOR (psi)	410	2,700	290
Thermal conductivity (W/m K)	<2	~20	<2
Thermal shock index (°C)	860	125	1,535
Weight density (g/cm^3)	0.46	0.85	0.72
Heat capacity per unit volume of filter @ 500°C (J/cm^3°C)	0.51	0.95	

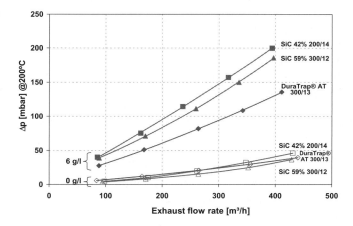

FIGURE 9.16. Pressure drop data for SiC and AT filters in clean versus soot-loaded state. Reprinted from Heibel et al. 2005, courtesy of SAE.

and AT filters. However, with 6-g/l soot loading, the AT filter shows excellent functionality due to its better connectivity (Heibel et al. 2005).

The chemical durability of the AT filter was measured by exposing it to the following environments:

1. Engine oil ash containing Fe, P, Ca, Zn, etc.
2. Iron and iron oxide debris in the exhaust system

TABLE 9.15. Comparison of salient characteristics of different filters.

Cordierite	Silicon carbide	Aluminum titanate
Low CTE	High CTE	Low CTE
Low heat capacity	High heat capacity	High heat capacity
Good thermal shock index	Good thermal shock index	Excellent thermal shock index
Susceptible to ash reaction	Potential issue of mechanical integrity	Not susceptible to ash reaction
Affordable	Higher cost	Minimal cost penalty

3. High-temperature oxidizing and reducing condition
4. Acidic solution resulting from SO_x and H_2O reactions in the presence of a catalyst

The results of these tests, detailed elsewhere (Ogunwumi et al. 2005), show that the AT filter passed all of the durability tests in a severe environment with no evidence of chemical reaction or decomposition. Table 9.15 compares the salient characteristics of cordierite, SiC, and AT filters.

The melting point of AT ($\sim$1,500 °C) is higher than that of cordierite ($\sim$1,460 °C), but it is less than the decomposition temperature for SiC ($\sim$2,400 °C). Although high melting temperature materials provide an added measure of system safety for the DPF itself, extreme temperatures should generally be avoided, as they may become a problem for the mat, the can, and other exhaust system components.

Cordierite has the lowest intrinsic material density of the three candidates, whereas SiC and AT have similar intrinsic material densities. However, in the configurations tested, the resulting bulk densities of AT filters were greater than cordierite but less than SiC.

AT has a lower specific heat than either SiC or cordierite. When the specific heat and density of the filter are combined, the heat capacity of the AT filter is found to be similar to that of the SiC filter, but it is considerably lower than the cordierite filter. A lower heat capacity filter (cordierite) can be heated quickly, resulting in faster regenerations. Faster regeneration generally means lower regeneration fuel penalty if raw fuel injection or additional engine power required to produce sufficient heat to initiate regular regeneration. Although a high heat capacity filter may be desirable for uncontrolled regeneration (to soak up excess heat), a high heat capacity also makes it more difficult to heat the filter for regular controlled regeneration.

Thus, the salient issue comes down to balancing the heat capacity of the filter with the material melting point and ash reaction temperature, such that the filter regenerates quickly and efficiently during controlled regeneration, while still having a sufficiently high melting point and/or ash reaction temperature to prevent pin holes and catastrophic failure during uncontrolled regeneration.

The thermal conductivity of AT is similar to that of cordierite (<2W/m-K), but it is much lower than the value measured for SiC (~20W/m-K at 500 °C). The thermal conductivity for all these materials drops as the temperature increases, such that their conductivity at ~1,300 °C is half the value at 500 °C. The value of a high thermal conductivity material is a matter of some debate, as the cooling effect due to high gas flow through the filter takes heat away from the hot spot much faster than the conductivity can draw the heat away from the hot spot.

The mean pore size (~13 μm) and pore size distributions for AT and SiC filters are similar. AT has a narrow distribution of highly connected pores as shown in Figure 9.15. The pore size, size distribution, and pore connectivity are highly tailorable for mean pore sizes between 5 and 20 microns. The percent porosity can also be adjusted. Tailoring the porosity characteristics has a profound effect on ultimate pressure drop, both clean and loaded.

The differences in thermal expansion coefficient, elastic modulus, and strength among the three materials translate into a different thermal shock index. Specifically, the very low CTE of cordierite material results in a very high thermal shock index (TSI) value. This is significant as cordierite is well known for its excellent thermal shock properties. Despite the high strength of silicon carbide, its high CTE and high elastic modulus lead to a low TSP (<200 °C). The segmentation of the commercial SiC may represent an effort to limit the distance over which thermal stresses can build. At the same time, however, segmentation may introduce strength-limiting flaws and reduce the axial MOR value.

The inferior thermal shock index of SiC has also led to the development of Si–SiC composite filters with improved thermal shock resistance and excellent ash and oxidation resistance up to 1,300 °C (Miwa 2001). In addition, their pore microstructures and permeability have been improved, thereby reducing the pressure drop during filtration. Table 9.16 compares the properties of porous SiC, Si–SiC, and cordierite materials. Indeed, the thermal shock index of the Si–SiC filter has increased by 15% due to its lower CTE and lower E-modulus.

9.7.3   Other Filter Materials

Several alternative filter materials offer partial filtration and command less of a pressure drop, with latter being critical for fuel economy. We will give a brief summary of a few of these alternative technologies; additional details may be found in the references cited in this section.

9.7.3.1   Partial Filter Technology (PFT). Although DOCs and DPFs are effective in removing 90% of particulates by mass and 99% by number, filter regeneration is dependent on an engine-out NO_x-to-PM ratio as well as exhaust gas temperature, which are difficult to achieve, particularly in old

TABLE 9.16. Properties of porous cordierite, SiC, and Si–SiC materials.

Property	SiC	Composite SiC: Si-metal and SiC	Cordierite
Wall porosity (%)	45	42	48
Mean pore size (μm)	10	8	13
Axial CTE ($10^{-7}\,°C^{-1}$)	43	40	3.3
Axial E-modulus (10^6 psi)	7.1	1.5	2.5
Axial MOR (psi)	7,700	1,740	1,500
Thermal shock index (°C)	250	290	1,820
Thermal conductivity (W/m-K)			
25 °C	61	13	0.8
400 °C	28	8	0.7
800 °C	21	6	1.0
Chemical resistance weight loss after 50 hr at 80 °C with 10% H_2SO_4	0.1	0.0	32.3
Oxidation resistance weight gain after 1,200 °C for 24 hr	3.0	3.9	—

engines with high PM emissions. Hence, a passive PM system known as PFT has evolved that combines DOC and DPF and reduces PM by up to 77% (Jacobs et al. 2006). It uses the NO_2:C reaction to oxidize a portion of soot, thereby regenerating the filter passively. A second advantage of PFT is that it does not accumulate significant amounts of lube oil ash, thereby minimizing the need for periodic ash cleaning.

Figure 9.17 shows the modular PFT system comprising a catalyzed DOC and a particulate filter. Figure 9.18 shows a cut-away of the filter portion. The exhaust gas flows through a metallic foil channel that contains a ramp or "shovel" providing a tortuous path for exhaust gas. The wall between the channels is made of porous sintered metal fleece material compressed between the foils. The "shovels" help force a portion of exhaust to pass through the fleece, thereby trapping a portion of soot that is then combusted by NO_2 produced by the upstream DOC. A typical temperature at the inlet of the PFT during the FTP cycle ranges from 250 °C for more than 40% of the cycle. The DOC reduces PM by about 25%, and the flow-through filter reduces it by an additional 25%.

9.7.3.2 Advanced Ceramic Material (ACM). The advanced ceramic material of mullite composition with high porosity and open connected pores has been developed by Dow Chemical Company (Pyzik et al. 2005). This material exhibits a unique microstructure compared with other ceramic substrates. The grains are fused to one another during sintering of standard ceramics, whereas the ACM microstructure exhibits three-dimensional interconnected mullite crystals with open connected pores. Such an acicular microstructure offers unique physical properties for filtration as well as long-term physical durability; see Figure 9.19.

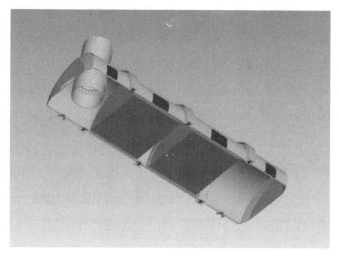

FIGURE 9.17. Modular PFT system. Reprinted from Jacobs et al. 2006, courtesy of SAE.

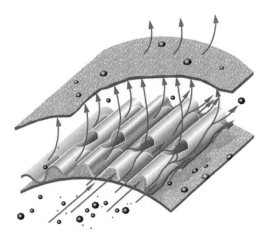

FIGURE 9.18. Foil substrate with "shovels" forcing soot into sintered metal fleece material. Reprinted from Jacobs et al. 2006, courtesy of SAE.

Both the composition and the manufacturing process can be tailored to yield the desired crystal structure, porosity, pore size distribution, grain morphology, and surface chemistry for high filtration efficiency, low back pressure, complete regeneration, as well as good mechanical and chemical durability. In addition, the acicular microstructure helps control the surface texture of channels, thereby allowing for good interaction between the catalyst and the trapped PM.

FIGURE 9.19. Acicular microstructure of mullite material for diesel filter application. Reprinted from Pyzik et al. 2005, courtesy of DEER Conference.

TABLE 9.17. Typical mechanical properties of Dow's ACM filter material (Pyzik et al. 2005).

Density	$3.17\,\text{g/cm}^3$
Wall porosity	60%
Bulk density	$0.52\,\text{g/cm}^3$
DPF weight	$1{,}300\,\text{g}$ ($5.66''$ diameter $\times$ $6''$ long)
Melting temperature	$>1{,}500\,°\text{C}$
Elastic modulus	$30\,\text{Gpa}$
Poisson's ratio	0.20
Modulus of rupture	$30\,\text{MPa}$
CTE	$27 \times 10^{-7}/°\text{C}$
Specific heat	$0.77\,\text{J/g K}$
Thermal conductivity	$1.30\,\text{W/m-K}$

By way of chemical durability, the weight loss suffered by ACM (due to etching) after 96 hours of exposure to 10% HNO_3 or 10% H_2SO_4 at $80\,°\text{C}$ was less than 1.2% comparable with that for SiC. Also, ACM demonstrated excellent resistance to ash components, including $Ca(OH)_2$, CeO_2, MgO, ZnO, NaCl, and Na_2SO_4, after 5 hours of exposure at $1{,}300\,°\text{C}$. For mechanical properties of ACM, see Table 9.17.

9.7.3.3   Ceramic Fibers and Cartridges. Filter cartridges can be assembled using high-temperature continuous ceramic fibers, typically polycrystalline metal oxide fibers made from alumina and silica. Nextel, a brand of synthetic fiber developed by 3M Company, is an example of a high-temperature fiber that has extensively been tested for diesel filter applications (Bloom 1995). Its properties are listed in Table 9.18.

TABLE 9.18. Physical properties of Nextel 312 fiber (Designer Guide, 3M Co., 1995).

Composition	Al_2O_3 62%, SiO_2 24%, B_2O_3 14%
Fiber diameter	10 to 12 μm
Fiber density	2.7 g/cm^3
Surface area	<1 m^2/g
Tensile strength	1.72 GPa
Elastic modulus	138 GPa
Use temperature	1,204 °C
Melting temperature	1,800 °C
Specific heat	1,047 J/kg K
Linear shrinkage	1.25%
CTE	$30 \times 10^{-7}/°C$

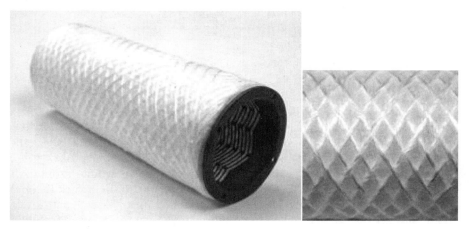

FIGURE 9.20. Wound fiber cartridge. Designer Guide, 3M Company 1995.

There are three types of filter cartridges made from Nextel fiber: basic cartridge, electric cartridge, and concentric tube pack. These cartridges are shown schematically in Figures 9.20, 9.21, and 9.22, respectively (Designer Guide, 3MCompany 1995). A brief description of each cartridge type follows (Majewski and Khair 2006).

The basic filter cartridge consists of a perforated support tube with ceramic fibers wound around it in a diamond-shaped pattern in many layers to achieve the desired fiber depth for filtration efficiency. The support tube is made of 1.22-mm-thick Type 304 stainless steel perforated with a 50% open area. The end sections of this tube, approximately 30 to 50 mm wide, are not perforated to form a gas seal during the winding process. One end of the tube is closed to force exhaust gas through the wound fiber layers. The flow of exhaust gas can be either outside-in or inside-out making the cartridge design more flexi-

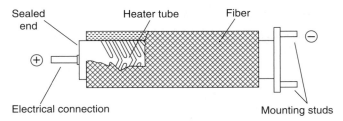

FIGURE 9.21. Fiber cartridge with internal heater. Designer Guide, 3M Company 1995.

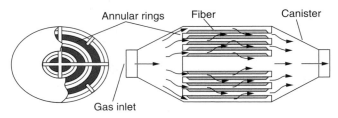

FIGURE 9.22. Concentric tube pack. Designer Guide, 3M Company 1995.

ble. These cartridges are available in 25 to 300 mm diameter and 75 to 1,500 mm length. The soot holding capacity and filtration efficiency depend on the acceptable pressure drop with typical values of 4 g/cartridge and 90% efficiency at a pressure drop of 14 kPa and flow rate of 3.4 m³/min.

In the case of an electric cartridge, the ceramic fiber is wound around an electrically resistive heater support structure made of a 0.46-mm-thick Inconel 600 nickel-based alloy sheet. The latter is punched in a "V-slot" pattern of about 50% open area dictated by the required electrical resistance. Much like basic cartridge design, the last 40- to 70-mm sections at each end of the material are not punched, allowing for the gas seal to form. The punched sheet is rolled into a cylinder and welded. The exhaust gas flows from inside to outside, thereby taking advantage of electrical heat for regeneration. The main exhaust gas stream bypasses the filter; only a controlled amount of air or exhaust gas (about 2.5% of nominal gas flow capacity of filter) is supplied to the cartridge. The electric cartridges are available in 38–76 mm diameter and 200 to 300 mm length. They can be operated on 12 or 24 V and require 400 to 1,000 W power.

The concentric tube pack (CTP) is made up of several concentric filter cartridges, each of slightly larger diameter than the previous one. They are separated by annular rings that block the space between concentric support tubes at alternate ends forcing exhaust gas to flow through the filter media and trapping particulate material in fiber layers. The canister is typically made from 1.6-mm-thick Type 409 stainless steel. The CTPs are available in outer diameters of up to 321 mm and lengths of up to 711 mm. The standard packs consist of three or five concentric cartridges and are capable of handling flow rates of 7 to 51 m^3/min.

Compared with wall-flow filters, the benchmark for diesel filter materials, the fiber filters offer the advantages of (1) resistance to thermal and mechanical stresses, (2) low risk of clogging with soot or ash for T_{max} < 900 C, (3) tolerance to additive ash accumulation if used with fuel additive, (4) good efficiency for removing nanoparticles, and (5) good noise attenuation. On the other hand, fiber filters suffer from (1) incompatibility with catalysts, thus requiring special catalyst formulations along with CVD (chemical vapor deposition) processing; (2) embrittlement of fibers caused by sintered ash deposits at T_{max} > 900 °C; (3) low filtration efficiency with susceptibility to blow off accumulated soot; (4) secondary fiber emissions from premature fiber breakage; (5) relatively large diameter compared with monolithic filters for equivalent filtration volume; and (6) relatively high cost.

9.7.4 New Filter Designs

As noted, one way to improve the filter's thermal durability is to reduce the peak regeneration temperature by increasing its thermal mass or heat capacity. This is most readily done by modifying the cell design, e.g., by increasing the cell density and wall thickness simultaneously. A series of regeneration tests were conducted on 2″ diameter × 6″ long (5.08 cm × 15.24 cm) EX-80 filters, with different cell designs, loaded with 9.6 g/l of soot, and the peak regeneration temperature was measured as a function of the filter's weight or heat capacity (Hickman 2000). These data, summarized in Figure 9.23, demonstrate

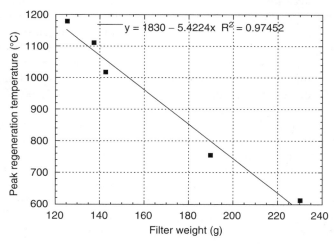

FIGURE 9.23. Effect of filter weight on peak regeneration temperature. Reprinted from Hickman 2000; courtesy of SAE.

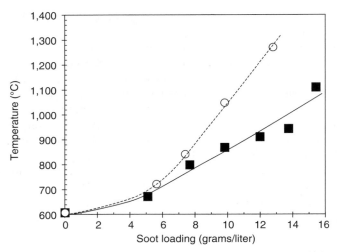

FIGURE 9.24. Maximum temperature in 14.4-cm × 15-cm filters of RC 200/19 (solid squares) and EX-80, 100/17 (open circles) during uncontrolled regeneration versus soot loading. Reprinted from Merkel et al. 2001a; courtesy of SAE.

that the peak temperature can be reduced by several hundred degrees by increasing the heat capacity via filter weight. A similar reduction in peak regeneration temperature was observed for the RC 200/19 filter whose heat capacity is 20% higher than that of the EX-80, 100/17 filter (see Figure 9.24) (Merkel et al. 2001a).

Another approach to improving thermal durability is to introduce stress relief slits in the center region of the filter that can reduce thermal stresses by 20% to 70% depending on slit dimensions and location as shown in Figure 9.25 (Miwa 2001). Of course, these slits must be filled with sealing material to prevent soot-laden exhaust gas from escaping. Regeneration tests on cordierite, SiC, and Si–SiC filters verified that both the improved material properties of the Si–SiC filter and the presence of stress-relief slits helped increase the failure temperature from 900 °C (for SiC) to 1,100 °C with soot loading as high as 22 g/l (see Figure 9.26).

The design approach can also be used to reduce pressure drop across the filter by increasing its diameter/length ratio while preserving the required filter volume and filtration area (Hickman 2000). Other design approaches for improving filter functionality and durability include modifying its channel geometry at inlet versus outlet faces. One such development by Corning Incorporated involves asymmetric cell technology (ACT) shown in Figure 9.27 where the open channels at inlet are larger than those at outlet compared with standard filter design with identical channel sizes (U.S. Patent 7,247,184). Such a channel design enables higher ash storage capacity combined with lower ash-loaded back pressure due to larger hydraulic diameter and higher open

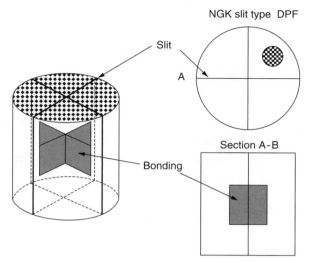

FIGURE 9.25. Filter design with stress-relief slits. Reprinted from Miwa 2001; courtesy of SAE.

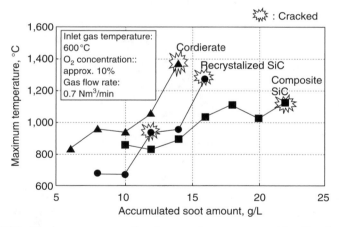

FIGURE 9.26. Maximum regeneration temperature versus soot loading for three different filters with and without stress-relief slits. Reprinted from Miwa 2001; courtesy of SAE.

volume at inlet. The ACT design also helps preserve the mechanical and thermal durability of filter.

Back pressure data, measured on an engine bench, help compare the functionality of a standard AT filter with that of an ACT–AT filter in Figure 9.28 at various soot and ash loadings (Heibel et al. 2007a). The slightly higher back pressure for the ACT-AT filter in clean state is most likely due to the lower

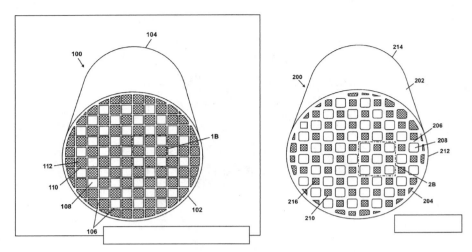

FIGURE 9.27. Schematic of standard and ACT channel geometries of a diesel filter. U.S. Patent 7,247,184.

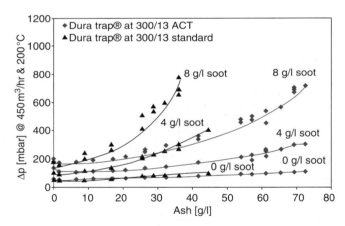

FIGURE 9.28. Comparison of pressure drop across standard AT versus ACT–AT filters for various ash and soot loadings. Reprinted from Heibel et al. 2007a, courtesy of SAE.

hydraulic diameter of open channels at outlet, which results in higher frictional and exit losses. Once the filter is loaded with ash and/or soot, the benefits of the ACT design become clear!

9.8 APPLICATIONS

Ceramic wall-flow filters have performed successfully since their introduction in the 1980s in passenger cars. Substantial numbers of filters have been installed to date and continue to meet emissions, back pressure, and durability requirements. With new developments in filter material, additives and catalyst technology, packaging designs, and regeneration techniques, ceramic wall-flow filters are being tested in new and more severe applications, including buses and trucks. The following examples help illustrate their design and durability.

9.8.1 Large Frontal Area Filter

As our first example, we analyze the thermal durability of an EX-54, 100/17 filter (10.63″ diameter × 12″ long or 27 cm × 30.48 cm) during regeneration. The thermal stresses during regeneration, which control filter durability, are governed by not only the filter properties and geometry but also by the regeneration conditions, e.g., flow rate, flow distribution, soot loading, % oxygen, and burner temperature. This example illustrates the effects of flow rate, burner temperature, and filter geometry on regeneration stresses.

The regeneration process was simulated by thermal cycling a clean filter with an electronically controlled burner and a centrifugal blower (for combustion air). The typical cycle consisted of 5 minutes of heating (with burner on) and 3 minutes of cooling (with burner off). Several engine conditions, namely full load, normal load, and idling, could be simulated by adjusting both the burner temperature and the flow rate. The radial temperature distribution during regeneration was obtained with the aid of 27 Type K chromel-alumel thermocouples at 5-s intervals (during the 8-minute cycle) as shown in Figure 9.29. With the burner temperature at 715 °C, the maximum radial gradient at midsection L3 occurred at t = 185 s with center and periphery temperatures approaching 930 °C and 300 °C, respectively (see Figure 9.30) (Gulati and Lambert 1991).

The thermal stresses were computed by the finite element code ANSYS (Swanson Systems, PA) taking full advantage of the axial symmetry of the filter (see Figure 9.31). Both the temperature-dependent physical properties of the EX-54 filter and the time-dependent thermocouple data were used as inputs to stress analysis. The maximum stresses in axial and tangential directions at the midsection are summarized in Table 9.19. It should be noted in Table 9.19 that the radial temperature gradient is the major contributor to thermal stresses; those due to axial gradient are less than 20%.

The axial stress approaches a value of 540 psi, far in excess of the high-temperature axial strength of 260 psi (see Table 9.7). Similarly, the tangential stress approaches 200 psi, which also exceeds the tangential strength of 120 psi. Figure 9.32 shows the variation of maximum axial stress at five different filter sections with time. It is clear that as the regeneration front moves from entry to exit sections, so does the dynamic stress wave from section L1 to L3 to L5. The maximum stress occurs at midsection and can result in ring-off failure if its magnitude and duration exceed the safe allowable values. Furthermore, the simulated regeneration with low flow rate and a burner temperature of 715 °C

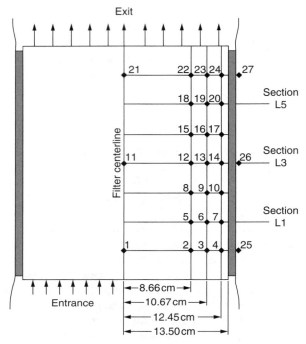

FIGURE 9.29. Thermocouple locations in the LFA filter. Reprinted from Gulati 1996b; courtesy of Marcel Dekker, New York.

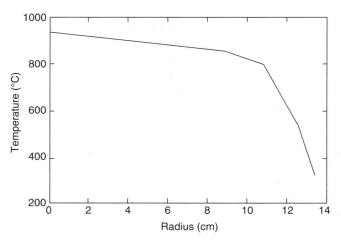

FIGURE 9.30. Regeneration temperature profile in radial direction at midsection ($T_{burner} = 715\,°C$, t = 185 s). Reprinted from Gulati 1996b; courtesy of Marcel Dekker, New York.

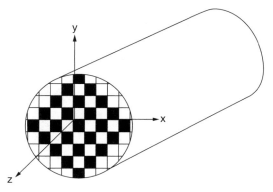

FIGURE 9.31. Coordinate system for computing thermal stresses in the diesel filter. Reprinted from Gulati 1996c; courtesy of Marcel Dekker, New York.

TABLE 9.19 Maximum tensile stress at midsection during simulated regeneration.

	Axial stress (psi)	Tangential stress (psi)
Radial gradient	450	200
Axial gradient	90	0
Total stress	540	200
$T_{burner} = 715\,^{\circ}C$		

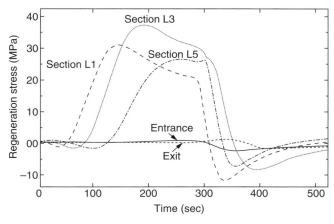

FIGURE 9.32. Dynamics of axial stress-time history during regeneration at $T_{burner} = 715\,^{\circ}C$. Reprinted from Gulati 1996b; courtesy of Marcel Dekker, New York.

is too severe for the specific LFA filter and hence unacceptable from a durability point of view.

The above stresses may be reduced by either lowering the burner temperature or using a smaller filter length. The former was verified by simulating the

TABLE 9.20. Effect of filter length on maximum axial stress at the midsection of the EX-54 filter.

Filter length (in)	Axial stress (psi)
12.0	333
10.6	319
9.4	290
8.3	254
7.1	218
6.0	167

regeneration process with a burner temperature of 600 °C. The maximum stress was reduced from 540 to 330 psi, and tangential stress was reduced from 200 to 125 psi. Thus, a 115 °C lower burner temperature reduced the maximum stresses by nearly 40%! Using the lower burner temperature, which proved highly beneficial, the filter length was varied from 12″ (30.48 cm) to 6″ (15.24 cm) in a stepwise fashion to reduce the axial stress further (see Table 9.20). It is clear from Table 9.20 that the axial stress can be halved to the safe allowable value (see Table 9.12) by using a lower burner temperature and reducing the filter length to 6″ (15.24 cm), thereby ensuring long-term durability.

Other ways to promote filter durability, which are currently being examined, include modification of filter properties via composition and process research and upgrading of packaging design via heavier mat, higher mount density, and stiffer shell (see Section 9.6).

9.8.2   Filter Durability for a 6.2-l Light-Duty Diesel Engine

In this second example, we examine the durability of another LFA filter of EX-47, 100/17 composition (7.5″ diameter × 8″ long or 19.05 cm × 20.32 cm) for a 6.2-l light-duty diesel engine with a burner bypass system (see Figure 9.33). In addition to 0.13-g/mile particulate emissions regulation, the filter must demonstrate a life durability of 120,000 miles. Based on the properties data alone (Tables 9.6–9.12), the EX-47 filter offers low CTE, high strength, small mean pore diameter, and high filtration efficiency.

This example demonstrates how the filter properties are modified by the regeneration process, thereby limiting the regeneration stresses to below the threshold value and ensuring filter reliability. These modifications are attributed to successive but controlled microcracking, which occurs during regeneration (Gulati et al. 1992a). The reliability of all components of the regeneration system, i.e., exhaust metering valve, air blower, fuel supply, fuel pump, solenoid valve, burner, and flow distributor, was further ensured by careful control of inlet gas conditions via air/fuel ratio, heating the gas mixture with an optimally designed fuel-powered burner, maintaining a certain minimum flow rate throughout the regeneration process, and limiting the soot loading to 20 g prior to regeneration. The typical regeneration took 10 minutes although shorter

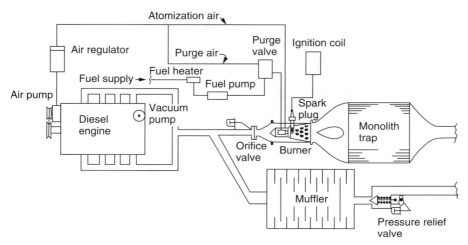

FIGURE 9.33. Burner bypass system for a 6.2-l light-duty diesel engine. Reprinted from Gulati 1996b; courtesy of Marcel Dekker, New York.

TABLE 9.21. Operating conditions during regeneration of 7.5-in-diameter × 8-in-long EX-47 filter.

Engine RPM	Engine torque (ft lb)	Flow rate through filter (lb/hr)	% O_2 in inlet gas
650	Idle	230	10.5%
1,200	90	246	8.9%
2,200	180	279	5.6%

durations also met Environmental Protection Agency (EPA) specifications for acceptable particulate level. The key advantages of shorter regeneration are significant fuel savings and minimal thermal fatigue of the ceramic filter, both of which are critical to the viability of a wall-flow particulate trap system. The soot filter was regenerated in the test cell under three different engine conditions to simulate the various driving conditions and terrains (see Table 9.21). Both the flow rate and the oxygen content of the inlet gas are also recorded in Table 9.21.

Of the three sources of stress, namely mounting, vibrations, and regeneration, the latter are most critical and are influenced by temperature gradients and physical properties. Figures 9.34 and 9.35 show thermocouple locations and the temperature-time history of center and peripheral regions defined by

$$T_c = \frac{1}{\pi r_o^2} \int_0^{2\pi} \int_0^{r_o} T(r) \, r \, d\theta \, dr \qquad (9.20)$$

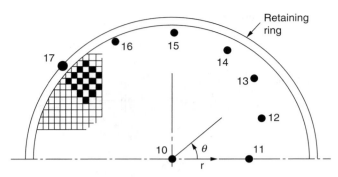

FIGURE 9.34. Thermocouple locations at midsection of filter during regeneration. Reprinted from Gulati 1996b; courtesy of Marcel Dekker, New York.

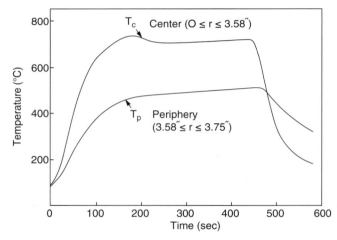

FIGURE 9.35. Temperature-time history at center and periphery of filter's midsection during 8-minute regeneration cycle at engine idle. Reprinted from Gulati 1996b; courtesy of Marcel Dekker, New York.

$$T_p = \frac{1}{\pi(R^2 - r_0^2)} \int_0^{2\pi} \int_{r_0}^{R} T(r) \, r \, d\theta \, dr \qquad (9.21)$$

where $r_0 = 3.58''$ and $R = 3.75''$ for this LFA filter. It is clear in Figure 9.35 that a radial gradient of ~200 °C is present for 5 of the 8-minute cycles and produces tensile stresses in the peripheral region in axial and tangential directions that, if excessive, can lead to thermal fatigue and/or fracture. When the burner is turned off at the end of regeneration, the temperature profile is reversed with the center becoming colder than the periphery. The thermal stresses also change sign and become tensile in the center region and compressive in the peripheral region. However, both the lower temperature and shorter duration

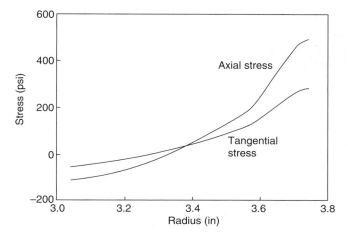

FIGURE 9.36. Radial variation of axial and hoop stresses at midsection at t = 200 s during regeneration of a new filter at engine idle. Reprinted from Gulati 1996b; courtesy of Marcel Dekker, New York.

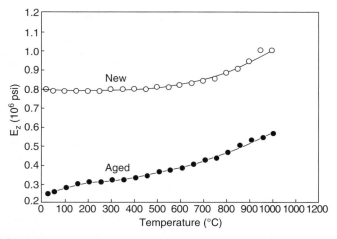

FIGURE 9.37. Variation of axial modulus of new and aged filters with temperature. Reprinted from Gulati 1996b; courtesy of Marcel Dekker, New York.

of the cooling cycle will reduce the stress magnitude and pose little concern about internal fracture.

Figure 9.36 shows the distribution of axial and tangential stresses in the peripheral region during regeneration. Both of these stresses exceed the MOR value in the region 3.6″ < r ≤ 3.75″ (9.144 cm to 9.525 cm) and produce macrocracking confined to that region. Consequently, the E-modulus of that region is reduced by nearly 60% with successive regenerations as indicated in Figure 9.37. Although the thermal expansion of this region is not affected, the regen-

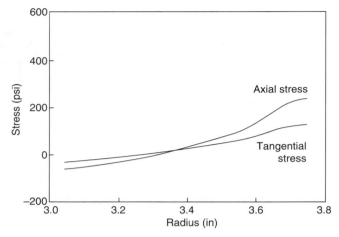

FIGURE 9.38. Radial variation of axial and hoop stresses at midsection at t = 200 s during regeneration of aged filter at engine idle. Reprinted from Gulati 1996b; courtesy of Marcel Dekker, New York.

TABLE 9.22. Maximum thermal stresses at midbed of new and aged EX-47 filters under various regeneration conditions.

Engine load speed (RPM)	Torque (ft lb)	Max. stress (psi)			
		New filter		Aged filter	
		σ_θ	σ_z	σ_θ	σ_z
750	idle	295	515	130	240
1,200	90	240	500	105	240
2,200	180	215	460	100	225

eration stresses are reduced by 50% due to lower E-modulus (see Figure 9.38). As a result, the aged filter experiences stresses lower than its threshold limit and continues to perform reliably over its required lifetime. Regeneration stresses at higher engine speeds and torque are 10–20% lower than those at engine idle (see Table 9.22). Thus, the likelihood of crack initiation and propagation is considerably lower under normal and full load conditions, which is good news from a total durability point of view since the majority of regenerations will occur under off-idle conditions.

Finally, the mounting design induces compressive stresses of 90 psi in axial direction and 75 psi in radial direction, which enhances the threshold limits of the EX-47 filter to 275 psi in the axial and 150 psi in the tangential direction. Comparing these with regeneration stresses in an aged filter, it is clear that the nature of macrocracking in the peripheral region will be stable and controlled due to low regeneration stresses. This points out the critical role that mounting plays in ensuring both the mechanical and the thermal durability of ceramic wall-flow filters.

9.8.3 High-Tech Filter with Optimum Performance

In this final example, we focus on performance and durability data for the EX-80 filter with optimized microstructure and physical properties. As noted in Sections 9.6 and 9.7, this high-tech composition offers low mean pore size for high filtration efficiency and low rate of pressure drop buildup at higher levels of soot accumulation. It has the lowest CTE values over the operating temperature range, which minimizes regeneration stresses. And it has the highest fatigue resistance, which permits higher threshold stresses desirable for long-term thermal durability. The above combination of filter properties, achieved by optimizing both the raw materials and the manufacturing process, was necessitated by more stringent functionality requirements and a 290,000-mile lifetime durability (Murtagh et al. 1994).

The filtration functionality of a 10.5″ diameter × 12″ long (26.67 cm × 30.48 cm) EX-80 filter, with a 100/17 cell structure, was measured by Ortech International (Ontario, Canada) using a 1989 Detroit Diesel 6V-92 TA DDEC II diesel engine and low sulfur diesel fuel (D-1 with 0.1% S). A steady-state engine speed-load condition of 300 ft lbs at 1,600 RPM was selected to achieve an engine flow of approximately 500 SCFM and an exhaust temperature of 260 °C. The engine soot output averaged 18 g/hr. All testing was performed with the engine's exhaust directed to, and diluted through, a dilution tunnel/PDP assembly of approximately 2000 SCFM. Forty-minute efficiency determinations were carried out until the engine's exhaust reached a back pressure of 5 in Hg. The pressure drop across the filter was measured at 1-min intervals during the test. Table 9.23 summarizes the collection efficiency and pressure drop data at three successive soot loadings, whereas Figure 9.39 captures the complete data.

Previous investigations have shown that filters with small pore size yield high efficiency and high back pressure, whereas those with large pore size yield low efficiency and low back pressure (Howitt and Montierth 1981; Kitagawa et al. 1992; Shinozaki et al. 1990). Although the EX-80 filter shows this trend at low soot levels, it soon shifts from conventional predictions due to formation of a soot membrane on the wall surface at higher loadings. During the early stages of filtration, the intrinsic pore size distribution of the filter wall is altered by soot penetration, which is minimal for the EX-80 filter due to its small pore size. Consequently, the bulk of the soot ends up as a soot membrane that behaves as a filtration medium and reduces the rate of pressure drop buildup due to absence of further alteration of the intrinsic pore structure of the filter wall. This dual pore-size filtration mechanism sets in at 15 g of soot loading in

TABLE 9.23. Filtration data for the LFA EX-80 filter.

Soot loading (g)	Collection efficiency (%)	Pressure drop (in. Hg)
5	87	1.6
15	88	2.0
50	92	3.1

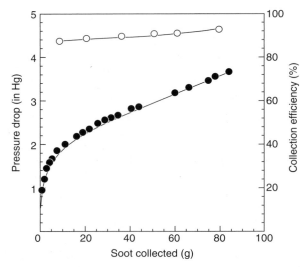

FIGURE 9.39. Pressure drop and filtration efficiency for the EX-80 filter as a function of soot loading. Reprinted from Gulati 1996b; courtesy of Marcel Dekker, New York.

the 10.5″ × 12″ (26.67 cm × 30.48 cm) filter tested, i.e., at 0.9 g/l of soot loading, and reduces the pressure drop buildup rate from 0.38 to 0.02 in Hg/g.

The effect of low CTE and high fatigue resistance of the EX-80 filter (see Tables 9.9 and 9.11) is best demonstrated by the excellent thermal shock behavior of a 11.25″ diameter × 12″ long (28.58 cm × 30.48 cm) filter under severe regeneration conditions produced by KHD Deutz Cologne, Germany (Murtagh et al. 1994). Figure 9.40 shows the midbed temperature profile during the 10-minute regeneration cycle. It should be noted that the center temperature T_c is nearly 500 °C higher than the periphery temperature T_p for the bulk of the regeneration cycle. The peak axial thermal stress, associated with these gradients, was estimated by finite element analysis to be 195 psi. The corresponding threshold strength of the EX-80 filter at $T_p = 200$ °C is 250 psi, well above the peak thermal stress. This implies no flaw growth during the 2175 regenerations, at 200-mile intervals, over the desired lifetime of 435,000 miles. Thus, the CTE, high MOR, and fatigue resistance ensure the physical durability of the EX-80 filter even under severe regeneration conditions.

9.9   SUMMARY

The stringent emissions legislation for diesel-powered vehicles has led to new developments in both the oxidation catalysts and the filters. These developments include new materials for catalyst supports and filters with higher heat capacity, filtration area, use temperature, and physical durability. In addition,

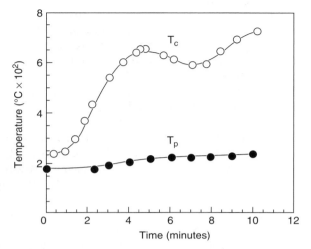

FIGURE 9.40. Thermal profile during severe regeneration as a function of regeneration time. Reprinted from Gulati 1996b; courtesy of Marcel Dekker, New York.

the advent of low sulfur fuels is helping develop better catalysts to meet the 120,000-mile vehicle durability. Similarly, improvements in engine design have reduced particulate material via more efficient fuel combustion. Furthermore, significant progress is being made in reducing oxides of nitrogen via NO_x adsorbers and $deNO_x$ catalysts. Also, a variety of fuel additives have been developed that help reduce the soot regeneration temperature, thereby reducing thermal stresses and enhancing the physical durability of diesel filters. Over 30 years of successful experience with ceramic catalyst supports for automotive application is proving valuable in designing a robust mounting system for diesel oxidation catalysts. In view of considerably lower operating temperature and longer physical durability requirement, relative to automotive catalysts, the intumescent mat used in packaging diesel oxidation catalysts has to be preheat-treated to ensure sufficient holding pressure on the catalyst against inertia, vibration, and road shock loads experienced in service.

Diesel particulate filters with a 200/12 cell structure are now available with twice the filtration area of a 100/17 cell structure, which require less space under the vehicle. However, their higher back pressure, accompanied by fuel penalty, must be weighed against space savings. Similarly, filters with 100/25 and 200/19 cell structures offer higher heat capacity, thereby reducing the peak regeneration temperature and the associated thermal stresses, which would enhance their physical durability. However, thicker cell walls would also lead to higher back pressure. Hence, certain trade-offs are necessary in selecting the optimum filter design. Improved engine controls along with reliable monitoring of back pressure and exhaust temperature would help limit the back pressure and trigger soot regeneration at more frequent intervals, thereby reducing the peak temperature. And yet, a certain critical mass of accumulated soot is necessary to sustain self-regeneration once initiated with the help of, for example, an electric heater. An alternative approach to safe regeneration would be to use fuel additives (organometallic compounds) that reduce the soot oxidation temperature to less than 500 °C.

Filter materials, which have a higher melting temperature than that of cordierite, have also been developed and are being used in commercial applications subject to more stringent emissions legislation. These materials include SiC, RC 200/19, and AT, which can withstand uncontrolled regeneration due to either their higher conductivity (SiC) or heat capacity (RC 200/19) or unique microstructure (AT). Similarly, new filter designs with stress relief slits have helped reduce regeneration stresses by 20% to 70%, thereby permitting the use of higher cell density or higher mass filters at high regeneration temperature. The high back pressure penalty associated with such designs is readily minimized by increasing the diameter/length ratio of these filters. In balancing the heat capacity and peak regeneration temperature, however, it must be kept in mind that the ash buildup does not react with filter material and render it weak or dysfunctional. An alternative approach for reducing back pressure is to modify channel geometry/size at inlet versus outlet, which also helps increase the soot and ash loading capacities of filter.

Although SiC offers higher thermal conductivity and melting temperature, which are desirable for uncontrolled soot regeneration, its order of magnitude higher thermal expansion coefficient can lead to inferior thermal shock resistance. An improved version, namely Si–SiC composite material, has recently been developed that offers a low thermal expansion coefficient and superior thermal shock resistance.

Other filter materials like sintered metal, mullite, and ceramic fiber cartridges are also available for both retrofit and new applications notwithstanding their cost.

Finally, the mounting system can play a major role in ensuring both the mechanical and thermal durability of diesel oxidation catalysts and filters notably for heavy-duty trucks with severe operating conditions and a 435,000-vehicle-mile durability requirement. Many of the robust packaging systems employed in automotive applications are equally applicable to both diesel oxidation catalysts and filters.

REFERENCES

Abdel-Rahman, A. A. "On the emission from internal combustion engines: A review, *International Journal of Energy Research*" 22: 483–513 (1998).

Amann, C., Stivender, D., Plee, S., and MacDonald, J. "Some rudiments of diesel particulate emission," SAE 800251 (1980).

Bloom, R. L. "The development of fiber wound diesel particulate filter cartridges," SAE 950152 (1995).

Brown, G. *Unit Operations*. Wiley and Sons, New York (1955).

Carman, P. *Flow of Gases through Porous Media*. Butterworth, London, England (1956).

Cooke, W. F., and Wilson, J. J. N. "A global black carbon aerosol model," *Journal of Geophysical Research*, 101: 19395–19410 (1996).

Cutler, W., and Merkel, G. "A new high temperature ceramic material for diesel particulate filter applications," SAE 2000-01-2844 (2000).

Designer Guide. "3M Nextel Diesel Filter Cartridges for Particulate Emission Control," 3M Company, St. Paul, MN (1995).

Eastwood, P. *Critical Topics in Exhaust Gas Aftertreatment*. Research Studies Press Ltd., Baldock, Hertsfordshire, England (2000).

Faiz, A., Weaver, C. S., and Walsh, M. P. "Air pollution from motor vehicles, standard and technology for controlling emission," World Bank, Washington, D.C. (1996).

Farleigh, A., and Kaplan, L. "Danger of diesel," U.S. Public Interest Research Group Education Fund (2000).

Farrauto, R., et al. "Reducing truck diesel emissions," *Automotive Engineering* (1992).

Gulati, S., and Helfinstine, J. "High temperature fatigue in ceramic wall-flow diesel filters," SAE 850010 (1985).

Gulati, S., and Kulkarni, N. "Catalytic converter durability," *Truck Engineering* (1992).

Gulati, S., and Lambert, D. "Fatigue-free performance of ceramic wall-flow diesel particulate filter," ENVICERAM '91; Saarbrücken, Germany (1991).

Gulati, S., and Sherwood, D. "Dynamic fatigue data for cordierite ceramic wall-flow diesel filters," SAE 910135 (1991).

Gulati, S. "Thermal stresses in ceramic wall-flow diesel filters," SAE 830079 (1983).

Gulati, S. Chap. 2. "Ceramic catalysts supports for gasoline fuel," in *Structured Catalysts and Reactors*. Editors Cybulski, A., and Moulijn, J. A., Marcel Dekker, New York (1996a).

Gulati, S. Chap. 18. "Ceramic catalysts, supports, and filters for diesel exhaust aftertreatment," in *Structured Catalysts and Reactors*, Editors Cybulski, A., and Moulijn, J. A., Marcel Dekker, New York (1996b).

Gulati, S., Lambert, D., Hoffman, M., and Tuteja, A. "Thermal durability of ceramic wall-flow diesel filter for light duty vehicles," SAE 920143 (1992a).

Gulati, S. "Strength and thermal shock resistance of segmented wall-flow diesel filters," SAE 860008 (1986).

Gulati, S. "Design considerations for diesel flow through converters," SAE 920145 (1992b).

Gulati, S., Sherwood, D., and Corn, S. H. "Robust packaging system for diesel/natural gas oxidation catalysts," SAE 960471 (1996c).

Health Effect Institute. "Understanding the health effects of components of the particulate matter mix: progress and next step," HEI Perspectives (2002).

Heibel, A., et al. "Performance and durability of the new corning duratrap[R] AT diesel particulate filter-results from engine bench and vehicle tests," Aachener Kolloquium Fahrzeug und Motorentechnik, Germany (2005).

Heibel, A., et al. "Performance evaluations of aluminum titanate diesel particulate filters," SAE 2007-01-0656 (2007a).

Heibel, A. "Advanced diesel particulate filter design for lifetime pressure drop solution in LD applications," SAE 2007-01-0042 (2007b).

Heywood, J. B. *Internal Combustion Engine Fundamentals.* McGraw-Hill, New York (1988).

Hickman, D. "Diesel particulate filter regeneration: Thermal management through filter design," SAE 2000-01-2847 (2000).

Howitt, J., Elliott, W., Morgan, J., and Dainty, E. "Application of a ceramic wall-flow filter to underground diesel emissions reduction," SAE 830181 (1983).

Howitt, J., and Montierth, M. "Cellular ceramic diesel particulate filter," SAE 810114 (1981).

Jacobs, T., et al. "Development of partial filter technology for HDD retrofit," SAE 2006-01-0213 (2006).

Johnson, T. "Diesel emission control technology in review," SAE 2001-01-0184 (2001).

Kitagawa, J., Asami, S., Ushara, K., and Hijikata, T. "Improvements of pore size distribution of wall flow type diesel particulate filter," SAE 920144 (1992).

Kitagawa, J., Hijikata, T., and Makino, M. "Analysis of thermal shock failure on large volume DPF," SAE 900113 (1990).

Lox, E., Engler, B., and Koberstein, E. CAPoC-II, Brussels, Germany (1990).

Ludecke, O., and Dimick, D. "Diesel exhaust particulate control system development," SAE 830085 (1983).

Majewski, W. A., and Khair, M. K. *Diesel Emissions and Their Control.* Second Edition. SAE International, Warrendale, PA (2006).

Mark, J., and Morey, C. *Diesel passenger vehicles and the environment,* Union of Concerned Scientists, Berkeley, CA, pp. 6–15 (1999).

Merkel, G., Beall, D., Hickman, D., and Vernacotola, M. "Effects of microstructure and cell geometry on performance of cordierite diesel particulate filters," SAE 2001-01-193 (2001a).

Merkel, G., Cutler, W., and Warren, C. "Thermal durability of wall-flow diesel particulate filters," SAE 2001-01-190 (2001b).

Miwa, S. "Diesel particulate filters made form newly developed sic and newly developed oxide composite material," SAE 2001-01-0192 (2001).

Murtagh, M. J., Sherwood, D., and Socha, L. "Development of a diesel particulate composition and its effect on thermal durability and filtration performance," SAE 940235 (1994).

Ogunwumi, S. B., et al. "Aluminum titanate compositions for diesel particulate filters," SAE 2005-01-0583 (2005).

Ohno, K., Shimato, K., Taoka, N., Santae, H., Ninomiya, T., Komori, T., and Salvat, O. "Characterization of SiC-DPF for passenger car," SAE 2000-01-0185 (2000).

Pyzik, A., et al. "Development of acicular mullite materials for diesel particulate filters application," DEER Conference, Chicago, IL (2005).

Rao, V. D., et al. "Advanced techniques for thermal and catalytic diesel particulate trap regeneration," SAE 850014 (1985).

Ritter, J. *Fracture mechanics of ceramics*, Vol. 4, Editor Bradt R. C., Plenum Press, New York (1978).

Shinozaki, O., et al. "Trapping performance of diesel particulate filters," SAE 900107 (1990).

Springer, K. VII SIMEA, Symp. for Auto. Engineers, Sao Paulo, Brazil (1993).

Stroom, P., et al. "Systems approach to packaging design for automotive catalytic converters," SAE 900500 (1990).

Swanson Systems. *ANSYS Finite Element Analysis*. Elizabeth, PA.

Tsien, A., Diaz-Sanxhez, D., Ma, J., and Saxon, A. "The organic component of diesel exhaust particles and phenanthrene, a major polyaromatic hydrocarbon constituent, enhances IgE production by IgE secreting EBV-transformed human B cells in vitro," *Applied Pharmaceutical Toxicology*, 142: 256–263 (1997).

U.S. Patent 7,247,184 (2007).

Vergeer, H., Gulati, S., Morgan J., and Dainty, E. "Electrical regeneration of ceramic wall flow diesel filter for underground mining application," SAE 850152 (1985).

Wade, W., et al. "Thermal and catalytic regeneration of diesel particulate trap," SAE 830083 (1983).

Wade, W., White, J., and Florek, J. "Diesel particulate trap regeneration techniques," SAE 810118 (1981).

Weaver, C. "Particulate control technology and particulate standards for heavy duty diesel engines," SAE 840174 (1984).

Wiederhorn, S. *Fracture mechanics of ceramics*, Vol. 2, Editor Bradt R. C., Plenum Press, New York (1974).

Yanowitz, J., McCormick, R., and Graboski, M. "In-use emissions from heavy duty diesel vehicles," *Environmental Science Technology* 34:729–740 (2000).

CHAPTER 9 QUESTIONS

1. What are the key requirements for a diesel particulate filter?
2. Name the critical properties of filter material to help meet the above requirements.
3. Consider an extruded ceramic filter, 5.66 in diameter × 8 in long, with a 200/12 square-cell structure and alternate plugging pattern on entry and exit faces; assuming uniform soot distribution, soot density of 0.05 g/cm^3, and plug depth of 0.4 in, calculate the following quantities for this filter:
   a) open frontal area (OFA)
   b) total open volume (TOV)
   c) soot filtration area (SFA)
   d) soot capacity of filter (grams of soot)

4. In the above example, assume a 3-l engine with a soot emission rate of 0.05 g/l; calculate the vehicle miles driven to reach the soot capacity of the above filter.

5. Automotive substrates can and do crack in service without affecting conversion efficiency; yet, diesel filters cannot afford to crack. Explain why?

6. Cracking in diesel filters can be mitigated by reducing regeneration stresses via controlled regeneration. The allowable stress must therefore be lower than the measured strength of filter to account for (a) acceptable failure probability, (b) stressed area of filter compared with that of strength specimen, and (c) loss of filter strength via fatigue after 2000 regenerations. Compute the allowable stress, as a fraction of its initial strength, for a cordierite filter by allowing for the above-mentioned losses.

Use the following equations for the failure probability factor (FPF), stressed area factor (SAF), and fatigue factor (FF):

$$FPF = [\ln\{1/(1-F)\}]^{1/m}$$
$$SAF = [A_s/A_f]^{1/m}$$
$$FF = [\tau_o/\{\tau_\ell(n+1)\}]^{1/n}$$

Assume

F = acceptable failure probability = 0.01 (i.e., 1%)

m = Weibull modulus of strength distribution = 12

n = fatigue constant of filter = 20

A_s = stressed area of strength specimen = 0.75 in^2

A_f = stressed area of filter = $d \times \ell$ (d = filter diameter = 9 in; ℓ = 12 in)

τ_o = equivalent time for measuring filter strength = 1 s

τ_ℓ = total duration of 2,000 regenerations @ 100 s each

  = 200,000 sec.

7. The typical filter materials used today are cordierite ceramic, silicon carbide, and aluminum titanate. Give a brief overview of their advantages and disadvantages.

10 Ozone Abatement within Jet Aircraft

10.1 INTRODUCTION

In the later part of the 1970s, flight personnel and passengers in commercial airliners frequently complained of headaches; irritation of the eyes, nose, and throat; and chest pains. It was also during this period that jet aircraft began to fly at higher altitudes to conserve fuel. The incidences were related to aircraft flying above 40,000 ft over polar routes where the ozone (O_3) concentration becomes significant. Fresh air containing small quantities of O_3 is brought through the air conditioning systems. This ozone is present in varying concentrations, depending on the location and altitude as shown in Figure 10.1 (Haldeman et al. 1977). Federal Aviation Administration (FAA) regulations now require that the airplane cabin ozone concentration (on a time weighted average) cannot exceed 0.1 vppm (sea level equivalent) (Wall Street Journal 1978; Federal Register 1980).

This application has matured to the point where publications on cabin air quality and safety now include a section on ozone and the methods to keep the air quality standard within the cabin for passenger safety using ozone catalytic converters (National Research Council 1986). Still there are ongoing studies to quantify the passenger exposure to ozone as evidenced from a recent publication where ozone concentrations were passively monitored in passenger cabins of commercial airliners flying domestic, Pacific, and southeast Asian routes. One hundred six flight segments were monitored for either the full duration and/or approximately 3 hrs during the middle portion of the flight for a total of 145 time-integrated measurements. This study showed that even in aircraft with catalytic ozone converters, passengers and flight crew may be exposed to increased ozone levels on domestic and international flights. Given the frequency of ozone excess, it is recommended that (1) ozone converters should be required equipment on all commercial passenger aircraft for mid- and high-latitude routes, (2) improved maintenance procedures should be required for catalytic converters (e.g., more frequent servicing/replacement), and (3) ozone should be routinely monitored on all mid- and high-latitude flights (Spengler et al. 2004).

Catalytic Air Pollution Control: Commercial Technology, by Ronald M. Heck and Robert J. Farrauto, with Suresh T. Gulati.
Copyright © 2009 John Wiley & Sons, Inc.

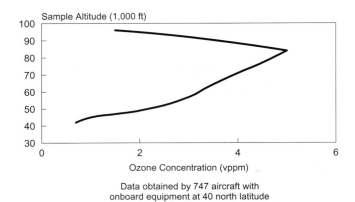

FIGURE 10.1. Ozone concentration levels at various altitudes. Reprinted with permission from Elsevier Science Publishers BV (Heck et al. 1992).

10.2   OZONE ABATEMENT

The airlines solicited several companies who sought solutions involving adsorption, thermal, and catalytic technologies to remove the ozone from the cabin. Ozone is a reactive substance and can be adsorbed or decomposed thermally or catalytically:

$$2O_3 \rightarrow 3O_2 \tag{10.1}$$

It was found that the required amount of adsorbent (e.g., carbon) would weigh too much for practical use. Thermal processes require higher temperatures and dictate that the air come from a higher stage of compression (which is hotter), but this would involve an efficiency penalty. Catalysts, on the other hand, permit the decomposition reaction to occur at much lower temperatures. Therefore, a lower stage of air compression (and associated lower temperatures) is sufficient. Figure 10.2 shows the catalytic and thermal decomposition of ozone. All aircraft equipped for ozone abatement, with the exception of the discontinued supersonic transport (SST), use a catalyst to decompose ozone.

The discontinued SST used thermal decomposition by taking the cabin air off a higher stage of compression. Compression of the air provides the energy to heat air for the jet engine. So depending on the stage of compression, the air is available at different isentropic temperatures. Of course, the higher the stage of compression for using the preheated air, the higher the parasitic power loss and the less efficient the jet engine.

Many precious metal and base metals catalytic materials were investigated (Carr and Chen 1982, 1983; Kent and Fein 1979; Chang 1980), and Figure 10.3 shows some candidate catalysts after short-term testing in the laboratory. Ozone is a powerful oxidant and will deactivate many catalytic materials after

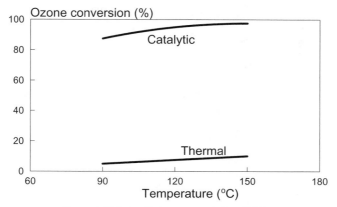

FIGURE 10.2. Catalytic conversion of ozone more efficient than the thermal process when measured at 18 milliseconds of residence time. Reprinted with permission from Elsevier Science Publishers BV (Heck et al. 1992).

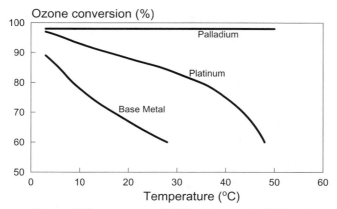

FIGURE 10.3. Pd catalyst most efficient for ozone conversion. Reprinted with permission from Elsevier Science Publishers BV (Heck et al. 1992).

a short time exposure. This caused considerable problems in early testing since many materials looked good on bench tests but failed in the aircraft after a relatively short time frame (Heck et al. 1992). This is demonstrated in Figure 10.4, where many materials are highly active at short times but deactivate after extended aging. The most effective catalytic material is about 1% Pd on γ-Al$_2$O$_3$ supported on a high cell density ceramic or metallic monolith. Beaded catalysts were tried, but their use was precluded by problems with reactor geometry and orientation, pressure drop, and the airplane operating environment (vibration and so forth).

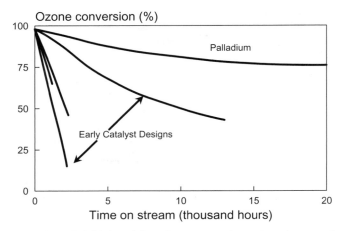

FIGURE 10.4. The high initial activity of ozone catalysts does not translate to long-term performance. Reprinted with permission from Elsevier Science Publishers BV (Heck et al. 1992).

TABLE 10.1. Range of design conditions for abaters in wide-body aircrafts.

Air flow (kg/s)	0.68–1.4
Temperature (°C)	120–200
Pressure (atm)	1.6–4.0
Allowable pressure drop (atm)	0.034–0.102
Vessel proof pressure (atm)	5–30
Required conversion (%)	83–93
Housing diameter (cm).	20–28
Length (cm)	40–60
Maximum weight (kg)	4.5–16
Shock and vibration	Individual aircraft manufacturer's specification

Source: Reprinted from Heck et al. (1992) with permission from Elsevier Science Publishers BV.

Typical operating condition ranges for wide-body jet aircraft are shown in Table 10.1. These catalysts operate at high space velocities of 200,000 to 500,000 hr^{-1}. Because of the high conversion requirements, the reactors are designed to operate in the bulk phase mass transfer control regime, so enhanced turbulence within the honeycomb chamber and geometric surface area are important design criteria. The reactor design uses segments of honeycomb usually 1 in (2.54 cm) in length for a total of five to seven segments. These segments interrupt the development of the boundary layer and provide for enhanced mass transfer and, hence, higher ozone performance. With reference to Chapter 4, this is equivalent to decreasing the L term in Figure 4.2 and, hence, to increasing the $N_{Re}d_{ch}/L$ term and the N_{Sh} or mass transfer coefficient K_g and therefore the mass transfer conversion. Diameters can range up to 12 in

FIGURE 10.5. Installation of ozone abater in underbody of jet aircraft. Reprinted with permission from Elsevier Science Publishers BV (Heck et al. 1992).

(30.5 cm). The major suppliers of aircraft ozone converters are BASF (formerly Engelhard) and Honeywell (BASF 1 2007; Honeywell 2008).

Because the ozone abater is located in the aircraft underbody in the ducting leading from the jet engine compressor to the passenger cabin, it must be designed to resist vibrations and other mechanical perturbations experienced in takeoff, flight, and landing. Still, it must be light enough to minimize the fuel penalty (Table 10.1). Figure 10.5 shows an ozone abater being installed in the underbody of a commercial jet aircraft. Each air handling system will contain an ozone abater (Bonacci and Heck 1983). Currently, several aircraft use catalytic converters, including the Boeing 747, 757, and 767; Lockheed L-1011; Douglas DC-10 and MD-11; Airbus A-300, A-310, A-319, A-320, A-321, A-330, and A-340; as well as the Gulfstream IV & V; Falcon 900 and 2000; Astra; Galaxy; and Global Express (Engelhard 1996). Because of the number of applications, the ozone abaters come in many different shapes and sizes as shown in Figure 10.6. An actual schematic of the ozone abatement system in the Boeing 767 has been published. The paper discusses the engineering aspects of a modern commercial jet airliner environmental control system (ECS), focusing on cabin air quality (Figure 10.7). A new 767 converter dissociates approximately 95% of the ozone entering the converter to oxygen. It has a useful life of about 12,000 flight hrs (Hunt et al. 1995).

The two major wide-body jet aircraft manufacturers, Boeing and Airbus, include in their sales brochures the use of catalysts for removing the ozone that comes into the air handling system (Nurcombe 2005; Hawk 2005). Boeing in the announcement of the new 787 Dreamliner actually has a diagram explaining the use of catalysts for the removal of ozone and maintaining cleaner healthier air in the airplane cabin (Hawk 2005).

The full cycle of replacement and repair of ozone abaters is now a mature business segment. For instance, Saywell International, one of the largest sup-

FIGURE 10.6. Ozone abaters for many aircraft applications.

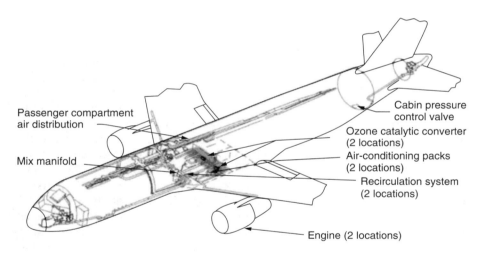

Typical components and system layout for the 767 enviromental control system (ecs)

FIGURE 10.7. System layout for 767 environmental control system showing ozone converters (Courtesy Boeing Inc.).

pliers of aircraft spares and components worldwide for aftermarket, is one distributor of ozone converters. They also publish an abbreviated component maintenance manual for the return, repair, replacement, and testing procedure for ozone converters (Saywell 2007; BASF 2007). AeroParts claims a unique procedure, which includes an overhaul process that restores the conversion efficiency of metallic converters to 96+% and ceramic converters to 98+% efficiency (AeroParts 2008). Over time, the effects of erosion and the accumulation of contaminants necessitate the replacement or overhaul of the ozone converter. The interval between overhaul and replacement depends on the type of aircraft and on the routes flown but usually falls between 10,000 and 14,000 hr. For best performance and service life, the ozone converter requires both cleaning and replacement of lost catalysts. AeroParts' unique procedure includes a thorough cleaning process that partially restores the performance of the ozone converter. A new catalyst may also be added. A major benefit of the AeroParts' procedure is their ability to overhaul the converter (for cleaning and addition of the catalyst) without opening the case or removing the cores. This ensures a short turnaround time resulting in significant cost savings.

Since the ozone converter is such a critical item for the aircraft operation, new and "off-the-shelf" exchange units are available 24 hr a day, 7 days a week, and both companies offer an aircraft on ground (AOG) program.

10.3   DEACTIVATION

The ozone catalysts sees many different environments, depending on the type of aircraft, airline maintenance procedure, flight routes, and time spent awaiting takeoff at different airports. The catalysts are operated at different inlet temperatures, depending on the compressor stage discharge temperature of the source air. The aircraft uses various lubricating oils and hydraulic fluids that can deposit on the catalyst. Flight routes determine the ambient exposure of the catalyst (e.g., flights to airports along the ocean involve exposure to salt water mist). Airports having high traffic volumes contain significant ground-level emissions, and the catalyst is exposed to the S and P present in fuels and lubricants. Dusts, halides from salt water, and so forth in the ambient environment all can contribute to deactivation.

To develop a replacement schedule for the ozone abater, an understanding of catalyst deactivation based on actual flight performance was required for development of a satisfactory catalyst. The palladium-based catalyst did not show any severe aging effects in the laboratory accelerated aging tests, which had feedstreams containing higher than anticipated concentrations of contaminants.

Full-scale tests were conducted on returned abaters for performance evaluation. Over a 10-year period, abaters were returned for evaluation as part of a joint effort with airline operators to develop service bulletins. One such study done for a fleet of Boeing 747 aircraft is shown in Figure 10.8. A total of 170 converters are shown. This figure contains the least-square lines for 50% and 85% confidence levels to meet a targeted performance at various flight hours. For example, at 10,000 hr, there is an 85% probability of meeting the targeted conversion, whereas the probability is 50% at 20,000 hr.

Figure 10.9 shows the performance of a fresh and aged Pd catalytic abater after 20,000 flight hr as a function of air preheat temperature. The initial per-

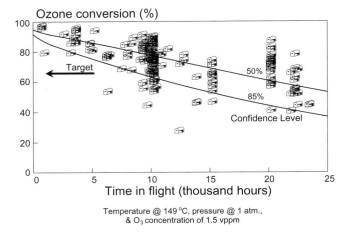

FIGURE 10.8. Prediction of ozone catalytic converter life as derived from a database of 170 converters from 747 aircraft. Reprinted with permission from Elsevier Science Publishers BV (Heck et al. 1992).

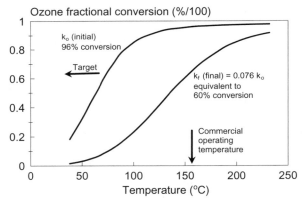

FIGURE 10.9. Ozone catalytic abater performance after 20,000 flight hr. Reprinted with permission from Elsevier Science Publishers BV (Heck et al. 1992).

formance is shown for comparison. Note that the fresh catalyst performance shows a sharp temperature response at low temperature, which is typical for catalytic reaction rates controlled by kinetics, whereas at operational temperatures (>150 °C), the rate is controlled by bulk mass transfer. The catalytic abater with 20,000 flight hr has lost considerable activity at low temperatures and never reaches bulk phase mass transfer controlled conversion. This response indicates the utilization of fewer active sites and an increase in pore diffusion resistance.

10.4   ANALYSIS OF IN-FLIGHT SAMPLES

In searching for the cause of a decline in performance, returned catalysts were analyzed using various surface analysis techniques. The results for a catalyst having 10,000 hr and 25,000 hr are shown in Table 10.2. This table contains the chemical analysis on the bulk aged catalyst samples done at seven different (uniformly spaced) axial locations along the abater. The elements analyzed are shown and compared with the fresh catalyst having no contamination.

Significant deposits of sulfur, phosphorous, and silica are noted (see Figures 10.10, 10.11, and 10.12). Furthermore, at 10,000 flight hr, there is an axial concentration profile for all three components. This concentration profile is much less pronounced for sulfur and phosphorus at 25,000 flight hr. The contaminant levels are much higher for phosphorus when comparing the different operating times, with only minor differences in the levels for silicon and sulfur. Apparently sulfur and silicon reach an equilibrium deposit concentration, whereas phosphorus continues to increase in concentration with flight hours.

Additional analysis was done to determine the nature of the phosphorus, silicon, and sulfur deposit on the washcoat of the catalyst. Elemental compositions of the mounted cross-section catalyst samples were determined with the scanning electron microscope and its energy-dispersive analyzer attachment. The cell flat wall and corner areas were scanned at 1,000–2,000 magnifications. Looking first at the catalyst with 10,000 flight hr, the presence of Si, S, P, Ca, and so forth with the major contaminants being silicon (9% maximum) and phosphorous (3% maximum) were detected preferentially at the surface of

TABLE 10.2   Elemental analysis of aged catalysts, 10,000 and 25,000 flight hr (weight %).

	C	S	P	Na	Fe	Ca	Ti	Si
Fresh	0.07	0.00	0.018	0.105	0.312	0.097	0.311	0.00
(a) 10,000 Flight hr								
1	0.05	0.60	0.050	0.122	0.324	0.079	0.329	1.73
2	0.04	0.54	0.047	0.094	0.331	0.096	0.326	1.49
3	0.03	0.45	0.040	0.084	0.311	0.085	0.321	1.26
4	0.04	0.41	0.038	0.142	0.295	0.093	0.358	1.5
5	0.04	0.32	0.037	0.100	0.346	0.079	0.350	1.59
6	0.04	0.37	0.034	0.088	0.326	0.090	0.339	1.27
7	0.04	0.31	0.033	0.092	0.345	0.085	0.338	1.29
(b) 25,000 Flight hr								
1	0.06	0.47	0.133	0.112	0.319	0.061	0.346	1.64
2	0.06	0.52	0.157	0.152	0.338	0.067	0.275	1.97
3	0.05	0.40	0.123	0.138	0.335	0.067	0.364	1.82
4	0.05	0.45	0.133	0.132	0.325	0.064	0.351	1.72
5	0.05	0.42	0.130	0.112	0.372	0.071	0.316	1.65
6	0.05	0.49	0.129	0.107	0.362	0.070	0.310	1.52
7	0.04	0.40	0.120	0.114	0.362	0.070	0.314	1.31

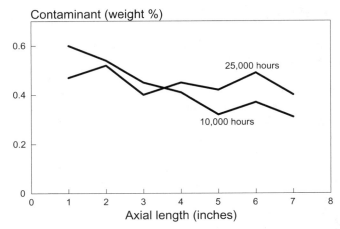

FIGURE 10.10. Sulfur content along the length of the ozone catalytic abater at 10,000 and 25,000 flight hours. Reprinted with permission from The Royal Society of Chemistry (Heck and Farrauto 1994).

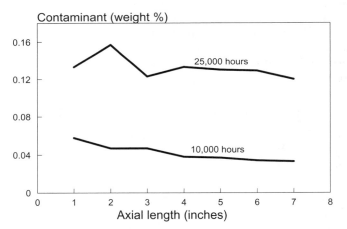

FIGURE 10.11. Phosphorus content along the length of the ozone catalytic abater at 10,000 and 25,000 flight hours. Reprinted with permission from The Royal Society of Chemistry (Heck and Farrauto 1994).

washcoat layers, whereas sulfur (4% maximum) was found to be uniformly distributed throughout the washcoat.

Microprobe analysis (see Figures 10.13, 10.14, and 10.15) on the cross-section samples of both aged catalysts indicated that silicon and phosphorus contaminations accumulate preferentially near the washcoat surface, whereas sulfur deposits throughout the layer without a radial gradient. For both aged catalysts, the absolute level of contamination increases with flight hours.

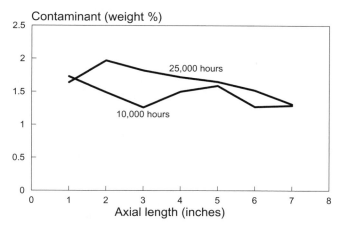

FIGURE 10.12. Silicon content along the length of the ozone catalytic abater at 10,000 and 25,000 flight hours. Reprinted with permission from The Royal Society of Chemistry (Heck and Farrauto 1994).

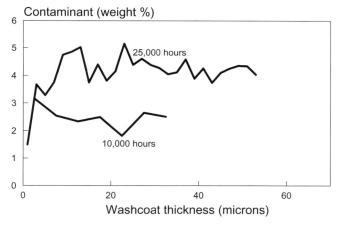

FIGURE 10.13. Sulfur content of catalyst washcoat at corner location of inlet section of ozone catalytic abater at 10,000 and 25,000 flight hr.

Using nitrogen (N_2) desorption data, the pore size distribution was derived for aged catalysts. The porosity data (see Table 10.3) of the bulk catalyst samples indicated that the catalyst with 25,000 flight hr contained about one half of the porosity of the catalyst having 10,000 flight hr (i.e., 0.0852 cc/g vs. 0.0318 cc/g porosity). The mean pore radius was also noted to be reduced from 53.2 Å to 37.9 Å as the catalyst aged. The pore size distribution profile (Figure 10.16) shows partial pore plugging and a subsequent increase in small mesopores (<50 Å) for the aged catalysts.

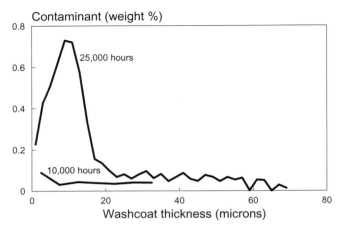

FIGURE 10.14. Phosphorus content of catalyst washcoat at corner location of inlet section of ozone catalytic abater at 10,000 and 25,000 flight hr.

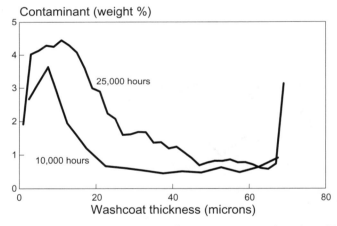

FIGURE 10.15. Silicon content of catalyst washcoat at corner location of inlet section of ozone catalytic abater at 10,000 and 25,000 flight hr.

Additional porosity data obtained with washcoat samples confirmed that a significant pore plugging occurred as the catalyst aging proceeded (Table 10.4). Aging at 10,000 hr caused a three-fold reduction in porosity, and extending the flight hours to 25,000 hr gave an eight-fold reduction. The Brunauer, Emmett, and Teller (BET) surface area showed a decline of three-fold at 10,000 hr and five-fold at 25,000 hr.

The effectiveness of the catalytic component, Pd, declined also as shown by carbon monoxide chemisorption measurements at 10,000 and 25,000 flight hr (Figure 10.17). The extra 15,000 hr of flight aging caused an average 33% reduction in palladium surface area.

TABLE 10.3. Pore distribution of aged catalysts (via N_2 desorption).

Catalyst Hours	Axial Location	Porosity (cm³/g)	Surface area (m²/g)	Mean pore Radius (Å)
10,000	1	0.02418	7.7	62.7
	2	0.02879	9.1	63.4
	3	0.03762	11.3	66.3
	4	0.02495	7.9	63.2
	5	0.01885	6.9	54.5
	6	0.03166	11.5	54.9
	7	0.01903	8.1	47.4
25,000	1	0.01336	8.0	34.3
	2	0.01506	6.0	50.1
	3	0.01466	5.5	53.1
	4	0.01680	8.8	39.4
	5	0.01358	6.9	39.7
	6	0.02076	11.0	38.4
	7	0.01583	7.3	44.1

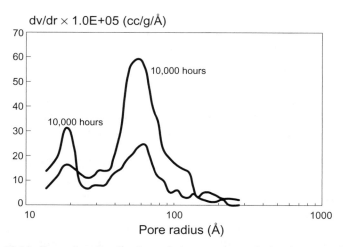

FIGURE 10.16. Pore size distribution of the ozone catalytic abater at 10,000 and 25,000 flight hr.

TABLE 10.4. Comparison of porosity of flight-aged catalyst.

Flight hours	0	10,000	25,000
BET desorption porosity (cm³/g)	0.25	0.085	0.032
BET area (m²/g)	98.8	32.6	18.7
Mean pore radius (Å)	49.4	53.2	37.9

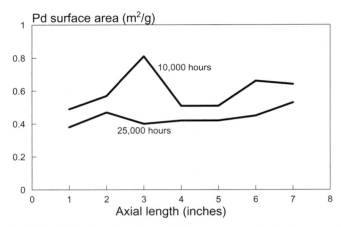

FIGURE 10.17. Pd CO chemisorption of the ozone catalytic abater at 10,000 and 25,000 flight hr.

Postanalysis of the catalyst with 10,000 and 25,000 flight hr indicates the following:

1. Silicon, phosphorus, and sulfur are the major contaminants.
2. There is a slight axial gradient for all three contaminants.
3. The concentrations of the major contaminants continue to increase from 10,000 flight hr to 25,000 flight hr.
4. Within the washcoat, phosphorus and silicon contamination is preferential at the surface, with phosphorus having a much more severe penetration gradient. This is possibly an aerosol-type deposition on the catalytic surface. The concentration of all three contaminants within the washcoat continues to increase in going from 10,000 to 25,000 flight hr.
5. Sulfur contamination seems to be uniform through the washcoat layer. This may indicate a gaseous compound deposition on the surface.

Based on the postanalysis of the catalyst, it is apparent that "masking" is the primary cause of deactivation. The "masking" causes significant pore plugging, reducing the accessibility of the ozone to the palladium catalyst. This pore plugging is caused by the phosphorus and silicon deposits. It is believed that the sulfur is a nonselective deposit, is chemically bound with the washcoat material, and does not deposit on the active sites.

To test this pore plugging theory, a proprietary mild chemical regeneration was tried using an all alkaline/acid-washing method. The catalyst performance was rejuvenated close to the fresh converter. Substantial reduction of the phosphorus was noted, with little change in the silicon and sulfur, which suggests that phosphorous was the primary cause of catalyst deactivation in this case.

The major source of sulfur is from the jet fuel and lubricating oils. Phosphorus is also from the lubricating oils and the hydraulic fluids used in various aircraft equipment.

Some catalysts also are exposed to de-icing fluids that may enter the air intake system, which then adsorb or condense within the catalyst washcoat. The silica-containing compounds have been traced to silicone parts in the air intake system. Dusts, halides from salt water, and so on also contribute to deactivation by masking the surface and preventing access to the catalytic sites. Typically, a converter can function for 10,000–20,000 flight hr before requiring replacement or regeneration. Depending on the nature of the deactivation and the age of the catalyst, washing solutions can be used to regenerate the activity. This, however, is not always necessary, because catalysts can be designed using different washcoat strategies (e.g., changing porosity or using inert overcoats) with tolerance to many poisons (Heck et al. 1992). Thermally induced deactivation modes such as sintering are not encountered simply because of the low temperature at which the catalyst functions.

10.5   NEW TECHNOLOGY

The airlines continue to request lighter weight material with less pressure drop, but of course with the same ozone conversion and long-term durability (Chang and Pluim 1982). Based on the experience of the first-generation ozone catalyst, two design advances were considered: lighter weight and lower pressure drop reactors and better catalyst durability. Since the space allocation within the aircraft is fixed, the geometric options for changing the reactor dimension are limited to reduce pressure drop. This means that the only design variable possible is to change the wall thickness of the honeycomb support. The normal support was 400 cpsi with a 6.5-mil wall thickness. Reducing the wall thickness will increase both the fractional open area (ε) and hydraulic diameter (d_{ch}) as well as the geometric surface area (a). [Refer to Chapter 4, Eq. (4.35).] The newest technology uses a thin-wall metallic honeycomb versus the previous ceramic honeycomb technology. In 1993, a new technology was introduced that uses a lightweight metallic support having a 400-cpsi cell density. This aluminum material has a 20% lower pressure drop and has reduced the weight of the abater by 20%. There is essentially no difference in the ozone conversion. This new lightweight design gives many design options depending on the requirements. One option is to maintain the same pressure drop but to reduce the converter cross section. In one example application, the cross section was reduced by approximately 10% by changing the support. Stainless steel converters are also in use with thin walls to reduce pressure drop; however, the reduction in weight is not as significant as the aluminum support. New catalyst technology was also developed that uses a base metal modified precious metal formulation catalyst to improve the overall catalyst performance (Carr and Chen 1983).

It is not unusual to be sitting in an aircraft on the ground and to experience a smell coming from the exhausts of aircraft in front of you in the taxi queue waiting to take off. The major odor-causing compounds in exhaust gases and fuels are from the family of compounds known as volatile organic compounds (VOCs). To remove these VOCs, another catalyst has been proposed that will not only remove the ozone but also any VOCs or hydrocarbons present on the same catalytic system (BASF 2 2007). New catalyst technology was developed to remove both ozone and VOCs. A new precious metal catalytic component was added to the existing Pd washcoat to allow for the simultaneous removal of both ozone and VOC odors with minimum installation costs and weight impact. The VOCs entering the bleed air supply via the engines may come from ground service vehicle exhausts or engine exhausts of other aircraft and mostly consist of small amounts of other compounds, such as sulfur. All of these compounds are present in very small amounts, but due to the sensitivity of the human nose, the cabin occupants can smell them, even down to levels of a couple of parts per trillion of the compound. The catalyst coating is applied to a core within the converter's body and oxidizes these odorous compounds, resulting in the formation of odorless water (H_2O) and carbon dioxide (CO_2) as reaction products. Since VOCs are present in low concentrations, the amount of water and carbon dioxide created by the conversion process is small. The combined converter has the same maintenance requirements as for the ozone converter, and the two converters are fully interchangeable (although of course the odor-removing function is lost if exchanged for a standard ozone converter). The combined ozone/VOC converter (VOZC) will be standard equipment for future Airbus programs (Nurcombe 2005).

REFERENCES

AeroParts. AeroParts Manufacturing & Repair, Inc., Rio Rancho, NM, 87124, http://aeroparts.net/index.shtml, (2008).
BASF 1, "Deoxo® Catalytic Ozone Converter, product information," http://www.catalysts.basf.com/Main/environmental/stationary_sources/indoor_air_quality/ozone_catalysts, (2007).
BASF 2, "Deoxo® Dual VOV/Ozone catalytic converter," http://www.catalysts.basf.com/Main/environmental/stationary_sources/indoor_air_quality/ozone_catalysts, (2007).
Bonacci, J., and Heck, R. "Air pollution control catalytic equipment," International Precious Metals Institute Seminar, Williamsburg, VA (1983).
Carr, W., and Chen, J. "Ozone abatement catalyst having improved durability and low temperature performance," U.S. Patent 4,343,776 (1982).
Carr, W., and Chen, J. "Ozone abatement catalyst having improved durability and low temperature performance," U.S. Patent 4,405,507 (1983).
Chang, J. "Catalyst for ozone decomposition," U.S. Patent 4,206,083 (1980).

Chang, J., and Pluim, A. "Catalytic converter for ozone removal in aircraft," U.S. Patent 4,348,360 (1982).

Engelhard, "Inspired performance in clearing cabin air of ozone," Engelhard Corporation EC6934 (1996).

Federal Register, "Airplane cabin ozone contamination," Code of Federal Register, 14 CFR Parts 25 and 121 (1980).

Haldeman, J., Nastrom, G., and Falconer, P. "Analysis of two years of GASP data," NASA Technical Memorandum (1977).

Hawk, J. "The Boeing 787 Dreamliner: More than an airplane," Director Certification, Government and Environment 787 Program (2005).

Heck, R., and Farrauto, R. "Catalysis for environmental control," pp. 120–138. In *Chemically Modified Surfaces* (special publication 139). Editors Perek, J., and Leigh, E. Royal Society of Chemistry Cambridge, England (1994).

Heck, R., Farrauto, R., and Lee, H. "Commercial development and experience with catalytic ozone abatement in jet aircraft," *Catalysis Today* 13: 43–85 (1992).

Honeywell. "Aerospace engineering green technology," http://www51.honeywell.com/aero/technology/key-technologies2/green-technology.html, (2008).

Hunt, E., Reid, D., Space, D., and Tilton, F. "Commercial airliner environmental control system engineering aspects of cabin air quality," Aerospace Medical Association annual meeting, Anaheim, CA (1995).

Kent, R., and Fein, M. "Catalyst composition for decomposing ozone," U.S. Patent 4,173,549 (1979).

National Research Council, Committee on Airliner Cabin Air Quality, "The airliner cabin environment: Air quality and safety," National Academic Press, Washington D.C. (1986).

Nurcombe, C. "Airbus quality: Still the best," Customized Systems, Airbus Deutschland, FAST Magazine (2005).

Saywell International, "Abbreviated component maintenance manual p/n's 40997001, 40997002, 40997003," BASF (Engelhard)—Tech Information, http://www.saywell.co.uk/pages/engelhard.htm, (2007).

Spengler, J., Ludwig, S., and Weker, R. "Ozone exposures during trans-continental and trans-pacific flights," *Indoor Air*, 14(s7): 67–73 (2004).

Wall Street Journal. "FAA proposes rule to restrict ozone intake of airlines." (1978).

CHAPTER 10 QUESTIONS

1. How does ozone come into the cabin of an aircraft? Why is it an issue?
2. Why do you think Pd is the best catalyst for ozone decomposition? Allow for some creative thinking and speculation in this answer.
3. Why is mass transfer so important in the design of an ozone reactor?
4. Based on your prior knowledge of catalyst deactivation in Chapter 5, explain the effects of the post-catalyst analysis on the ozone reactor performance. Some good scientific speculation should be allowed here.

PART III
Stationary Sources

11 Volatile Organic Compounds

11.1 INTRODUCTION

One of the key elements in any new manufactured product is the life-cycle analysis, and a key element of the analysis is the effect on the environment. This includes both the sources of contaminants to the water system as well as the air we breathe. Of course it is difficult to manufacture products without some waste streams. Wastewater treatment, a prime consideration in the 1960s and 1970s in the United States, has met with success in cleaning up many rivers and lakes. It is being revisited in the United States and certainly is receiving worldwide attention as it is becoming a serious issue again on a global basis. Probably more common, but many times less noticeable, is the emissions of organic compounds from manufacturing the multitude of consumer products used everyday. In most manufacturing processes, either for the raw materials, intermediates, or the finished product, organic materials are present as chemicals, solvents, release agents, coatings, decomposition products, pigments, and so on that eventually must be disposed. In such manufacturing, there is usually a gaseous effluent that contains low concentrations of organic that is vented to the atmosphere. Examples of commercial process having volatile organic (VOC) emissions are as follows:

- Chemical plants
- Petroleum refineries
- Pharmaceutical plants
- Automobile manufacturers
- Airplane manufacturers
- Food processors
- Fiber manufacturers
- Textiles manufacturers
- Printing plants
- Can coating plants
- Wire enameling plants

Catalytic Air Pollution Control: Commercial Technology, by Ronald M. Heck and Robert J. Farrauto, with Suresh T. Gulati.
Copyright © 2009 John Wiley & Sons, Inc.

- Electronic component plants
- Painting facilities
- Wood stoves

The organics present depend on the process and can include ketones, aldehydes, aromatics, paraffins, olefins, acids, chlorinated hydrocarbons, fluorinated hydrocarbons, and higher molecular weight organics, which are often present as aerosols.

Several methods are available for abating these emissions, including:

- Liquid absorbents
- Solid adsorbents (pressure swing, temperature swing, etc.)
- Scrubbing
- Precipitation
- Capture devices (filters, etc.)
- Condensation (refrigeration, cryogenic, etc.)
- Thermal incinerators
- Catalytic incinerators
- Membranes
- Biodegradation
- Plasma

The emissions regulations governing the VOC output from the various industrial applications depend on the application or point source, and these regulations are documented in the Federal Register in the United States (CFR 2007).

To complement the enforcement of the point source regulations established by the Environmental Protection Agency (EPA), the Clean Air Act (CAA) establishes a pair of programs, known as New Source Review (NSR), that regulate the construction and modification of large stationary sources of air pollution. The NSR requires a permit before construction or modification may begin on a stationary source that has the potential to emit more than a specified level of emissions. The permit must indicate that the construction or modification will include advanced emission controls (NSR 2006). Recent studies of ambient air quality have highlighted the issues with VOCs in the atmosphere contributing to the indoor concentrations of various hydrocarbons. These materials include benzene, ethylbenzene, xylenes, and methyl tertiary butyl ether (MTBE), as well as some of the chlorinated compounds, tetrachloroethylene, 1,1,1-trichloroethane, and carbon tetrachloride, all of which had significant contributions from outdoor sources (Sax et al. 2004). So it is not just exposure to VOCs in the working environment but in the home as well from industrial sources. Regulation of VOC emissions has also reached the state level in the United States. The California Air Resource Board (CARB) and several northeastern states that are members of the Ozone

Transport Commission (OTC) are now regulating the VOC levels of many consumer products with the goal of improving air quality. It is anticipated that all OTC states will ultimately regulate the VOC content of consumer products (OTC 2007).

11.2 CATALYTIC INCINERATION

The catalytic incineration method has become most popular because, in many cases, it is more versatile and economical for the low concentrations of organic emissions (i.e., <5,000 vppm). The basic catalytic oxidation reaction of an organic molecule (HC) is shown as follows:

$$HC + Air\ (O_2) \xrightarrow{\text{catalyst}} CO_2 + H_2O \qquad (11.1)$$

The actual operating temperature and amount of preheat varies, depending on the organic molecule, space velocity, composition of feed (i.e., contaminants, water vapor, and so forth), and organic concentration. Typical examples of operating temperatures are given in Table 11.1 (Bonacci et al. 1989).

One way of comparing thermal versus catalytic abatement is to look at the energy required (air preheat temperature) to obtain quantitative removal of a given hydrocarbon species. The operating temperatures shown in Table 11.1 are well below the corresponding temperatures necessary to initiate thermal (noncatalytic) oxidation (Bonacci et al. 1989). The catalyst initiates a reaction at lower temperatures by lowering the activation energy. This demonstrates the major advantage of catalyzed processes, which is that they proceed faster than noncatalytic reactions, allowing lower temperatures for the same amount of conversion. This translates directly to improved economics for fuel use and less expensive reactor construction materials, since corrosion is greatly reduced.

Selection of the catalytic material for various organic pollutants has been the subject of many studies. Both base metal oxides and precious metals as well as combinations of them are used for both hydrocarbons and chlorinated hydrocarbons (Martin et al. 1992; Spivey 1987; Spivey and Butt 1992; Noordally et al. 1993; Stein et al. 1962; Yao 1980; Cohn and Haly 1965; Hardison and Dowd 1977; Pisarczyk et al. 1996; Heneghan et al. 2004). As a rule, precious metals (especially platinum and/or palladium dispersed on carriers) are preferred because of their activity, resistance to deactivation, and ability to be regenerated. Platinum seems to be the preferred catalysts for saturated hydrocarbons and higher molecular weight species (Yao 1980). Palladium is preferred for methane and low-molecular-weight olefins (Stein et al. 1962). Base metal oxides are less frequently used except for those conditions where the feed gas is relatively free of contaminants such as sulfur. An exception is the use of a Cr_2O_3 catalyst in a fluidized bed process in which abrasion renews the catalytic surface and minimizes the retention of poisons (Hardison and Dowd 1977).

TABLE 11.1. Operating temperatures for catalytic abatement of organic compounds.

Name of constituent	Chemical formula	Formula weight	Temperature rise at 1,000vppm (°C)	Operating temperature (°C)	Concentration before treatment (vppm)
Styrene	$C_6H_8CHCH_2$	104.15	138	250	310
Acetaldehyde	CH_8CHO	44.05	35	350	240
Benzene	C_6H_8	78.12	103	210	380
Toluene	$C_6H_5CH_3$	92.14	123	210	320
m-Xylene	$C_6H_4(CH_3)_2$	106.17	143	210	270
Phenol	C_6H_5OH	94.11	101	300	380
Formaldehyde	HCHO	30.03	17	150	410
Acrolein	CH_2CHCHO	56.06	51	180	500
Acetic acid	CH_2COOH	60.05	26	350	590
Bultyric acid	C_3H_7COOH	88.11	66	250	370
Acetone	CH_3COCH_3	58.08	57	350	410
Methyl ethyl ketone	$CH_3COC_2H_5$	72.11	74	220	380
Methyl isobutyl ketone	$CH_3COC_4H_8$	100.16	116	250	270
Ethyl acetate	$CH_3COOC_2H_5$	88.11	68	350	350
Butyl acetate	$CH_3COOC_4H_9$	116.16	108	350	480
Methyl alcohol	CH_3OH	32.04	21	150	830
Ethyl alcohol	C_2H_5OH	46.07	44	350	550
Isopropyl alcohol	C_3H_7OH	60.01	64	280	230
Butyl alcohol	C_4H_8OH	74.12	84	260	330
Carbon monoxide	CO	28.01	9	150	4,000
Methyl cellosolve	$HOCH_2CH_2OCH_3$	76.01	55	300	110
Ethyl cellosolve	$HOCH_2CH_2OC_2H_5$	90.12	76	300	80
Butyl cellosolve	$HOCH_2CH_2OC_4H_9$	118.18	118	300	50

Source: From the *Encyclopedia of Environmental Control Technology*, Volume 1. Copyright 1989 by Gulf Publishing (Bonacci et al. 1989).

Figures 11.1 and 11.2 show typical response curves using a standard (range of 0.1% to 0.5%) Pt/γ-Al₂O₃ catalyst for the oxidation of olefins/paraffins and solvents, respectively.

The preferred support for organic abatement catalysts is the monolith or honeycomb with 200 to 400 channels per square inch (cpsi), either ceramic (cordierite) or metallic (aluminum or stainless steel). The honeycomb

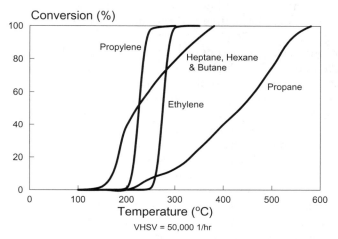

FIGURE 11.1. Performance of Pt monolithic catalyst for olefins and paraffins. Reprinted from the *Encyclopedia of Environmental Control Technology*, Volume 1. Copyright 1989 by Gulf Publishing (Bonacci et al. 1989).

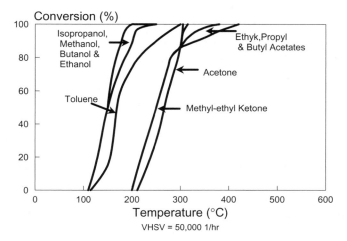

FIGURE 11.2. Performance of Pt monolithic catalyst for various industrial solvents. Reprinted from the *Encyclopedia of Environmental Control Technology*, Volume 1. Copyright 1989 by Gulf Publishing (Bonacci et al. 1989).

structure offers the major advantage of low pressure drop due to its high open frontal area (i.e., 70% for ceramic, up to 90% for metallic).

For VOC catalysts, the best operating region is where the reaction is controlled by bulk phase mass transfer. Here the most important design parameters are the Reynolds number, the geometric surface area of the support, and the reactor cross-sectional area (see Chapter 4). These all must be selected at the lowest possible pressure drop. The honeycomb support satisfies these criteria. In some limited cases, other catalyst structures such as screens and beads are used. One process has a fluid bed system using a base metal bead catalyst (Hardison and Dowd 1977).

Specific catalyst compositions are kept proprietary; however, most commonly, 0.1% to 0.5% Pt on high-surface-area γ-Al_2O_3 is used, either as a bead or a washcoat on a monolith. In limited cases, small amounts of Pd or Rh are added to promote a particular reaction where the Pt may be deficient. Honeycomb geometries vary depending on the nature of the feed and pressure drop restraints. For example, feed gases containing dust usually require larger diameter honeycomb channels (i.e., 100 cpsi). Larger channels decrease pressure drop but also have lower geometric surface areas per given volume, which decreases bulk mass transfer conversion. Therefore, some compensation in size of the total honeycomb is required to give a specific conversion. The volume of catalyst used also depends on the degree of conversion desired. For example, 30,000 1/hr of space velocity will typically produce 99% conversion for a 200-cpsi honeycomb, whereas 60,000 1/hr produces 90%. Trade-offs of this type require that the plant engineers work closely with the catalyst manufacturer.

For feeds that are known to contain large amounts of sulfur (i.e., greater than 50 ppm S), less reactive carriers such as TiO_2 or α-Al_2O_3 are used for the Pt. Those carriers are relatively inert to the formation of sulfates compared with high-area γ-Al_2O_3. For feeds without sulfur, non-precious metal catalysts such as CuO or Co_2O_3 are also used on the γ-Al_2O_3 carrier. Of course, it is also possible to operate the catalyst at high enough temperatures so that the sulfur does not adsorb on the catalytic sites, thus preventing deactivation by poisoning. However, this requires additional energy and a fuel penalty. Furthermore, base metal oxides such as CuO, Co_2O_3, and so on undergo sintering, oxidation state change, and reaction with the carrier at elevated temperatures (Shoup et al. 1975).

Catalytic researchers are always looking for catalysts that will operate at lower temperatures and be resistant to catalyst deactivation. Several patents have addressed this concept for lower temperature operation for destruction of VOCs in an oxygen-containing gas. The new catalyst uses Ce and Zr as the key compositions. Other catalytic materials are added along with precious metals (Roark and White 2002). Operating temperatures of 150°C to 200°C are claimed for complete oxidation. The performance for complete oxidation of VOCs is shown to be better than the conventional Pt/Al_2O_3-type catalyst. Commercial catalysts are available as honeycomb catalysts for all VOC applications (Eltron 2007).

There are many suppliers of VOC catalysts, including BASF, Johnson Matthey, DCL International, Eltron, Advanced Catalyst Systems, Sud Chemie, and Catalytic Combustion.

11.3 HALOGENATED HYDROCARBONS

Several exhaust streams contain halogenated hydrocarbons either as the vapor from a solvent or as a material that was directly added in the process.

With the emphasis on cleanup of water-contaminated sites, which in many cases contain chlorinated hydrocarbons, there has developed a need for catalysts that are active toward the oxidation and destruction of these compounds (eq. 11.2 not balanced).

$$C_xH_yCl + O_2 \rightarrow CO_2 + H_2O + HCl \tag{11.2}$$

The production of HCl requires the addition of a scrubber and sometimes corrosion-resistant materials for reactor construction. Water vapor must be present in this reaction for HCl to be formed and to prevent Cl_2 gas formation. Conventional Pt- and Pd-based catalysts are inhibited by the presence of the chlorine (Spivey and Butt 1992; Wang et al. 1992; Yu et al. 1992; Simone et al. 1991). Catalysts have been developed containing Cr_2O_3, V_2O_5 + Pt, and other compounds to overcome this activity loss (Nguyen et al. 1994; Agarwal et al. 1992; Berty 1991; 1992; Muller et al. 1993; Freidel et al. 1993). Figure 11.3 shows a comparison of a standard VOC catalyst composed of a standard Pt/γ-Al_2O_3, a Cr_2O_3/γ-Al_2O_3, and a newly commercialized Pt catalyst containing

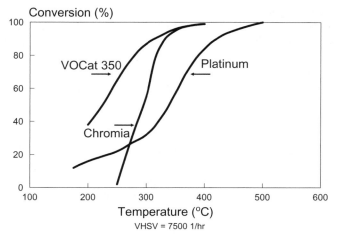

FIGURE 11.3. Catalytic destruction of trichloroethylene over various monolithic catalysts.

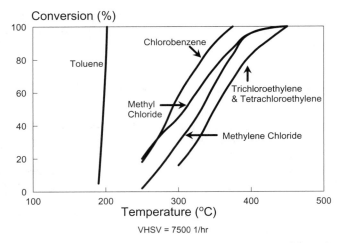

FIGURE 11.4. Catalytic destruction of various chlorocarbons with an improved Pt-based monolithic catalyst.

proprietary base metal oxide promoters (designated VOCAT350) (Nguyen et al. 1994).

Performance curves using this new improved catalyst for various chlorinated hydrocarbons are shown in Figure 11.4. Chlorobenzene is the most reactive chlorinated species. Note the ease of conversion of toluene compared with the chlorine-containing molecules.

Many new catalysts and processes capable of treating streams containing chlorinated hydrocarbons have been commercialized in the last 2 years. In general, many have not disclosed the details of the catalyst, but some performance data are available. One such catalyst and process is now on stream in the purification of an off-gas containing a variety of chlorinated hydrocarbons produced as by-products in the production of vinyl chloride (Muller et al. 1993). A bi-metallic Pt and Pd catalyst on γ-Al$_2$O$_3$ yields a 99% conversion of chlorinated hydrocarbons, generating an outlet temperature of between 580 °C and 680 °C at a space velocity of 10,000 1/hr. An activated Al$_2$O$_3$, specially treated to resist Cl$^-$, is used in a pre-bed at an inlet temperature of 350–380 °C, to aid in the decomposition of the aromatic hydrocarbon solvent absorbent and some of the chlorinated hydrocarbons, and to protect the Pt and Pd catalyst by filtering particulates. The physical structure of the catalyst is not given. Another catalyst is described as comprising vanadium oxide, zirconium oxide, and at least one oxide of manganese, cerium, or cobalt (Nguyen et al. 1994). Platinum group metals and tungsten oxide may also be added.

A catalyst and process has been successfully used to abate hydrocarbons and brominated hydrocarbons from a terephthalic acid plant using a noble metal catalyst supported on a ceramic honeycomb (Chen and Tran 2001). Purified terephthalic acid (PTA) is a key raw material for production of polyester fibers and polyethylene terephthalate (PET) bottle resin and film. The world-wide PTA manufacturing capacity has expanded substantially over the past 10 years to meet the increasing PTA demand. The vent gases from PTA plants often contain low concentrations of CO, methyl bromide (MeBr), and various VOCs, including methyl acetate (MeAc), xylene, toluene, benzene, and acetic acid. The gas composition in this vent is as follows:

- CO (3,000–7,000 ppm)
- MeBr (CH_3Br) (5–50 ppm)
- VOCs (500–1,000 ppm) (xylene, toluene, benzene, methanol (MeOH), acetates, acetic acid)
- O_2 (3–5%), H_2O (2%)

Depending on the application, the compounds to be removed vary as follows:

- CO removal only
- CO and VOC removal (excluding MeBr)
- CO, VOC, and MeBr removal

These compounds need to be removed to as high as 98% before the vent gas can be emitted from the process. The possible catalytic reactions are as follows:

$$C_xHy + (x + y/2)O_2 \rightarrow xCO_2 + yH_2O \qquad (11.3)$$

$$CO + 1/2\,O_2 \rightarrow CO_2 \qquad (11.4)$$

$$3Ch_3Br + 5O_2 \rightarrow 3CO_2 + 4H_2O + HBr + Br_2 \qquad (11.5)$$

Since being introduced in 1991, catalytic oxidation has been accepted as the most effective commercial control technology for these applications. Commercially, catalytic oxidation systems (CATOXs) have been installed either at the process vent (low-pressure operation) or before the expander (high-pressure operation).

Most PTA plants have very high throughput, in the range of 50,000 to 100,000 SCFM. The energy requirement to preheat the gas to the catalyst oxidation temperature is the major operational cost for a vent gas control system. A catalyst reduces the oxidation temperature, which thus reduces energy consumption. Through continuing development efforts by catalyst suppliers for new catalysts, the catalyst oxidation temperature has been reduced from 350 °C to 450 °C initially to the 250–300 °C temperature range. This development has resulted in greater than 40+% fuel saving. Also, the increase of the catalyst activity has reduced the initial capital cost for catalyst and reactor.

Among various catalysts investigated, precious metal catalysts, such as Pt and Pd, have shown to be most active and are widely used commercially. These catalysts have demonstrated to achieve 98% removal of these pollutants for more than 5 years of continuing operations. In the presence of bromine-containing environments, the activities of these catalysts are strongly influenced by the support materials. Optimal combinations of the catalyst and support materials are essential to operate catalytic oxidation at low temperatures. From a study of support properties and catalyst compositions, new catalyst formulations have been commercialized for PTA vent gas control to cover the broad range of operating temperatures and performance requirements regarding VOC and MeBr conversion. Initial formulations of PTA vent gas catalyst had an operating temperature around 375 °C, which was 75 °C lower in comparison with a conventional Pt/Al$_2$O$_3$ catalyst. The performance of VOCat 300H about 375 °C is reflected in Figure 11.5. Improvements in the catalyst technology were directed at lowering the operating temperature. VOCat PTA LT catalyst was developed to abate CO and most VOCs selectively at temperatures as low as 225 °C without significantly converting MeBr as shown in Figure 11.6. VOCat 450H has been developed to remove all these compounds within a temperature range of 250 °C to 300 °C without forming bromobenzene (see Figure 11.7). In contrast to VOCat 450H is catalyst VOCat 500H, which is designed to remove MeBr over the same temperature range as

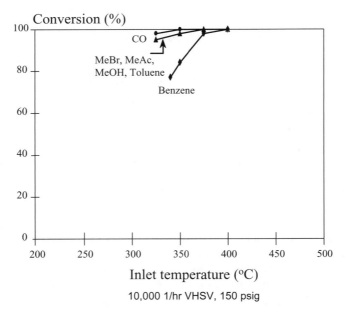

FIGURE 11.5. VOCat 300H achieves 98+% CO and VOC conversions at 280 °C+ temperature for PTA applications (Chen and Tran 2001).

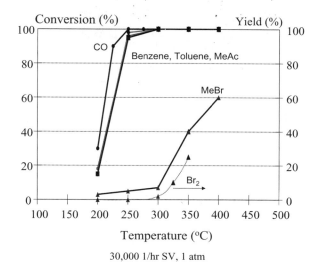

FIGURE 11.6. VOCat PTA LT catalyst designed to abate CO and most VOCs selectively at temperatures as low as 225 °C without significantly converting MeBr (Chen and Tran 2001).

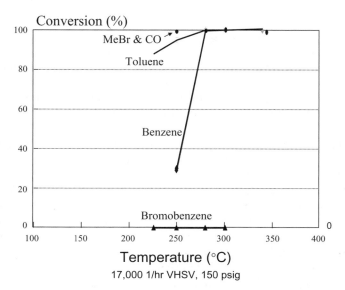

FIGURE 11.7. VOCat 450H has been developed to abate CO and most VOCs within a temperature range of 250 °C to 300 °C without forming bromobenzene (Chen and Tran 2001).

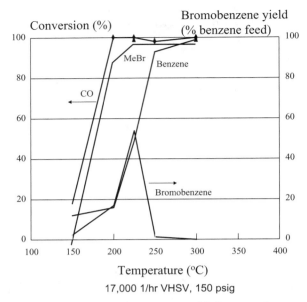

FIGURE 11.8. VOCat 500H designed to remove MeBr over the temperature range of 250 °C to 300 °C (Chen and Tran 2001).

shown in Figure 11.8. So by modifying the catalytic components, two different selectivities for the MeBr reaction [Eq. (11.5)] can be obtained.

Another study (Freidel et al. 1993) showed data from pilot plant studies that indicated greater than 95% conversion of CO and HC in the presence of 80-ppm CH_3Br, at a space velocity of 33,000 1/hr and inlet temperatures of 250–300 °C. The same authors report the success of this catalytic process in destruction of dioxins.

An interesting process has been developed (Berty 1991; 1992) that uses base metal oxides of copper and manganese deposited on sodium carbonate for the catalytic oxidation of chlorinated hydrocarbons and freons. The chlorine generated reacts with the carrier, forming NaCl and eliminating the need for a downstream scrubber and corrosion-resistant materials of construction. The process, with a reported 98% destruction efficiency, operates at a space velocity of 6,700 1/hr and at an inlet temperature of 250–400 °C. The upflow reactor contains multiple layers of catalyst, which are periodically removed as breakthrough is observed by analytical sensors located between the layers. The inlet layer is the first to be replaced. Reference is made to regeneration of the catalyst, but no details are provided.

The semiconductor manufacturing industry is operating under a Memorandum of Understanding (MOU) with the EPA. Under this MOU, the industry is targeting emissions reductions to 10% below 1995 levels by 2010 (Brown et al. 2001). Perfluoro compounds (PFCs), examples of which include CF_4, C_2F_6, and SF_6 (sulfur hexafluoride), are used in the manufacture of semiconductor materials. Significant emissions reductions have already been achieved through process optimization and alternative chemistries (e.g., adaptation to NF_3), but these will not be sufficient to meet future PFC emissions needs. Although the emissions of these materials are small in terms of total mass emitted, the emissions are significant in terms of global warming potential (GWP). This is because the atmospheric lifetimes of PFCs are on the order of many thousands of years, resulting in a GWP of many times that of CO_2. Catalytic abatement processes have been applied to the control of PFC emissions. The catalytic process involves passing the PFC-laden stream through a catalyst bed at an elevated temperature with the following reactions:

$$C_2F_6 + 3H_2O \rightarrow CO + CO_2 + 6HF \tag{11.6}$$

$$CO + 1/2\,O_2 \rightarrow CO_2 \tag{11.7}$$

The catalyst consists of Pt/ZrO_2-SO_4 (Feaver et al. 1999). The catalytic decomposition of PFCs proceeds according to a catalyzed hydrolysis reaction. Therefore, water, rather than oxygen, is necessary for the decomposition reaction to proceed. The catalytic destruction of various fluorinated compounds is listed in Table 11.2.

About 60,000 wafers were processed over the course of the 5-month evaluation. During this time, the system achieved removals of >99.5% for CF_4 (tetrafluoromethane), 99% for C_2F_6 (hexafluoroethane), and 97% for c-C_4F_8 (octafluorocyclobutane). No decrease in the catalytic activity was observed during the evaluation. Note that there was a pre-scrubber upstream of the catalytic reactor to filter SiF_4 from the tool exhaust. In the reactor effluent only, CO_2 and HF were detected and no F_2, OF_2, CO, or products of incomplete oxidation were detected.

An accelerated aging test was developed that indicated that after 5 months of operation, the catalyst could maintain >80% of its initial lifetime. This indicates that the actual lifetime of the catalyst may be in excess of 2 years. Also it was determined that the catalyst was not a hazardous material after

TABLE 11.2. Listing of T_{95} (°C) for selected fluorine-containing compounds (Brown et al. 2001).

Compound	T_{95} (°C)
NF_3	375
CHF_3	425
SF_6	510
CF_4	650
C_2F_6	690
C_3F_8	690
c-C_4F_8	705

exposure to the exhaust. Therefore, no special handling of the used catalyst is required.

11.4 FOOD PROCESSING

Restaurant cooking emits a significant amount of visible fine particulate matter (i.e., smoke) and VOCs into the air. VOCs can react with oxides of nitrogen (NO_x) in the presence of sunlight to generate ozone. Ozone as well as fine particulate matter (PM_{10}) can cause health effects on the respiratory system. According to a recent study by the California South Coast Air Quality Management District (SCAQMD 1997), just the restaurants in the south coast basin alone generate approximately 11.6 tons/day of particulate matter and 1.6 tons/day of VOCs. These alarming numbers raised public concern on the impact of cooking emissions on air quality and human health, and they have recently prompted southern California to implement the nation's first environmental regulation to reduce cooking emissions from restaurants (SCAQMD Rule 1138).

Restaurant equipment that contributes to these emissions includes charbroilers, griddles, and deep fat fryers. According to a study conducted by the University of California Riverside, College of Engineering, Council for Environmental Research and Technology (CE-CERT) (Norbeck and Welch 1997), charbroiling is the major contributor to emissions from restaurants and contributes more than 80% of all the cooking emissions (i.e., VOC and particulate matter). Currently, Rule 1138 only requires the addition of an emission control device for the chain-driven charbroiled. An extensive study was conducted by the SCAQMD to evaluate the performance of various available and emerging control techniques for chain-driven charbroilers. Among all the assessed technologies that are capable of reducing both PM and VOC (i.e., catalytic oxidizer, fiber-bed filter, and incineration), the catalytic oxidizer is the only device that is both cost-effective and meets the Best Available Control Technology (BACT) guidelines. Based on this study, SCAQMD concluded that the catalytic oxidizer (Hoke et al. 1998) is the only acceptable emission control method for chain-driven charbroilers (SCAMD).

Unlike the traditional catalytic VOC abatement technology, the effect of both catalyst and converter design on control of cooking emissions is not well studied. The catalytic converter is normally mounted onto a ventilation shroud that connects the converter to the top of the broiler. The hot cooking emissions flow through the ventilation shroud by natural convection upward to the converter. Figure 11.9 shows a schematic for this food processing with a catalytic system.

The catalytic converter unit operates at a nominal temperature range of 530–650 °C while cooking food, and 370–430 °C during idle or warm-up mode. The treated cooking exhaust is emitted from the converter, mixes with room air, and is then vented outside the room through the ventilation hood installed

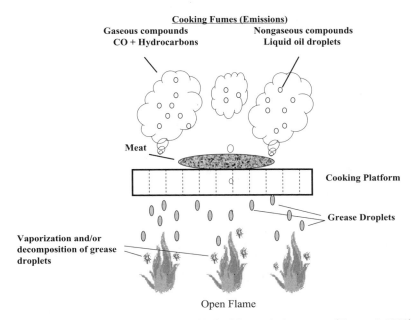

FIGURE 11.9. Schematic of charbroiled with catalytic system (Fu et al. 1999).

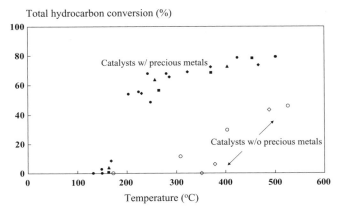

FIGURE 11.10. Precious metal catalyst reduces emissions from charbroiled (Fu et al. 1999).

above the broiler. Figure 11.10 compares the total hydrocarbon emission from a chain-driven charbroiled with the installation of two types of catalytic converters. Note that a precious metal catalyst formulation is compared with a base metal catalyst formulation. The precious metal gives superior performance.

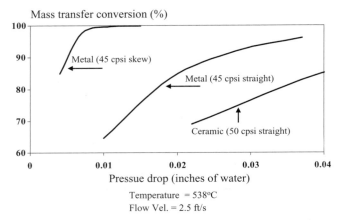

FIGURE 11.11. Changing monolith channel shape and flow pattern improves performance for abating emissions from charbroilers (Fu et al. 1999).

The system design for application of catalysts was mentioned in Chapter 5 in regard to the trade-off between mass transfer and pressure drop. This is the classic catalytic reactor design problem, and the optimum converter design for a charbroiled requires high VOC and particulate removal efficiency with a minimum pressure drop loss. The effects of substrate design, as well as catalyst formulation, were evaluated. It was found that metal substrate with interrupted channel design gives a better mass transfer-limited conversion rate with lower pressure drop loss. The skew channels cause a disruption of the boundary layer; therefore, higher mass transfer performance occurs as well as a concurrently higher pressure drop (Fu et al. 1999).

A photograph of the actual catalyzed metallic monolith insert in the duct of a commercial charbroiler operation is shown in Figure 11.12. This is designed as an insert to fit in the duct from the charbroiler assembly.

The actual performance of the catalytic system on a charbroiler installation is shown in Figure 11.13 and compares the uncontrolled emissions with the controlled emissions using a catalyst. Reductions of emission around 80% are achieved.

11.5 WOOD STOVES

An alternative application for VOC catalysts is the catalytic oxidation of distillable organics and CO present in the exhaust gases from wood stoves. Catalytic oxidation results in improved energy efficiency, safety, and cleaner emissions. In more rural settings like Colorado, Vermont, and Washington, this is common practice for wood stove installations. Because of the particulate loading in the exhaust from the wood stove, low cell density honeycombs (e.g., 64 cpsi) with a Pt/Al_2O_3 catalyst are used (Farrauto and Heck 1992). The stove

FIGURE 11.12. Catalyzed metallic monolith insert in the duct of a commercial charbroiler operation (Fu et al. 1999).

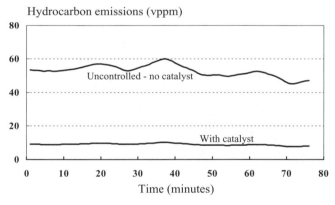

FIGURE 11.13. Catalyst substantially reduces emissions from food process of charbroiling (Fu et al. 1999).

operates with a deficiency of air to prevent the wood from combusting too rapidly. The combustion products are therefore rich in HC, CO, and wood-derived distillable gaseous components. A monolithic catalyst is placed above a baffle near the top of the stove where secondary air is injected at the inlet face generating an oxidizing environment to complete the oxidation of the emissions. The exotherm generated provides additional heat for the room and maintains a cleaner chimney since the distillable wood-derived components are catalytically oxidized before they can condense on the colder outer walls of the chimney. The maximum temperature in the catalyst is typically around

900 °C, so Al_2O_3 and Pt sintering are primary modes of deactivation. A catalyst life of two heating seasons is expected.

11.6 PROCESS DESIGN

In designing an organic abatement system, several components usually constitute the entire system. Of course, this is highly dependent on the needs of the user. Typically a system comprises the following:

- Filter (or guard bed)
- Catalyst
- Reactor
- Preheat burner
- Instrumentation and controls
- Analytical equipment
- Heat exchanger(s)
- Auxiliary equipment (piping, valves, etc.)

The total package is usually provided as a compact unit on a structural frame that can be readily implemented in the manufacturing process. Figures 11.14a) and 11.14b) show two standard configurations for a catalytic abatement unit with and without heat recovery (Farrauto and Heck 1992). Substantially more detail on design and sizing of organic abatement units is available in the literature (Bonacci et al. 1989; Bonacci and Heck 1983; Grzanka 2007; EPA 2002; Vatavuk et al. 2000) and Chapter 4.

Several system design engineering firms in the United States are engaged in assembling catalytic abatement systems (CSM; Durr Systems, Met-Pro, and Anguil, to name a few). Many case studies of VOC abatement system design and applications are listed on the Institute of Clean Air Companies website (ICAC 2001).

11.7 DEACTIVATION

Sintering is not usually a problem for VOC applications, since modern catalyst materials have been developed from mobile source applications (i.e., automotive catalysts) to resist high-temperature degradation. Deactivation usually occurs by a mechanism called *fouling* or *masking* in which a residue from the process stream deposits on the catalyst surface. This residue may be a dust from the manufacturing process or an organic char. Another example of this mechanism is via aerosols that contain a metallo-organic compound that, when decomposed, leaves a residue of inorganic material. A common source is lubricating oil, which deposit and then decompose, leaving Zn, P, and Ca compounds. When these materials deposit on the catalytic surface, they can block

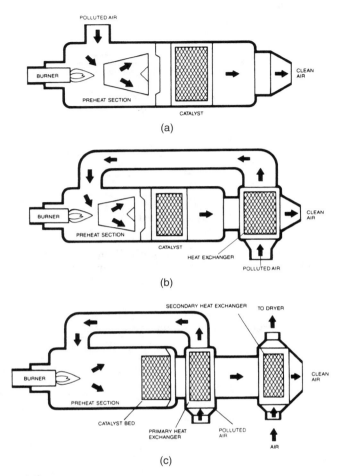

FIGURE 11.14. a) VOC abatement system with no heat exchange equipment. b) VOC abatement system with heat exchange equipment. c) VOC abatement system with secondary heat exchange equipment. Reprinted by courtesy of Marcel Dekker Inc. (Farrauto and Heck 1992).

access to the pores and/or react with the catalytic sites, rendering them less active. A physical representation of these phenomena is shown in Figures 5.9 and 5.10 (see Chapter 5).

11.8   REGENERATION OF DEACTIVATED CATALYSTS

Various catalyst regeneration techniques can be used for precious metal-based catalysts. However, the most practiced involves chemical washing, which selectively dissolves impurities without disturbing the basic catalyst material. In many cases, the catalyst surface and the activity are restored to their initial

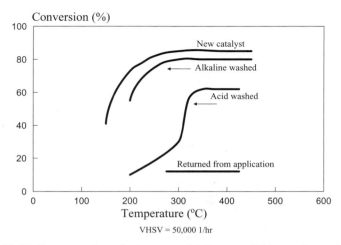

FIGURE 11.15. Regeneration of a commercial Pt monolithic catalyst using various washing procedures. *Source*: Reprinted with permission from Elsevier Science Publishers BV (Chen et al. 1992)

TABLE 11.3. Analysis of contaminants on the returned platinum honeycomb catalyst from metal strip coating operation.*

| | Weight percent (%) | | | |
| | As received | After acid Cleaning | After alkaline Cleaning | Fresh |
Contaminant				
Pb	0.007	0.007	0.001	0.003
Sn	0.3	0.2	0.03	0.001
Na	2	2	0.4	0.3
P	0.3	0.3	0.03	0.03
Cu	0.02	0.02	0.003	0.003
BET surface area (m^2/g)	3.55	5.09	9.60	15.0

*Analytical method—optical emission spectroscopy.

state. The composition of the regeneration solutions and the precise methods employed are proprietary, but they include careful treatment with mild acids, bases, or chelating agents. By using the proper reactor, operating conditions, and periodic maintenance with chemical cleaning, a useful catalyst life of 5 to 10 years is common.

The usual procedure in developing regeneration technology is to remove a catalyst sample from the commercial reactor and to test for activity in a laboratory unit. Using toluene as a model compound, the conversion versus operating temperature is determined, and then various chemical washing procedures are tried. Figure 11.15 shows one such laboratory study for a metal strip-coating operation (Chen et al. 1992). For this application, the catalyst was in operation for 3 years and was severely deactivated. The alkaline wash was

TABLE 11.4. In-field regeneration of a catalyst at commercial installations (Heck et al. 1988).

	Typical Solvents, Organics, and/or Miscellaneous Compounds	Cleaning Method Effective	Years on Stream	Average Performance (%)			Catalyst Temperature (°C)	
				X_D	X_i	X_R	Inlet	Exit
Can coating	Mineral spirits, methylisobutylketone	Acid	9	66	90	86	350	470
		Acid	5	60	95	95	315	370
	Iso-phorone, diisobutylketone	Acid	5	65	95	85	315	370
	Butyl cellosolve	Acid	5	70	95	94	315	370
		Alkaline	12	85	95	95	340	400
		Acid/alkaline	8	73	90	90	315	425
Metal coating	Methylethylketone, toluene, methylisobutylketone	Alkaline	4	75	95	95	315	370
	Iso-butanol	Physical, thermal, acid	8	70	90	90	315	370
Automotive paint bake	Toluene, xylene, methylketone	Acid	4	71	95	95	315	370
	Isopropyl alcohol	Acid	4	71	95	93	315	370
		Acid	3	91	95	95	400	455
Glove manufacture	Formaldehyde, phenolics	Physical	10	90	95	91	290	345
		Air lance	3	85	90	87	455	510
Phthalic anhydride	Phthalic, maleic anhydride	Acid	12	70	95	92	345–370	400
Oil production	H$_2$S condensate gas	Alkaline	1	50	—	85	315	—
Synthetic fibers	Teflon coating, titron, tars	Alkaline	3	50	—	99	315	425
Plastic products	Methanol, ethanol	Alkaline	4	13	—	99	315	425
Ceramic products	Vermiculate, nylon waste	Alkaline/acid	2	—	—	90	315	425
Synthetic fabrics	Scotchguard, thermosol dye	Alkaline/acid	4	80	—	90	315	425

the most effective in restoring the catalytic activity close to a nearly fresh condition. Analysis of the catalyst using optical emission spectroscopy, as given in Table 11.3, showed deposits of Sn, Na, and P and a substantial reduction in surface area from 15 to $3.55\,m^2/g$. The deposits were masking the pores of the catalytic surface, and the alkaline wash substantially removed them and restored the surface area to $9.60\,m^2/g$ or 65% of the original value. Using such regeneration procedures has resulted in a substantially longer life for organic abatement catalysts. Some regeneration data for various commercial installations is given in Table 11.4 (Heck et al. 1988).

Similar principles have been applied to catalysts used in the food industry (Fu and Chen 2001). Like many other catalytic technologies, the activity of the food service catalyst can be affected by fouling. A reliable and efficient cleaning method is needed to rejuvenate the catalyst routinely for better performance. The most common deposits found on food service catalysts are unburned grease particles and/or partially burned organic char. The grease particles are normally found on areas where the catalyst temperature is not high enough to initiate oxidation. The organic char is the result of the dehydrogenation of the deposited grease particles. As the grease builds up on the catalyst surface, it hinders the diffusion of oxygen to the catalyst surface. This favors "coking" of the grease particles rather than catalytic oxidation since the oxygen is blocked from reaching the catalytic surface. The deposits on a fouled food service catalyst surface from a chain-driven char broiler were analyzed by X-ray photoelectron spectroscopy (XPS), and several inorganic compounds such as P, K, Na, and Si were found. These compounds are believed to be from various sources, such as the ingredients used for food preparation, chemical agents used to clean the cooking appliances, and decomposition of the phospholipids contained in the meat. These major contaminants on the catalyst

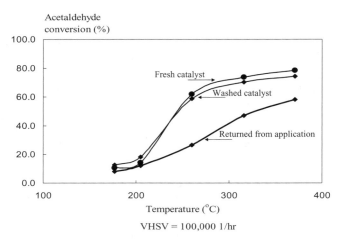

FIGURE 11.16. Regeneration of a commercial Pt monolithic catalyst used for abatement in food processing (Fu and Chen 2001).

surface cause catalyst deactivation. Using acetaldehyde as a model compound, the conversion versus operating temperature is determined, and then various chemical washing procedures are tried. An in-house washing method was then developed using a mild alkaline solution. The method was tested and proved to be effective in restoring food service catalyst activity without hurting the catalyst as shown in Figure 11.16.

Companies such as BASF and Catalytic Combustion offer services for routine cleaning and regeneration of the VOC catalyst.

REFERENCES

Agarwal, S., Spivey, J., and Butt, J. "Catalyst deactivation during deep oxidation of chlorocarbons," *Applied Catalysis A: General* 82: 259–275 (1992).

Berty, J. "Catalyst for destruction of toxic organic chemicals," U.S. Patent 5,021,383 (1991).

Berty, J. "Catalyst for destruction of toxic organic chemicals," U.S. Patent 5,114,692 (1992).

Bonacci, J., Farrauto, R., and Heck, R. "Catalytic incineration of hazardous wastes," pp 130–178. *Encyclopedia of Environmental Control Technology*. Editor Cherminisoff, P., Gulf Publishing, Houston, TX (1989).

Bonacci, J., and Heck, R. "Air pollution control catalytic equipment," IPMI Meeting Proceedings, Williamsburg, VA (1983).

Brown, R., Rossin, J., and Thomas, C. "Catalytic process for control of PFC emissions," Semiconductor International (2001).

CFR, Code of Federal Regulations, Title 40, Protection of the Environment, Chapter 1, Environmental Protection Agency (2007).

Chen, J., Heck, R., and Farrauto, R. "Deactivation, regeneration and poison resistant catalysts: Commercial experience in stationary pollution abatement," *Catalysis Today* 11: 517–545 (1992).

Chen, J., and Tran, P. "Catalytic oxidation of PTA vent gases," ICMAT 2001, MRS (Singapore), Paper I10-01-IN (2001).

CFR, Code of Federal Regulations. "Standards of performance for new stationary sources," 40(60) (1998).

Cohn, J., and Haly, A. "Process for effecting combustion reactions of oxygen-containing gases," Canadian Patent 709,313 (1965).

Eltron. "Catalysts for complete oxidation of volatile organic compounds (VOCs)," Eltron Research and Development TECH BRIEF, http://www.eltronresearch.com, (2007).

EPA, "The EPA air pollution control cost manual: Section 3 VOC controls," EPA/452/B-02-001 (2002).

Farrauto, R., and Heck, R. "Precious metal catalysis," pp 218–274. *Encyclopedia of Chemical Processing and Design*. Editor McKetta, J., Marcel Dekker, New York (1992).

Feaver, W., and Rossin, J. "The catalytic decomposition of CHF_3 over ZrO_2-SO_4," *Catalysis Today* 54(1): 13–22 (1999).

Freidel, I., Frost, A., Herbert, K., Meyer, F., and Summers, J. "New catalyst technologies for the destruction of halogenated hydrocarbons and volatile organics," *Catalysis Today* 17: 367–382 (1993).

Fu, J., and Chen, J. "Catalyst fouling and cleaning in food service applications," Paper No. 155, 94th Annual Meeting of Air & Waste Management Association (2001).

Fu, J., Czarnecki, L., Patellis, C., Bouney, A., and Whittenberger, W. "Catalytic method for controlling restaurant emissions," Paper 99-TP217, 92th Annual Meeting of Air & Waste Management Association (1999).

Grzanka, R. "Operating cost reduction strategies for oxidizers," Chem Show (2007).

Hardison, L., and Dowd, E. "Air pollution control: Emission control via fluidized bed oxidation," *Chemical Engineering Progress* 75(8): 31–35 (1977).

Heck, R., Durilla, M., Bouney, A., and Chen, J. "Air pollution control-ten years operating experience with commercial catalyst regeneration," Paper 88-83B.11, 81st APCA Annual Meeting, Dallas, TX (1988).

Heneghan, C., Hutchings, G., and Taylor, S. "The destruction of volatile organic compounds by heterogeneous catalytic oxidation," in *Catalysis*. Volume 17 The Royal Society of Chemistry, London, England (2004).

Hoke, J., Larkin, M., Farrauto, R., Voss, K., Whitely, R., and Quick, M. "System and method for abating food cooking fumes," U.S. Patent, 5,756,053 (1998).

ICAC, "Case Studies—Institute of Clean Air Companies, VOCs and HAPs Control," http://www.icac.com/i4a/pages/Index.cfm?pageid=3357, (2001).

Lester, G. "Catalytic destruction of hazardous halogenated organic compounds," International Patent Number 90/13352, 82nd Annual Air and Waste Management Association Meeting, Anaheim, CA (1989).

Martin, A., Nolan, S., Gess, P., and Baesen, T. "Control of odors from CPI facilities," *Chemical Engineering Progress* 88: 53–61 (1992).

Muller, H., Deller, K., Despeyroux, B., Peldszus, E., Kammerhofer, P., Kuhn, W., Spielmannleitner, R., and Stoger, M. "Catalytic purification of waste gases containing chlorinated hydrocarbons with precious metal catalysts," *Catalysis Today* 17: 383–390 (1993).

Nguyen, P., Stern, E., Deeba, M., and Burk, P. "Oxidative destruction of chlorocarbons," U.S. Patent 5,238,041 (1994).

Noordally, E., Richmond, J., and Tahir S. "Destruction of volatile organic compounds by catalytic oxidation," *Catalysis Today* 17: 359–366 (1993).

Norbeck, J., and Welch, W. "Further development of emission test methods and development of emission factors for various commercial cooking operations—final report," SCAQMD Contract No. 96027, College of Engineering—Center for Environmental Research and Technology, University of California, Riverside (1997).

NSR. "New source review for stationary sources of air pollution," Board on Environmental Studies and Toxicology (BEST) (2006).

Ojala, S. "Catalytic oxidation of volatile organic compounds and malodorous organic compounds," University of Oulu, Academic Dissertation, Finland (2005).

OTC. "Volatile organic compounds (VOCs) what are they and what are their effects?" http://www.parish-supply.com/volatile_organic_compounds.htm, (2007).

Pisarczyk, K., Singh, N., Sigmund, J., and Leonard, R. "The dynamics of the oxidation reaction of volatile organic compounds promoted by a novel monolithic support base metal catalyst," 96-MP4A.06, AWMA Meeting (1996).

Roark, S., and White, J. "Catalysts for low-temperature destruction of volatile organic compounds in air," U.S. Patent 6,458,741 (2002).

Sax, S., Bennett, D., Chillrud, S., Kinney, P., and Spengler, J. "Differences in source emission rates of volatile organic compounds in inner-city residences of New York City and Los Angeles," *Journal of Exposure Analysis and Environmental Epidemiology* 14: S95–S109 (2004).

SCAQMD. "Summary of proposed rule 1138," Report from South Coast Air Quality Management District (1997).

Shoup, R., Hoekstra, K., and Farrauto, R. "Thermal stability of a copper chromite auto exhaust catalyst," *American Ceramic Society Bulletin* 54(6): 565–568 (1975).

Simone, D., Kennelly, T., Brungard, N., and Farrauto, R. "Reversible poisoning of palladium catalysts for methane oxidation," *Applied Catalysis* 70: 87–100 (1991).

Spivey, J. "Complete catalytic oxidation of volatile organics," *Industrial and Engineering Research* 26: 2165–2180 (1987).

Spivey, J., and Butt, J. "Literature review, deactivation of catalysts in the oxidation of volatile organic compounds," *Catalysis Today* 11: 465–500 (1992).

Stein, K., Feenan, J., Hofer, L., and anderson, R. "Catalytic oxidation of hydrocarbons," *Bureau of Mines Bulletin* 608: (1962).

Vatavuk, W., van der Vaart, D., and Spivey, J. "Section 3, VOC controls, section 3.2, VOC destruction controls, chapter 2, incinerators," EPA/452/B-02-001 (2000).

Wang, Y., Shaw, H., and Farrauto, R. "Catalytic oxidation of trace concentrations of trichloroethylene over 1.5% Pt on γ-Al$_2$O$_3$," *ACS Symposium Series* 45: 125–140 (1992).

Yao, Y-F. "Oxidation of alkanes over noble metal catalysts," *I&EC Product Research and Development* 19(3): 293–298 (1980).

Yu, T., Shaw, H., and Farrauto, R. "Catalytic oxidation of trichloroethylene over PdO catalysts," *ACS Symposium Series* 45: 141–152 (1992).

CHAPTER 11 QUESTIONS

1. Mention and describe six approaches to removing VOCs from the commercial process.

2. When is the use of a catalyst a viable option?

3. Design a catalytic process for removing a VOC from an exhaust gas. List the major equipment, and draw a flow diagram to describe the process. The process is to remove 1,000 vppm of toluene in an exhaust gas of 10,000 scfm. The exhaust gas is available at 120 °C. Use 400 cpsi of honeycomb. Assume the reaction is mass transfer controlled and the design space velocity is 50,000 hr^{-1}.

4. You are hired by a major company that coats wires with organic-silicones that electrically insulate the wire. The emissions of organic and inorganic components of the wire coating operations are present in large excesses of air. The plant manager, who has no experience with catalysis, sees that you have taken such a course in catalysis and thinks perhaps you can improve the performance by using catalytic technology compared with their current pollution abatement methods of filtration and scrubbing.

   Here is a great chance for you to show that you are a highly trained chemical engineer so you recommend that an R&D program with multiple phases be established to determine the feasibility of a catalytic approach to reduce toxic emissions. You are selected as the project leader, with a small staff of technicians, and must now develop a 3-year research program. The company has absolutely no catalyst testing equipment available, but this is so important for their business they give you $500,000 to purchase whatever you need.

   You decide that only catalyst testing equipment will be purchased and that any required characterization will be outsourced to a local university that is well equipped with the type of equipment necessary. Furthermore, any catalyst materials will be purchased from catalyst companies. Prepare your approach plan in phases with specific objectives and goals. Some phases may be done in parallel and some in series. Indicate which phases will be performed in series and parallel.

5. Refer to a restaurant catalyst from U.S. Patent 5,580,535:

   a. How was thermal gravimetric analysis (TGA)/differential thermal analysis (DTA) used to evaluate candidate catalysts?

   b. What is the significance of Table II?

   c. Would you suggest a ceramic or metal monolith for the abatement catalysts and why?

   d. Contrast Claims 1–7. How are they different from each other?

   e. Contrast Claims 1, 8, and 9.

12 Reduction of NO$_x$

12.1 INTRODUCTION

A major source of nitric oxide (oxides of nitrogen) emissions is through combustion of fuels in engines and power plants. NO$_x$ emissions also can be significant in chemical operations such as nitric acid plants. More recently, the emissions of nitrous oxides (N$_2$O) from fiber-producing plants has received attention because of its global warming effects. The nitric oxide emissions and the SO$_x$ emissions are components of acid rain since, when mixed with water vapor in the clouds, they form nitric and sulfuric acid, respectively. Furthermore, NO$_x$ participates in photochemical ozone (smog) generation by reaction with hydrocarbons.

NO$_x$ is formed thermally in the combustion process by combination of the N$_2$ and O$_2$ present in the air. At temperatures greater than 1,500 °C, this reaction proceeds at appreciable rates through a well-characterized mechanism called the Zeldovich equation (Zeldovich 1946). If a compound in the combustion process has bound nitrogen (e.g., pyridine), NO$_x$ is readily formed at much lower temperatures through an oxidation process (Fenimore 1972). The NO$_x$ formed from bound nitrogen can be controlled by staged rich combustion. Several possible combustion modifications, including low NO$_x$ burners, overfire air, reburning, and water or steam injection can substantially lower the thermal NO$_x$ emissions (EPA 1999; Srivastava et al. 2005). In addition, direct injection of ammonia or urea (selective noncatalytic reduction of NO$_x$ or SNCR) into the flue or exhaust gas is also used to remove 30% to 60% of the NO$_x$ (Smith 2005). However, for high removal rates, in a lean environment, catalytic aftertreatment using ammonia or urea is required to meet federal regulations. These regulations contained in Title 40 Part 60 Subpart D of the Code of Federal Regulations vary for each point source (Code of Federal Regulation 2007). Some examples of point source regulations follow:

Point Source	Regulation
Utility boilers	40 CFR, Part 60, Subpart DB
Industrial boilers	40 CFR, Part 60, Subpart DB
Gas turbines	40 CFR, Part 60, Subpart GG
Stationary engines	40 CFR, Part 60, Subpart DC
Nitric acid production	40 CFR, Part 60, Subpart G

Catalytic Air Pollution Control: Commercial Technology, by Ronald M. Heck and Robert J. Farrauto, with Suresh T. Gulati.
Copyright © 2009 John Wiley & Sons, Inc.

An excellent resource for the standards of performance for new stationary sources is the Texas Natural Resource Conversation Commission (TNRCC 2007).

12.2   NONSELECTIVE CATALYTIC REDUCTION OF NO$_x$

One of the earliest techniques used to abate NO$_x$ emissions from engines and nitric acid plants was first to deplete the oxygen by operating the engine near stoichiometric or by adding a hydrocarbon or purge gas to deplete the oxygen via a chemical reaction in the exhaust.

For a stationary engine operation, the engine is normally operated near stoichiometric conditions, whereby the catalytic chemistry is similar to automotive three-way catalyst technology (see Chapter 6). A typical response of engine emissions from a natural gas-fired stationary engine is given in Figure 12.1.

The main differences from this application relative to automotive exhaust control are in the operating conditions (temperature, steady-state operation, and so forth) and the aging phenomena. A typical application of engine non-selective catalytic reduction (NSCR) of NO$_x$ would be the natural gas recompression engines used on the gas pipelines throughout the Unites States or for small cogeneration engines in hospitals, universities, and other institutions. The NO$_x$, CO, and sometimes HC emissions are to be controlled. In situations where high conversions are required, the engine will be equipped with a feedback control loop to maintain the engine operation near stoichiometric. Also, a second catalyst may be added with air injection to reduce CO and HC emissions further.

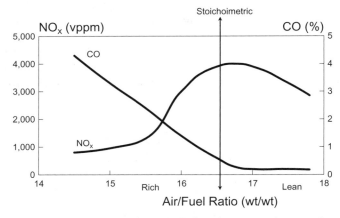

FIGURE 12.1. Emissions of natural gas-fueled stationary engine as a function of air/fuel ratio. Reprinted with permission from ASME (Burns et al. 1983).

The major reactions involved in NSCR NO$_x$ are as follows:

$$CO + 1/2\,O_2 \rightarrow CO_2 \tag{12.1}$$

$$H_2 + 1/2\,O_2 \rightarrow H_2O \tag{12.2}$$

$$HC + O_2 \rightarrow CO_2 + H_2O \tag{12.3}$$

$$NO_x + CO \rightarrow CO_2 + N_2 \tag{12.4}$$

$$NO_x + H_2 \rightarrow H_2O + N_2 \tag{12.5}$$

$$NO_x + HC \rightarrow CO_2 + H_2O + N_2 \tag{12.6}$$

A typical commercial catalyst ranges from 0.1% to 0.5% platinum, plus rhodium supported on a high-surface-area γ-Al$_2$O$_3$ washcoated onto a 200- to 400-cpsi ceramic or metallic honeycomb. There are, however, limited cases where beaded catalyst beds of similar catalysts compositions are in operation. Usually, the reactor operates at 50,000 hr^{-1} to 100,000 hr^{-1} space velocity. A schematic of an NSCR NO$_x$ system for engines is given in Figure 12.2. The performance of an NSCR NO$_x$ catalyst for a given engine operating condition is shown in Figure 12.3 (Burns et al. 1983).

The operating temperatures range from 450 °C to 900 °C, with the cogeneration applications having the higher range. Thermal deactivation is rare but does occur, mainly through engine maladjustments or some failure mode such as miss-fueling (e.g., bad fuel injector), miss-fire (e.g., bad spark plug), and so on. Normal operating conditions are well within the range for the stability of the catalytic elements of Pt and Rh and the components used to stabilize the surface area. The major mode of deactivation is masking of the catalyst with materials from the engine (e.g., lubricating oil, turbocharger coolant, and so on).

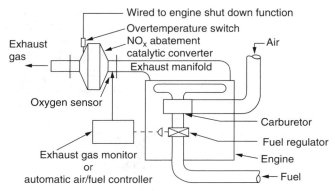

FIGURE 12.2. Major equipment in NSCR system for stationary engines. Reprinted with permission from ASME (Burns et al. 1983).

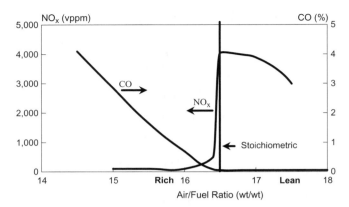

FIGURE 12.3. Emissions of natural gas-fueled stationary engine with an NSCR system. Reprinted with permission from ASME (Burns et al. 1983).

Phosphorus from the engine lubricating oil is believed to be the major source of deactivation, and specifications have been placed on its contents in the lubricating oil. Several studies have been conducted to find lube oil formulations more forgiving to the catalyst's longevity (Smrcka 1991). Recall that these engines operate continuously, and a 1-year exposure is equivalent to about 300,000 miles of operation of an automobile. Thus, the exposure to impurities from the engine operation is many times more severe than the catalyst sees in an automotive application. Depending on the nature of the contamination, the NSCR NO$_x$ can be regenerated using an air lance or air jets for cleaning, or a washing technique to remove the compounds masking the catalytic sites.

The other major application of NSCR technology is in the abatement of the exhaust gas from nitric acid plants (Marzo and Fernandez 1980; Adlhart et al. 1971; Gillespie et al. 1972). This so-called *tail gas* abatement is one of the earliest techniques to abate NO$_x$ emissions from oxidizing environments. Nitric oxide emissions produced during the manufacture of nitric acid by high-temperature catalytic oxidation of NH$_3$ can be conveniently abated by a non-selective process. Reddish-brown NO$_2$ is emitted to the atmosphere in the presence of excess air. Since the environment is oxidizing, and NO$_2$ must be reduced to N$_2$, it is necessary in NSCR to first consume all the excess O$_2$ by combustion by adding a fuel such as CH$_4$ (natural gas), liquid petroleum gas (LPG), or purge gas. The NO$_2$ is then catalytically reduced to N$_2$ by the residual fuel or its byproducts (i.e., CH$_4$, CO, and so on). Tail gas abatement is accomplished through a three-step process as shown by the following reactions:

$$HC + NO_2 \rightarrow NO + CO_2 + H_2O \qquad (12.7)$$

As evident from the reddish-brown plume, the tail gas contains predominantly NO$_2$. In fact, in the 1960s, this was the only reaction that was used. The effluent was bleached or visibly clear, and the NO was diluted quickly so that the visible color on reoxidation in the ambient air went unnoticed.

$$HC + O_2 \rightarrow CO_2 + H_2O \text{ (depletion of } O_2\text{)} \qquad (12.8)$$

$$HC, CO, H_2 + NO \rightarrow N_2 + CO_2 + H_2O \text{ (NO reduction)} \qquad (12.9)$$

The catalyst ranges from 0.3% to 0.5% platinum, plus small amounts of rhodium deposited on an alumina carrier on a ceramic honeycomb structure. Again, in some instances, a beaded catalyst is used. The operating temperature varies, depending on the fuel used to deplete the oxygen. Purge gas, which contains some H_2 and CO, operates at 300–350 °C, whereas natural gas requires a higher temperature of 500–550 °C. LPG usually requires 350–450 °C. Normally, these reactors operate at 50,000 to 100,000 hr^{-1} space velocity.

The reactor can be axial flow or radial flow for honeycomb catalysts, and it may be single stage or two stage, depending on the oxygen content of the tail gas from the nitric acid plant. For the high-pressure operating plants typical in the United States, the effluent from the reactor is used to drive a gas turbine that operates the air compressor of the air oxidation plant, thus supplying all the power needed for compression.

Deactivation usually occurs because of thermal sintering of the precious metal and/or Al_2O_3 after continuous exposure to temperature close to 900 °C caused by the exothermic nature of the process. Deposition of iron from upstream corrosion of piping can also deactivate the catalyst. Phosphorous from the compressor lubricating oils is also a common poison that masks the catalytic sites.

A hybrid low NO_x process is commercially available. It is composed of a thermal O_2 burn-out step, followed by NO_x reduction using added hydrocarbon fuel, and a final oxidation step to remove excess hydrocarbon and CO (Freidel et al. 1993).

12.3   SELECTIVE CATALYTIC REDUCTION OF NO_x

12.3.1   Introduction

Selective catalytic reduction of NO_x using NH_3 was first discovered in 1957 (Cohn et al. 1961). The discussion that follows is an historical perspective of the development of the SCR technology. It was discovered that NH_3 can react selectively with NO_x, producing elemental N_2 over a Pt catalyst in excess amounts of oxygen. The major desired reactions are as follows:

$$4NH_3 + 4NO + O_2 \rightarrow 4N_2 + 6H_2O \qquad (12.10)$$

$$4NH_3 + 2NO_2 + O_2 \rightarrow 3N_2 + 6H_2O \qquad (12.11)$$

One undesirable reaction produces N$_2$O that, due to its strong infrared absorptivity, is considered to be a powerful greenhouse gas:

$$2NH_3 + 2O_2 \rightarrow N_2O + 3H_2O \tag{12.12}$$

The injected ammonia can be wasted by catalytic partial oxidation to elemental nitrogen. This is a nonselective reaction.

$$4NH_3 + 3O_2 \rightarrow N_2 + 6H_2O \tag{12.13}$$

It can also be completely oxidized to NO, another nonselective reaction.

$$4NH_3 + 5O_2 \rightarrow 4NO + 6H_2O \tag{12.14}$$

At temperatures below about 100–200 °C, the ammonia can also react with the NO$_2$ present in the process gas producing explosive NH$_4$NO$_3$.

$$2NH_3 + 2NO_2 + H_2O \rightarrow NH_4NO_3 + NH_4NO_2 \tag{12.15}$$

This reaction can be avoided by never allowing the temperature to fall below about 200 °C. The tendency for formation of NH$_4$NO$_3$ can also be minimized by metering into the gas stream less than the precise amount of NH$_3$ necessary to react stoichiometrically with the NO$_x$ (1-to-1 mole ratio). By doing so, there is no little excess NH$_3$ that can "slip" out of the reactor.

As an example of the responses of an SCR catalyst to temperature and NH$_3$/NO$_x$ feed conditions, Figures 12.4 and 12.5 demonstrate the NO$_x$ conver-

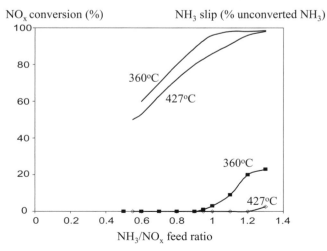

FIGURE 12.4. Effect of NH$_3$/NO$_x$ feed ratio on NO$_x$ conversion and NH$_3$ slip for V$_2$O$_5$/TiO$_2$ SCR catalyst on 200 cpsi.

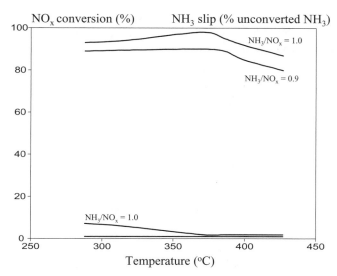

FIGURE 12.5. Effect of temperature on NO$_x$ conversion and NH$_3$ slip for V$_2$O$_5$/TiO$_2$ SCR catalyst on 200 cpsi.

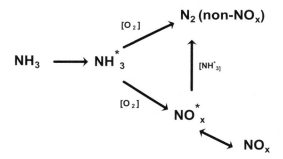

FIGURE 12.6. Reaction network—catalytic reaction scheme of NH$_3$, NO$_x$, and O$_2$.

sion and NH$_3$ slip as a function of the NH$_3$/NO$_x$ ratio and temperature for a V$_2$O$_5$/TiO$_2$ monolithic catalyst.

It is apparent that ratios much greater than 1 result in significant NH$_3$ slip. In all applications, there is always a specification on permitted NH$_3$ slip. Frequently this is <5–10 vppm.

The major reactions involved in SCR NO$_x$ reduction are depicted schematically in Figure 12.6. An excellent discussion of the kinetics of NO$_x$ reduction and selective NH$_3$ reaction over a honeycomb catalyst is given in Chen et al. 1990.

When sulfur is present in the flue gas, such as in coal-fired boilers and/or power plants, or petroleum-derived liquid fuels such as distillate or diesel, the oxidation to SO$_2$ to SO$_3$ [reactions (12.16) and (12.17)] results in formation of H$_2$SO$_4$ upon reaction with H$_2$O. Obviously, this results in condensation downstream and excessive corrosion of process equipment.

$$2SO_2 + O_2 \rightarrow 2SO_3 \tag{12.16}$$

$$SO_3 + H_2O \rightarrow H_2SO_4 \tag{12.17}$$

The reaction of NH$_3$ with SO$_3$ also results in formation of (NH$_4$)$_2$SO$_4$ and/or NH$_4$HSO$_4$ [reactions (12.18) and (12.19)], which deposits on and fouls downstream process equipment such as heat exchangers causing a loss in thermal efficiencies (Matsuda 1980; Kobayashi et al. 1988).

$$NH_3 + SO_3 + H_2O \rightarrow NH_4HSO_4 \tag{12.18}$$

$$2NH_3 + SO_3 + H_2O \rightarrow (NH_4)_2SO_4 \tag{12.19}$$

Few applications of SCR NO$_x$ catalysts existed until the early 1970s, when reduction of the emission of NO$_x$ became an important control issue for stationary power sources in Japan. The Pt technology was not applicable in this exhaust temperature region (i.e., >250 °C) due to its poor selectivity for NO$_x$ reduction at these higher temperatures, so it was during this period that the base metal catalysts were found to be effective. Figure 12.7 shows a comparison of the operating temperature range for the various catalyst technologies available for SCR NO$_x$ (Heck et al. 1994).

Note that the Pt catalysts lose selectivity above 250 °C. At >250 °C, the V$_2$O$_5$/Al$_2$O$_3$ catalyst first was used. However, its use was restricted to sulfur-free exhaust gases because the alumina reacted with SO$_3$ forming Al$_2$(SO$_4$)$_3$ and deactivated the catalyst. This problem led to another key development—

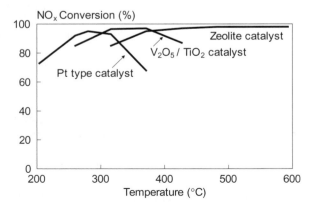

FIGURE 12.7. Operating temperature window for different SCR catalyst formulations. Reprinted with permission from Heck et al. Copyright 1994 American Chemical Society.

the use of a nonsulfating TiO$_2$ carrier for the V$_2$O$_5$, which then became the catalyst of choice (Heck et al. 1987; Bosch and Janssen 1988). These catalysts functioned at higher temperatures and over a broader range than Pt (see Figure 12.7) (Heck et al. 1994).

Finally, zeolite-based catalysts have been developed that function at higher temperatures, thus extending the operating temperature range for SCR NO$_x$ (Byrne et al. 1992). In addition there has been a flurry of activity in the diesel emission control area looking at new zeolite materials that cover a broad operating range equivalent to the V$_2$O$_5$/TiO$_2$-type technology. Please refer to Chapter 8 of this textbook for more information.

As discussed in Chapter 1 on catalyst fundamentals, the SCR NO$_x$ reaction with NH$_3$ is a classic example of selectivity. It is because of the competing reactions that the temperature operating window is narrow and the conversion response curve resembles a bell-shaped curve, as depicted in Figure 12.8. This is a similar catalytic reaction phenomenon as described in Chapter 6 for a lean NO$_x$ automotive catalyst.

SCR catalysts can be prepared in many different geometric structures. (Figure 12.9) Some are extruded into pellets or homogeneous monoliths, whereas others are supported on parallel metal plates or ceramic honeycomb structures; still others are fixed on a wire mesh. The appropriate structure depends on the end-use application as discussed below.

Since each general class of catalyst materials has different temperature performance characteristics, the engineer has considerable design flexibility to select the most cost-effective catalyst composition, structure, and operating system to optimize the abatement process (Heck 1999). The active catalytic component and temperature ranges may be classified as indicated:

- Low temperature (175–250 °C): Platinum
- Medium temperature (300–450 °C): Vanadium
- High temperature (350–600 °C): Zeolite

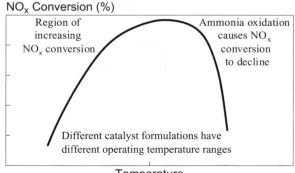

FIGURE 12.8. SCR NO$_x$ is a classic example of catalytic selectivity.

TABLE 12.1. Original participants in U.S. market during 1990s.

Company	Catalyst description	Operating Temperature (°C)
ARI Industries	Metal oxide/extruded monolith	290–370
	Zeolite/extruded monolith	315–480
Babcock Hitachi	V/Ti/metal plate	240–415
Camet	Precious metal/metal monolith	225–275
Cormetech	V/Ti/extruded monolith	200–450
Engelhard	Precious metal/ceramic monolith	175–320
(BASF)	V/Ti/ceramic monoltih	300–425
	Zeolite/ceramic monolith	315–605
Hitachi Zosen	V/Ti/extruded monolith	330–420
	V/Ti/wire mesh	330–420
IHI	V/Ti/extruded monolith	200–400
JMI	V/Ti/metal monolith	340–425
KHI	V/Ti/extruded monolith	300–400
MHI	V/Ti/extruded monolith	200–400
Norton	Zeolite/extruded monolith	220–520
Steuler	Zeolite/extruded monolith	300–520
UBE	V/Ti/extruded monolith	250–400

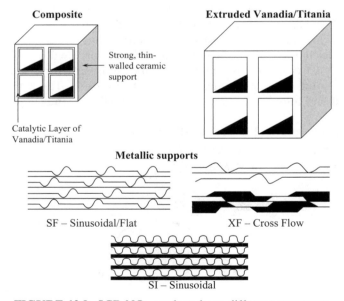

FIGURE 12.9. SCR NO$_x$ catalysts have different structures.

A more complete list of catalyst manufacturers, general catalyst compositions, catalyst support structures, and temperatures of operation is shown in Table 12.1. These were the early participants in this technology in the early 1990s and many do not participate (or actually exist) in the present market. An excellent review of the fundamentals of SCR is also available (Bosch and

Janssen 1988) as well as commercial operating conditions. The major SCR NO$_x$ catalyst suppliers going into the twenty-first century are BASF (formerly Engelhard), Cormetech, Haldor Topsoe, Siemens, Mitsubishi, and Babcock Hitachi. An extensive cost analysis of the various SCR catalysts approaches as well as alternative approaches has been completed (DOE 1999).

12.3.2 Low-Temperature SCR

At low temperatures, the SCR reactions (12.10) and (12.11) dominate, so NO$_x$ conversion increases with increasing temperature as shown in Figure 12.10. At about 225–250 °C, the rates of reaction (12.13) and the oxidation of NH$_3$ to NO$_x$ and H$_2$O (12.14) becomes so dominant that the conversion versus temperature plot reaches a maximum and begins to fall. Notice that the NH$_3$ slip or unconverted NH$_3$ is very low. To use such a Pt-based catalyst, one must control the temperature of the process gas to above about 200 °C to avoid NH$_4$NO$_3$ formation [reaction (12.15)] but not to exceed about 275 °C, where the catalyst loses its selectivity toward the NO$_x$ reduction reaction. This narrow window for temperature control adds a great deal of expense and complexity to the overall process design. Also, the formation of N$_2$O as a possible by-product needs to be determined for any Pt catalyst. Consequently, this technology is not commonly used today.

A new low-temperature process has emerged for NO$_x$ abatement of tail gas from nitric acid plants. This catalyst operates at inlet temperatures of 180–200 °C. The process uses NH$_3$ injection with four beds of catalyst. Two beds are composed of oxides of Cu and Ni on γ-Al$_2$O$_3$ cylinders. The two other beds are composed of Pt on γ-Al$_2$O$_3$. The space velocity is reported to be 15,000 1/hr and reduces the NO$_x$ by 92–95% (Blanco et al. 1993).

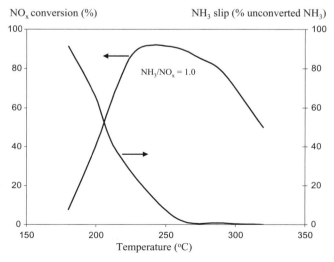

FIGURE 12.10. NO$_x$ conversion and NH$_3$ slip of low-temperature SCR catalyst.

A Pt catalyst supported on a stainless steel substrate with 320 cpsi has also been reported for gas turbine applications. It simultaneously removes NO$_x$ by reduction with injected NH$_3$ and oxidizes CO and HC at a catalyst inlet temperature of less than 280 °C (Pereira et al. 1988).

The Shell DeNO$_x$ System differs from conventional honeycomb SCR systems in two important aspects: the catalyst and the catalyst reactor module (Brundrett et al. 2001). The catalyst is in the form of porous particles (extrudates) and can be produced in a range of sizes and shapes to meet specific performance requirements. Due to the high activity of the catalyst, high NO$_x$ removal efficiencies with simultaneous control of NH$_3$ slip can be obtained at relatively low temperatures. The catalyst reactor module is based on the so-called lateral flow reactor (LFR) principal. The LFR is a packed-bed-type reactor that offers the advantage of low pressure drop, even at high space velocities. The reactor is basically cross-flow, or radial flow, which is usually required for low pressure drop fixed-bed reactors using a particulate catalyst as shown below in Figure 12.11.

A low pressure drop over the SCR reactor reduces power consumption or power loss depending on the application. Furthermore, the LFR design makes possible the use of relatively small catalyst pellets, which reduces the amount of catalyst required and facilitates fast loading and unloading of the catalyst from the reactor. Development of the LFR technology has resulted in a modular construction system, providing a high degree of flexibility in the

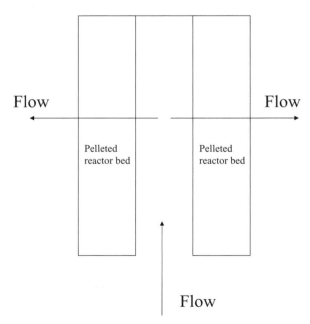

FIGURE 12.11. Schematic of radial flow reactor with catalytic particles.

design of SCR systems for specific applications, particularly retrofit. The pelleted SCR catalyst has a typical application of 160–230 °C and a range of 150–375 °C.

12.3.3 Medium-Temperature Operation

Medium-temperature V$_2$O$_5$-containing catalysts operate best in a temperature range between 260 °C and 450 °C. This has the obvious advantage of a broader operational range than Pt. However, as shown in Figure 12.7, it too shows a maximum followed by a decline in NO$_x$ conversion where the catalyst loses selectivity. NO$_x$ conversion initiates at about 225 °C, continues to rise to a plateau at about 400 °C, and then falls as the rate of ammonia oxidation and decomposition [reactions (12.13) and (12.14), respectively] begin to dominate.

From the discussion above, it is clear that selectivity is lost above about 425 °C. The exposure temperature of this catalyst must not exceed about 450–475 °C. The active anatase phase of TiO$_2$ with a surface area of 80–120 m^2/g irreversibly converts to rutile with a surface area of less than 10 m^2/g.

Recent advances in the vanadium technology have sought to improve the operating performance by

- Reducing the V$_2$O$_5$ level to reduce formation of SO$_3$ (Hums et al. 1997)
- Addition of WO$_3$ and MoO$_3$ to improve the redox properties and thus reactivity (Casgrande et al. 1999)
- Replacement of WO$_3$ with MoO$_3$ to be more tolerant to As poisoning (Spitznagel et al. 1994)

The addition of W was shown to give better high-temperature stability, and further modification of the catalyst improved the higher temperature selectivity, thus reducing the loss of NH$_3$ to the nonselective reactions to N$_2$ and NO$_x$. Figure 12.12 shows the performance of an improved catalyst technology that extends the higher operating temperature of the vanadia catalyst and shows improved durbility.

Studies on modifications of the V$_2$O$_5$/TiO$_2$ catalyst formulation have continued looking at the effect of WO$_3$ and SiO$_2$ and sulfated SiO$_2$. These studies were addressed at increasing the V$_2$O$_5$ concentration to increase the SCR catalyst activity and yet minimize SO$_2$ oxidation to SO$_3$. With the higher intrinsic activity, it may be possible to decrease the catalyst volume (Kobayashi et al. 2005; Kobayashi and Hagi 2006).

12.3.4 High-Temperature Operation

The suitability of zeolite catalysts for SCR above 450 °C has been known since the 1970s, when the zeolite mordenite was identified as an active SCR catalyst (Pence and Thomas 1980). Mordenite has a well-defined crystalline structure

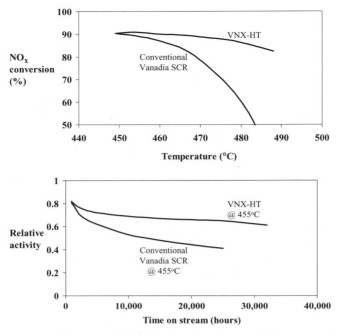

FIGURE 12.12. Modified vanadia/titania catalysts have a higher operating temperature and improved durability.

composed of a SiO$_2$-to-Al$_2$O$_3$ ratio of about 10. It has two intersecting pore structures, one containing 12 oxygens in the elliptical-shaped aperture with a size of 6.7×7.0 Å, and the other having 8 oxygens with a pore of 2.9×5.7 Å. Hydroxyl groups (OH$^-$) are bonded to each Al$^+$ in the pore structure, giving rise to Bronsted acidity, which is believed to be the site for catalytic activity. The H$^+$ of the OH$^-$ can be ion-exchanged by other cations while maintaining electrical neutrality, altering the chemistry of the active sites. Manufacturers do not usually reveal the precise chemical composition of the zeolite, so it is not possible to describe them in detail. Commercially available zeolite SCR catalysts can operate at temperatures as high as about 600 °C. When NO$_x$ is present, this catalyst does not oxidize ammonia to NO$_x$ according to reaction (10.14). Therefore, unlike the Pt and V$_2$O$_5$ catalysts, its selectivity toward NO$_x$ conversion continually increases with temperature as shown in Figure 12.13. (See NMR in Chapter 3 on catalyst characterization.)

At exposure temperatures approaching 600 °C, in a high water content process stream, zeolites tend to deactivate by a process called *de-alumination* where the Al^{+3} in the SiO$_2$-Al$_2$O$_3$ framework migrates out of the structure. This leads to permanent deactivation and, in the extreme case, collapse of the crystalline structure.

New technology advances have continued with the zeolite-based SCR NO$_x$ catalyst. New zeolite-based materials have been invented that extend the

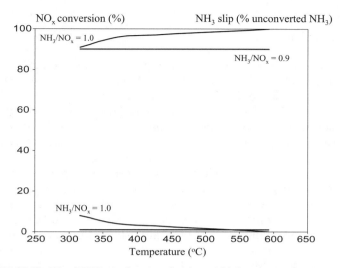

FIGURE 12.13. The ZNX catalyst is selective at higher operating temperatures.

operating temperature range for NO_x reduction. The zeolite material called ETZ is the newer technology that is being offered for higher temperature applications (around 580 °C) for extended operation. Some patent literature describes an iron-promoted beta zeolite commonly used for SCR from combustion gases (Tran et al. 2006).

12.4   COMMERCIAL EXPERIENCE

Selection of the proper catalyst composition and structure depends very substantially on the fuel used in the front end of the process. In many power plants, coal containing high levels of ash and sulfur is used, which creates an exhaust with high particulate levels. Gas turbines generate a much cleaner exhaust since they use liquid or gaseous fuels. Fortunately, a wide variety of catalyst compositions and structures exist to meet modern NO_x abatement requirements. Listed below are several applications for SCR NO_x:

- Gas-fired utility boilers
- Coal-fired boilers
- Oil-fired boilers
- Process heaters
- Gas turbines
- Stationary engines
- Nitric acid plants

- Steel mills
- Chemical plants

Because of the variant exhaust gas compositions, particulate loading, and contaminants, there are different catalyst support structures as shown in Figure 12.9. In any application of SCR NO$_x$ catalyst technology, the most important design element is the ammonia injection grid (AIG). If this is not designed properly, then the NH$_3$/NO$_x$ ratio entering the catalyst will not be uniform. This can result in lower NO$_x$ conversion and NH$_3$ leak in the catalyst exhaust. An excellent review of commercial experience in the twentieth century was presented showing the issues with catalyst design and performance (Hums 1998), whereas many current publications on commercial experience in the United States and worldwide can be found at the National Energy Technology Laboratory website at http://www.netl.doe.gov. The extruded catalyst and the metallic support are typically used in high dust conditions and have low cell densities (10 to 100 cells/in^2 or cpsi). The composite catalyst (either on a metallic or a ceramic monolith) is used in low dust conditions and has a higher cell density (from 64 cpsi to 400 cpsi).

12.4.1   Coal/Petroleum-Fired Power Plants or Boilers

The position of the SCR reactor in the process effluent strongly influences the decision as to what catalyst composition and physical structure should be used (Heck 1999). For a high ash, high sulphur containing exhaust, there are three possible positions for the SCR catalyst (Janssen and Meijer 1993; ICAC 1997):

1) High dust
2) Low dust
3) Tail end

A typical installation for an SCR NO$_x$ unit is shown in the schematic in Figure 12.14. Note that the critical design parameters are the location of the SCR reactor structure and the design of the ammonia (or urea) injection grid. The mixing of the ammonia must be uniform to assure a set NH$_3$/NO$_x$ ratio, which will then dictate the amount of NO$_x$ removal. There are several design options for SCR installations in coal-fired utility boilers as shown in Figure 12.15 (ICAC 1997).

One commercial catalyst supplier has a unique approach for selecting the honeycomb configuration as shown in Figure 12.16 for different applications classified as coal, oil, and gas as the source for the fuel used in the combustion process (Cormetech 2007).

The extruded titania-based honeycomb structure and cell pitch is determined for each customer application after an engineered analysis of system performance requirements.

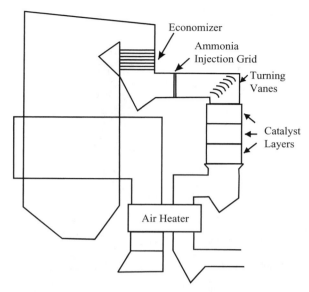

FIGURE 12.14. Typical high dust schematic for SCR in a utility boiler. Courtesy ICAC (ICAC 1997).

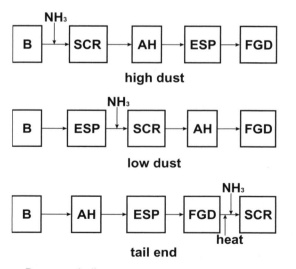

B   =   boiler
AH   =   air preheater
ESP =   electrostatic precipitator (or other dust collector)
SCR =   selective catalyst reduction
FGD =   flue gas desulphurisation

FIGURE 12.15. Design options available for operation in utility boiler applications (ICAC 1997).

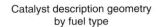

Catalyst description geometry
by fuel type

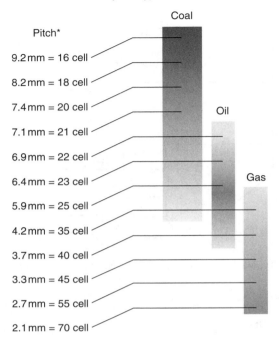

Coal

Pitch*

9.2 mm = 16 cell

8.2 mm = 18 cell

7.4 mm = 20 cell   Oil

7.1 mm = 21 cell

6.9 mm = 22 cell

6.4 mm = 23 cell   Gas

5.9 mm = 25 cell

4.2 mm = 35 cell

3.7 mm = 40 cell

3.3 mm = 45 cell

2.7 mm = 55 cell

2.1 mm = 70 cell

*Pitch defines cell size from
center line of the walls.

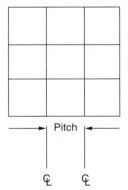

Pitch

FIGURE 12.16. Precise catalyst pitch is tailored for each gas path application (Cormetech 2007).

12.4.1.1 High Dust. In general for high dust/coal-fired boiler applications, a V_2O_5/TiO_2 extruded catalyst of low cell density (9 to 11 cpsi) operating at 350 °C to 400 °C and 3,000 to 5,000 1/hr space velocity is preferred. The TiO_2 catalyst can also be bonded to parallel stainless steel plate supports (Hums 1998). This steel is necessary since it resists corrosion by the SO_3 present. The SCR catalyst is located before any flue gas particulate or scrubbing operations so it must have large channels or holes to avoid plugging and excessive back pressure (i.e., high pressure drop) buildup. It is common practice to have a vertical stack of three to four catalyst beds with a dummy bed at the inlet. The vertical orientation favors low accumulation of ash in the channels of the honeycomb. Flow of the flue gas is usually downflow through the catalyst beds. The dummy or blank honeycomb helps to straighten the flow to minimize erosion of the subsequent catalyst beds caused by the high level of particulates present. In the "high dust" configuration, particulate levels can vary between 1 and 30 g/m^3 depending on the type of boiler and the quality of the fuel. Sometimes processes operate with a soot blower at one or more catalyst beds to purge the channels periodically of unwanted dust and ash deposits. A typical installation for high dust location in utility boilers is shown in Figure 12.14. A comparison of a module of extruded titania-based catalyst and the commercial bed of modules is shown in Figure 12.17. Note the size of the reactor structure for a power plant installation.

These pictures show that some individual elements are in special trays that can be removed and sent to be tested for catalyst activity. Coal (and gas) modules are custom designed and are not standardized. Individual SCR catalyst elements are typically 150-mm square, and the length is determined by multiple parameters. Each project has different requirements and in turn modules of different dimensions. Typically, an SCR catalyst converts over 80% of the inlet NO_x with less than 5 ppm of NH_3 slip.

Catalyst deactivation occurs primarily by the accumulation of fly ash, containing alkali and alkaline earth metal oxides, and sulfur compounds on the surface or within the pore structure of the catalyst. Sulfates of Ca, Mg, and Ba are frequently found blocking 50–65% of the pore volume depending on the specific bed location (Prins and Nuninga 1993; Chatterjee et al. 1993). With certain fuels irreversible selective poisoning of the catalyst by As occurs (Hums 1998).

There have been many years of operating experience in the so-called high dust applications, and knowledge has been acquired to specify routine catalyst maintenance. For instance, one catalyst manufacturer recommends SCR catalyst system inspections and evaluations performed at least once per year. This should include physical inspection of the catalyst, reactor, and ammonia injection system. Samples of the catalyst are also taken for analysis and evaluation in the laboratory (Pritchard 2006). Experience has also shown that various catalyst cleaning procedures can be effective where excessive accumulation of particulate matter or dust has accumulated, decreasing catalyst performance and increasing pressure drop. These cleaning methods can be classified as dry

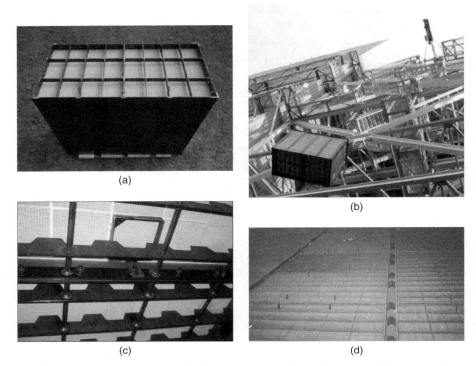

(a)

(b)

(c)   (d)

FIGURE 12.17. SCR NO$_x$ installation for power plant. Courtesy of Cormetech Inc. a) Module containing extruded V$_2$O$_5$/TiO$_2$ SCR NO$_x$ catalyst b) Lifting module into SCR NO$_x$ reactor structure c) SCR NO$_x$ modules in place d) Wall of SCR NO$_x$ catalyst.

or wet. The dry process is typically some sort of vacuum procedure, whereas the wet process involves using an aqueous solution after a drying process. The cleanings or catalyst rejuvenations can be done *in situ*, on-site or off-site. In addition to the SCR catalyst suppliers, some companies actually provide SCR catalyst regeneration technologies and management services to power plant operators (Hartenstein and Gutberlet 2001; Bullock and Hartenstein 2003; Focus on Catalysis 2007).

European experience with SCR reflects that 15% of the catalyst is replaced annually. Since most of the early high dust SCR experience was in Europe with the typical coals available locally, there was a question of whether applications in the United States would be successful where low-sulfur, high-calcium sub-bituminous coals are used at many power plants to comply with SO$_2$ regulations. There was some concern that firing of such coals may lead to deposits on SCR catalysts, which in turn may experience an accelerated deactivation range for commercial use (Srivastava et al. 2005). Initial operating experience has demonstrated SCR systems capable of 93% NO$_x$ removal with a maximum of 3-ppm NH$_3$ slip. Another issue was possible catalyst arsenic poisoning on dry bottom boilers firing coals from western Pennsylvania and West Virginia. These coals unusually have low contents of free calcium oxide (CaO) in the fly ash. CaO acts to scavenge gaseous forms of arsenic to form calcium arsenide. To address this issue, some facilities have accelerated their catalyst management plans, and others are adding small amounts (1–2% of coal feed rate) of pulverized limestone to their coal (Srivastava et al. 2005). It is becoming apparent that with more in-field operating experience, the SCR systems are working satisfactory in the United States and with proper maintenance procedures and are achieving >85% NO_x removal.

A projected catalyst life of about 5 to 9 years is typical with replacement of the front catalyst sections with provisions for soot blowing and various maintenance procedures.

12.4.1.2 *Low Dust.* The SCR catalyst can be positioned downstream of an electrostatic precipitator providing the catalyst with a low dust environment (see Figure 12.15). Extruded homogeneous honeycombs with cell geometries between 11 and 50 cpsi and composed of either V_2O_5/TiO_2 are used. Temperatures are typically between 300 °C and 450 °C dependent on the need to control SO_3 production and overall NO_x conversion. Space velocities vary depending on the feed composition, but 5,000 to 10,000 1/hr is typical. In the "low dust" and "tail end" configuration, particulate levels are much lower, typically less than 100 mg/m³.

Since the flue gas still contains high levels of sulfur in the form of SO_2, the catalyst and process conditions, i.e., lower temperatures, must be selected to minimize production of SO_3. In some applications, promoters such as Mo and W are added to the V_2O_5/TiO_2 catalyst to decrease the formation of SO_3 and to extend catalyst life (Gutberlet and Schallert 1993).

Fine dust that passes through the precipitator will deposit on the horizontally mounted catalyst bed and result in pore plugging by alkali sulfates. Catalysts in this environment have a typical life of about 5 to 9 years.

12.4.1.3 *Tail End Location.* This location provides the cleanest feed gas to the catalyst since it has passed through an electrostatic precipitator and a flue gas desulfurizer (see Figure 12.15). Consequently, honeycombs with cell densities of 200 cpsi can be used with little worry of channel plugging. Extruded homogeneous V_2O_5/TiO_2 catalyst or ceramic or metal substrates washcoated with V_2O_5/TiO_2 can be used. Relatively low temperatures of 300–350 °C are used to minimize the amount of energy needed to reheat the flue gas from the flue gas desulfurizer (FGD) to the required SCR temperature. Conversions in excess of 95% of the inlet NO_x with less than 10 ppm of NH_3 slip are achieved (Lee 1993).

The catalyst life of 5 to 9 years is typical with deactivation due to slow poisoning by feed contaminants and sintering of the anatase structure of TiO_2 to the less active, low-surface-area rutile structure.

12.4.2   Gas Turbines

The flue gases produced from power plants or boilers fueled by natural gas or light distillate gases are free of ash and sulphur and thus present a clean process gas to the SCR catalyst. For such applications, high cell densities up to 200 cpsi can be used since plugging is not an issue. The limiting factor on cell density is pressure drop, which is limited to about a 4″ (10.1 cm) H$_2$O column for gas turbines. Higher cell density substrates translate to substantially smaller catalyst volumes and thus to higher space velocities of 20,000 to 40,000 hr^{-1}. V$_2$O$_5$/TiO$_2$ deposited on a cordierite or metal support function at temperatures close to 400 °C, while zeolites, either homogeneously extruded or supported on a ceramic, can be used over 500 °C to ensure conversions of greater than 95% (less than 9-ppm NO$_x$ at the outlet) with less than 10 ppm of NH$_3$ slip (Byrne et al. 1992). Figure 12.18 contains an equipment schematic of an SCR NO$_x$ catalyst installed in the ducting downstream of a gas turbine. Figure 12.19 contains the equipment layout for the SCR catalyst and the AIG. The AIG design contains the ancillary equipment needed for delivering the NH$_3$ and the mixing device required for providing a uniform NH$_3$/NO$_x$ ratio to the catalyst. Notice that the catalyst is located between heat exchangers in the heat recovery section to obtain the optimum operating temperature for the SCR NO$_x$ catalyst. Since the combined cycle gas turbines do not have a heat exchange section, the exhaust temperatures are higher and the zeolite technology must be used. Catalyst lives of 5 to 6 years are common since deactivation due to process impurities is a minimum. Lubricating oils containing P and Zn from seals and compressors seem to be responsible for the slow deactivations observed in the field.

As more experience with SCR for gas turbines has been accumulated, it has become apparent that there is a substantial cost penalty for operating SCR

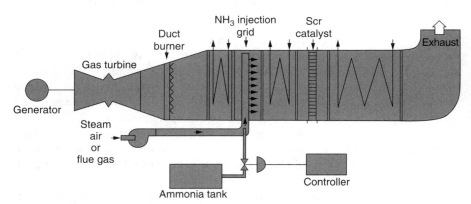

FIGURE 12.18. Schematic of typical installation of SCR NO$_x$ unit in cogeneration applications.

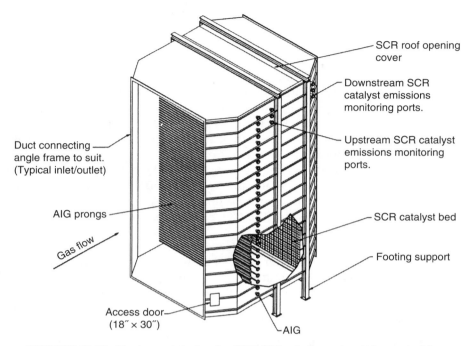

FIGURE 12.19. Equipment design for SCR NO$_x$ abatement unit in gas turbine.

systems above 80% conversion (Carpenter 2002). This is because of the increased quantity of catalyst, increased back pressure, increased quantity of ammonia, and potential increase in ammonia slip. It is therefore recommended that for very high NO$_x$ removal rates, that modification to the gas turbine operation or combustion process be done to reduce the NO$_x$ when possible; otherwise, the SCR process cost increase will be required.

12.4.3 Chemical Plants

Publications on this application are sparse. One application using a V$_2$O$_5$/TiO$_2$-type catalyst was for removing the NO$_x$ from an automotive catalyst plant off-gas. In the manufacturer of automotive catalysts, various inorganic rare earth nitrate compounds are included in various catalyst formulations. Calcination of an automotive catalyst causes these compounds to decompose, producing oxides of nitrogen (NO$_x$). Other by-products include particulate and volatile organic compounds (VOCs).

$$2\text{La}(\text{NO}_3)_3 \rightarrow \text{La}_2\text{O}_3 + 3\text{NO} + 3\text{NO}_2 + 3\text{O}_2 \qquad (12.20)$$

$$\text{Sr}(\text{NO}_3)_2 \rightarrow \text{SrO} + \text{NO} + \text{NO}_2 + \text{O}_2 \qquad (12.21)$$

$$Ce(NO_3)_3 \rightarrow CeO_2 + NO + 2NO_2 + O_2 \qquad (12.22)$$

$$2Nd(NO_3)_3 \rightarrow Nd_2O_3 + 3NO + 3NO_2 + 3O_2 \qquad (12.23)$$

The concentration of NO$_x$ and other pollutants purged from the calciner depend on the quantity of nitrates and on other compounds used in the catalyst formulation and production rate. Concentrations of NO$_x$ up to 10,000 ppm$_v$ are possible with VOC levels reaching 500 ppm. Dust concentration from the oven is 0.003 grains per dry standard cubic foot with a particle size of 90% < 4 microns. The gas temperature averages 335 °C. Local environmental regulations limit the combined maximum discharge of NO$_x$ to 200 ppm (Heck and Womelsdorf 2002). Since dust is also present in the off-gas, there is a ceramic filter in the exhaust stream to remove the dust before passing to the composite 100 cpsi composite SCR catalyst. The NO$_x$ removal efficiency for this application is >99%.

Another chemical plant application for SCR is the production of hydrogen from the steam reforming of methane. The off-gas from the steam methane reformer (SMR) contains NO$_x$ that is formed in the combustion in the reformer furnace. Past experience using SCR at SMR installations showed chromium-oxide species deposits on the catalyst, which causes activity loss. This phenomenon has also been observed for SCRs in ethylene plants. This direct experience with hydrogen-plant flue gas is important for initial catalyst sizing and prediction of run length with the presence of chromium-oxide species in the flue gas. The loss of catalyst activity experienced has been attributed to a masking of the active catalyst surface by an ongoing deposition of some form of chromium, believed to have been evaporated at parts-per-billion (ppb) to ppm concentrations from alloy metals in the reformer tubes in contact with hot flue gas (Kunz et al. 2003; O'Leary et al. 2002). This masking layer manifests itself as a discoloration of the SCR catalyst surface, heavy at the inlet and becoming gradually lighter approaching the outlet. The NO$_x$ removal efficiency is nominally 80% to meet the outlet NO$_x$ permit levels.

Another application is to reduce the NO$_x$ emissions from fired heaters that are used for the vaporization of liquefied natural gas (LNG) at a receiving terminal (Focus on Catalysts May 2007).

12.4.4   Combined DeNO$_x$ and DeSO$_x$

A two-stage catalytic process that combines NO$_x$ reduction and SO$_x$ removal by conversion of SO$_2$ to H$_2$SO$_4$ has been developed for high sulfur-containing process effluents. Ammonia is injected into a NO$_x$ particulate free gas stream. The process gas contains large and recoverable amounts of sulfur. The stream is passed through a honeycomb where the SCR catalyst reduces the NO$_x$ to N$_2$ at 450 °C. The exit gas then passes into a second bed comprising an extruded V$_2$O$_5$ honeycomb catalyst known to be active for converting SO$_2$ to SO$_3$. The latter is then scrubbed producing a salable sulfuric acid solution (Ohlms 1993).

A similar process has been described (Blumrich and Engler 1993) in which CO and HC are also removed in addition to NO_x and SO_x.

12.4.5   Fluidized Bed SCR

A novel process was commercialized in mid-1992 for the combined reduction of NO_x, HCs, and CO from a chemical manufacturing site (Addison 1992; Hardison et al. 2006). The process uses a fluid bed catalyst composed of a proprietary metal oxide on an abrasion-resistant ceramic sphere about 1/8 in (0.32 cm) in diameter. The fluidization process slowly abrades feed gas salt deposits from the surface of the sphere providing a renewed surface for catalysis. The feed gas contained about 4,000 ppm of NO_x, and reductions of >90% with an NH_3 slip of less than 10 ppm was achieved at an NH_3-to-NO_x ratio of close to 1. The system operates with an effective space velocity of 4,000–16,000 1/hr at a temperature of 275–350 °C. It is reported that 99% of VOCs are reduced and almost 90% of the CO. These SCR systems are used extensively in Japan and western Germany to eliminate 80–90% of NO_x emissions from utility boilers and industrial furnaces. Costs have been lowered considerably over the past 10 years, and this approach is considered to be a simple system attractive for refinery and industrial process heaters, boilers, and gas turbines.

12.5   NITROUS OXIDE (N₂O)

Because of its contribution to global warming or "the greenhouse effect," N_2O emissions are now being abated from nylon manufacturing plants (Thiemens and Trogler 1990; Reimer et al. 1993) and nitric acid plants (Focus on Catalysts Jan. 2007; Focus on Catalysts Sept. 2007). N_2O is formed and emitted into the atmosphere in significant concentrations from several industrial chemical processes, in particular the production of nitric acid, caprolactam, and adipic acid. These emissions have been growing at about 0.3% per year and increase the O_3 depletion rate by 6%, thus contributing 4% to the greenhouse effect (Byrne 1995). Recent data show this increase continues but may have slowed some with about a 2% increase from the 1990 levels as measured in 2004 (EPA 2007). Technologies using either precious metals or Co or Cu supported on a variety of zeolites, including ZSM-5, have been found to promote the following decomposition reaction (Li and Armor 1992; Riley and Richmond 1993):

$$N_2O \rightarrow N_2 + 1/2 O_2 \tag{12.24}$$

Figure 12.20 shows the response for this decomposition reaction using various catalysts on beta zeolite. Note that this reaction does not require a chemical reductant to be added to the effluent stream since it is purely a decomposition

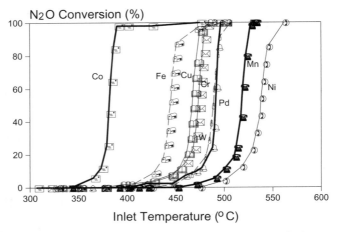

FIGURE 12.20. Co seems to be the most active catalyst for N$_2$O decomposition on beta zeolite (Byrne 1995).

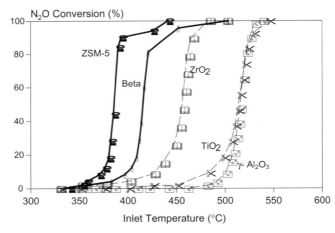

FIGURE 12.21. The catalyst support has a key function in N$_2$O decomposition (Byrne 1995).

reaction. Co seems to be the most reactive catalyst, and its performance on various washcoats is shown in Figure 12.21. Both ZSM-5 and beta zeolite supports can be used for this application.

Technologies and the need for N$_2$O abatement continue with many applications throughout the world (Schimizu et al. 2000).

12.6 CATALYTICALLY SUPPORTED THERMAL COMBUSTION

12.6.1 Introduction

This breakthrough technology would be directly applicable to combustion processes for heat generation. It differs from all other commercial processes currently being practiced in that it prevents the formation of pollutants rather than cleaning up that which has been produced. In its simplest form, it uses a catalyst to replace the burner used in fuel combustion.

The operational theory points directly to a major advantage of a catalytic system: The fuel-air composition can be below that of a flammable mixture and still initiate and sustain an oxidation process. In operation, the lean fuel–air mixture is passed through a monolithic catalyst bed and (provided the inlet temperature exceeds lightoff) the oxidation reaction commences. The conversion versus monolith length profile is depicted in Figure 12.22.

The heat generated at the catalytic surface is continuously transferred to the bulk gas as the oxidation reaction proceeds through the bed, continuously raising the bulk gas temperature. Shortly after the catalytic conversion reaches mass transfer, the temperature and composition of the bulk gas are sufficient to initiate homogeneous or thermal reaction, which completes the combustion of the fuel. It is the initiation of the homogeneous reaction that differentiates this technology from traditional catalysis for which the maximum conversion is limited to mass transfer. The maximum temperature obtained is controlled by the initial fuel–air composition. Since this mixture is lean relative to that of a flammable mixture, the maximum temperature is less than $1,500\,°C$, which minimizes formation of thermal NO_x and still completes the oxidation of the

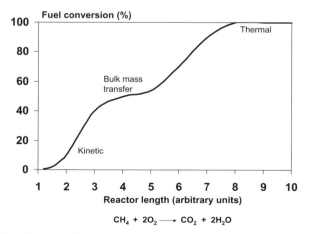

$$CH_4 + 2O_2 \longrightarrow CO_2 + 2H_2O$$

FIGURE 12.22. Catalytically supported thermal combustion. A thermal reaction is initiated by a monolithic catalyst, and combustion is completed with ultra-low emission levels.

fuel to CO$_2$ and H$_2$O with no other pollutants generated. This gas temperature is more than sufficient to operate modern gas turbines.

Conventional gas burners require a fuel–air composition close to the flammable range in order to support a flame. The resultant adiabatic flame temperature is far above the 1,500 °C where nitrogen fixation producing NO$_x$ becomes appreciable. Furthermore, because of the inhomogeneity of the flame, significant amounts of carbon monoxide and difficult-to-oxidize hydrocarbons are emitted. All three of these pollutants then must be reduced with aftertreatment technology. The inlet temperature to modern gas-fired turbines is between 1,300 °C and 1,400 °C, so the combusted gas must be cooled prior to impinging the blades of the turbine. Thus, a high temperature is first generated, creating high levels of NO$_x$, and then it is cooled by the addition of secondary air for turbine operation.

If catalytically supported thermal combustion could be used in gas-fired combustion processes, it would create virtually no pollutants eliminating the need for downstream aftertreatment, such as SCR of nitric oxides, and carbon monoxide and hydrocarbon abatement technology for gas-fired burners.

The kinetics for methane combustion in excess air over Pd catalysts are generally understood (McCarty 1995):

$$CH_4 + O_2 \rightarrow CO_2 + 2H_2O \qquad (12.25)$$

The reaction is first order in methane, essentially zero order in O$_2$ and CO$_2$, and −0.8 negative order in H$_2$O indicating strong inhibition. The activation energy is about 37 kcal/mole with water present and about 21 kcal/mole dry (Van Giezin 1997).

Catalytic combustion models in monoliths are also well established (Hayes and Kolaczkowsky 1997).

12.6.2   The Technical Problems

Since catalytic combustion was discovered and patented in the mid-1970s (Pfefferle 1976), companies have recognized the commercial and environmental significance and have investigated many aspects of the process and material requirements. The catalytic problems that had to be solved were formidable. Monolithic materials are necessary to accommodate the flow rates needed (Pfefferle and Pfefferle 1987) with minimum pressure drop. The catalyzed washcoat must continuously initiate oxidation or lightoff of the fuel–air mixture at compressor discharge temperatures less than 350 °C at 13 atmospheres. Catalyst compositions are available to lightoff naphtha, diesel, LPG, low BTU gases, and so on, but not for the preferred fuel, natural gas. The latter is composed primarily of methane, which is extremely difficult to oxidize with or without a catalyst. Under transient conditions of startup, shutdown, and/or load changes, the catalyst would experience temperatures up to 1,300–1,400 °C, but it must do so without losing its high initial activity for lightoff. These conditions go far beyond what is required of a catalyst for any other application, and thus, new technology for both the catalyst and the washcoat had to be developed. (Pfefferle and Pfefferle 1987; Farrauto and Heck 1992).

The monolithic support materials would have to resist deformation and/or melting after experiencing 1,300 °C to 1,400 °C. Cordierite, used as a monolithic substrate for automobile catalytic converters, melts at about 1,300 °C, so it would not be an acceptable material. Furthermore, when an emergency shutdown in a power plant occurs, the fuel is immediately shut off rapidly cooling the catalyst surface from 1,300 °C to about 350 °C (compressor discharge temperature for 13 atmospheres). The substrate surface cools more rapidly than its bulk, establishing expansion gradients, which leads to cracking or thermal shock.

Since its discovery almost 30 years ago, research has focused on finding new materials and process operating conditions to bring this technology to commercial reality. It seems there is now reason to be confident that this has finally been accomplished, albeit in a limited way.

12.6.3 Technical Background

The search for new catalysts, monolithic substrate materials, and process conditions needed to operate under the harsh conditions of the gas turbine has met with some success. A family of new proprietary catalyst formulations has been developed. Some formulations provide excellent low-temperature light-off activity for natural gas, whereas other formulations are capable of withstanding temperature surges during transient operation up to 1,300–1,400 °C. All of these materials are based on PdO-containing catalysts and will be discussed later.

The key finding is that Pd, in its oxidized form, is the most active catalytic species for the high-temperature oxidation of methane (Farrauto et al. 1990, 1992, 1993). During atmospheric combustion, the catalyst surface temperature rises above 800 °C, where PdO on γ-Al$_2$O$_3$ decomposes, as shown in Eq. (12.22), to less active Pd metal, resulting in a loss of activity and a cooling of the surface:

$$\underset{\text{methane active}}{PdO} \rightarrow \underset{\text{methane inactive}}{Pd + 1/2\,O_2} \tag{12.22}$$

When the catalyst surface temperature decreases below about 600 °C, the Pd surface reoxidizes and redisperses, and the activity is regenerated as indicated in Eq. (12.23):

$$1/2\,O_2 + Pd \rightarrow PdO \tag{12.23}$$

The decomposition of the PdO and its reformation upon cooling in air is also dependent on the support (Farrauto et al. 1995). The effect of operating pressure on the kinetics of the CH$_4$ oxidation reaction has also been studied (Kuper et al. 1999).

Expanding on the relationships between PdO and its activity for natural gas conversion, a family of new materials was developed that resist PdO decomposition up to 1,300 °C and retain high activity for methane oxidation. These materials are produced by reacting various oxides, such as La$_2$O$_3$, CeO$_2$ (Kennelly and Farrauto 1993), and/or Pr$_2$O$_3$ (Chou et al. 1992) with Pd salts during the preparation process. The La$_2$O$_3$–PdO compounds raise the stability of PdO decomposition to over 900 °C, whereas Pr$_2$O$_3$–PdO increases the temperature for decomposition up to 1,300 °C. As the stability of the PdO compound is increased, the activity is decreased. Therefore, these materials represent suitable catalysts for a segmented bed in which the most active is used as an ignition section, whereas the least active, but more stable, is used for the final segment that is expected to see temperatures close to 1,300 °C.

Laboratory aging studies using methane as a fuel have been performed on these catalyst formulations supported on proprietary monolithic materials and using novel reactor designs (Farrauto et al. 1994). In this study, they used the most highly active PdO/γ–Al$_2$O$_3$ catalyst for the inlet section and the most stable but least active catalyst for the exit section. The highest temperatures were experienced downstream from the exit of the catalyst in the homogeneous zone, but during startup and shutdown, transients did raise the temperature within the bed to 1,300 °C. Thus, it is necessary to have materials capable of withstanding these excursions even for a short time. The system behaved with stable performance for hundreds of hours with emissions of NO$_x$, CO, and HC below 2 ppm.

Thermally stable metal substituted hexaluminate materials have been suggested as alternative catalysts to Pd compounds for the high-temperature sections of the combustor (Machida et al. 1987). A catalyst composed of BaMn$_2$Al$_{10}$O$_{19}$ was evaluated (Groppii et al. 1996) for the exit section of the combustor; however, it was concluded that it was neither sufficiently active nor stable for combusting natural gas.

An alternative approach uses a process designed to protect the catalyst and monolith against high-temperature exposure. The fuel addition is staged to keep the catalyst and monolith temperature below 1,000 °C. The process is referred to as a hybrid because a portion of the fuel is catalytically oxidized, after which additional fuel is added at the catalyst exit (Furuya et al. 1987). The authors claim it is necessary to use an igniter to initiate the homogeneous reaction at the exit of the catalyst bed where the secondary fuel is added. By restricting the catalyst bed temperature below 1,000 °C, monolithic materials such as cordierite and combinations of more traditional palladium and platinum catalysts could be used (Kawakami et al. 1989). The process is reported to be feasible for turbine inlet temperatures up to 1,300 °C, based on laboratory scale tests, with ultra-low emissions.

The most successful design to date is the combustor design of Catalytica Energy Systems of Palo Alto, California (Anson 2000). This technology was sold to Kawasaki in 2006 (Eco Stock Edge 2006). It uses a metal monolithic substrate that acts as a heat exchanger. The metal is an alloy composed of iron, chrome, aluminum, and yttrium with a use temperature of about 1,100 °C. The catalyzed washcoat is coated onto only alternating channels of the metal wall (Dalla Betta et al. 1993). The PdO/ZrO$_2$ containing washcoat is applied before the metal is wound into the honeycomb shape. An example from one catalyst manufacturer shows that one side of each foil leaf is coated with catalyst and that the other side is left uncoated. The layers are arranged so that coated sides face each other to form catalyzed channels, and uncoated sides face each other to form uncoated channels. Combustion goes to completion in the coated channels. However, no combustion takes place in the adjacent uncoated channels (Whittenberger and Retallick 2003). The net effect is that only half of the fuel is burned (assuming the channels are of equal size), and the bulk temperature rise is limited to half of the adiabatic rise. Catalytic reactions occur in the catalyzed channel, and a portion of the reaction heat generated is transferred through the metal wall to the uncoated channel. By maintaining the catalyst surface temperature below the temperature for PdO decomposition, the activity is maintained. If, however, transients excursions occur and the PdO decomposes, the gas temperature decreases protecting the catalyst system against permanent degradation. The activity is regained when the Pd metal is reoxidized at about 600 °C as predicted from previous work (Farrauto et al. 1990). This temperature control mechanism is not fool proof since start-up transients have been observed to occur faster than the PdO/Pd transition (Beratta et al. 1999; Forzatti 2000).

The catalyst washcoat is deposited in gradients with decreasing amounts from inlet to outlet (Dalla Betta et al. 1993). This provides sufficient catalyst to initiate the oxidation at the inlet but limits the amount of reaction downstream. The design controls the maximum temperature in the monolith. The outlet temperature is sufficiently high to initiate homogeneous combustion downstream of the catalyst. Pilot plant studies were performed at 11 atmospheres with combustor outlet temperatures of 1,300 °C with NO$_x$, CO, and HC all below 2 ppm.

Another technology that is being tested under government contracts uses a concept originally proposed in the 1970s whereby a catalyst is operated fuel rich and air is added to burn the effluent from the catalytic stage (Etemad et al. 2004; Smith et al. 2005). The two-stage combustion process consists of a first-stage rich-catalytic lean-burn (RCL^{TM}) catalytic reactor, wherein a fuel-rich mixture contacts the catalyst and reacts while final and excess combustion air cools the catalyst. The second stage is a gas-phase combustor, wherein the catalyst cooling air mixes with the catalytic reactor effluent to provide for final gas-phase burnout and dilution to fuel-lean combustion products. The first-stage reactor is an air-cooled tubular heat exchanger with the catalyst coating on the outside of the tubes and cooling air flowing through the inside of the tubes. The tubular catalytic reactor operates at a rich air-to-fuel ratio. Full-scale tests have been conducted with coal-derived syngas (Etemad et al. 2004) and are now being conducted under government contract using a hydrogen/nitrogen blend on a Solar gas turbine (NETL 2006).

Research into lean-burn catalytic combustion has continued since the mid-1970s, but to date, no significant commercial installations exist. One main issue is the absence of a sufficiently active methane oxidation catalyst that will lightoff natural gas at compressor discharge temperature in a gas turbine of about 350 °C at 13 atmospheres. The technology has now shifted toward a fuel-rich lightoff followed by injection of additional air downstream, but to date, no significant progress has been made to move this intriguing technology closer to commercialization.

REFERENCES

Addison, G. "First commercial operation of an innovative SCR process," ARI Technologies Inc., 600 N. First Bank Dr. Palantine, IL 60067 (1992).

Adlhart, O., Hindin, S., and Kenson, R. "Processing nitric acid tail gas," *Chemical Engineering Progress* 67(2): 73–78 (1971).

Anson, O. "Low emission gas turbine combustor field demonstration," Final report for California Air Resources Board, Contract No. 96-337 (2000).

Beratta, A., Baiardi, D., Prina, D., and Forzatti, P. "Analysis of a catalytic annular reactor for short contact times," *Chemical Enginering Science* 54: 765 (1999).

Blanco, J., Avila, P., and Marzo L. "Low temperature multibed SCR process for tail gas treatment in a nitric acid plant," *Catalysis Today* 17: 325–332 (1993).

Blumrich, S., and Engler, B. "The DESO$_x$NO$_x$-process for flue gas cleaning," *Catalysis Today* 17: 301–310 (1993).

Bosch, H., and Janssen, F. "Catalytic reduction of nitric oxides—a review of the fundamentals and technology," *Catalysis Today* 2: 369–521 (1988).

Brundrett, C., Maaskant, O., and Genty, N. "Application and operation of the shell low temperature SCR technology on ethylene cracker furnaces," 13th Annual Ethylene Producers Conference (2001).

Bullock, D., and Hartenstein, H. "Full-scale catalyst regeneration experience of a coal-fired U.S. merchant plant," 2003 Conference on Selective Catalytic Reduction (SCR) and Selective Non-Catalytic Reduction (SNCR) for NO$_x$ Control (2003).

Burns, K., Collins, M., and Heck, R. "Catalytic control of NO$_x$ emissions from stationary rich—burning natural gas engines," ASME 83 D6P-12 (1983).

Byrne, J., Chen, J., and Speronello, B. "Selective catalytic reduction of NO$_x$ using zeolites for high temperature applications," *Catalysis Today* 13(1): 33–42 (1992).

Byrne, J. "Zeolites as catalysts for nitrous oxide emissions reduction," Materials Research Society meeting, Canjun, Mexico (1995).

Carpenter, K. "NOx emission solutions for gas turbines," NETL 2002 Conference on Selective Catalytic Reduction (SCR) and Selective Non-Catalytic Reduction (SNCR) for NOx Control (2002).

Casgrande, L., Liette, I., Nora, P., Forzatti, P., and Baiber, A. "SCR of NO$_x$ by NH$_3$ over TiO$_2$-supported V$_2$O$_5$-MoO$_3$ catalysts: reactivity and activity behavior," *Applied Catalysis B: Environmental* 22: 63–77 (1999).

Chatterjee, S., Lee, H., and Rolando, S. "An analysis of the parameters effecting the application of SCR catalysts to high sulphur coal-fired boilers," Norton Chemical Process Products, Akron, OH (1993).

Chen, J., Speronello, B., Byrne, J., and Heck, R. "Kinetics of NO_x reduction and selective NH_3 oxidation over a honeycomb catalyst," AIChE Summer National Meeting, San Diego, CA (1990).

Chou, T., Kennelly, T., and Farrauto, R. "Praseodymium-palladium binary oxide catalyst compositions containing the same and methods of use," U.S. Patent 5,102,639 (1992).

Cobb, D., Glatch, L., Ruud, J., and Snyder, S. "Application of selective catalytic reduction (SCR) technology for NOx reduction from refinery combustion sources," *Environmental Progress* 10(1): 49–59 (1991).

Code of Federal Regulations, office of the Federal Register, U.S. Government Printing Office (2007).

Cohn, G., Steele, D., and andersen, H. "Nitric acid tail gas abatement," U.S. Patent 2,975,025 (1961).

Cormetech. "Catalyst Description Geometry," http://www.cormetech.com/catalystoverview.htm (2007).

Dalla Betta, R., Tsurumi, K., and Shoji, T. "Graded palladium-containing partial combustion catalyst and a process for using it," U.S. Patent 5,248,251 (1993).

Dalla Betta, R., Ribeiro, F., Shoji, T., Tsurumi, K., Ezawa N., and Nickolas, S. "Catalyst structure having integral heat exchange," U.S. Patent 5,250,489 (1993).

DOE. "Cost analysis of NOx control alternatives for stationary gas turbines," Contract No. DE-FC02-97CHIO877, U.S. Department of Energy Environmental Programs, Chicago Operations Office, Chicago, IL ONSITE SYCOM Energy Corporation, Carlsbad, CA (1999).

Eco Stock Edge. "Kawasaki to acquire xonon cool combustion gas turbine technology from catalytica energy systems," (2006).

EPA. "Technical bulletin: Nitrogen oxides (NOx)—why and how they are controlled," Clean Air Technology Center, EPA/F-99-006R (1999).

EPA. "Inventory of U.S. greenhouse gas emissions and sinks: 1990–2005," USEPA #430-R-07–002, http://www.epa.gov/globalwarming/publications/emissions/, (2007).

Etemad, S., Smith, L., and Burns, K. "System study of rich catalytic/lean burn (RCL®) catalytic combustion for natural gas and coal-derived syngas combustion Turbines: Final report," DOE Contract No. DE-FG26-02NT41521 (2004).

Farrauto, R., and Heck, R. "Precious metal catalysis," pp 218–274. *Encyclopedia for Chemical Processing and Design*. Editor. McKetta. J. Marcel Dekker, New York (1992).

Farrauto, R., Hobson, M., Kennelly, T., and Waterman, E. "Catalytic chemistry of supported palladium for combustion of methane," *Applied Catalysis A: General* 81: 227–234 (1992).

Farrauto, R., Kennelly, T., and Waterman, E. "Process conditions for operation of ignition catalyst for natural gas combustion," U.S. Patent 4,893,465 (1990).

Farrauto, R., Hobson, M., Kennelly, T., and Waterman, E. "Catalytic chemistry of supported palladium for combustion of methane," *Applied Catalysis A: General*, 81: 227–237 (1992).

Farrauto, R., Kennelly, T., Waterman, E., and Hobson, M. "Process conditions for operation of ignition catalyst for natural gas combustion," U.S. Patent 5,214,912 (1993).

Farrauto, R., Larkin, M., Fu, J., and Feeley, J. "Catalytic combustion for ultra-low emissions," Spring Meeting of the Materials Research Society, San Francisco, CA (1994).

Farrauto, R., Lampert, J., Hobson, M., and Waterman, E. Thermal decomposition and reformation of PdO catalysts: Support effects," *Applied Catalysis B: Environmental* 6: 263 (1995).

Fenimore, C. "Formation of nitric oxide from fuel nitrogen in ethylene flames," *Combustion and Flame* 19: 289–296 (1972).

Focus on Catalysts. "Heehaw and Mitsui jointly execute N$_2$O reduction project," Focus on Catalyst, p 7, January (2007).

Focus on Catalysts. "Peerless Mfg awarded \$5 M contract for SCR equipment for LNG facility," p 7, May (2007).

Focus on Catalysts. "Sasol to receive carbon credits," Focus on Catalysts, p 7, September (2007).

Focus on Catalysts. "SCR-Tech announces SCR catalyst regeneration orders totaling \$1.0 M," Focus on Catalysts, p 4, July (2007).

Forzatti, P. "Environmental catalysis for stationary applications," *Catalysis Today* 62: 51 (2000).

Freidel, I., Frost, A., Herbert, K., Meyer, F., and Summers, J. "New catalyst technologies for the destruction of halogenated hydrocarbons and volatile organics," *Catalysis Today* 17: 367–382 (1993).

Furuya, T., Yamanaka, S., Hayata, T., Koezuka, J., Yoshine, T., and Ohkoshi, A., "Hybrid catalytic combustion for stationary gas turbine-concept and small scale test results," Gas Turbine Conference and Exhibition, Anaheim, CA (1987).

Gillespie, G., Boyum, A., and Collins, M. "Nitric acid: Catalytic purification of tail gas," *Chemical Engineering Progress* 68(4): 72–74 (1972).

Gutberlet, H., and Schallert, B. "Selective catalytic reduction of NO$_x$ from coal fired plants," *Catalysis Today* 16: 207–236 (1993).

Groppi, G., Tranconi, E., and Forzatti, P. "Investigation of catalytic combustion for gas turbine applications through mathematical model analysis," *Applied Catalysis A* 138: 177 (1996).

Hardison, L., Nagl, G., and Addison, G. "NOx reduction by the Econ-NOx™ SCR process," *Environmental Progress* 10(4): 28 (2006).

Hartenstein, H., and Gutberlet, H. "Catalyst regeneration—an integral part of proper catalyst management," 2001 EPRI Workshop on Selective Catalytic Reduction, Baltimore MD (2001).

Hayes, R., and Kolaczkowsky, S. "Introduction to catalytic combustion," Gordon and Breach Science Publishers, London, England (1997).

Heck, R. "Catalytic abatement of nitrogen oxides-stationary applications," *Catalysis Today* 53: 519–523 (1999).

Heck, R., Bonacci, J., and Chen, J. "Catalytic air pollution controls—commercial development of selective catalytic reduction of NO$_x$," Paper 87-523, 80th Annual Meeting of APCA, New York (1987).

Heck, R., Chen, J., Speronello, B., and Morris, L. "Family of versatile catalyst technologies for NO$_x$ removal in power plant applications," ACS Symposium Series 552, Washington D.C. (1994).

Heck, R., and Womelsdorf, R. "Environmental catalysis in the chemical industry," *Chemical Engineering* (2002).

Hums, E. "Is advanced SCR technology at a standstill? A provocation for the academic community and catalyst manufacturers," *Catalysis Today* 42: 25–35 (1998).

Hums, E. "Understanding of deactivation behavior of DeNOx catalysts: A key to advanced catalyst applications," *Kinetics and Catalysis* 39(5): 603–606 (1998).

Hums, E., Sigling, R., and Spielman, H. "Tailored selective reduction application, and operating experience in coal- and gas-fired power plants," Chemistry *for Sustainable Development* 5: 303–313 (1997).

Hums, E. "Is advanced SCR technology at a standstill? A provocation for the academic community and catalyst manufacturers," *Catalysis Today* 42: 25–35 (1998).

ICAC, "White paper: Selective catalytic reduction (SCR) control of NOx emissions," SCR Committee of Institute of Clean Air Companies (1997).

Janssen, F., and Meijer, R. "Quality control of DeNO$_x$ catalysts," *Catalysis Today* 16: 157–185 (1993).

Kawakami, T., Furuya, T., Sasaki, Y., Yoshine, T., Furuse, Y., and Hoshino, M. "Feasibility study on honeycomb ceramics for catalytic combustor," Gas Turbine and Aeroengine Congress and Exposition, Toronto, Canada (1989).

Kennelly, T., and Farrauto, R. "Catalytic combustion process using supported palladium oxide catalysts," U.S. Patent 5,216,875 (1993).

Kobayashi, N., Sayoma, K., and Miyazowa, M. "Dry catalytic process gives promising results," *Modern Power Systems* 37–39 (1988).

Kobayashi, M., Kuma, R., Masaki, S., and Sugishima, N. "TiO$_2$-SiO$_2$ and V$_2$O$_5$/TiO2-SiO$_2$ catalyst: Physico-chemical characteristics and catalytic behavior in selective catalytic reduction of NO by NH$_3$," *Applied Catalysis B: Environmental* 60: 173–179 (2005).

Kobayashi. M., and Hagi, M. "V$_2$O$_5$-WO$_3$/TiO$_2$-SiO$_2$-SO$_4^{2-}$ catalysts: Influence of active components," *Applied Catalysis B: Environmental* 63: 104–113 (2006).

Kunz, R., Hefele, D., Jordan, R., and Lash, F. "Use of SCR in a hydrogen plant integrated with a stationary gas turbine—case study: the port arthur steam-methane reformer," AWMA 96 Annual Meeting and Exhibition, Paper # 70093 (2003).

Kuper, W., Blaauw, M., Van der Berg, F., and Graaf, G. "Catalytic combustion concept for gas turbines," *Catalysis Today* 47: 377–389 (1999).

Lee, H. Recent developments in SCR NO$_x$ control technology, Norton Chemical Process Products, Akron, OH (1993).

Li, Y., and Armor, J. "Catalytic decomposition of N$_2$O," U.S. Patent 5,171,553 (1992).

Li, L., and Armor, J. "Catalytic decomposition of nitrous oxide on metal exchanged zeolites," *Applied Catalysis B: Environmental* 1: L21–L29 (1992).

Machida, M. Eguchi, K., and Arai, H. "Effect of additions on the surface area of oxide supports for catalytic combustion," *Journal of Catalysis* 103: 385 (1987).

Marzo, L., and Fernandez, L. "Destroy NO$_x$ catalytically," *Hydrocarbon Processing* 87–89 (1980).

Matsuda, S. "Selective catalytic reduction of nitrogen oxides with ammonia in presence of sulfur trioxide-deposition of ammonium bisulfate," 80-6.1. 73rd Annual Meeting of AIChE, Montreal, Canada (1980).

McCarty, J. "Kinetics of PdO combustion catalysts," *Catalysis Today* 26: 283 (1995).

NETL. "Catalytic combustion for ultra-low NOx hydrogen turbines," DE-FC26-05NT42647 (2006).

Ohlms, N. "DeSO$_x$NO$_x$ process for flue gas cleaning," *Catalysis Today* 16: 247–261 (1993).

O'Leary, J., Kunz, R., and von Alten, T. "Selective catalytic reduction (SCR) performance in steam-methane reformer service: The chromium problem," Paper ENV-02–178 2002 NPRA Environmental Conference, New Orleans, LA (2002).

Pence, D., and Thomas, T. "Reduction of nitrogen oxides with catalytic resistant aluminosilicate molecular sieves and ammonia," U.S. Patent 4,220,632 (1980).

Pereira, C., Plumlee, K., and Evans, M. "Camet metal monolith catalyst system for cogen applications," *2nd International Symposium on Turbomachinery, Combined Cycle Technologies and Cogeneration-IGTI*. Editors, Serovy, G., and Fransson, T. Book No. 100270, American Society of Mechanical Engineers, New York (1988).

Pfefferle, W. "Catalytically-supported thermal combustion," U.S. Patent 3,928,961 (1976).

Pfefferle, L., and Pfefferle, W. "Catalytic combustion," *Catalysis Review* 29: 219 (1987).

Prins, W., and Nuninga, Z. "Design and experience with catalytic reactors for SCR-DeNO$_x$," *Catalysis Today* 16: 187–205 (1993).

Pritchard, S. "Long term catalyst health care," *Power Magazine* 150: 1–5 (2006).

Reimer, R., Slaten, C., Seapan, M., Lower, M., and Tomlinson, P. "Abatement of N$_2$O emissions in the adipic acid industry," Paper No. 73A, Summer National Meeting of AIChE, Seattle, WA (1993).

Riley, B., and Richmond, J. "A catalytic process for the decomposition of nitrous oxide," *Catalysis Today* 17: 277–284 (1993).

Schmizu, A., Tanaka, K., and Fujimori. "Abatement technologies for N$_2$O emissions in the adipic acid industry," *Chemosphere—Global Change Science*, 2(3–4): 425–434 (2000).

Smith, L., Karim, H., Castaldi, M., Etemad, S., Pfefferle, W., Khanna, V., and Smith, K. "Rich-catalytic lean-burn combustion for low-single-digit NOx gas turbines," *Journal of Engineering for Gas Turbines and Power* 127: 27–35 (2005).

Smith, T. "Industrial and mobile NOx control practices and options," Geographic Strategies Group, U.S. EPA's Office of Air Quality Planning and Standards, Research Triangle Park, NC (2005).

Smrcka, N. "Development of a new-generation low-ask oil additive package for natural gas engines," ASME 91-ICE-A, Energy-sources Technology Conference and Exhibition, Houston, TX (1991).

Spitznagel, G., Huttenhofer, K., and Beer, J. *Environmental Catalysis*. Editor Armor, J. ACS, Washington D.C. p. 172 (1994).

Srivastava, R., Hall, R., Khan, S., Lani, B., and Culligan, K. "Nitrogen oxides emission control options for coal-fired electric utility boilers," *Journal of Air and Waste management Association* 55: 1367–1388 (2005).

Thiemens, M., and Trogler, W. "Nylon production: An unknown source of atmospheric nitrous oxide," *Science* 251: 932–934 (1990).

TNRCC. Texas Natural Resource Conversation Commission, http://www.tnrcc.state. tx.us/permitting/airperm/opd/60/60hmpg.htm, (2007).

Tran, P., Liu, X., Chen, J., Lapadula, G., and Furbeck, H. "Hydrothermally stable metal promoted zeolite beta for NOx reduction," U.S. Patent 7,118,722 (2006).

Van Giezin. "Catalytic combustion of methane," PhD thesis, University of Utrecht, the Netherlands (1997).

Whittenberger, W., and Retallick, W. "Controlled catalytic combustion module (CCM)," http://catacel.com/Papers/Whittenberger-CCMJuly2003.pdf, (2003).

Zeldovich, J. "The oxidation of nitrogen in combustion and explosions," *Acta Physiochimica, USSR* 21: 577 (1946).

Zwinkels, M., Jaras, S., Menon, P., and Griffin, T. "Catalytic materials for high-temperature combustion," *Catalysis Reviews, Science and Engineering* 35(3): 319–358 (1991).

CHAPTER 12 QUESTIONS

1. Describe how SCR NO_x is a classic example of selectivity. Include Figure 12.8 in your discussion.

2. List the many possible contaminants that may affect a NO_x catalyst performance, and describe the source and the possible mechanism for deactivation.

3. Why is W and Mo added to the SCR NO_x catalyst? Review and comment on references Casgrande et al. 1999 and Spitznagel et al. 1995 for additional insight.

4. Why is SNCR used in combination with SCR for NO_x reduction greater than 90%?

5. Consider the selective reduction of NO_x (SCR) with NH_3 using a V_2O_5/TiO_2//monolith catalyst.

   a. The rising portion (T < 250 °C) of the conversion vs. temperature profile

   b. The maximum portion of the profile (at about 350 °C)

   c. The descending portion of the profile (T > 400 °C)

   Describe the chemical reactions occurring and the relative activation energies and selectivities for the rising and declining conversions for V_2O_5.

6. If it is determined that the rate of the undesirable reaction is limited by chemical kinetics while the desired selective catalytic reduction reaction is limited by pore diffusion, suggest how you would design the monolith-supported catalyst and the process conditions?

13 Carbon Monoxide and Hydrocarbon Abatement from Gas Turbines

13.1 INTRODUCTION

Combustion turbines that burn fuel to generate heat are widely used for the cogeneration of electricity and steam. Since the 1980s, a common practice in operation of these gas turbines has been to use water or steam injection to lower the combustion flame temperature to reduce the NO_x emissions formed from N_2 and O_2 at elevated temperatures. The quenching of the flame results in an increase in CO and hydrocarbon (HC) emissions. Using water or steam injection can typically reduce the NO_x emissions from 150 vppm to about 40 vppm, with a concurrent increase in CO emissions from 10 vppm to as high as 400 vppm (and an increase in hydrocarbon emissions). Water injection is not without problems (e.g., corrosion, water purification, and so forth), and alternative burner technologies such as staged combustion are being developed. However, water injection is still viable technology, and in some cases, it is used in combination with the new burner developments. Several gas turbine classifications exist, including simple cycle, regenerative cycle, cogeneration cycle, and combined cycle (State of New Jersey 2004).

The first "cogen" systems were fueled by natural gas or a mixture of natural gas and refining gas. More recently, abatement systems have been required to operate with dirtier liquid fuel feeds, and some operating permits have included requirements for destruction of HC and limitations on the maximum allowable SO_3 (sulfuric acid emissions). To meet these requirements, a family of catalysts has been developed for clean and dirty applications in CO and HC removal (Speronello et al. 1992; Chen et al. 1993; Pereira et al. 1990). The catalysts requirements are as follows:

- Oxidize CO and HC.
- Minimize the oxidation SO_2 to SO_3.
- Resist poisons in exhaust gas.

Catalytic Air Pollution Control: Commercial Technology, by Ronald M. Heck and Robert J. Farrauto, with Suresh T. Gulati.

13.2 CATALYST FOR CO ABATEMENT

The basic performance of a CO removal catalyst (Speronello et al. 1992; Chen et al. 1993) in the exhaust of a natural gas-fueled turbine is shown in Figure 13.1. The catalyst used can range from 0.05% to 0.5% Pt catalyst dispersed on $\gamma\text{-Al}_2\text{O}_3$ on a ceramic or metallic straight flow-through honeycomb with 100 to 400 cpsi or metallic honeycombs with herringbone or skew channel shapes (BASF 2007). Since this reaction is basically bulk mass transfer controlled above 150 °C, the catalyst efficiency can be increased by increasing the linear velocity, the reactor volume, and the number of cells (cpsi) of the straight flow-through ceramic or metallic honeycomb, or as in the case of the metallic with herringbone and skew channels increasing the mass transfer rate through increased turbulence (Berndt and Landri 2002). Of course what always has to be balanced is the increased pressure drop of the reactor. Increasing the space velocity diminishes the maximum conversion obtainable, as shown in Figure 13.1.

A schematic of the installation of a catalyst system for CO abatement in a gas turbine is shown in Figure 13.2. Note that the oxidation catalyst can be located anywhere downstream in the gas turbine exhaust as long as the temperature is above 100 °C (preferably above 250 °C) to minimize catalyst requirements (e.g., volume). These catalytic systems are assembled together in modules (e.g., 2 ft × 2 ft) and fabricated into walls of catalyst as shown in Figure 13.3. Typical catalytic depths range from 3 to 6 in (7.6 to 15.2 cm). These catalyst walls have been fabricated up to 30 ft (9.4 meters) wide. Depending on the installation, space velocities range from 100,000 to 300,000 hr^{-1} and cell geometries from 100 cpsi to 400 cpsi. Pressure drops of 1 (2.54) to 10 in (25.4 cm) of water are typical. Either ceramic or metallic substrates can be used; however,

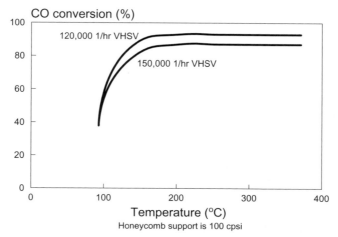

FIGURE 13.1. Space velocity as a key design parameter for CO abatement in exhaust from a gas turbine using a 100-cpsi catalyzed monolith.

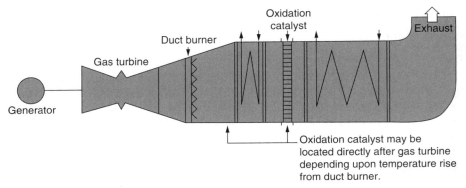

FIGURE 13.2. Schematic of the CO abatement catalyst in a heat recovery unit from a gas turbine.

(a)   (b)

FIGURE 13.3. Installation of CO catalyst abatement ceramic modules in the exhaust from a gas turbine engine. a) Installation of a ceramic catalyst module. b) Wall of catalyst modules.

most modern commercial installations now use metal substrates because of the lower pressure drop and, hence, a smaller cross section of the catalytic reactor that is placed in the heat recovery unit (Pereira et al. 1988; Gulian et al. 1991) (Figure 13.4).

CO abatement systems for fuels containing sulfur require a different catalyst formulation and process conditions that minimize the $SO_2 \rightarrow SO_3$ reaction and reduce the poisoning effect of both SO_2 and SO_3 on the catalyst. SO_2 chemisorbs onto the Pt sites at temperatures below about 300 °C, resulting in inhibition of the CO oxidation reaction. Above about 300–350 °C, the SO_2 is converted to SO_3, which can react with the γ-Al_2O_3 washcoat forming $Al_2(SO_4)_3$. This leads to deactivation by pore blockage. Figure 13.5 shows the performance of a standard Pt/γ-Al_2O_3 on a honeycomb after exposure to a mixture of SO_2 and SO_3 at about 400 °C. Also shown is an improved catalyst tolerant

(a)   (b)

FIGURE 13.4. Installation of CO catalyst abatement metallic modules in the exhaust from a gas turbine engine. a) Installation of a metallic catalyst module. b) Wall of catalyst modules (Courtery BASF).

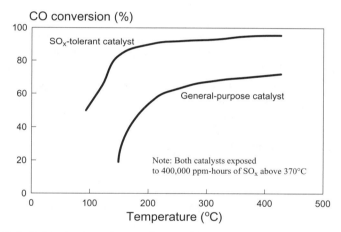

FIGURE 13.5. Sulfur in the exhaust of gas turbine reduces activity and requires SO_x-tolerant catalyst formulation.

to the SO_2/SO_3 environment. This proprietary formulation has a washcoat that is unreactive toward SO_3 and additives that suppress the activity of Pt toward producing SO_3 at the reaction conditions.

13.3 NON-METHANE HYDROCARBON (NMHC) REMOVAL

Air quality regulations currently do not limit methane emissions. This is because it is not toxic and largely considered unreactive for the atmospheric processes that cause photochemical air pollution. Thus, hydrocarbon reduction efficiency usually concerns the destruction of non-methane hydrocarbons (NMHC). The catalytic oxidation activity of catalysts for hydrocarbon removal, however, depends strongly on hydrocarbon type. For example, unsaturated or substituted hydrocarbons, such as alkenes (e.g., propylene), aromatics (e.g., toluene), and aldehydes (e.g., formaldehyde), are much easier to oxidize (i.e., require lower temperature) than saturated paraffins such as propane. Among paraffinic compounds, those with lower carbon numbers are more difficult to convert than those with higher carbon numbers (e.g., propane is much more difficult to convert than hexane).

Hydrocarbons in the exhaust of combustion turbines are composed of "combustion-derived" and "fuel-derived" products. Fuel-derived hydrocarbons are uncombusted hydrocarbons from the fuel that have reached the turbine exhaust, whereas combustion-derived hydrocarbons are compounds that have undergone a partial chemical change in the turbine combustor that falls short of complete conversion to carbon dioxide and water.

Depending on the fuel source, the exhaust HC species and composition can vary greatly. It has been reported that, for aircraft turbine engines operating with jet A fuel, combustion-derived ethylene, acetylene, propylene, and formaldehyde accounted for a major portion (22% to 43%) of the non-methane hydrocarbons (Spicer et al. 1990). Fuel-derived, C_7–C_9 hydrocarbons represent a large portion of the remaining NMHC in the exhaust. On the other hand, the NMHC in the exhaust of natural gas-fired turbines are usually reported to consist of fuel-derived ethane and propane, with trace levels of combustion-derived formaldehyde, benzene, and other substances. The requirements of a catalytic NMHC emissions abatement system can vary widely as a result of these distinctly different NMHC compositions for turbines using different fuels.

13.4   OXIDATION OF REACTIVE HYDROCARBONS

Except for C_2–C_5 paraffins (i.e., ethane, propane, butane, and pentane), the conversion rates of NMHC compounds are controlled by the rate of gas phase mass transfer to the catalyst surface when operating at temperatures greater than 250 °C (see Chapter 4). These reactive, often combustion-derived, hydrocarbons include alkenes, alkynes, aromatics, C_6^+ paraffins, and partially oxygenated hydrocarbons. Their conversion efficiency will depend primarily on their gas phase diffusivities and, like CO oxidation, on the geometric surface area of the honeycomb catalyst.

The conversion rates of reactive hydrocarbons will always increase with increasing catalyst cell density, because the geometric surface area for reaction increases. However, the absolute conversion level for each species will depend on its diffusion rate in the exhaust gas. In general, larger, heavier molecules (like C_8 and C_9 molecules) will diffuse more slowly than smaller, lighter molecules such as ethylene. Table 13.1 gives the conversions of several reactive

TABLE 13.1. Conversion of reactive hydrocarbons over standard Pt catalyst.

Hydrocarbon compound	Conversion (%)
Carbon monoxide (CO)	90
Acetylene (C_2H_2)	86
Ethylene (C_2H_4)	85
Formaldehyde (CH_2O)	77
Benzene (C_6H_6)	72
Toluene (C_7H_8)	71

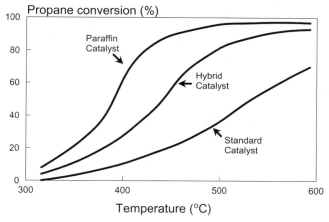

FIGURE 13.6. Propane conversion can be increased using a catalyst optimized for light paraffins.

hydrocarbons over an abatement system designed for 90% CO removal. As the size of the hydrocarbon molecule increases, hydrocarbon conversion decreases due to decreased gas diffusivity. However, Table 13.1 shows that high conversion of reactive hydrocarbons can be achieved using the typical CO abatement system designs with no additional catalyst.

13.5   OXIDATION OF UNREACTIVE LIGHT PARAFFINS

The primary types of hydrocarbons in the exhaust of a natural gas-fired combustion turbine are light paraffins, and these are among the least reactive molecules for oxidation. Methane, ethane, propane, butane, and to a lessor extent pentane require either special catalysts, higher temperatures, or both before they can be destroyed using practical volumes of catalyst in a combustion turbine exhaust.

Figure 13.6 is a graph of propane conversion as a function of temperature for three proprietary catalyst formulations, all of which contain different amounts and type of precious metals. For more difficult-to-oxidize hydrocarbons, a combination of Pt and Pd is preferred to optimize the hydrocarbon conversion in the exhaust.

There has been one limitation on the application of the hybrid and paraffin catalyst formulations in combustion turbine exhausts: their sensitivity to sulfur compounds. Most natural gas contains negligible sulfur (1 to 2 vppm), but natural gas specifications typically allow up to 1 grain S/100 ft^3 (about 30 vppm of sulfur). This level is similar to that found in sweetened refinery gas. Figure 13.7 shows propane conversion versus temperature over a standard Pt/γ-Al$_2$O$_3$-type catalyst for both 0 and 30 vppm of SO$_2$ in the exhaust gas. This sulfur level is equivalent to much higher levels than might be present in natural gas turbine exhaust, but it was used in a laboratory study to accelerate the aging. This type of test has successfully predicted long-term performance in the exhaust of a turbine with much lower sulfur contents. Figure 13.7 shows that sulfur reduces the propane conversion from >40% at 425 °C to about 5%.

Based on these experiences, more sulfur-tolerant catalysts have been developed for the control of light paraffins in natural gas-fired combustion turbine exhausts. These catalysts have washcoats formulated to be less reactive with SO$_3$ than γ-Al$_2$O$_3$ and contain proprietary additives that suppress the oxidation of SO$_2$ to SO$_3$. In one such case, the addition of Rh or Pd to the Pt was found to suppress the SO$_3$ formation (Cordonna et al. 1989). Figure 13.8 compares propane conversion activity for the improved and standard Pt/γ-Al$_2$O$_3$ catalyst.

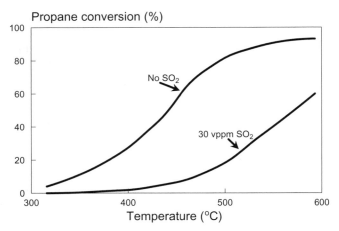

FIGURE 13.7. Sulfur in gas turbine exhaust affects the performance of the catalyst for propane conversion.

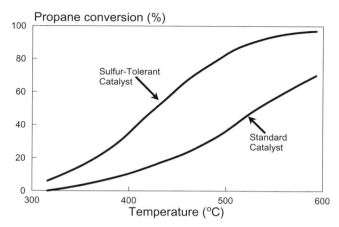

FIGURE 13.8. Sulfur-tolerant catalyst needed for propane conversion of gas turbine exhaust.

13.6 CATALYST DEACTIVATION

There are three primary sources of catalyst poisons and contaminants in the exhausts of cogeneration combustion turbines: fuel contaminants, boiler leaks, and turbine lubricants. Of the three, sulfur oxides from liquid fuels are the most common. Poisoning from boiler feed additives and turbine lubricants can be much more severe.

Sulfur in the exhaust of a combustion turbine interacts with a catalyst in two ways:

1) At temperatures below about 300 °C, SO_2 chemisorbs on to the active metal sites and inhibits the oxidation of both CO and HC. It is for this reason that most processes operate above 300 °C.

2) Excessive amounts of SO_3 are formed, which react with γ-Al_2O_3 and lead to formation of $Al_2(SO_4)_3$, which deactivates the catalyst. Catalysts are formulated to minimize SO_3 production by adding a species that suppresses activity of the metal toward this reaction without dramatically decreasing activity toward CO and HC. New washcoats have been developed that adequately disperse the active catalytic metals but are unreactive toward SO_3.

Catalyst contamination from sources like turbine lubricant and boiler feed water additives (e.g., Zn, P, Ca, and so on) is usually much more severe than deactivation by sulfur compounds in the turbine exhaust. This is because these materials mask the catalyst surface, and this is not reversible during operation of the gas turbine. Catalyst formulation can be modified by adjusting the

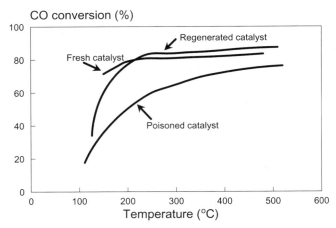

FIGURE 13.9. Proper regeneration technique can restore fresh activity to a severely poisoned catalyst.

micro- and macroporosity to improve poison tolerance, but no catalyst is immune to a contaminant that coats or masks its surface and prevents CO or HC to access the active sites. Therefore, catalyst regeneration remains the best option for maintaining good conversion in the dirtiest environments. Commercial experience led to the development of catalyst regeneration procedures that have been capable of restoring essentially fresh catalyst activity in every case of deactivation caused by masking. The principles of this regeneration procedure are similar to those procedures described in Chapter 11.

Regeneration strategies have been developed for both ceramic (cordierite) honeycomb supports and metal monoliths (Johnson Matthey 2006). Some regeneration treatments require that the catalyst withstand both strong acid and strong base exposure at elevated temperatures. Metal monoliths may have to be made of stainless steels for some regeneration procedures.

Figure 13.9 is an example of a catalyst that was poisoned during operating in a cogeneration system. Several years of operation in a "dirty" turbine exhaust caused this catalyst to accumulate major amounts of an inorganic poison that both neutralized the active sites and masked the catalyst surface. As a result, catalyst activity fell from the fresh catalyst levels to that shown for the poisoned catalyst. The activity restoration after use of the proper washing solutions was successful (Heck et al. 1990).

REFERENCES

BASF. "CAMET oxidation catalysts," BASF Camet product sheet, www.basf-catalysts. com, (2007).

Berndt, M., and Landri, P. "An overview about Engelhard approach to non-standard environmental catalysis," *Catalysis Today* 75: 17–22 (2002).

Chen, J., Speronello, B., and Heck, R. "Catalytic control of unburned hydrocarbon emissions in combustion gas turbine exhausts," 93-mp-7.03, 86th Annual AWMA Meeting, Denver, CO (1993).

Cordonna, G., Kosanovich, M., and Becker, E. "Gas turbine emission control: Platinum and platinum-palladium catalysts for carbon monoxide and hydrocarbon oxidation," *Platinum Metals Review* 33(2): 46–54 (1989).

Gulian, F., Rieck, J., and Periera, C. "Camet oxidation catalyst for cogeneration applications," Industrial and Engineering Chemistry *Research* 30: 122–126 (1991).

Heck, R., Chen, J., and Collins, M. "Oxidation catalyst for cogeneration applications-regeneration of commercial catalyst," 90-105.1, 83rd Annual AWMA Meeting, Pittsburgh, PA (1990).

Johnson Matthey, "Gas turbine oxidation catalysts for control of CO, VOC and HAPs," www.jmssec.com, (2006).

Pereira, C., Plumblee, K., and Evans, M. "Camet metal monolith catalyst system for cogen applications," Editors, Servoy, G., and Fransson, T. *2nd International Symposium on Turbomachinery, Combined Cycle Technologies and Cogeneration- IGTI-Vol 3*, 1002705 (1988).

Pereira, C., Gulian, F., Czarnecki, L., and Rieck, J. "*Dual function catalyst for clean gas applications*," 90-105.6, 83rd Annual AWMA Meeting, Pittsburgh, PA (1990).

Speronello, B., Chen, J., and Heck, R. "A family of versatile catalyst technologies for NO_x and CO removal in co-generation," 92-109.06, 85th Annual AWMA Meeting, Kansas City, MO (1992).

Spicer, C., Holdren, M., and Smith, D. "Chemical composition of exhaust from aircaft engines." Gas Turbine and Aeroengine Congress and Exposition, Brussels, Belgium, (1990).

State of New Jersey. "State of the art (SOTA) manual for stationary combustion turbines," Second Revision: (2004).

CHAPTER 13 QUESTIONS

1. In the exhaust of a gas turbine, list all the potential oxidation reactions and describe the selectivity in these reactions.

2. Compare the different properties between a ceramic and a metallic honeycomb. Why is a metallic better for gas turbine applications?

14 Small Engines

14.1 INTRODUCTION

Emission regulations have been put into place worldwide for small engines. These regulations include the so-called handheld engines (leaf blowers, chainsaws, etc.), non-handheld engines (riding lawn mowers, etc.), all-terrain vehicles (ATVs), marine engines, snowmobiles, and the so-called two wheelers or motorcycles. For instance, in the United States, these small engines are classified by the Environmental Protection Agency (EPA) as nonroad spark-ignition engines, equipment, and vessels; small spark-ignition handheld engines; motorcycles; ATVs; and snowmobiles. Spark-ignition (SI) nonroad engines rated below 25 horsepower (19 kW) are used in household and commercial applications, including lawn and garden equipment, utility vehicles, generators, and a variety of other construction, farm, and industrial equipment. Spark-ignition engines are also used in marine vessels, including outboard engines, personal watercraft, and sterndrive/inboard engines.

14.2 EMISSIONS

These engines are usually of the two-stroke or four-stroke design, and the emissions tend to be high compared with automotive applications. The two-stroke engine is much smaller and lighter than its counterpart and thus is more suitable for a handheld or device.

There are two major differences between the design of the combustion chambers and the combustion process. The two-stroke engine vents the combustion products during the power stroke while fresh fuel is being added in contrast to the four-stroke engine that vents the combustion products in a separate stroke. Second, lubricating oil is premixed with the gasoline because there is no provision for delivering oil separately based on the design of the crankcase of the two-stroke engine. In the case of four-stroke engines, there is a separate compartment that delivers oil to the combustion chamber. These two differences have a profound effect on the emissions, which for two-stroke engines have more fuel-derived hydrocarbons and more oil-derived pollutants.

Catalytic Air Pollution Control: Commercial Technology, by Ronald M. Heck and Robert J. Farrauto, with Suresh T. Gulati.
Copyright © 2009 John Wiley & Sons, Inc.

Furthermore, two-stroke engines tend to operate rich of the stoichiometric ($\lambda < 1$) air-to-fuel ratio, and thus, the exhaust is cooler and richer in fuel-derived hydrocarbons and carbon monoxide than a stoichiometric engine. The rich operating mode allows for less expensive materials of construction for the device since the maximum temperatures are lower. Thus, catalytic aftertreatment will abate higher hydrocarbon (HC) and CO emissions with greater ash deposition and its poisoning effect of Zn and P due to the larger consumption of oil. Although two-stroke engines have higher emissions than the four-stroke engines, the latter have higher combustion efficiencies. Typical pre-2000 emissions cover the following range (Mooney et al. 1994):

CO (%)	HC (vppm)	NO_x (vppm)	O_2 (%)	Exhaust temperature (°C)
1.0–9.0	1,500–5,000	50–400	0.2–1.0	300–600

A complete breakdown of the hydrocarbon speciation for the exhaust emissions from ten four-stroke lawn mower engines has been completed for both a 1990 national average blend and a reformulated gasoline (Gabele 1997). Reformulated gasoline usage resulted in lower organic and carbon monoxide emissions that contributed to lower photochemical reactivity factor or ozone-forming potential. Benzene emissions were also lower. The NO_x emissions increased with the reformulated gasoline since the combustion efficiency was higher and thus higher combustion temperatures. (See Chapter 7 for NO_x formation in combustion processes.)

14.3 ENVIRONMENTAL PROTECTION AGENCY (EPA) REGULATIONS

Currently, the EPA has regulations for many small-engine and nonroad applications as shown in Table 14.1 (EPA 1 2006).

The EPA divides the engines into classes as shown in Table 14.2. The reproposed standards for handheld engines are given in Table 14.3 (EPA 2 2007). As this book was going to press, the EPA has finalized these standards as of September 2008 (EPA 2008.)

For small nonroad, nonhandheld engines, the EPA is proposing HC + NO_x exhaust emission standards of 10 g/kW-hr for Class I engines starting in the 2012 model year and 8 g/kW-hr for Class II engines starting in the 2011 model year. No changes in the exhaust emission standards have been proposed for handheld emissions. For spark-ignition engines used in marine generators, a more stringent Phase 3 CO emission standard of 5 g/kW-hr is proposed. This would apply equally to all sizes of engines subject to the small SI standards.

For marine SI engines and vessels, the EPA is proposing a more stringent level of emission standards for outboard and personal watercraft engines starting with the 2009 model year. The proposed standards for engines above 40 kW

TABLE 14.1. EPA nonroad engine program.

Locomotives engines	40 CFR Part 92	April 16, 1998	63 FR 18978	Exhaust
Marine diesel engines	40 CFR Part 94	December 29, 1999	64 FR 73300	Exhaust
Other nonroad diesel engines	40 CFR Parts 89, 1039	June 29, 2004	69 FR 38958	Exhaust
Marine SI engines	40 CFR Part 91	October 4, 1996	61 FR 52088	Exhaust
Recreational vehicle SI engines	40 CFR Part 1051	November 8, 2002	67 FR 68242	Exhaust & Evaporative
Small SI engines (SI engines ≤ 19 kW [or ≤30 kW if total displacement is ≤1l)]	40 CFR Part 90			Exhaust
a. Handheld (HH)		a. January 12, 2004	a. 69 FR 1824	
b. Nonhandheld (NHH)		b. March 30, 1999	b. 64 FR 15208	
Large SI engines (SI engines >19 kW (or >30 kW if total displacement is ≤1 liter)	40 CFR Part 1048	November 8, 2002	67 FR 68242	Exhaust & Evaporative

TABLE 14.2. EPA engine classes.

Nonhandheld				Handheld		
Class I-A	Class I-B	Class I	Class II	Class III	Class IV	Class V
<66 cc	66 to <100 cc	100 to <225 cc	≥225 cc	<20 cc	20 cc to <50 cc	≥50 cc

are 16 g/kW-hr for $HC + NO_x$ and 200 g/kW-hr for CO. For engines below 40 kW, the standards increase gradually based on the engine's maximum power. They are proposing new exhaust emission standards for sterndrive and inboard marine engines. The proposed standards are 5 g/kW-hr for $HC + NO_x$ and 75 g/kW-hr for CO starting with the 2009 model year. For sterndrive and inboard marine engines above 373 kW with high-performance characteristics, they are proposing a CO standard of 350 g/kW-hr.

TABLE 14.3. EPA proposed standards for handheld engine classes.

Engine class	Reproposed emission standards for handheld engines by model year HC + NO$_x$ (g/kW-hr)			
	2005	2006	2007	2008 and later
Class III	100	50	50	50
Class IV	89	50	50	50
Class V	129	110	91	72

In California, the regulations for motorcycles have been revised based on advances in catalyst and engine technologies. As a result, the California Air Resources Board (CARB) adopted a new set of standards that applied to 280 cc and larger motorcycles starting with the 2004 model year with additional reductions required in the 2008 model year. HC and NO$_x$ emissions are combined into a single new standard that will give manufacturers additional flexibility to lower emissions and provide motorcycles that meet consumer needs. HC plus NO$_x$ emissions is required to be reduced to 1.4 g/km for the 2004 model year and 0.8 g/km for the 2008 model year. This represents a significant reduction over the current standards. CARB is also discussing with manufacturers a proposal to offer an incentive plan encouraging the introduction of cleaner motorcycles prior to the 2008 model year for a type of retrofit program.

Table 14.4 shows the current CARB standards for small spark-ignition engines based on model year and displacement (CARB 1 2004). Small off-road engines are equal to or less than 19 kW (25 horsepower) and include both handheld equipment (such as string trimmers and chain saws) and nonhandheld equipment (such as lawn mowers and generators as well as industrial equipment).

CARB recently has proposed new so-called Tier 3 standards for engines above 80 cc (CARB 2007). The proposed standards are based on the use of a catalyst that would reduce HC + NO$_x$ by 50% at the end of useful life. For engines >80 cc to <225 cc, the proposed Tier 3 standard is 8 g/kW-hr HC + NO$_x$ at the end of useful life. For engines 225 cc or above, the proposed Tier 3 standard is 6 g/kW-hr HC + NO$_x$ at the end of useful life. No change is proposed to the current CO emission standard. In previous documents released to the public, staff initially proposed an implementation date of 2006 for the proposed Tier 3 standards. Proposed implementation was for the 2007 model year for engines between 80 and 225 cc and with the 2008 model year for engines 225 cc and above as shown in Table 14.5.

In addition, voluntary optional low exhaust emission standards for small engines are proposed. Engines meeting these standards will be classified as a "Blue Sky Series" engine. The optional standards are presented in Table 14.6.

TABLE 14.4. Exhaust emission standards for spark-ignition engines (grams per kilowatt-hour).

Model year	Displacement category	Durability period (hours)	Hydrocarbons plus oxides of nitrogen	Carbon monoxide	Particulate
2006 and subsequent	<50 cc	50/125/300	50	536	2.0
	50 cc to 80 cc inclusive	50/125/300	72	536	2.0
2005	>80 cc to <225 cc Horizontal shaft engine	50/125/300	16.1	549	
	>80 cc to <225 cc Vertical shaft engine	N/A	16.1	467	
	≥225 cc	125/250/500	12.1	549	
2006	>80 cc to <225 cc	125/250/500	16.1	549	
	≥225 cc	125/250/500	12.1	549	
2007	>80 cc to <225 cc	125/250/500	10.0	549	
	≥225 cc	125/250/500	12.1	549	
2008 and subsequent	>80 cc to <225 cc	125/250/500	10.0	549	
	≥225 cc	125/250/ 500/1,000	8.0	549	

TABLE 14.5. Proposed exhaust emission standards for spark-ignition engines.

Year	Displacement	Standards g/kW-hr [g/bhp-hr]	
		HC + NO$_x$	CO
2002–2005	>65 to <225 cc Horizontal Shaft	16.1 [12.0]	549 [410]
	>65 to <225 cc Vertical Shaft	16.1 [12.0]	467 [350]
	≥225 cc	12.1 [9.0]	549 [410]
2006 and later	>65 to <225 cc	16.1 [12.0]	549 [410]
	≥225 cc	12.1 [9.0]	549 [410]
2007 and later *(Proposed)*	>80 to <225 cc	8.0 [6.0]	549 [410]
2008 and later *(Proposed)*	≥225 cc	6.0 [4.5]	549 [410]

TABLE 14.6. "Blue Sky Series" engine emission standards g/kW-hr [g/bhp-hr].

Model Year	Displacement	HC + NO$_x$	CO	PM*
2005 and later	<50 cc	25 [18.5]	536 [400]	2.0 [1.5]
2005 and later	≥50 to ≤80 cc	36 [26.9]	536 [400]	2.0 [1.5]
2007 and later	>80 to <225 cc	4.0 [3]	549 [410]	N/A
2008 and later	>225 cc	3.0 [2.3]	549 [410]	N/A

*The PM standard is applicable to all two-stroke engines.

14.4   CATALYSTS FOR HANDHELD AND NONHANDHELD ENGINES

Catalyst supports for small-engine applications have ranged from monoliths to foams, either ceramic or metallic (Dettling et al. 2001; Chaudhari et al. 1999). Also, catalyst coatings have been applied to exhaust manifold/muffler parts to effect the removal of emissions. In some cases, wires are coated and used as a catalyst. The impact of restricted space within the muffler for the catalyst and the pulse flow characteristics challenges conventional catalyst sizing methods as well as the methods of catalyst mounting. Space velocities range from 250,000 1/hr to 1,500,000 1/hr for small off-road engines and motorcycles (Mooney et al. 1994; Dettling et al. 2001). Since many of theses small engines are calibrated on the rich side of stoichiometric, air is injected into the exhaust to promote complete oxidation and the catalyst can see up to a 500 °C increase in the surface temperatures due to the adiabatic temperature rise of oxidation of CO and unburned hydrocanbon (UHCs). The implementation of small-engine catalysts is proceeding worldwide for both two and three wheelers and small off-road engines. Many technical challenges remain, but the consumer of future small engines will operate much cleaner devices, which will aid in the reduction of ambient pollution and hence smog (mainly ozone) formation.

14.4.1   Handheld and Nonhandheld Engines

Most of the small off-road spark-ignition engines (SOREs) run on gasoline, and some use alternative fuel such as liquefied petroleum gas (LPG) or compressed natural gas (CNG). These engines are rated at or below 19 kW (25 horsepower). Small off-road engines are used to power a broad range of lawn and garden equipment, including lawn mowers, leaf blowers, and lawn tractors, as well as generators and small industrial equipment.

In an attempt to meet emission standards, manufacturers are looking at engine modifications and the use of catalysts. The handheld equipment market is dominated by the two-stroke engine, because of its high power-to-weight ratios, multipositional operation, simple construction, lower manufacturing costs, and low maintenance requirements. The inherent design of a two-stroke engine allows a portion of unburned fuel that enters the combustion chamber to escape to the atmosphere. This process, known as scavenging, results in excessive exhaust HC emissions. To meet the more stringent emission standards, some manufacturers are expected to incorporate internal engine redesign coupled with the use of a catalyst. A modified two-stroke engine designed to reduce scavenging will minimize the deterioration of the catalyst by significantly reducing the catalyst's exposure to "escaping" fuel and oil (CARB 2003). The four-stroke engine is the primary internal combustion engine design used in personal transportation and nonhandheld equipment applications. Compared with a typical two-stroke engine, a four-stroke engine can achieve as much as a 30% improvement in fuel economy and emit significantly less HC emissions. In the four-stroke engine design, consumers do not need to pre-mix fuel with oil. Although four-stroke engines require periodic oil changes, and are thought to be "too heavy" when used with larger sized handheld equipment, recent advances in small four-stroke engine design has allowed the four-stroke engine to be an attractive alternative to its two-stroke counterpart.

One of the issues of concern with implementation of a catalyst for emission control from small engines was the safety concerns with these devices in regard to exposure to higher than normal operating temperatures, which may cause skin burns to the operator. The EPA conducted an extensive study on the safety of emission controls for nonroad spark-ignition engines less than 50 horsepower (EPA 1 2006). A total of 19 nonhandheld engines were selected for laboratory and field tests. Twelve engines were Class I engines and were evenly split between side-valve and overhead valve engine designs from four different engine families. Eight engines were Class II with an overhead valve engine design from three different engine families. This study focused on five areas:

- Engineering analysis and emission testing of current technology Class I and Class II engines and Class I and Class II engines with properly designed emission control systems capable of achieving exhaust emission reductions beyond the Phase 2 standards (advanced prototype systems).
- Exhaust emission and safety assessment testing of Class I and Class II engines in both a stock configuration and equipped with advanced prototype emission control systems. Engines were tested both in the laboratory and in the field over a broad range of operating conditions; external exhaust system surface temperatures were measured using infrared thermal imaging, whereas temperatures for lubricant, cylinder head, and exhaust gases were measured using thermocouple probes.
- Laboratory analysis of significant off-nominal operating conditions that were identified by engine manufacturers, original equipment manufacturers (OEMs), and EPA staff.
- Assessment of the potential safety impacts of evaporative emission control requirements.
- The completion of design failure mode and effects analyses (FMEA) for Class I and Class II engines used in walk-behind and ride-on mowers and three process FMEAs for consumer use of lawn equipment. These studies were conducted as an additional tool for identifying potential safety concerns in going from Phase 2 to potential Phase 3 standards.

As a result of this study, the EPA concluded that based on the technical work and subsequent analysis of all of the data and information, there was a strong indication that catalyst-based standards can be implemented without

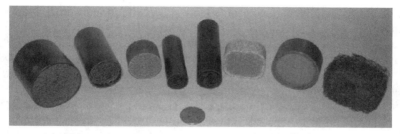

FIGURE 14.1. Different types of catalyst supports for small-engine applications (EPA 1 2006).

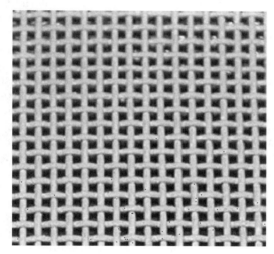

FIGURE 14.2. Screen catalyst support for small engines. (Courtesy Engelhard).

an incremental increase in the risk of fire or burn to the consumer. In many cases, the designs used for catalyst-based technology lead to an incremental decrease in such risk.

In the EPA study, many different types of catalyst and catalyst designs were incorporated that represent the breadth of catalyst support types being considered for small-engine applications as shown in Figure 14.1. In general loadings were between 50 and 70 g/ft^3, and Pt/Pd/Rh loading ratios varied from 0:5:1 to 5:0:1. Catalyst volume varied from approximately 50% to 55% of the engine displacement.

One catalyst manufacturer uses a coated screen as shown in Figure 14.2. These screen catalysts are coated on large, flat sheets that can be cut to desired shapes, stacked, or rolled into radial flow or heat tube configurations without significant washcoat loss or damage (Adomaitis 2004; Kumar and Galligan 2005). Stacking the screens results in higher performance.

A durability study was conducted on six nonhandhelds in low-emission configurations (Lela and White 2004). Four engines were used in walk-behind mower applications; one was used in a riding mower, and one was used in constant-speed/generator applications. The goal was to reduce the tailpipe-out HC plus oxides of nitrogen (NO_x) emissions to 50% or less of the current useful life standard of 12 g/hp-hr for Class I engines (65 cc < displacement < 225 cc) or 9 g/hp-hr for Class II engines (displacement ≥ 225 cc). Low-emission engines were developed using three-way catalytic converters, passive secondary-air induction systems, and enleanment, when needed. Commercially available catalysts were provided for this study. The main conclusions of this work were that catalyst technology can be successfully applied to small off-road engines; that such applications are durable; and that HC + NO_x reductions of 50–70% were demonstrated over the useful lives of several small engines. Useful engine life ranged from 125 hr to 500 hr.

Another study considered five 6.5-hp, single-cylinder, side-valve, 190-cc engines that were modified to have ceramic honeycomb catalytic converters inside the muffler (Doll and Reisel 2007). Ports were drilled into the muffler to provide secondary air for use in the converters since all of these engines were operated fuel rich. Because of the uncontrolled operation of these carbureted engines, there was substantial variability in emissions and in the air-to-fuel ratio. In addition, the engine-out hydrocarbon emissions increased with engine age. Tests were conducted at various intervals up to 187.5 hr, and decreases in conversion were noted. The results indicated that soot buildup, ash deposits, and physical deterioration of some of ceramic honeycombs contributed to this decline. The HC + NO_x conversion varied from 60% to around 30% over the test period for the five engines tested. Postanalysis indicated deposits of zinc and phosphorus on catalyst most likely from the engine lubricating oil. It was unclear from this study why three of the catalysts suffered significant wear.

The operating space velocities of small-engine catalysts are significantly higher than automobile catalyst applications and typically can range from 200,000 hr^{-1} up to 1,000,000 hr^{-1}.

Regarding the useful life of engines for use with catalysts for emissions control, manufacturers are required to declare the applicable useful life category for each engine family at the time of certification (EPA 3 2007). For nonhandheld engines, manufacturers shall select a useful life category as follows in Table 14.7. For handheld engines, manufacturers shall select a useful life category as follows in Table 14.8. Engines with gross power output greater than 19 kW have an engine displacement less than or equal to 1 L that must certify to a useful life period of 1,000 hr. The manufacturer must provide specific data to support the choice of the useful life category.

A small-engine test procedure has been specified by the EPA beginning with the 2005 model year. The test consists of prescribed sequences of engine operating conditions to be conducted on an engine dynamometer or equivalent load and speed measurement device (EPA 3 2007). The exhaust gases

TABLE 14.7. Useful life categories for nonhandheld engines [hours].

Class I	125, 250, 500
Class II	250, 500, 1,000
Class I-A	50, 125, 300
Class I-B	125, 250, 500

TABLE 14.8. Useful life categories for handheld engines (hours).

Class III, IV, V	50, 125, 300

generated during engine operation are sampled either raw or dilute, and specific components are analyzed through the analytical system. The test consists of three different test cycles that are application specific for engines spanning the typical operating range of nonroad spark-ignition engines. Two cycles exist for Class I-B, I and II engines, and one is for Class I-A, III, IV, and V engines. The test cycles for Class I-B, I, and II engines consist of one idle mode and five power modes at one speed (rated or intermediate). The test cycle for Class I-A, III, IV, and V engines consists of one idle mode at idle speed and one power mode at rated speed. These procedures require the determination of the concentration of each pollutant, fuel flow, and the power output during each mode. The measured values are weighted and used to calculate the grams of each pollutant emitted per brake kilowatt hour (g/kW-hr).

14.4.2 Marine Engines

Spark-ignition engines used in marine vessels include outboard engines, personal watercraft, and sterndrive/inboard (SD/I) engines. Various feasibility and safety studies have been conducted on these engines. One study performed laboratory testing of six catalyst designs on an SD/I engine. This included in-water testing of a boat with catalysts over a severe operation using both fresh and salt water. Laboratory testing was conducted for six catalyst designs on an SD/I engine. Results from this testing suggest that significant emission reductions can be achieved from SD/I engines through the use of catalysts. The fresh water durability study showed that these emission reductions can be achieved over the lifetime of the engine. The salt water testing is in progress (EPA 2 2006).

Another study looked at twin 1.7-l catalysts installed on a 7.4-l marine engine rated at 310 hp at 4600 rpm with oxygen-feedback stoichiometric air–fuel control (CARB 2001). A composite emission rate of 3.2 g/kW-hr of $HC + NO_x$ was achieved compared with the baseline without a catalyst of 12.9 g/kW-hr $HC + NO_x$. The engine was modified, and EGR was added to

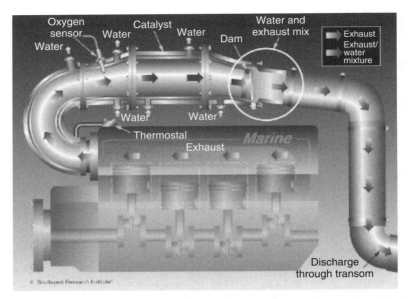

FIGURE 14.3. Schematic of the catalyst application in a marine engine (CARB 2 2004).

lower the engine baseline NO_x; 2.6 g/kW-hr $HC + NO_x$ was achieved. The compact cylindrical 0.8-l catalysts were tested in the exhaust manifold riser position, well upstream of the water mixing point. Other catalyst volumes and locations were tested. Regardless of the catalyst-system design, a catalyst near the exhaust–water mixing point, or a catalyst well upstream of the exhaust–water mixing point, the emission test results were below the proposed standard of 5 g/kW-hr. The catalyst resulted in a maximum-power decrease of the engine of 6 kW (about 3%). EGR was not required to meet the proposed standards. A representation of one catalyst design for a marine engine is given in Figure 14.3 (CARB 2 2004).

The EPA recently announced that one of the manufacturers involved in their catalyst test programs, Indmar Marine Engines, successfully completed its own catalyst development and durability testing program in parallel with the EPA studies (EPA 2 2006). They are now selling inboard marine engines equipped with catalysts, and they report excellent emission performance without any loss in performance (Indmar 2007).

14.4.3 Motorcycles

Perhaps the motorcycle mode of transportation has received the most attention for controlling small-engine emissions. It is estimated that there are over a billion motorcycles on the road worldwide (Walsh 2004) with approximately 25 million manufactured annually (Yavuz et al. 2004). The engines are two and four stroke with the majority being two stroke. Since the late 1990s, the motorcycle industry has shifted its production from two stroke to a cleaner four stroke for all engine classes with the exception of 50-cc mopeds. Because such small-utility engines are weight sensitive, the lighter two-stroke engine remains preferred for these applications. The base calibration of these engines is fuel rich to obtain more power output. Many studies have shown that improvements in the engines, such as frequent servicing, and engine modifications, such as operation closer to stoichiometric, cannot meet the proposed emission limits in any country worldwide so that catalytic approaches are needed in combination with these enhancements. Typical conversions of CO and UHCs are shown in Figure 14.4 for an oxidation catalyst. The increase in engine speed can be related to an increase in operating space velocity for the catalyst. The catalyst was a ceramic converter 40 mm in diameter containing two equal lengths of 4.94 g/l tri-metal catalysts (each 34 mm in length) and operated on a four-stroke engine.

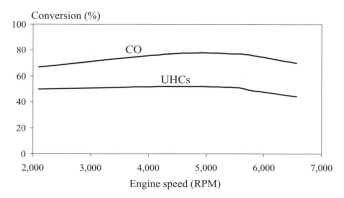

FIGURE 14.4. Performance of a tri-metal catalyst for motorcycle emission control (Hwang et al. 1997).

Catalyst technology is not new to emission control from motorcycles as it has been used successfully on motorcycles and mopeds for over 15 years. During that time, over 15 million two- and three-wheel vehicles have been equipped with catalyst systems (MECA 2002). Catalyst technology applied to two-stroke motorcycles and mopeds has demonstrated a capability of reducing emissions in the range of 50–60% for HC and 50–80% for CO. If secondary air injection is used, control efficiencies in excess of 90% for both HC and CO can be achieved. For four-stroke engines, a three-way catalyst can achieve in excess of 90% HC + NO_x and CO emissions. Catalyst aging studies have been conducted out to 60,000 km with no change in performance for oxidation catalysts (Palke 2004).

Several design approaches are being considered depending on the engine design, including:

- Rich engine calibration to reduce NO_x followed by secondary air injection to oxidize CO and UHC over a catalyst
- Rich engine calibration followed by dual catalyst (i.e., reduction catalyst followed by an oxidation catalyst) system with secondary air injection
- Closed-loop system with stoichiometric engine operation with electronic controller and oxygen sensor and a TWC catalyst
- Closed-loop system with lightoff catalyst (i.e., close-coupled catalyst) with electronic controller and oxygen sensor and a TWC catalyst

Studies showed that cold-start and high-speed operations are the high-emission contributors requiring a lightoff catalyst and engine control strategies to meet projected Euro III emissions and beyond (Kumar et al. 2006). Catalyst loadings from 50 to $150 g/ft^3$ of precious metals have been proposed with 200-cpsi to 400-cpsi supports. The close-coupled catalyst may be 100 cpsi.

14.4.4 Snowmobiles

Snowmobiles have become quite an issue in the last 10 years due to there popularity and use in national parks and other public recreational areas. It has been shown that it would take almost 100 automobiles to emit the combined HC, CO, and NO_x emissions of one snowmobile (EPA 2002). Emission rates have been measured as shown in Table 14.9.

TABLE 14.9. Emissions from various snowmobile engine configurations.

	Snowmobile emissions (g/bhp-hr)			
Engine category	HC	CO	NO_x	PM
Baseline two stroke	111	296	1	2.7
Recalibrated two stroke	54	147	1	2.7
Direct Injection two stroke	22	90	3	0.6
Four stroke	8	123	9	0.2

Most snowmobiles currently use two-stroke engines, but stricter emissions standards will require the use of cleaner technology in the future. A two-stroke engine requires oil to be mixed with the gas for lubrication purposes. This results in burned and unburned oil in the exhaust. Also, these engines are inherently inefficient in completely burning gas and have high amounts of unburned hydrocarbons in the exhaust. EPA regulations and the demand from more environmentally conscious consumers have created a need for a more environmentally friendly snowmobiles using a four-stroke engine (Bennis et al. 2003/2004). These snowmobiles have significantly lower emission of unburned hydrocarbons, carbon monoxide, and NO_x.

In 1999, in response to increasing concern about snowmobile noise and air pollution in environmentally sensitive areas, Wyoming officials worked with the Society of Automotive Engineers (SAE) to form and organize a new intercollegiate design competition, the SAE Clean Snowmobile Challenge 2000 (CSC2000). The goal of the competition was to develop a snowmobile with improved emissions and noise characteristics that did not sacrifice performance. Modifications were expected to be cost-effective and practical (Montana 2000). Many useful innovations in design and use of both oxidation and three-way catalyst derived from automotive technology have substantially reduced the baseline emissions of the snowmobile and the tailpipe emissions using catalyst technology.

14.5   CATAYST DURABILITY

Catalyst deactivation for small-engine applications can occur through a thermal mechanism, masking of the catalyst surface, and poisoning by sulfur (McDowell et al. 1998; Tyo and Palke 1999; Palke et al. 1999; O'Sullivan 2001). Since the converter is located inside the muffler, the peak temperatures can severely stress the catalytic active washcoat and the monolithic material. Catalysts using the segregated washcoat concept have been employed to minimize the thermal effects. This process can be used to separate precious metals from each other or to add high-temperature stabilizing agents to particularly high-temperature-sensitive components (Hwang et al. 1997).

Since the two-stroke engines use lubricating oil in the gasoline mixture, the ash components in the lubricating additive pass through the exhaust to the catalyst. Four-stroke engines also have lubricating oil in the exhaust due to bypass around the piston rings and valve guide seals. Another contributor to oil consumption is cylinder bore distortion when the engine is hot. This problem is more severe with side-valve engines than with overhead-valve engines because a side-valve engine's exhaust port is adjacent to the cylinder and more difficult to cool. The industry trend toward overhead-valve engines will help to alleviate this problem. Other approaches include tighter manufacturing tolerances and the use of improved seals, which limit the oil available to the valve guides. Analysis of an engine-aged catalyst reveals that the ash material from the lubricating oil is depositing on the catalyst surface resulting in a loss of the effective mass transfer area through pore plugging (Mooney et al. 1994). Electron microprobe analysis revealed that a considerable amount of phosphorus and zinc had accumulated on the top surface of the catalyzed washcoat, whereas sulfur and calcium tended to be deposited throughout the washcoat layer. Catalyst manufacturers are aware of the effects of lubrication oil contamination and have designed catalysts that resist it for other applications.

In 2002, it was estimated that more than 15 million motorcycles and mopeds worldwide have been equipped with catalysts, with most of these units being powered by two-stroke engines. In addition, many two-stroke engines equipped with catalysts have been certified to California's current exhaust emission standards. These applications have been successful despite that two-stroke engines burn lubricating oil and the concentrations of oil contaminants in the exhaust are significantly higher than typical automotive (or lawnmower) exhaust (MECA 2002).

REFERENCES

Adomaitis, J. "Motorcycle catalyst presentation: Meeting the euro-3 challenge for 4-stroke motorcycles," AVECC 2004 Beijing, China (2004).

Bennis, D., Byrd, J., Higgins, J., and Bushnell, C. "4 stroke emission control," MEE 387 Design III (2003/2004).

CARB 1. "Rulemaking on small off-road utility engine regulations," 1998 Hearing 1999, http://arbis.arb.ca.gov/homepage.htm, (1999).

CARB 2. "California exhaust emission standards and test procedures for 1995 and later small off-road engines," California Air Resources Board (1999).

CARB. "State of California air resources board staff report, public hearing to consider adoption of emission standards and test procedures for new 2003 and later spark-ignition inboard and sterndrive engines," http://o3.arb.ca.gov/regact/marine01/isor.pdf, (2001).

CARB. "Staff report: Initial statement of reasons for proposed rulemaking, public hearing to consider the adoption of exhaust and evaporative emission control requirements for small off-road equipment and engines less than or equal to 19 kilowatts," (2003).

CARB 1. "California exhaust emission standards and test procedures for 2005 and later small off-road engines," (2004).

CARB 2. "Status of inboard/sterndrive catalyst testing," (2004).

CARB. "Final regulation order: Chapter 9, division 3, title 13, California code of regulations, Chapter 9: Off-road vehicles and engines pollution control devices, Article 1. Small off-road engines," http://www.arb.ca.gov/regact/sore03/2fro.pdf, (2007).

Chaudhari, M., Jose, T., Vora, K., Dias, C., Reck, A., and Diewald, R. "An investigation on application of metallic substrates and emitubes for a two-stroke three-wheeler," SAE 990033 (1999).

Dettling, J., Larkin, M., Adomaitis, J., and Galligan, M. "Emission control strategies for 2 and 4 stroke motorcycles in India," SAE 200-01-0002 (2001).

Doll, N., and Reisel, J. "Catalyst deterioration over the lifetime of small utility engines," *Journal of Air and Waste Management Association*, 57:1223–1233 (2007).

EPA, Federal Register, 40 CFR Parts 60, 63, et al. "Control of emissions from nonroad sparks-ignition engines and equipment: proposed rule," Friday, May 18, 2007.

EPA, "Environmental impacts of newly regulated nonroad engines: Frequently asked questions," Office of Transportation and Air Quality (2002).

EPA 1. "Proposed emission standards for new nonroad spark-ignition engines, equipment, and vessels," http://www.epa.gov/otaq/equip-ld.htm, (2007).

EPA 2. "Regulatory announcement: Reproposed phase 2 standards for small spark-ignition handheld engines," http://www.epa.gov/otaq/regs/nonroad/equip-ld/hhsnprm/f99027.htm, (2007).

EPA 3. "Title 40: Protection of environment, part 90, control of emissions from nonroad spark-ignition engines, at or below 19 kilowatts. Subpart B-emissions standards and certification provisions," http://ecfr.gpoaccess.gov/cgi/t/text/textdx?c=ecfr&sid=06a483f87ec5dd4faa5fa04cac19d835&rgn=div8&view=text&node=40:20.0.1.1.4.2.1.5&idno=40, (2007).

EPA "EPA finalizes emission standards for nonroad spark-ignition engines, equipment and vessels," EPA420-F-08-0 13, September 2008.

EPA 1. "EPA technical study on the safety of emission controls for nonroad spark-ignition engines <50 horsepower," EPA420-R-06-006 (2006).

EPA 2. "Marine engine manufacturer develops low emission inboard marine engines," EPA420-F-06-057, (2006).

Gabele. "Exhaust emissing from four-stroke lawnmower engines," *Journal of Air and Waste Management* 47: 945–952, September 1997.

Hwang, H., Dettling, J., and Mooney, J. "Catalytic converter development for motorcycle emission control," SAE 972143 or JSAE 9734764 (1997).

Indmar Marine Engines, ETX/CAT, http://www.indmar.com/Innovations/ETX-CAT/index.html, (2007).

Kumar, S., and Galligan, M. "Advanced emission control for single cylinder engines," SAE 2005-26-017 (2005).

Kumar, S., Adomaitis, J., and Alive, K. "Advanced emission control for motorcycles, scooters and recreational vehicles," SAE 2006-01-0020, JSAE200666520 (2006).

Lela, C., and White, J. "Durability of low emissions small off-road engines," Final Report Prepared for California Resources Board, SwRI 08.05734 (2004).

MECA. "Emission control of two-and three-wheel vehicles," Manufacturers of Emission Controls Association (1999).

MECA. "Statement of the manufacturers of emission controls association on the U.S. Environmental Protection Agency proposed rulemaking on control of emissions from highway motorcycles," Docket No. A-2000-02, (2002).

McDowell, A., Douglas, R., McCullough, G., and Kee, R. "Catalyst deactivation on a two-stroke engine," SAE 982015 (1998).

Montana, Department of Environmental Quality. "Solutions—clean snowmobile challenge," http://www.deq.state.mt.us/CleanSnowmobile/solutions/challenge/index.asp, (2000).

Mooney, J., Hwang, S., Daby, K., and Winberg, J. "Exhaust emission control of small 4-stroke air cooled utility engines an initial R&D report," SAE 941807 (1994).

O'Sullivan, R. "Two-stroke exhaust catalyst durability on indian 2-wheeler and comparison with catalysts aged on an engine bench," SAE 2001-01-003 (2001).

Palke, D., Mital, R., Dillon, J., and Hopmann, M. "Catalysts durability and physicochemical changes due to extended vehicle operation and their impact on catalyst performance—a case study," SAE 990001 (1999).

Palke, D. "Catalytic aftertreatment and 2- and 3-wheel vehicles," Asian Vehicle Emission Control Conference (2004).

Tyo, M., and Palke, D. "Thermal and poisoning effects on the performance of motorcycle emission control catalysts," SAE 1999-01-3301 (1999).

Walsh, M. "Status of worldwide environmental regulations," 20th Annual Mobile Sources Clean Air Conference (2004).

Yavuz, B., Adomaitis, J., and Galligan, M. "A catalytic two-stroke of genius," Pollution Engineering (2004).

CHAPTER 14 QUESTIONS

1. Why was it necessary to study the safety aspects of adding a catalyst to the exhaust of small-engines to reduce emissions?

2. What are the benefits of moving from two-cycle engines to four-cycle engines? The drawbacks?

PART IV
New and Emerging Technologies

15 Ambient Air Cleanup

15.1 INTRODUCTION

Emission control technology related to implementation of the Clean Air Act has concentrated on exhaust controls and better engine combustion technology for mobile source and pollution prevention through better processing for stationary source (Dartt and Davis 1994). This evolution has been true for automotive vehicles as well as industrial processes. The development of the three-way catalyst and computer-controlled multipoint fuel injection are two examples in the evolution of today's low-emission vehicles. Both exhaust controls and pollution prevention have had successes and major accomplishments, and in many cases, more controls in these areas will certainly be implemented.

Entirely new approaches are required to achieve the substantial improvement that will be needed to meet the health standard for ozone in several areas such as Southern California, Houston, Atlanta, Chicago, and New York. New approaches are required to achieve significant air quality progress without limiting an area's economic competitiveness. PremAir® catalyst technology represents one of these new approaches. This concept involves the actual removal of pollutants from ambient air (Hoke et al. 1996; 1999). "PremAir" is the trade name for a family of catalysts that are capable of reducing air pollutants.

The California Air Resources Board (CARB) has recently adopted direct ozone reduction (DOR) technologies as an emission control alternative. The application of DOR technologies on motor vehicles allows an automaker to receive non-methane organic gas (NMOG) emission credits, which may be applied to offset vehicle tailpipe emissions or evaporative emissions from fuel tanks. Additionally, DOR technologies can enhance the "green image" of an automaker. DOR devices involve catalyst coatings on radiators or other surfaces in such a way that the amount of ozone in the ambient air passing through such surfaces is reduced (CARB 1999).

Catalytic Air Pollution Control: Commercial Technology, by Ronald M. Heck and Robert J. Farrauto, with Suresh T. Gulati.
Copyright © 2009 John Wiley & Sons, Inc.

15.2   PREMAIR® CATALYST SYSTEMS

The PremAir® technology involves the application of a special catalyst coating to heat exchange surfaces such as the radiator of an automobile, an air conditioner condenser, or other heat-exchange surfaces encountered in stationary applications. Since the PremAir® catalyst is usually applied to a heat-exchange surface over which ambient air is already being moved or processed, no new equipment is needed. In general, the catalyst coating can be applied to existing surfaces, which can be plastic, ceramic, or metallic.

15.2.1   Mobile Applications

The hydrocarbons (HCs) and NO_x pollutants emitted from industrial and biogenic processes can react photochemically in the presence of sunlight to form ozone, the major component of smog. Therefore, the NO_x and UHC (unburned hydrocarbons) emissions from an automobile contribute to the ozone formation cycle. By equipping a vehicle with a radiator coated with an ozone-reducing catalyst, the ozone in the air can be removed catalytically through the following reaction.

$$2O_3 \rightarrow 3O_2 \tag{15.1}$$

As a result, the ozone removed by the catalyzed radiator can be related back to the tailpipe emissions through the photochemical ozone formation cycle (Okayama et al. 1999). In fact, atmospheric modeling calculations have been able to correlate vehicle tailpipe NMOG emissions with ozone removal by a catalyst-coated radiator (Greger et al. 1998; Hoke et al. 1999; CARB 1999). Since CARB has agreed to grant emissions credits for DOR technologies such as an ozone-destroying radiator catalyst, automakers now have an alternative tool to help meet specific vehicle emission standards (CARB 1999). This becomes particularly significant in the case of the super ultra-low-emission vehicle (SULEV), which must meet an emission standard of 0.01-g/mile NMOG hydrocarbon (see Chapter 7). Atmospheric modeling calculations have shown that DOR radiator-catalyst technology can achieve potential emission credits of ca. 0.01-g/mile equivalent hydrocarbons or 10 mg/mile (CARB 1999). Therefore, these UHC equivalent numbers are within the range of HC tailpipe emission regulations for SULEV vehicles (i.e., 10 mg/mile). The partial zero-emissions vehicle (PZEV) vehicle has stringent evaporative emission standards and the same tailpipe emissions as the SULEV vehicle. In the case of the PZEV, the equivalent HC removed by the catalyst-coated radiator can be applied to both tailpipe and evaporative emission standards.

Figure 15.1 shows the typical performance for ozone destruction over a catalyst-coated radiator as a function of flowrate (space velocity) and temperature. Note that a car radiator operating temperature is typically above 75 °C. The energy of activation for this catalyzed reaction under the radiator condi-

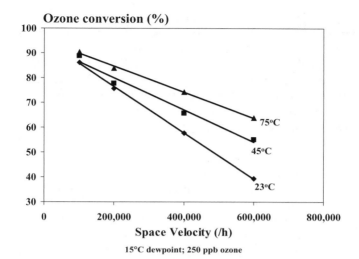

FIGURE 15.1. PremAir catalyst performance at various operating conditions.

tions of about $80\,°C$ is around $2\,kcal/g\text{-}mol$, which means the reaction is essentially bulk mass transfer controlled.

Several studies have been conducted on various catalyst formulations (Oh et al. 1998, Wu and Kelly 1998), but the most active and stable catalytic material is a MnO_2-containing catalyst (Hoke et al. 1999; 2001a; 2001b; 2006; Allen et al. 2003). On-road studies have been conducted with several vehicles under many driving conditions at different locations in Europe and the United States. In these evaluations, different vehicles were outfitted with catalyst-coated radiators. Additionally, analytical instrumentation and hardware were installed on the vehicles to measure ozone concentrations, radiator airflows, and radiator exhaust air temperature during on-road driving. On-road data were usually collected over a period of several weeks. The following results summarize the average performance measured on several vehicles during all driving conditions:

- Chevrolet Caprice
  78% average conversion
- Ford Contour
  80% average conversion
- Volvo S70
  47% average conversion
- Volvo S70 Turbo
  75% average conversion
- Volvo S90
  70% average conversion

Since the radiator is a geometric support having similar characteristics as a honeycomb, it can be characterized by a cell or fin density, depth, cross-sectional area, and volume. Also, radiators have louvers or punched slots along the channel length that increases turbulence and prevents the flow from becoming developed (i.e., no boundary layer). Therefore, by knowing these parameters, a correlation can be developed to predict performance using basic mass transfer principles established in Chapter 5. Some cars will be equipped with radiators that provide high ozone conversion (high-performance radiators), whereas others will be equipped with radiators that provide lower ozone conversion (low-performance radiators). In the future, the radiator design may be adjusted to maximize the performance.

Initial use of the PremAir® catalyst has occurred on Volvo vehicles primarily as a "green" image (New York Times 2000). Nissan also installed catalyst-coated radiators on their initial production of their Sentra CA PZEV vehicle (Nishizawa et al. 2000). None of these installations were used for credits to offset tailpipe and/or evaporative emissions, which is the prime purpose of this new technology. Another application of this technology for the "green" image was at Sunline Bus Company in Palm Springs, California. This particular fleet of buses was powered by natural gas (Hoke et al. 2001).

To date, seven car companies have used PremAir® catalysts, with some car makers using the technology to achieve PZEV certification credits from CARB. Since its launch, more than 3 million PremAir®-equipped cars have been sold in the United States, Europe, and Asia. The technology is featured on most Volvo cars and on certain BMW, Mercedes, and Hyundai models (BASF 2007).

15.2.2   On-Road, Long-Term Performance

Several studies have been conducted to evaluate the on-road performance of the PremAir® catalyst system (Greger et al. 1998; Peterson et al. 2000; Hoke et al. 1999). These studies have been conducted with several different vehicle models at several different worldwide locations and during all seasons of the year. As a result, the catalyzed radiators have been exposed to varied conditions. During these studies, greater than 2.5 million miles were accumulated by all test vehicles. Figures 15.2 and 15.3 show the results of one study completed in Sweden by the Volvo Car Corporation using taxi fleet vehicles. Mileage accumulation reached 120,000 miles (Peterson et al. 2000). The following conclusions can be drawn from these studies:

- Catalyst performance can be maintained for 120,000 miles.
- The level of catalyst-stable performance is a function of the radiator design.
- The level of catalyst performance is also a function of the catalyst formulation.

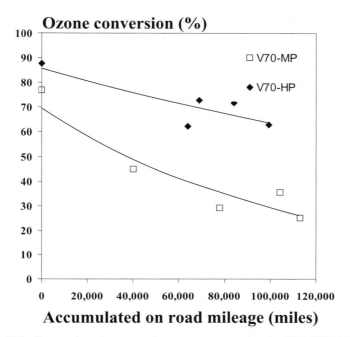

FIGURE 15.2. On-road performance for ozone conversion for V70-HP and V70-MP taxi radiators. Reprinted with permission, © 2000 Society of Automotive Engineers, Inc. (Peterson et al. 2000).

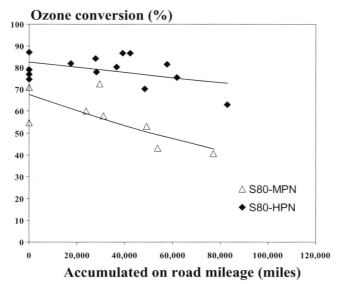

FIGURE 15.3. On-road performance for ozone conversion for S80-HPN and S80-MPN taxi radiators. Reprinted with permission, © 2000 Society of Automotive Engineers, Inc. (Peterson et al. 2000).

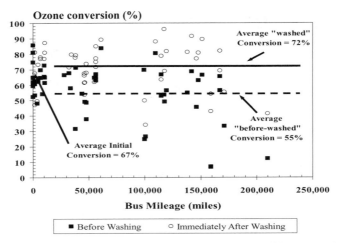

FIGURE 15.4. Idle ozone conversion for coated bus air conditioner condensers as a function of accumulated mileage (Hoke et al. 2001).

- Preliminary evidence indicates that the primary mode of decline in catalyst performance is the result of masking of the catalyst surface by microscopic airborne particulate matter (APM).

Another long-term on-road study was conducted for the bus application in Palm Springs, California. In this study, 33 transit buses in the Sunline fleet were equipped with radiators and condensers coated with a pre-production formulation of the catalyst beginning in June 1997. The purpose of this study was to evaluate the long-term durability characteristics of the catalyst coating. In addition, 15 of these buses were outfitted with sampling tubes and thermocouples mounted on the radiators and condensers to allow for periodic off-road catalyst performance testing. Figure 15.4 shows the performance on the coated condensers of the buses.

For both the radiators and the condensers, long-term mileage accumulation up to 200,000 miles per bus showed excellent coating durability and coating retention. In general, maintenance of catalytic activity was better for the coated condensers due to their protected location in a recessed well on the bus roof. A significant drop in catalyst activity over time was observed for the coated radiators although performance could be significantly improved with washing. The drop in performance was believed to result from large amounts of particulate contamination on the catalyst surface. Advanced coating technologies are under development to minimize the effect of particulate contamination on long-term catalyst performance.

15.2.3 Catalyst Deactivation

As part of the credit calculation for determining the hydrocarbon-reducing benefit of the catalyst-coated radiator (CARB 1999), the deactivation factor relating to end-of-life performance to fresh performance must be known. Detailed studies have been conducted in the laboratory and on the road to understand the possible mechanism for long-term changes in the catalyst performance. These studies have shown that there are many possible mechanisms for a decline in catalyst performance as measured by ozone conversion. Possible mechanisms include the following:

- Masking by particulate matter (PM_{10} and $PM_{2.5}$)
- Poisoning by reactive ambient gases (SO_x and NO_x)
- Poisoning by water-soluble salts (NaCl and $CaCl_2$)
- Masking by debris (mud, feathers, sticks, and so forth)
- Morphological changes to coating (crystallite growth and porosity loss)
- Coating loss (flaking)

The contribution from each possible mechanism has been quantified through on-road tests and subsequent catalyst analysis and activity tests. The major deactivation is due to deposits of airborne particulate matter (APM). This APM comes from the PM_{10} and $PM_{2.5}$ that are suspended in the atmosphere. The actual mode of deactivation is shown schematically in the Figure 15.5. The sources and types of APM have been identified by many studies of airborne particulates (EPA 1998). The APMs consist of particles containing the following elements: Ca, Zn, Al, Si, Fe, S, Cu, Cl, Ti, Mn, C, N, and K. The APM deposits on the surface and within the pores of the MnO_2-containing catalyst coating, thus reducing the BET surface area and porosity of the catalyst. The extent of

Fresh PremAir® Coating

Aged Coating

Normal deactivation
- buildup of APM contaminants
- penetration of APM contaminants
- masking of APM contaminants
- can be reversible

FIGURE 15.5. Schematic showing mode of deactivation of PremAir® catalyst.

deactivation is very sensitive to the total geometric surface area and depth of the vehicle radiator. Higher geometric surface area and deeper radiators that operate at lower space velocities have a design that promotes excellent durability and a much lower deactivation factor. New catalyst preparations have also been developed to minimize the effect of APM deposits on the lower geometric surface area (low performance) radiator designs (Hoke et al. 2001).

CARB has published guidelines that specify procedures for calculating the NMOG credit and certifying vehicles. This document also outlines requirements for catalyst durability, emission warranty, and on-board diagnostics (OBD). In calculating the allowable credit, coated radiator frontal surface area, airflow, ozone conversion, and deactivation factor (e.g., after 150,000-mile aging for PZEV vehicles) are all key parameters. Although one method for determining the long-term catalyst deactivation factor is to age a vehicle on-road for 150,000 miles, the cost and time to accomplish this make it an unattractive option. A lab-scale accelerated aging test that allows accurate prediction of catalyst performance after 150,000 miles of on-road driving has been developed (Heck et al. 2003).

Because APM is not localized to roadways, off-road accelerated aging of catalyst samples is possible. Stationary aging of catalyst samples under continuous ambient airflow is one possible method. Although this method exposes the catalyst to the same type of contaminants that occur on the roadway, relatively long exposure times are required to achieve sufficient deactivation. Although this process is an improvement compared with actual on-road aging, a shorter duration aging method is more desirable for certification purposes. The exposure of catalyst samples to synthetic particulate contamination in a lab-scale reactor is the preferred method for accelerated aging. Commercially available aerosol generators are well suited for this application since they can reproducibly generate highly concentrated mists of potential airborne and roadway contaminants. The method simulates real-world particulate contamination, and it differentiates the long-term performance characteristics of various radiator geometries and catalyst formulations. Validation of the method has been demonstrated by comparing accelerated aging results after short exposure times to actual 150,000-mile, on-road aging data (Heck et al. 2003).

15.2.4 Stationary Applications

In addition to mobile heat-exchange devices such as those installed on cars, trucks, and buses, the ozone-destroying catalyst technology is also applicable to stationary heat-exchange devices that come into contact with ambient air. Commercial and residential air conditioning (A/C) condensers are a particularly suited application since these devices process huge volumes of air and their peak operating period is during the summer months when ambient ozone levels are at their highest (Rudy et al. 1996; Hoke and Heck 2000). Ozone is destroyed by the catalyst coated onto the condenser, and the purified air is

TABLE 15.1. Ozone conversion depends on specific condenser design (Hoke and Heck 2000).

	Goodman	Carrier	Trane
Condenser depth (in)	1.25	0.75	2.75
Condenser fin density (fins per in)	19	16	14
Total condenser frontal surface area (ft^2)	11.9	9.4	31.1
Rated air flow (ft^3/min)	2,000	2,000	12,400
Ozone conversion (%)	77	66	88

then exhausted through the top of the A/C housing. Although A/C condensers typically operate at significantly lower temperatures than automobile radiators (e.g., 40 °C vs. 75 °C), the catalyst is sufficiently active to destroy ozone at high conversion levels. Again, in these instances, the ambient air is already passing through the heat transfer surface, which has high geometric surface area and good heat transfer properties and, hence, good mass transfer properties.

Similar to the vehicle application, the in-field performance of the catalyst technology in the air conditioning application is dependent on the specific design of the condenser coil. Table 15.1 illustrates fresh ozone destruction performance for several different types of air conditioning condensers.

Although the Goodman and Carrier are both 3-ton residential air conditioners, their different designs give rise to different measured ozone conversions. The Goodman condenser is thicker, it has a larger frontal surface area, it has a higher fin spacing, and it operates at a lower space velocity when compared with the Carrier unit. As a result, the measured ozone conversion of the Goodman is larger than that measured on the Carrier. The highest conversion was observed with the Trane air conditioner. This is a much larger 20-ton light commercial unit. Although the fin spacing is the smallest of the three condenser types, its low operating space velocity and its internal fin structure gives rise to excellent ozone destruction performance.

To determine the long-term efficacy of the catalyst technology in a residential A/C application, a commercially available, 3-ton Goodman air conditioner (model CK361B) was purchased from a local distributor, and the condenser coil was removed and spray coated with catalyst. The coated condenser was installed back into the air conditioner housing, and the unit was subsequently outfitted with Teflon sampling lines at several locations on the inlet and outlet faces of the condenser coil and on the exhaust air fan guard. This was done to measure ozone concentration before and after the condenser and subsequently to calculate the ozone conversion efficiency across the condenser coil. The unit was put into service in central New Jersey during July 1996, and ozone conversion was measured periodically over a 3-year period (summer 1996 through summer 1999) to determine long-term performance. Except for the first season, the air conditioner operated continuously (24 hr per day) from approximately

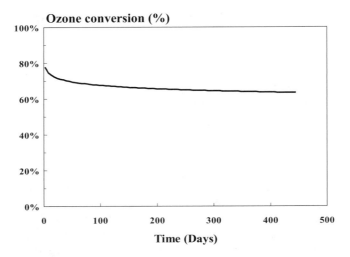

Goodman 3 ton residential A/C (Model CK361B); New Jersey unit
operated continuously during summer 1996, 1997, 1998, and 1999.

FIGURE 15.6. Initial results show good long-term performance for ozone conversion of A/C condensers (Hoke and Heck 2000).

mid-May through mid-October. As part of a typical annual maintenance program, the coil was treated with a mild detergent solution and then rinsed with tap water at the beginning of each summer season. Results are summarized in Figure 15.6. Even after three operating seasons, the catalyst has shown excellent long-term activity maintenance.

15.3 OTHER APPROACHES

Cleaning the ambient air has received attention in Japan for removing ambient air pollutants using photocatalyst. These catalysts rely on the properties of titania that is activated by sunlight to carry out chemical reactions. The applications have been directed to remove NO_x and odorous materials in the ambient air. In one application, photocatalysts are being sprayed on the road surface in Japan as a cement mixture. The cement incorporates a 15% titanium dioxide photocatalyst. This is designed to be used on road surfaces to decompose the nitrogen oxides emitted by motor vehicles. The catalysts destroyed 25% of the nitrogen oxides emitted in simulation trials (Japan Chemical Week 1999). Another application using a new type of concrete paving block has been developed and tested and has been shown to purify the air around roads by removing nitrous oxide from the atmosphere. The interlocking paving block contains titanium dioxide in its uppermost layer that can act as a photocatalyst. The catalyst, which is activated by sunlight and rain, promotes the oxidation of nitrous oxide to nitrate ions that are washed away by rain or neutralized by the alkalinity of the concrete. Tests indicate an 80% cut in pollution levels under nitrous oxide concentrations of 0.05–1.0 ppm and normal weather conditions. The blocks are permeable to water (European Chemical News 1998).

The use of titanium dioxide incorporated into an air filter to remove harmful gases through a photocatalytic reaction has been developed. Initially, the air filter will be applied to household situations, but the company is also expecting automotive applications. The gas purification capability of the new material is around three times that of conventional titanium dioxide. The filter consists of a honeycomb structure made from a paste of titanium dioxide and active carbon, plus a special oxidation co-catalyst that increases the speed of decomposition of pollutants, such as nitrogen oxides and formaldehyde (Japan Chemical Week 1998).

Photocatalytic coatings have been applied to tunnels to remove N_2O and CO. Reduction over 40% have been achieved (Focus on Catalysts 2007). The coating is a photochemical cement that is reactive in the presence of light (artificial or natural) and air. A similar type concept uses a concrete that is designed to remove automatically smog-forming compounds for buildings (Bigham 2007). The final product uses light to accelerate natural oxidation through the process known as photocatalysis to degrade NO_x and SO_2. A test section of treated concrete road was tested and demonstrated that levels of NO_x were reduced by as much as 70%.

This area of photocatalysis is in its infancy, and future studies are needed to prove the long-term efficacy of this approach.

REFERENCES

Allen, F., Blosser, P., Heck, R., Hoke, J., and Poles, T. "Catalytic material for treating pollutant-containing gases," U.S. Patent 6,586,359 (2003).

BASF. "Premair® ozone-eating technology featured on high end Ferrari supercar," http://www.investor.basf.com/en/presse/mitteilungen/automotive/pm.htm?pmid=2690&n=50&id=V00-AF2pABTbWbcp-H2, (2007).

Bigham, R. "10 top technologies for 2007," *Pollution Engineering* pp 20–21, (2007).

CARB. State of California Air Resources Board. "Manufacturer's advisory correspondence 9906a: Certification procedure for direct ozone reduction (DOR) technologies," (1999).

Dartt, C., and Davis, M. "Catalysis for environmentally benign processing," *Industrial and Engineering Chemistry Research* 33: 2887–2889 (1994).

EPA. "$PM_{2.5}$, A fine particle standard," *Environmental Protection Agency* 1: 36 (1998).

European Chemical News. 69(1831): 28 (1998).

Focus on Catalysts. "Beating the smog in Shangai," (2007).

Greger, L., Berequist, M., Gottberg, I., Wirmark, G., Heck, R., Hoke, J., Anderson, D., Rudy, W., and Adomaitis, J. "PremAir® catalyst system," SAE Paper 982728 (1998).

Heck, R., Hoke, J., and Allen, F. "PremAir catalyst system on-road durability—a mechanistic case study," 17th North American Catalysis Society Meeting, Toronto, Canada (2001).

Heck, R., Hoke, J., and Buelow, M. "PremAir® catalyst system: Accelerated aging method," 18th North American Catalysis Society Meeting, Cancun, Mexico (2003).

Hoke, J., White, D., anderson, D., Heck, R., and Rudy, W. "PremAir® catalyst system: Fleet test results—sunline transit agency," 94th Annual Meeting of Air and Waste Management Association Paper 1139 Orlando, FL (2001).

Hoke, J., Anderson, D., Heck, R., Poles, T., and Steger, J. "New technical approach for ambient pollution control—premair™ catalyst systems," SAE 960799 (1996).

Hoke, J., and Heck, R. "PremAir® catalyst systems—technology update," 93th Annual Meeting of Air and Waste Management Association, Paper 849 Salt Lake City, UT, (2000).

Hoke, J., Heck, R., and Poles, T. "PremAir® catalyst system—a new approach to cleaning the air," SAE 1999-01-3677 (1999).

Hoke, J., Heck, R., and Allen, F. "Method and device for cleaning the atmosphere," U.S. Patent 6,190,627, 2001.

Hoke, J., Novak, J., Steger, J., Poles, T., Quick, M., Heck, R., Hu, Z., Durilla, M., "Method and apparatus for treating the atmosphere," U.S. Patent 6,214,303, (2001).

Hoke, J., Allen, F., Blosser, P., Hu, Z., and Heck, R. "Vehicle having atmospheric treating surfaces," U.S. Patent 7,083,829, (2006).

Japan Chemical week, 1 Jan 1998, 39, 7, (1957).

Japan Chemical week, 8 Apr 1999, 40, 5, (2020).

New York Times. "Carmakers to put "smog-eating" radiator in some models", The New York Times, January 14, 2000, page F1.

Nishizawa, K., Momoshima, S., Koga, M., Tsuchida, H., and Yamamoto, S. "Development of new technologies targeting zero emissions for gasoline engines", SAE 2000-01-0890 (2000).

Oh, S., Sinkevitch, Baker, J., and Nichols, G. "Use of catalytic monoliths for on-road ozone destruction", SAE 980677 (1998).

Okayama, T., Hatcho, S., Yamamoto, Y., Kishi, N., Hayashi, T., Myers, T., and Whitten, G., "Estimation of ozone reduction catalyst effect at on-road condition", in 1999 Global Powertrain Congress, Stuttgart, Germany, Global Powertrain Congress, Ltd. (1999).

Peterson, M., Bergqvist, M., Gottberg, I., Hoke, J., and Heck, R. "PremAir® catalyst system—long-term on-road aging results," SAE paper 2000-01-2925, (2000).

Rudy, W., Ober, R., Durilla, M., Anderson, D., Hoke, J., and Heck, R. "A new approach for ambient pollution reduction—premAir™ catalyst systems field experience," 89th Annual Meeting of Air and Waste Management Association, Paper 96-MP4A.08 Nashville, TN, (1996).

Wu, M.-C., and Kelly, N. "Clean-air catalyst system for on-road applications I: Evaluation of potential catalysts," *Applied Catalysis B: Environmental* 18(1–2): 79–91 (1998).

Wu, M.-C., and Kelly N. "Clean-air catalyst system for on-road applications II: Mechanistic studies of pollutant removal," *Applied Catalysis B: Environmental* 18(1–2): 93–104 (1998).

CHAPTER 15 QUESTIONS

1. How is ground-level ozone formed, and why is it such an issue?
2. How can the removal of ground-level ozone be equated to a vehicle's tailpipe emissions?
3. What are the major hurdles in implementing a catalytic process for ambient air purification?

16 Fuel Cells and Hydrogen Generation

16.1 INTRODUCTION

This book describes the most modern catalytic technologies commercially in use to abate emissions generated by combustion of hydrocarbon-based fuels from stationary and mobile sources. This is referred to as exhaust emission control or aftertreatment or "end of pipe" with the catalyst positioned in the exhaust system to convert the unburned CO, hydrocarbon (HC), and NO_x. Throughout the world, emission standards are becoming ever more stringent, thus demanding more from the catalyst and engineering design of the engine and power plant. Compounding the problem, fossil fuel combustion generates large quantities of greenhouse gases such as CO_2. As pollutants are removed to extremely low levels and fossil fuel supplies are not viewed as sustainable, it is logical to investigate alternative power sources that will not require emission controls. The need for alternative energy sources that are free from the geopolitics and the supply issues associated with fossil fuels, but are environmentally friendly with a low carbon fingerprint, is now more than ever important for all of us (Rifkin 2002). We are now progressing toward a completely new technological era of the hydrogen economy and the hydrogen fuel cell (Farrauto 2005).

Only the basics of materials for the hydrogen economy will be addressed in this chapter, but advances are being made daily so breakthroughs can be expected. Clearly this entire book addresses the success catalysis has had in cleaning our environment of unwanted pollutants. Now it is time for catalysis to play a key role in developing clean and cost-effective technology for energy generation.

The authors have chosen to concentrate on the low-temperature proton exchange membrane (PEM) fuel cell, its electro-catalysts, and the technology being developed for generating hydrogen since these involve the greatest use of catalysts. Other fuel cells will also be discussed since they too will have a strong impact on the environment and energy for the future.

Catalytic Air Pollution Control: Commercial Technology, by Ronald M. Heck and Robert J. Farrauto, with Suresh T. Gulati.
Copyright © 2009 John Wiley & Sons, Inc.

16.1.1 A Comparison of Power Generation with Fossil Fuel-Powered Heat Engines and Fuel Cells

Why are fuel cells a promising alternative to fossil fuel combustion for transportation and stationary power generation currently used in conventional power generating plants? A hydrogen–oxygen (air) fuel cell directly converts its chemical energy to electricity and heat with efficiencies ranging from 40% to 80%. The lower efficiency is for electricity only, whereas the high efficiency is obtained by utilization of the heat. Figure 16.1 presents a cartoon comparison of the major power generation technologies compared with the fuel cell.

The top section shows the combustion of coal (or any other carbon-based fuel) and the consequent generation of heat and CO, HC, NO_x, particulates, and CO_2. The heat generated vaporizes high-pressure steam in a boiler that is directed to a steam turbine, which spins a magnet in the field of a metal coil and induces a current by the electromagnetic induction effect. The efficiency of a steam turbine ranges from 45% for a single-cycle power plant to about 55% for a combined-cycle power plant. The middle frame shows the familiar internal combustion engine, which combusts a fuel in a piston that drives the work stroke spinning the fly wheel turning an axle producing motion to the wheels of the vehicle. It also generates primary pollutants due to incomplete

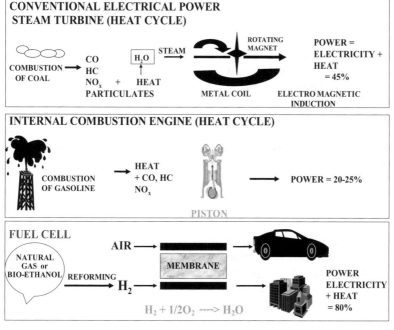

FIGURE 16.1. A comparison of power generation for a coal-fired power plant, the internal combustion engine, and a H_2–O_2 low-temperature fuel cell.

combustion (CO and HC) and nitrogen fixation (NO_x). For a modern diesel passenger car, the efficiency is no greater than 20–25%. In each of these heat engines, fossil fuel is combusted and pollutants are formed. The heat and high pressure are converted to mechanical energy to generate power. The thermodynamics of this heat engine cycle limits efficiency.

The fuel cell requires hydrogen; however, no infrastructure exists for its distribution. Thus, a convenient and local source is to convert a readily available infrastructure fuel to hydrogen. It must be repeated that the use of natural gas (mostly CH_4) is only a transitional fuel, whereas other natural sources of energy are being developed. Natural gas is readily available in pipelines in high-population areas of the world, is rich in H_2 (25% relative to gasoline, which is about 10–12% H_2), and is obtained from more politically stable regions of the world. The natural gas is to be converted to hydrogen via a hydrocarbon reforming process described later. Hydrogen would then be available for fuel cell vehicles at service stations. Alternatively, H_2 could be produced by reforming bio-fuel blends such as E85 (85% ethanol and 15% gasoline).

Returning to the bottom frame, it is observed that the fuel cell converts H_2 and O_2 to power without combustion. Since this operation has no mechanical steps, higher efficiencies than the traditional power generation systems are possible. The amount of greenhouse CO_2 produced during the reforming reactions, to be described later, is less than the heat engines described in frames 1 and 2.

With the current state of cost and power efficiency, some argue against the fuel cell. Some studies suggest, from an energy point of view, it is more feasible to use natural gas directly as a fuel for transportation as opposed to converting it to H_2 for a fuel cell-powered vehicle (Hetland and Mulder 2007). The authors indicate there is an advantage for H_2 regarding lower CO_2 emissions; however, it is important to recognize that combustion of any hydrocarbon-based fuel generates primary pollutants such as carbon monoxide, unburned hydrocarbons, oxides of nitrogen, and particulates. Clearly we need a strategy to transition to a hydrogen economy from one that uses fossil fuels that makes economic, political, and environmental sense. The key is ultimately to develop H_2 from renewable sources.

16.1.2 General Markets for Fuel Cells

Transportation applications: An ultimate goal for the hydrogen economy is to develop a fuel cell-powered vehicle with H_2 stored on-board safely and with a reasonable driving range approaching 350 miles. The only fuel cell that shows promise for this application is the low-temperature, Nafion®-based Proton Exchange Membrane (PEM) simply because of high-power density relative to the other fuel cells (Acres et al. 1997; St. Pierre and Wilkinson 2001; Appleby 1999). Simply stated, a 50–75-kW system can potentially fit under the hood of the vehicle. An additional advantage is its low operating temperature of 80 °C that will allow relative rapid start-up. For this reason, many car companies are aggressively developing and testing PEM fuel cell systems.

A major cost factor is the Pt electro-catalyst necessary for the fuel cell electrodes. The cost of such a system is extremely challenging with automobile manufacturers targeting no more than 15–20 g of Pt per vehicle depending on vehicle size. In comparison, a typical automobile has 2–5 g of precious metal in the catalytic converter. Some estimate that the current low-temperature vehicle fuel cell contains 50–60 g of Pt, so cost reduction without sacrificing reliability is a key issue. Since there will be no aftertreatment required, the catalytic converter can in theory be eliminated. A secondary automotive application is auxiliary power in which a fuel cell will generate power for the vehicle's lights, heating, and entertainment systems.

High-pressure tanks lined with carbon or glass fibers (Funck 2003) are one choice for liquid hydrogen storage on the vehicle (Wolf 2003; Cooper 2007). Storage of tanked hydrogen is practiced today in demonstration fuel cell vehicles as well as in hydrogen hybrid engines as practiced by BMW. Most automobile companies envision the local filling station generating high-pressure H_2 from natural gas (or possibly an ethanol-based fuel such as E85) that will charge either the high-pressure tank or hydrogen cartridge in your car (Ogden 2003). Also, much activity in chemical storage materials is composed of in a wide variety of metal hydrides and intermetallic compounds such as LiH. $LiNH_2$, $NaAlH_4$, MnH_2, NiH_2, PdH, MgH_2, and so on (Sandrock 2003; Cooper 2007) are claimed to absorb up to their own weight of H_2. Mazda and Hiroshima University have announced they have developed a new alloy of Pd–Mg permitting a low-temperature operation ($<100\,°C$) for filling and discharging the H_2, but no details are available. Carbon nanofibers have also received attention for storage (Chambers et al. 1998); however, other studies indicate that considerably more research and development is necessary before useful H_2 storage devices can be made of carbon fibers (Schutz 2000). Mueller et al. (2006) recently published data on the use of molecular organic framework materials (MOFs) with high H_2 storage capacity with rapid fill and discharge capability.

The reader is advised to check the websites of the auto companies continuously for updates on progress (check the websites for fuel cell vehicles). Frequently, there are bold new announcements of demonstrations from automobile companies of their plans for commercializing fuel cell vehicles, the most optimistic being 2012–2015 It is, however, not expected that they will penetrate the mass market before 2020.

Construction of H_2 fueling stations is slowly progressing all over the world in preparation for the introduction of commercially viable fuel cell vehicles (Nguyen et al. 2005). The reader is encouraged to search the Web for H_2 fueling stations. California has a particularly aggressive program (CaH_2Net), but there are many others throughout the world (Copper 2007). Clearly there are major economic and logistical challenges to making such an infrastructure available.

Stationary applications: One additional advantage is that some fuel cells lend themselves to a distributed model whereby power (electricity and hot water) can be generated in the home or commercial building independent of the existing centralized power station. The hydrogen can be produced from natural gas for which extensive infrastructures exist. Centralized power generating plants have large losses in transmission and are subject to brownouts and blackouts due to peak demands. Furthermore, they are subject to weather conditions and are expensive to transmit to homes or offices in remote locations. As new buildings are built, power companies must absorb the cost of adding new generating capacity. Clearly fuel cells have a place in today's energy hungry world provided they can be implemented with high reliability at competitive prices.

Stationary power can be met with different fuel cells depending on the power demands. For homes requiring less than about 5 Kw$_e$, the Nafion®-based PEM fuel cell system is the most popular for combined heat and power using natural gas, liquid petroleum gas (LPG), or kerosene as the reformate fuel. Some lower temperature solid oxide fuel cells are also be considered for home use. Large installations such as hospitals, schools, businesses, and centralized housing developments favor phosphoric acid, molten carbonate, or solid oxide systems.

Portable power applications: Portable power is likely to be the first major application for fuel cells primarily to replace or recharge batteries. The market is first-responder emergency as well as military and business communications, none of which are as cost sensitive as products for the mass market. Many conditions exist in which grid power is not available, and thus, fuel cell power offers an alternative to batteries. Many companies are developing fuel cell systems for portable applications based on either direct methanol systems in which the methanol is directly fed to the fuel cell anode and then electrochemically oxidized or reformed methanol fuels that deliver H_2 to the anode.

16.1.3   Types of Fuel Cells

Four basic fuel cells systems are available. The solid polymer electrolyte or the low-temperature PEM has an electrical efficiency of close to 40% (when used in a combined heat and power mode, the efficiency approaches 80%). The alkaline fuel cell with an electrical efficiency of 50–60% is not suitable for terrestrial applications due to its sensitivity to CO_2 in the atmosphere neutralizing the electrolyte. The electrical efficiency of a phosphoric acid (PAFC) fuel cell is 35–40%. The high-temperature fuels cells such as molten carbonate (MCFC) and solid oxide (SOFC) both have electrical efficiencies between 40% and 80% depending on the mode of operation and heat recovery (Copper 2007). Each has its own particular application in diverse markets. The basic technologies for all four will be only described in very simple terms in this text later. For more detail, the reader should consult the *Handbook of Fuel Cells* (Vielstich et al. 2003).

The first major uses of fuel cells came in the 1960s for the space program. The acid-based PEM fuel cell was briefly used in the Gemini space program but was short lived because of the slow kinetics of the reaction relative to the alkaline fuel cell. It is now, however, the primary choice for terrestrial applications (stationary, portable power, and vehicles) due to high current densities and tolerance to CO_2, which neutralizes the electrolyte in the alkaline fuel cell (Kordesch and Simader 1996).

Here alkaline electrolyte systems were used operating with liquefied H_2 and O_2 on-board. The use of phosphoric acid fuel cells followed in early 1990 by United Technologies (International Fuel cells and ONSI) who commercialized the PC25 with some systems up to 11 mw for large power plants such as that built by Tokyo Electric in 1993. Currently, there about 30 large power plants in Japan, the United States, and Europe using fuel cells. Approximately 250 units of 50–500-kW throughout the world are used for stationary applications such as schools, apartments, commercial building, and so on. So fuel cells have been commercialized but for principally niche markets. The desire now is to produce them broadly for the mass market.

The PEM fuel cell operates at 75 °C and is the number one choice for automobile companies for vehicle applications. The pioneering work of Dialmer-Benz Chrysler-Ballard demonstrated NECAR 1 in 1994 at 50 kW with a weight per kW ratio of 21 kg/kW (Hirschenhofer et al. 1998). NECAR 2 was demonstrated in 1996 at 6 kg/kW, NECAR 3 in 1997, and NECAR 4 in 2000. These vehicles were operated with an on-board methanol reformer to generate the H_2. On-board methanol is no longer considered viable for on-board vehicle applications given its toxicity and solubility in water. Many of these companies are no longer in existence specifically for fuel cell technology, but they represent a rich history of the pioneering work that has evolved to the fuel cell vehicles anticipated for the future.

A market expected before fuel cell vehicles are commercialized is that of residential or distributed power for homes. The concept is to reform fuels for which an infrastructure exists, such as natural gas, LPG, and/or kerosene, providing H_2 for the anode of the PEM fuel cell for home and business use. Companies such as Plug Power have pioneered in developing this technology for American homes and businesses delivering 5 Kw$_e$. In Japan, companies such as Tokyo Gas, Nippon Oil, Osaka Gas, Sanyo, Aisin, and Mitsubishi Heavy Industries have targeted 1-Kw$_e$ systems designed to provide combined hot water and electricity. An additional complication is their model to start and stop once per day adding additional complications to the materials and control strategies for both the fuel cell and the integrated fuel processor.

Molten carbonate fuel cells operate between 600 °C and 700 °C and are being demonstrated in many locations. FuelCell Energy demonstrated its Direct Fuel Cell® (DFC ®) 2-mW power plants in Santa Ana, California, in 1995. This same company, in cooperation with PPL Energy Plus, completed a demonstration in July 2000 of a 250-kW system for over 12,000 hr in Danbury, Connecticut, generating 1.9 million kW-hr of power. Additional demonstrations of 250-kW units will be going to Alabama (Mercedes-Benz), Louisiana, Asia, Germany, and 1 mW in Washington (Kings County) operating on digester gas in late 2002. Efficiencies of 50–60% are achieved by cogeneration (combined cycle) of electricity and heat. MTU is demonstrating a 250-kW cogenerating system in a hospital in Germany.

Solid oxide fuel cell systems, some of which operate around 1,000 °C, will likely be commercialized for centralized power generation above about 200 kW. In April 2000, Siemens-Westinghouse manufactured a natural gas-fueled, 220-kW hybrid SOFC and micro-turbine for use in Irvine, California. It provides electricity for 200 homes. The hot exhaust from the fuel cell drives the micro-turbine such that 55% efficiency is realized. Efficiencies approaching 70% are predicted. The anode fuel is CO and H_2 produced by partial oxidation and reforming of the hydrocarbon fuel conducted within or adjacent to the anode compartment. The anode is relatively thick, whereas the electrolyte and cathode are very thin. The cathode fuel is air. Their market is stationary 1–25-kW and 3–5-kW auxiliary power (i.e., air conditioning or music center) for vehicles. Ceres Power, in the United Kingdom, in concert with British utilities, is investigating a new SOFC system that operates at 500 °C for residential applications.

The need to replace batteries in cellular phones, portable laptop computers, digital equipment battery chargers, and so on strongly suggests the value of a fuel cell operated on direct or reformed liquid fuel such as methanol, which can easily be changed simply by replacing a small lightweight cartridge. Direct methanol fuel cell portable power devices are available but in limited use. Companies such as Ultracell, Dupont, Casio, MTI, Toshiba, SMART, and Samsung are actively engaged in research, so the reader is encouraged to check their websites periodically for updates on commercialization.

16.2   LOW-TEMPERATURE PEM FUEL CELL TECHNOLOGY

16.2.1   The Electrochemical Reactions for H_2 Fueled Systems

The fuel cell can be thought of as a Galvanic cell in which spontaneous oxidation of a species occurs at the anode and reduction of another species at the cathode. Figure 16.2 shows such a cell in which a zinc anode is spontaneously oxidized to zinc ions with electrons traveling through an external circuit to reduce the Au ions in the gold plating of a utensil (the cathode). The voltage driving the reaction is the spontaneous oxidation of zinc metal to zinc ions and Au ions spontaneously reducing to its metallic state.

$$\text{Anode}: \quad 3Zn - 6e^- \rightarrow 3Zn^{+2} \quad E^0 = +0.76\,V \qquad (16.1)$$

$$\text{Cathode}: \quad 2Au^{+3} + 6e^- \rightarrow 2Au \quad E^0 = +1.50\,V \qquad (16.2)$$

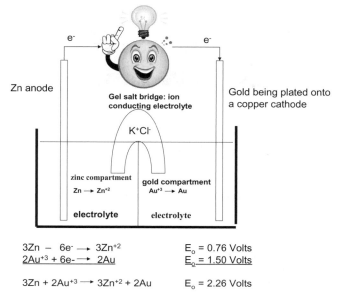

$3Zn - 6e^- \rightarrow 3Zn^{+2}$ $E_o = 0.76$ Volts
$2Au^{+3} + 6e- \rightarrow 2Au$ $E_o = 1.50$ Volts

$3Zn + 2Au^{+3} \rightarrow 3Zn^{+2} + 2Au$ $E_o = 2.26$ Volts

FIGURE 16.2. Galvanic cell of Zn and Au.

E_{cell}^o (Anode + Cathode single half-cell voltages) $E_{cell}^o = +2.26\,V$ (16.3)

The positive voltage (E_{cell}^o) is equated to the total cell free energy by $\Delta G_{cell}^o = -nFE_{cell}^0$, a negative value indicating a spontaneous reaction. Here n is the number of electrons transferred and the Faraday constant (F) is the number of coulombs per 1 mole of electrons.

The two half-cells are separated by an ion charge-carrying gel salt bridge that transfers cations (K^+ as shown in Figure 16.2) to maintain charge neutrality as Zn ions are generated at the anode and Au ions are reduced at the cathode. It also serves to minimize mixing of the Zn ions with the Au ions.

The H_2–O_2 fuel cell (Figure 16.3) operates by the electro-catalytic-oxidation of H_2 and reduction of O_2 to form H_2O, electricity, and heat (Voss 1999). It will continue to supply power provided H_2 and O_2 continue to be supplied and the electrocatalysts retain their activity.

Both anode and cathode reactions are catalyzed by Pt on carbon; however, their respective compositions are different.

$$H_2 \xrightarrow{\text{Pt/C}} 2H^+ + 2e^- \quad E^\circ = 0.00\,V \qquad (16.4)$$

$$O_2 + 4H^+ + 4e \xrightarrow{\text{Pt/C}} 2H_2O \quad E^\circ = 1.23\,V \qquad (16.5)$$

One undesirable side reaction is the reduction of O_2-forming hydrogen peroxide with a negative voltage that must be subtracted from the single cell voltage of O_2 reduction to H_2O (16.2):

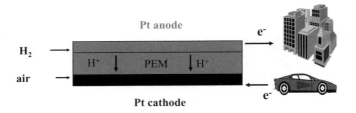

Anode:	$H_2 - 2e^- \longrightarrow 2H^+$
Cathode:	$1/2\,O_2 + 2H^+ + 2e^- \longrightarrow H_2O$
Net reaction :	$H_2 + 1/2\,O_2 \longrightarrow H_2O$

FIGURE 16.3. A single cell of the PEM fuel cell.

$$O_2 + H_2O + 2e^- \rightarrow HO_2^- + OH^- \quad E^\circ = -0.07\,V \tag{16.6}$$

$$\text{Net reaction} \quad H_2 + 1/2\,O_2 \rightarrow H_2O \quad E^\circ_{cell} = 1.16\,V \tag{16.7}$$

$$\Delta H^\circ = -57\,Kcal/mol\,(-242\,kJ/mol)$$

The net voltage of the cell (E_{cell}) is related to standard state cell voltage (E°_{cell}) and to the partial pressures of the reactants and products through the Nernst equation:

$$E_{cell} = E^\circ_{cell} + RT/2F \ln (P_{H_2})/(P_{H_2O}) + RT/2F \ln (P_{O_2})^{1/2} \tag{16.8}$$

The anode and cathode compartments are separated by a solid polymer membrane (analogous to the salt bridge in Figure 16.2). The membrane prevents the H_2 and O_2 from mixing and provides a path for H^+ to migrate from anode where they participate in the cathodic reaction. The membrane will be discussed in more detail later.

16.2.2 Mechanistic Principles of the PEM Fuel Cell

The H_2 is first dissociatively chemisorbed onto the Pt electro-catalyst followed by electro-catalytic oxidation to protons and electrons. This is a relatively easy and fast reaction. The reduction of O_2 at the Pt/C cathode is the slower of the two reactions and controls the rate and power output of the fuel cell (Gottesfeld and Zawodzinski 1998) at low current densities. Thus, the reaction is initially electrochemically controlled, which is favored by higher Pt loadings. A corrosion-resistant carbon is also required given the potentials experienced

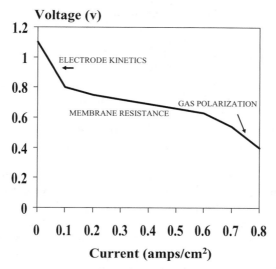

FIGURE 16.4. Voltage-current profile for the PEM fuel cell. The area under the curve represents the power output.

at the cathode. The rate-limiting steps, however, change with increasing current drawn from the cell as discussed below.

Figure 16.4 shows a voltage-current profile for a typical low-temperature PEM fuel cell. The open-circuit voltage ($E°$) is the theoretical value of about 1.16 V. When current is drawn, there is a drop in voltage and power (power = voltage × current) controlled by the slow O_2 reduction at the cathode. As current is further increased, the voltage-current output more slowly decreases caused by the transition of the rate-limiting step to H^+ migration through the membrane. One can envision this caused by increased H^+ traffic through the membrane. Finally, as the current required further increases, the output drops considerably due to polarization or lack of sufficient supply of gaseous O_2 to the cathode due to buildup of the water layer at the surface of the electrode. The O_2 (air) must undergo bulk mass transfer diffusion through the H_2O layer generated at the cathode. Sweeping away the water by the incoming air is a critical issue that must be engineered into the air input flow field. Thus, there are three distinct steps in the fuel cell operation, each of which can limit the rate of reaction analogous to those in heterogeneous gas phase reactions.

A key factor to note is that the fuel cell converts chemical energy directly to electrical energy and heat with no mechanical steps involved and free of thermodynamic inefficiencies associated with traditional heat cycles in power generation. For this and its environmental benefits, there continues to be tremendous interest in developing cost-effective fuel cell systems as an alternative technology for conventional fossil fuel-powered energy generation.

16.2.3   The Membrane Electrode Assembly (MEA)

The electrodes consist of highly dispersed Pt (20–40 Å (2–4 mm)) deposited (about 30 w%) on nonporous conductive carbon powders (300 Å (30 mm) in diameter). The layer of electro-catalyst about 50 um thick is admixed with an optimized amount of a Nafion® solution to enhance conductivity (Raistrick 1986). It is deposited onto the surface of the membrane (50–175 um) by spraying, painting, or filtration. The Pt loading is about $0.25\,mg/cm^2$, although it is desirable to reduce it to less than $0.1\,mg/cm^2$. The electrodes are hot pressed onto each side (anode and cathode) of the membrane to ensure intimate contact.

Pt on carbon is currently the only viable active electro-catalyst for both anode and cathode for hydrogen-oxygen acid electrolyte fuel cells (Ralph 1999; Gasteiger et al. 2005). The reduction of O_2 on the cathode is a slow step so about two to three times as much Pt is usually used than in the anode. The cathode must use a carbon resistant to corrosion especially on air exposure open-circuit conditions (system off) where the electrode potential exceeds about 1.1 V. Ruthenium is sometimes present in the Pt anode when the CO levels in the H_2 exceed about 10 ppm, but cleaner H_2 from reformers will eliminate the need for it. Direct methanol fuel cells require Ru alloyed with the Pt in the anode to broaden the useful voltage-current range.

The anode and cathode gases are dispersed through gas diffusion layers (GDLs) positioned on top of each electrode. The diffusion layer is also permeable to allow the cathode product water to escape. The GDL is composed of electrically conductive carbon cloth (300–400 um) woven from carbon fibers that are melt coated with Teflon (40–70%) to render them hydrophobic to prevent flooding by water. The main channels must be kept open for gas permeability (Paganin et al. 1996). The combination of GDL, electrodes, and solid polymer electrolyte is called the membrane electrode assembly or MEA.

Electrically conductive nonporous graphite bipolar plates have grooved microchannels on their surface to allow the gases to be delivered uniformly to their respective GDLs (Tonkovich et al. 1999). Also, polymer-based, bipolar plates are being developed (Chemical and Engineering News 2007). Sandwiched between these conductive plates is the single-cell MEA with an open-circuit voltage output of about 1.16 V. These are stacked in series to increase the power or voltage output. For this reason, the opposite side of the bipolar graphite plate is also grooved to permit the other reactant gas to flow as shown in the Figure 16.5.

16.2.4   The solid polymer membrane

The most well-known solid polymer electrolyte used is polyperfluorosulfonic acid (PFSA) Nafion® (Dupont trade name) developed in 1960 by Dupont (Mauritz and Moore 2004). It has a

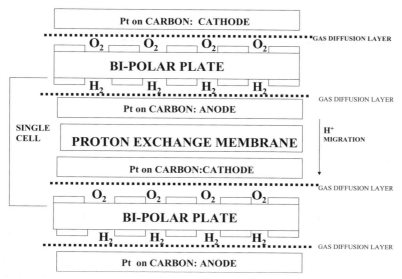

FIGURE 16.5. A PEM single cell arranged in a "stack." Each cell is separated by a electrically conductive impermeable bipolar plate that serves as a gas manifold for the cells connected in series.

$$CF_2 = CFOCF_2CFO(CF_3)CF_2CF_2SO_3^-H^+$$

hydrophobic Teflon® backbone with side chains of strongly acidic–$SO_3^-H^+$ that dissociate forming hydronium ions, H_3O^+. The thickness of the membrane varies between 50 and 175 microns. The membrane is conductive to protons provided sufficient moisture is present to permit ionization of the sulfonic acid group but is impermeable to gases (Grot 1989; Singh and Shaki 1998). Inverted spherical micelles are formed due to the interface between the hydrophobic fluorocarbon structure and the hydronium ions forming channels 10–20 Å (1–2 mm) in size. Through these channels, ion transport occurs, which produces the ion conductivity of the membrane. This model is known as the "cluster network," and it was proposed by Hsu and Gierke (1982).

The membrane swells with water uptake and has high conductivity at a humidity close to 100% at 80 °C (Summer et al. 1998). Such a low temperature is attractive for rapid electrochemical start-up for transportation and residential fuel cell applications, but low temperatures make the anode more susceptible to poisons such as CO always present in reformate (Gottesfeld and Pafford 1988). At low humidity or higher temperatures, the membrane acts as an insulator (Stone et al. 1996; Malhotra and Datta 1997; Thampan et al. 2000). Because the membrane has a low permeability of hydrogen and oxygen, it prevents their mixing across its interface (Gottesfeld and Zawodzinski 1998).

It is used extensively as the membrane for the chlor-alkali industry (Clapper 1980) where Cl ions are oxidized to Cl_2 and water is reduced to hydroxide at the cathode.

Alternative fluorine-containing membranes are also being developed by Dupont (Doyle and Rajendran 2003), Aciplex-S by Asahi Glass (Nakao and Yoshitake 2003), Gore, 3M, and Ballard. Nonfluorine membranes are also being developed (Kreuer 2003). A difference between membranes is related to the number of CF_2 groups (i.e., n) in the backbone of the solid polymer. A comparison of cell power outputs associated with proton conductivity can be found (Kordesch and Simader 1996; Jones and Roziere 2003; Lin et al. 2003).

Membranes based on polybenzimidazole (PBI) that operate at temperatures up to 200 °C have also been developed (Wainright et al. 2003; Wainright et al. 1995). Phosphoric acid is used as the ionically electrically conducting media as opposed to sulfonic acid groups in Nafion®. Unlike F-containing polymer membranes, they require no humidification and operate at temperatures above 180 °C. PEMEAS (now BASF) has developed a family of new higher temperature membranes also based on PBI under the trade name Celtec P 1000 that are commercially available. The performance of these membranes has been reported (Xiao et al. 2005a, 2005b). Because of their higher temperature operation, the Pt anodes can tolerate more CO than in low-temperatures systems. If successful, this can have far reaching, positive consequences in simplifying the reformer since the water gas shift reactor (14.7.3 water gas shift) can be reduced in size and the CO clean-up reactor (16.7.4.PROX) can be eliminated.

16.2.5 PEM Fuel Cells Based on Direct Methanol (DMFC)

This system also uses PEM technology but is differentiated from the hydrogen fueled system in that it operates directly on methanol with no need to reform the fuel to H_2 (Metkemeijer and Achard 1994; Doyle and Rajendran 2003). The operating temperature is 40–80 °C.

$$\text{ANODE:}\quad CH_3OH + H_2O \rightarrow CO_2 + 6H + 6e^- \quad E_o = 0.029\,V \quad (16.9)$$

$$\text{CATHODE:}\quad 1\tfrac{1}{2}O_2 + 6H^+ + 6e^- \rightarrow 3H_2O \quad E_o = 1.229 \quad (16.10)$$

$$\text{NET:}\quad CH_3OH + 1\tfrac{1}{2}O_2 \rightarrow CO_2 + 2H_2O \quad E_o = 1.26\,V \quad (16.11)$$

The electrolyte is an acidic solid polymer (i.e., Nafion®), but it must be impermeable so the methanol in the anode does not cross over to the cathode. Permeability or methanol cross-over remains a problem minimizing the power output. Both electrodes contain large amounts of expensive Pt dispersed on conductive carbons, but the anode is a PtRu alloy on carbon to minimize its deactivation caused by partially oxidized products such as aldehydes, carboxylic acids, and so on that results in poisoning of the reaction. The electrochemical oxidation of methanol involves a six-electron transfer. The rate-limiting

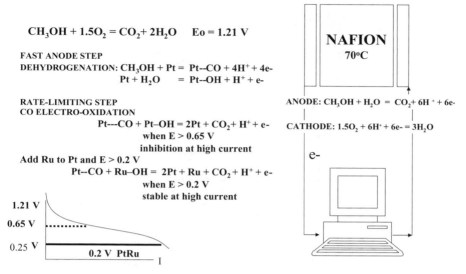

FIGURE 16.6. Cartoon showing the DMFC fuel cell system.

step for Pt only is reaction of adsorbed CO on Pt reacting with adsorbed OH. A voltage greater than about 0.8 V is needed to overcome this step, but this limits the current to low values. The presence of Ru, alloyed to the Pt, catalyzes the adsorbed CO and OH reaction extending the voltage to 0.25 allowing higher current densities. A cartoon is shown in Figure 16.6 for a DMFC system.

In addition to the methanol cross-over problem, the retention of Ru in the alloy, which leaches into the acidic solution, remains a problem. Evidence shows the Ru migrates through the membrane onto the cathode creating a mixed potential that reduces the effectiveness of the cathode.

Commercial products from SMART and Dupont are available for specialized low-power output applications.

16.3   THE IDEAL HYDROGEN ECONOMY

We are on the road to a hydrogen economy. Success will free us from substantial dependence on petroleum-based fossil fuels, will render combustion for power generation no longer necessary, and thus will make obsolete the need for primary pollution abatement of CO, HC, NO_x, and particulates. Agrawal et al. (2007) envisions a sustainable fuel economy in what is referred to as H_2CAR. His model assumes all carbon fuel will be supplied by biomass with H_2 combusted to gasify biomass. The H_2 will be generated from nuclear energy or solar photovoltaic energy electrolyzing water. The gasifier product will be primarily synthesis gas (H_2 and CO) used to produce diesel-type transporta-

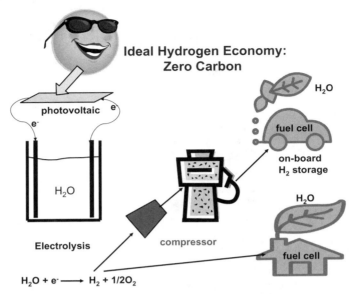

FIGURE 16.7. Ideal H_2 economy with the sun providing energy for a photovoltaic device generating sufficient voltage to electrolyze water to H_2 and O_2.

tion fuels via catalytic Fischer–Tropsch processing. The CO_2 generated is recycled to the gasifier where it participates in reverse water gas shift generating more CO. This is an excellent example of an ideal hydrogen economy that frees us from fossil fuel with no primary pollutants nor greenhouse gas emissions.

Over time we will slowly see the replacement of heat engines that require fossil fuels with hydrogen or renewable fuels. We are talking about an ideal hydrogen economy that can operate a fuel cell. This idealistic description is represented by the cartoon in Figure 16.7.

Here the sun is the source of energy generating a voltage through a photovoltaic device sufficient to electrolyze water. The released H_2 is fed to the anode of a fuel cell cleanly generating heat and electricity with the only product being water.

16.4   CONVENTIONAL HYDROGEN GENERATION

16.4.1   Hydrogen Production in the Chemical Industry

One major issue is the fuel and process to be used for generating H_2 for the PEM fuel cell anode. The most important choices are fuels for which an infrastructure exists, such as gasoline, diesel, natural gas, LPG, and kerosene. Therefore, it is necessary to convert readily available fuels to H_2.

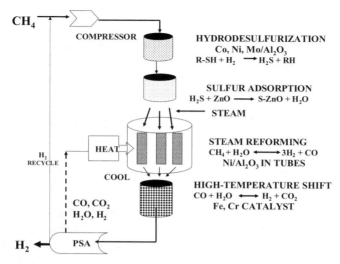

FIGURE 16.8. Traditional unit operations for the generation of H_2 for chemical applications (gas compositions are dry) with pressure swing adsorption.

Producing H_2 from hydrocarbons is currently practiced in the chemical industry (Bartholomew and Farrauto 2006; Rostrup-Nielsen 1993; Armor 1999; Brown 2001a) for production of ammonia and alcohol under steady-state conditions with carefully controlled catalytic unit operations. A schematic of the major unit operations is shown in Figure 16.8.

The first step in the chemical production of pure H_2 is desulfurization because sulfur compounds are poisons to all downstream catalysts. The hydrodesulfurization process is carried out at pressures exceeding 500 psig and 500 °C using Co, Mo/Al_2O_3 particulate catalysts shown in reaction (16.12). Obviously this step is not required for methanol but would be necessary for petroleum-based fuels.

$$HC\text{-}S + H_2 \rightarrow H_2S + HC \tag{16.12}$$

The H_2S produced is removed by adsorption [see Eq. (16.13)] on ZnO extrudates at about 400 °C.

$$H_2S + ZnO \rightarrow ZnS + H_2O \tag{16.13}$$

The next step is nickel (Ni/Al_2O_3) catalyzed steam reforming [see Eq. (16.14)] that is highly endothermic and requires high-energy input (inlet temperatures exceed 800 °C) depending on the fuel. The reaction below shows the stream reforming of methane, the major component in natural gas, as the most common fuel used. Conversions are favored by thermodynamics at high temperature and steam, but materials of construction dictate the operation of the reformer.

$$CH_4 + H_2O \leftrightarrow CO + 3H_2 \quad \Delta H = 49\,Kcal/mole \qquad (16.14)$$

The product gas exiting from the reformer contains about 10–12% CO (dry gas), which is fed to a high-temperature water gas shift (WGS) reactor [reaction (16.15)]. This slightly exothermic reaction is thermodynamically favored at low temperatures where the kinetics are slow. For this reason, WGS is conducted as a two-stage process. The first stage uses an extruded Fe, Cr, Al at about 350 °C, which converts about 80% of the CO and enriches the stream by an equivalent amount of H_2.

$$CO + H_2O \leftrightarrow H_2 + CO_2 \quad \Delta H = -10\,Kcal/mole \qquad (16.15)$$

The exit gas is cooled to below 200 °C, where a Cu, Zn Al_2O_3 composite extruded catalyst is used to reduce the CO well below 0.5% CO.

The remaining CO, if present, poisons downstream catalysts, so it is removed by pressure swing adsorption or methanation [Eq. (16.16)] over Ni- or Ru-based catalysts at about 250 °C.

$$CO + 3H_2 \rightarrow CH_4 + H_2O \quad \Delta H = -49.3\,Kcal/mole \qquad (16.16)$$

Weight, size, and transient operations are not critical for H_2 plants, but they are important for fuel cell systems that will be present in the home, building, portable digital device, or under the hood of a vehicle. Furthermore, safety issues (given the consumer use of the fuel cell) cannot in any way be compromised. Therefore, it is necessary to review the process details carefully and to determine their applicability for the fuel cell system. What will become obvious is that it is not possible simply to decrease the size of a traditional H_2 plant and to apply it a fuel cell system. Thus, new catalysts and process conditions along with considerable catalytic engineering must be found to satisfy this demanding application. Listed below are some limitations of the traditional H_2 plant as applied to the fuel cell system.

1) Desulfurization to the ppb level is required to protect the highly sulfur-sensitive Ni steam-reforming catalyst and downstream catalysts from being poisoned. The traditional hydrodesulfurization process is practiced at pressures and temperatures in excess of that which is practical for most fuel cell systems. Furthermore, recycle of H_2 to the front end of the process adds complication.
2) The steam-reforming process requires complicated engineering to provide sufficient heat to drive the endothermic reaction at appreciable rates.
3) The active Ni steam-reforming catalysts are highly pyrophoric (or self-igniting) upon exposure to air as indicated in [Eq. (16.17)].

$$Ni + O_2 \rightarrow NiO + Heat \qquad (16.17)$$

This condition is a major safety concern since the system may be subject to air exposure as a consequence of an accident or by an untrained consumer attempting to perform maintenance on his new power generator.

4) The two-shift catalysts, Fe, Cr, Al and Cu, Zn, Al are extremely pyrophoric when activated (reduced), and therefore, safety from runaway heat generation and fires cannot be ensured upon air exposure.

5) Pressure swing adsorption is the preferred technology for producing H_2 for chemical synthesis.

16.5 HYDROGEN GENERATION FROM NATURAL GAS FOR PEM FUEL CELLS

16.5.1 Organo-Sulfur Removal for Gaseous and Liquid Fuels

Removing sulfur for fuel cell applications will not typically use hydrodesulfurization (HDS) as performed in large-scale, traditional hydrogen generation processes given the absence of pressure as required for HDS. The technology of choice is adsorption. For natural gas, up to 10 ppm of odorants such as mercaptans, di-sulfides, and tetrahydothiophene (THT) are intentionally added for safety purposes. Consequently, technologies for their removal and the establishment of maintenance schedules for replacement or environmentally acceptable regeneration procedures of the traps need to be developed. Zeolites, activated carbon, and high-surface-area amorphous materials are the primary adsorbent candidates for desulfurizing natural gas provided they can function close to ambient conditions of temperature and pressure (Satokawa et al. 2005; Farrauto et al. 2003; Dicks 1996). The composition of hydrocarbon and sulfur compounds of natural gas and LPG depends on the location. For example, in the United States, mercaptans and di-sulfides are frequently used as odorants, whereas in Europe, THT is added to the gas, which may already contain varying amounts of COS and H_2S. LPG varies considerably in hydrocarbon composition and sulfur content. It is not unusual for LPG to contain propylene that competes for adsorption sites with the many different organo-sulfur compounds present. Adsorbents can be designed to remove 99% of the sulfur, but this still will leave 100–200 ppb of sulfur in the gas stream. New methods to reduce this to <10 ppb will likely be required to ensure an adequate life for all fuel processing and fuel cell catalysts.

The amount of organo-sulfur compounds that will be present in the near future in liquid fuels is still an open question being debated by oil companies, and it is likely to be mandated by government regulations. It is likely that sulfur levels in liquid fuels will be reduced further to less than 30 ppm; however, even at these greatly reduced levels, their removal to acceptable levels for the processor and fuel cell is still challenging. Liquid fuels such as gasoline, diesel, heating oil, and so on that contain large molecular weight organo-sulfur compounds may require adsorbents as well as new technologies (Maldonado et al. 2005; Ma et al. 2005; Ng et al. 2005). Naturally this problem does not exist for sulfur-free methanol; however, the lack of an adequate distribution or infrastructure and toxicity issues may limit the widespread use of methanol. Most major automobile companies are investigating liquid fuel and have eliminated their investigations of methanol.

16.5.2 Reforming

16.5.2.1 Steam Reforming. Primary steam reforming is the most efficient process for H_2 generation because it maximizes the amount of H_2 produced. For some applications, the higher yield of H_2 achieved by direct reforming is critical for efficiency of the fuel cell. It is commercially practiced with a tubular reactor composed of a packed bed of Ni pellets surrounded by combustion gases to provide the endothermic heat of reaction.

16.5.2.2 Autothermal Reforming. An alternative technology to steam forming, especially for fuels more complicated than natural gas such as LPG or liquid fuels, is autothermal reforming (ATR), which was first patented in 1985 (Hwang et al. 1985) and demonstrated for natural gas by Hochmuth (1992). In the latter case, only monoliths were used as catalyst supports. Recent advances have been made for natural gas, LPG, diesel, and gasoline (Cuzens et al. 2000; Hwang and Farrauto 2002; 2005; Farrauto et al. 2003).

ATR combines an exothermic catalytic partial oxidation (CPO) [reaction (16.18)] with an endothermic catalytic steam-reforming (SR) [see Eq. (16.19)] reaction to produce a hydrogen- and carbon-monoxide-rich product. Relative to conventional steam reforming, it reduces the size of the reactor and allows for a more rapid response to the transient operation. The example below illustrates the processing of natural gas:

$$CH_4 + 1/2\,O_2 \rightarrow 2H_2 + CO \tag{16.18}$$

$$CH_4 + H_2O \rightarrow 3H_2 + CO \tag{16.19}$$

The key is to limit the amount of O_2 to minimize the amount of partial oxidation and to maximize the amount of steam reforming since the latter generates the $3H_2$ and CO.

The balance between the heat generated and that adsorbed by the endothermic steam-reforming reactions is critical for adiabatic operation. This process simplifies the heat transfer problems typical of the traditional endothermic steam-reforming process. The catalyst technologies use a precious metal catalyst for the CPO and another catalyst specifically designed for maximum steam reforming (Hwang et al. 1985; Hwang and Farrauto 2002; 2005). The proper combination of precious metals are tolerant to sulfur and convert organo-sulfur compounds to H_2S, which can be adsorbed downstream using conventional ZnO. This advantage is important since Ni catalysts are intolerant to sulfur. Other catalysts have been reported (Krumpelt et al. 1999; Gray and Petch 2000; Wieland, 2001).

A significant advantage of ATR over traditional steam reforming is a more simplistic reactor design. Heat is generated internally, eliminating the need for external heat transfer. The penalty is a lower H_2 yield because of the addition of air, which oxidizes some fuel and dilutes the product with N_2.

16.5.3 Water Gas Shift

The product gas from a natural gas steam reformer (which depends on the $H_2O/$C ratio and the outlet temperature) is typically about 8% CO + 50% H_2 + 27% H_2O + 15% CO_2. It is cooled to 350 °C where in the high temperature shift reaction [Eq. (16.15)], the CO is reduced further to about 2–3%. This is usually sufficient for a fueling station that operates at a pressure exceeding about 8–10 atmospheres. In pressure swing adsorption (PSA), the reformate enters a chamber where all the molecules are condensed in the adsorbent except for about 90% of the H_2, which is now high purity and suitable for further use in a fuel cell or to be stored for a fuel cell vehicle. The pressure is released vaporizing the tail gas, which has sufficient heating value that it can be catalytically combusted providing the heat for the steam-reforming reaction. For fueling station applications, it is preferred to use a precious metal on a monolith catalyst that operates about five to ten times the space velocity of a traditional Cu, Zn, Al bed (Farrauto et al. 2007) with virtually no pressure drop and high attrition resistance without air and liquid water sensitivity.

When the hydrogen is to be used in a PEM fuel cell that operates at atmospheric pressure, a second WGS reactor operating at about 180 °C is required to reduce the CO to less than 1% CO. The shift reactions are thermodynamically favored at low temperatures, but the kinetics are so slow that large volumes (i.e., low space velocities <2000 h^{-1}) are needed when using Cu-containing particulates (Ladebeck and Wagner 2003; Li et al. 2000). Precious metal monoliths operate at 5,000–10,000 h^{-1} (Farrauto et al. 2007).

$$CO + H_2O \leftrightarrow H_2 + CO_2 + HEAT \qquad (16.15)$$

Another critical issue is the inherent self-heating properties of the Fe, Cr high-temperature shift (HTS) and the Cu, Zn low-temperature shift (LTS) catalysts when in their active states. Exposure to air as a result of an accident or a maintenance mistake will render the system completely unsafe since an exotherm of up to 650 °C are typical. Not surprisingly, this would be a catastrophic event for transportation applications. Furthermore, because of the exothermic nature of the reduction of the commercial catalyst and its sensitivity to sintering, the activation must be carried out with very slow heating and careful temperature control, which is time consuming, dangerous, and therefore impractical in the field. Discharging a pyrophoric catalyst from the processor also requires careful passivation with dilute air to avoid excessive and unsafe temperatures.

16.5.4 Preferential Oxidation of CO (PROX)

The PROX reactor is used as the final step in fuel processing of a hydrocarbon to rid the reformate of CO, which poisons the anode of a fuel cell operating <100 °C. Excessive amounts of CO adsorb preferentially on the Pt anode and block the sites. It is not necessary for a PSA system, which rejects CO, nor for a fuel cell, which uses a membrane (i.e., PBI), that operates above 180 °C where CO no longer adsorbs on Pt.

The principal technology is to use a highly selective catalyst that will catalyze the oxidation of the CO without oxidizing excessive amounts of the H_2 or methanating the CO or CO_2 present in the reformate. This is challenging since the CO is typically no more than 0.2–0.5 Vol. % compared with 50–80% H_2 (dry gas basis). Controlled amounts of air are injected into the stream.

$$CO + 1/2 O_2 \rightarrow CO_2 \quad \Delta H = -67.6 \, Kcal/mole \qquad (16.20)$$

$$H_2 + 1/2 O_2 \rightarrow H_2O \quad \Delta H = -57.8 \, Kcal/mole \qquad (16.21)$$

Several particulate-supported catalysts have been studied for this reaction, including precious metals (Oh and Sinkevitch 1993; Brown et al. 1960; Cohn 1965; Bonacci et al. 1980; Kahlich et al. 1997; Korotkikh and Farrauto 2000; Liu et al. 2003). Interestingly Au has shown reasonable activity and selectivity for this reaction (Bethe and Kung 2000; Sanchez et al. 1997; Kahlich et al. 1999; Teng et al. 1999). There are also reports of the advantages of CeO_2 as a support for PROX reactions (Liu and Stephanopoulis 1995; Avgouropoulos et al. 2001), but to date, no gold catalysts have been commercialized due to lack of stability.

New PROX publications using ceramic monoliths (Korotkikh and Farrauto 2000; Liu et al. 2003; Shore and Farrauto 2003; Roberts et al. 2003), foams (Chin et al. 2006), and heat exchangers have appeared (Farrauto et al. 2007).

16.5.5 Anode Tail Gas Oxidizer

Approximately 5–10% of the inlet H_2 to the anode is unreacted and exits the fuel cell. Air is injected into the stream, and the H_2 is catalytically oxidized. The heat of reaction is sometimes recovered for use within the processor. Typically the catalyst is designed to oxidize H_2 but may also function as a start-up, so it must also oxidize the hydrocarbon feed. It is often deposited on a monolith or a heat exchanger when heat recovery is necessary.

Fuel processors integrated to low-temperature fuel cells will operate at atmospheric pressure with typically zeolite-based particulates to remove the

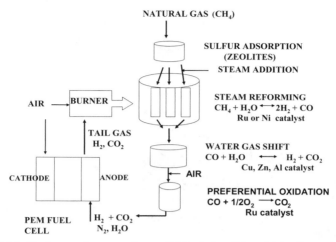

FIGURE 16.9. Reformer integrated to a PEM fuel cell in which all particulate catalysts are used.

odor-bearing sulfur compounds present in natural gas. A typical design is shown in Figure 16.9. After hydrodesulfurization, the natural gas is converted to a H_2- and CO-rich gas by the catalytic steam reforming usually with a Ni- or Ru-based particulate catalyst. The CO is then processed further via water gas shift using a Cu, Zn, Al particulate catalyst to more H_2 and CO_2. The CO is generally reduced to less than 0.5%, which must further be reduced to avoid excessive poisoning of the Pt anode of the Nafion®-based MEA fuel cell. The CO can be reduced with a Ru-based particulate catalyst by the PROX reaction, which reduces the CO to less than 10 ppm, which is of sufficient quality for the fuel cell. The anode system is designed to permit utilization of 85–90% of the H_2, the residual of which is combusted in a burner to generate the heat necessary for the endothermic steam-reforming reaction.

Farrauto et al. (2007) have commercialized catalyst technology for steam-reforming natural gas in which the process side of a heat exchanger is washcoated with a precious metal on stabilized Al_2O_3, whereas the opposite side is washcoated with a precious metal on stabilized Al_2O_3 catalyst, which combusts tail gas from a PSA to provide the heat for the endothermic steam-reforming reaction. This design minimizes heat transfer resistance, allowing space velocities up to ten times that of a packed bed translating to significant decreases in volume, lower pressure drop, and enhanced mechanical stability. The technology is currently commercialized for hydrogen service stations on the hydrogen highway. This is shown as Figure 16.10.

16.5.6   PEM Fuel Cells Based on Reformed Methanol

The electrochemical technology is based on H_2–O_2, however, the H_2 is derived from methanol rather than from hydrocarbon fuel. A great deal of this tech-

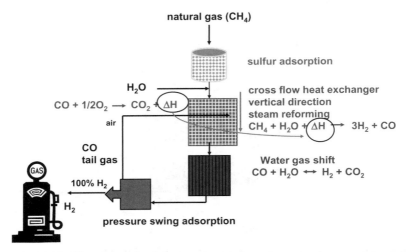

FIGURE 16.10. Heat integrated fuel processor based on precious metal catalyzed monolith and heat exchanger technology for service stations.

nology was developed for on-board reforming (16.22) for transportation applications mainly by Dailmer-Benz-Ballard (Xcellsis) using a Cu, Zn, Al particulate catalyst supplied by BASF.

$$CH_3OH + H_2O \rightarrow 3H_2 + CO_2 \tag{16.22}$$

It is a convenient fuel in that it contains no sulfur and can be reformed at modest temperatures of 200–250 °C and ideally produces 3 moles of H_2 and 1 mole of CO_2. The idea of on-board reforming of methanol (Trimm and Onsan 2001) was abandoned in the early 2000s due mainly to toxicity and water solubility issues. However, this work still represents a key milestone in the development of a hydrogen economy and the use of fuel cell technology. Steam reforming of methanol (Pettersson and Westerholm 2001; Ranganathan et al. 2005) to produce a H_2 and CO_2 mixture is required for portable power fuel cell and other stationary applications. High selectivity to CO_2 is required to minimize the poisoning affect of CO on the anode performance.

Methanol steam reforming is now finding potential use for portable power applications as a convenient source of H_2 for a high-temperature PEM fuel cell system based on a polybenzimidazole or PBI membrane. Advanced reforming catalysts based on promoted PdZn supported alloys (Castellano et al. 2007) are providing advantages over traditional Cu-based materials with enhanced thermal stability and resistance to deactivation by air and liquid water exposure. The market drive is to charge batteries for portable power applications. A water methanol 50–50 mole % mixture is fed to a microchannel reactor containing a selective reforming PdZn washcoated catalyst (Castellano et al. 2007). Using a high-temperature PBI membrane (>180 °C) catalyzed with a Pt on carbon anode, there is no need to reduce the CO further since it

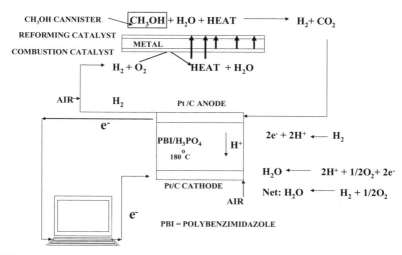

FIGURE 16.11. Cartoon of a methanol reformer integrated to a PEM membrane made of PBI.

does not adsorb appreciably at its operating temperature. Heat is externally supplied to drive the endothermic steam-reforming reaction. The H_2-rich reformate can be directly sent to the anode or through a Pd membrane for further purification. When H_2 purification membranes are used, the CO does not permeate so a Nafion®-type MEA can be used.

Protonex and UltraCell Power are working on an advanced products.

In Figure 16.11, a cartoon is shown using reformed methanol integrated into a fuel cell system. The water–methanol mixture is passed over a washcoat catalyst on a heat exchanger designed for reaction between 230 °C and 400 °C. The endothermic heat of the reaction is provided by the combustion of the tail gas from the anode or from a Pd membrane on the opposite side of the heat exchanger.

16.6 OTHER FUEL CELL SYSTEMS

16.6.1 Alkaline Fuel Cell

The Apollo Space Program (1960–1968) used the alkaline fuel cell system because it had better kinetics and delivered a higher voltage than the acid-based system; also, there is no peroxide intermediate to lower the power output (Kinoshita 1992). The reactions are shown below:

$$\text{ANODE:} \quad 2H_2 + 4OH^- \rightarrow 4H_2O + 4e^- \quad Eo_{anode} = -0.828\,V \quad (16.23)$$

$$\text{CATHODE:} \quad O_2 + 2H_2O + 4e^- \rightarrow 4OH^- \quad Eo_{cathode} = 0.401\,V \quad (16.24)$$

$$\text{NET:} \quad H_2 + 1/2\,O_2 \rightarrow 2H_2O \quad Eo_{\text{cathode}} - Eo_{\text{anode}} = 1.23\,\text{V} \quad (16.25)$$

Although its lifetime is only 2,000–5,000 hours maximum, this was sufficient for early space exploration. The water produced was used for drinking by the astronauts. Since liquid H_2 and O_2 were used, there was no concern for CO_2, which will neutralize the alkaline electrolyte. This limitation renders the alkaline fuel cell impractical for any application where CO_2 is present as in the case of reformate and ground level air. For these cases, the PEM is preferred.

The electrolyte for the 1981 space shuttle was KOH/asbestos with a mixture of Pt + Pd on carbon bonded by polytetrafluoroethylene (PTFE). The cathode was predominately Au promoted with a small amount of Pt on a nickel grid. Other possible electrode materials are Ni–Ti and Pt–Pd for the anode as well as Ag and perovskites for the cathode (Cornils et al. 2000).

16.6.2 Phosphoric Acid Fuel Cell

The first commercialized fuel cells (PC25) were manufactured by United Technologies. Systems were supplied to Tokyo Electric in 1993 for an 11-mW power plant. Currently there are about 30 large power plants in Japan, the United States, and Europe. There are approximately 250 units of 50–500 kW throughout the world used for stationary applications such as schools, apartments, commercial building, and so on.

The reactions are identical to the PEM fuel cell in that H_2 is oxidized at the anode and O_2 from the air is reduced at the cathode but at about 200 °C. Because of the higher operating temperatures, the Pt anode can tolerate larger concentrations of CO and, consequently, no PROX is required: when less than 1% CO. The electrolyte is 100% phosphoric acid adsorbed on SiC and is sufficiently conductive at 200 °C. The anode is 0.1 mg/cm^2 of Pt dispersed on carbon black (i.e., Vulcan XC –72), which is admixed with a Teflon polymer such as PTFE to render it hydrophobic to minimize flooding by water. The electrode is then printed via a doctor blade onto porous graphite paper composed of graphite fibers bonded with phenolic resins. The cathode is Pt on a corrosion-resistant carbon, but it requires more Pt (i.e., 0.5 mg/cm^2) since the kinetics of the electrocatalytic reduction of O_2 are much slower than the anode reaction (Stonehart 1990; Petrow and Allen 1976; Appleby 1985).

Each electrode assembly, composed of the electrodes, electrolyte, and gas dispersion graphite paper, is stacked similar to the PEM with cooling plates every four to six stacks. They are stacked using bipolar grooved or channeled conductive plates (necessary for gas flow) bonded on each side by the anode of one assembly and the cathode of the next producing a series stack for increased power output.

The major problem is cost ($4,000/kW), relatively low current densities, and longer start-up times compared with the PEM that operates at a lower temperature. For this reason, they are no longer considered the most attractive system for residential or vehicular applications.

The major source of deactivation is corrosion and dissolution of the carbon and sintering of the Pt. Corrosion can be controlled by densifying the carbon black by treating in an inert atmosphere at elevated temperatures (i.e., 90 °C).

16.6.3 Molten Carbonate Fuel Cell

The electrolyte is conductive for carbonate ions between 600 °C and 700 °C. It is typically 50% each of Li and K carbonates stabilized on a γ-LiAlO$_2$ with additives of particles and/or fibers of α-Al$_2$O$_3$ to give mechanical strength. The selection of Li and K carbonates and the amount of each depends on resistance to solubility of the Ni-containing cathode as well as on resistance to corrosion. Increasing the amount of Li carbonate increases basicity, which decreases solubility (Appleby and Nicholson 1980). The presence of K carbonates decreases corrosion. Other additives such as Ca, Ba, or Sr also decrease solubility. The exact compositions are highly proprietary. Thicknesses are typically 0.5 mm and are prepared by tape casting (Brdar and Farooque 2005).

The anode is porous sintered Ni (1 m^2/g, 50–70% porosity) with about 10% Cr to stabilize against excessive thermal sintering (Cornils et al. 2000). Typically the thickness of the catalyst layer is 1.5 mm. Frequently a small amount of Li is added to decrease sintering further.

The cathode is Ni doped with small amounts of Li to minimize the solubility of the Ni in the electrolyte. The surface area is less than 0.5 m^2/g (65% porosity) and about 0.75 mm thick. An alternative cathode material is LiCoO$_2$ that in limited tests performs essentially comparable with Ni cathodes but is less likely to corrode (Makkus et al. 1992). A schematic of the cell reactions is shown in Figure 16.12.

Additional O$_2$ is brought into the cathode compartment to combine with the CO$_2$ for the main electrochemical reduction reaction. The carbonate pro-

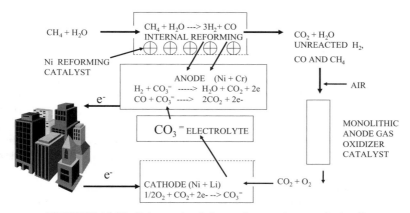

FIGURE 16.12. Schematic of the molten carbonate fuel cell.

duced carries the current to the anode and participates in the electrochemical oxidation of the H_2. Each cell has a manifold for water, fuel, and air.

$$\text{ANODE:}\quad H_2 + CO_3^{-2} \rightarrow H_2O + CO_2 + 2e^- \tag{16.26}$$

$$\text{CATHODE}\quad 1/2\,O_2 + CO_2 + 2e^- \rightarrow CO_3^{-2} \tag{16.27}$$

$$\text{NET:}\quad H_2 + 1/2\,O_2 + CO_{2\,\text{cathode}} \rightarrow H_2O + CO_{2(\text{anode})} \tag{16.28}$$

$$E = Eo + RT/2F \ln{(PH_2)}/PH_2O)(CO_2)_{\text{anode}} + RT/2F \ln{(PO_2)}^{1/2}(PCO_2)_{\text{cathode}} \tag{16.29}$$

The voltage per cell varies between about 0.7 V and 1.0 V depending on the current drawn. Individual cells are stacked, making a repeating series generating proportionally higher voltage and power. For a typical megawatt power plant, about 340 cells are stacked. Steam, fuel, and air are added to each cell. Typically each cell is about 2 by 3 or 4 ft. Each stack is targeted to have a 40,000-hr life, but it is not unusual to replace components periodically due to corrosion, fatigue, and catalyst deactivation at the severe operating conditions (Hoffmann et al. 2003).

A unique feature of both the MCFC and the SOFC (see the next section) is the internal reforming process. Due the high temperatures necessary for these fuel cells, a fossil fuel can be reformed in the anode chamber with the endothermic energy provided by the fuel cell. The hydrocarbon fuel is mixed with steam (about 2–3 moles of H_2O per mole of C) at about 650 °C. The reforming catalyst Ni/Al_2O_3 is positioned adjacent to the anode but usually in a separate compartment management (Dicks 1998) so the fuel is internally reformed with the endothermic heat of reaction provided by the fuel cell. This eliminates all of the unit operations associated with the external reformers for the lower temperature fuel cells. It also puts less demand on the cooling equipment provided heat integration is successfully included in the design [as shown in Eqs. (16.14) and (16.15) and displayed again below]:

$$CH_4 + H_2O \leftrightarrow 3H_2 + CO \quad \Delta H = 49\,\text{Kcal/mole}$$

$$CO + H_2O \leftrightarrow H_2 + CO_2 \quad \Delta H = -10\,\text{Kcal/mole}$$

In most modern molten carbonate and solid oxide fuel cell designs, some precondition of the fuel to be either steam reforming or catalytic partial oxidaton of the hydrocarbon has been found to improve the performance and life of the SOFC. It is improved by minimizing the coke formation in the anode as well as by minimizing the amount of cooling occuring in the anode during internal reforming, which causes thermal stresses in the fuel cell compartment.

As H_2 is produced, it is oxidized, electrochemically shifting the equilibrium for both the reforming and the water gas shift reactions, producing more H_2 and CO_2. For this reason, little CO is present in the effluent. The anode exhaust gas composed of mainly H_2, some unreacted hydrocarbon, and CO_2 and H_2O (from the reaction at the anode) is mixed with some air and passed through a catalytic oxidizer designed to convert the hydrocarbons and H_2 to H_2O and CO_2. The exhaust is then passed to the cathode compartment to provide the CO_2 necessary for the cathodic reaction. Due to the molten salt vapor pressure, some is carried into the anode exhaust and is deposited on the anode oxidizer, leading to its deactivation. Some scrubbing of the alkali carbonates is designed into the process loop to minimize deposition onto the catalysts.

The steam-reforming catalyst is a specially designed Ni-based material not unlike what is conventionally used in standard steam-reforming plants practiced in the chemical industry. The higher temperature operation permits cogeneration of electricity and heat, at about 450 °C, which improves the overall system efficiency to almost 60%.

It is subject to deactivation by coke formation, sintering, and poisoning by impurities in the fuel, i.e., sulfur compounds, as well as by alkali carbonate from the electrolyte. Shields of SiC and other ceramic membranes have been developed that minimize the poisoning effect of the electrolyte (Passalaqua et al. 1996).

The current density is much lower than for the PEM so its major market is for large-scale power plants for buildings, and so on.

16.6.4 Solid Oxide Fuel Cell

The electrolyte is typically a material, such as 10% Y_2O_3 stabilized ZrO_2. The Y^{+3} replaces a Zr^{+4} in the lattice, freeing an O^{-2} for conduction from the cathode to the anode. Conduction occurs at a reasonable rate at 1,000 °C. The anode is 30% porous and is composed of Ni/ZrO_2 (150 microns thick). The cathode is $LaMnO_3$, doped with about 30% Sr (1 mm thick). Recent developments are available (Kawada and Mizusaki 2003). The schematic of the electrochemical reactions is shown in Figure 16.13.

The anode reactions are shown in Eqs. (16.30) and (16.31):

$$H_2 + O^{-2} \rightarrow H_2O + 2e^- \tag{16.30}$$

$$CO + O^{-2} \rightarrow CO_2 + 2e^- \tag{16.31}$$

The cathode reaction is shown in Eq. (16.32):

$$O_2 + 4e^- \rightarrow 2O^{-2} \tag{16.32}$$

The net reaction is shown in Eq. (16.33):

$$O_2 + H_2 + CO \rightarrow H_2O + CO_2 \tag{16.33}$$

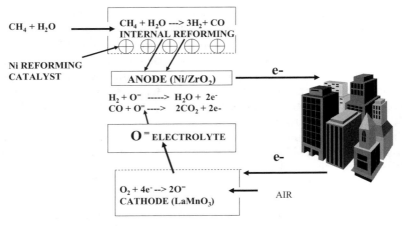

FIGURE 16.13. Solid oxide fuel cell.

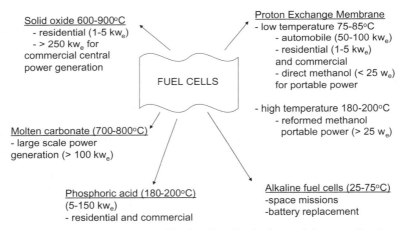

FIGURE 16.14. A summary of fuel cell technologies and their applications.

Typical cell voltage = $0.8\,V$ at $1\,amp/cm^2$ at $1,000\,°C$. The high operating temperatures cause a decrease in the thermodynamic free energy of formation of H_2O (a product of the anode reaction), leading to a 100-mV loss relative to other fuel cells.

It has a reported efficiency of over 60–80% for cogeneration systems.

Since the SOFC operates at $1,000\,°C$, less active but more stable catalysts can be used. Furthermore, the catalysts (electrocatalysts) are much less sensitive to impurities in the fuel, so for this reason, it is especially attractive for fuels generated from coal-based gasification plants. It is still desirable to remove most of the sulfur to preserve their lives.

As with the MCFC, the Ni-based reforming catalyst must be in close contact with the anode for heat management (Dicks 1998). No CO_2 recycling is required as it is in the molten carbonate fuel cell. There are no flooding issues or electrolyte migration since the electrolyte is an O^{-2} conductive solid oxide.

A potential problem is the thermal stress from expansion differences between the anode and the cathode and the solid electrolyte that leads to delamination. Altair has reported some new electrode/electrolyte designs that minimize this problem (Press Release 2001).

A summary of fuel cell technologies and their applications is shown in Figure 16.14.

REFERENCES

Acres, G., Frost, J., Hards, R., Potter, S., Thompson, D., Burstein, G., and Hutchins, G. "Electrocatalysts in fuel cells," *Catalysis Today*, 38: 393 (1997).

Appleby, A. "The electrochemical engine for vehicles," *Scientific America* 281(1): 74 (1999).

Appleby, A. *Fuel Cells: Trends in Research and Applications*, Springer-Verlag, Berlin, Germany (1985).

Appleby, A., and Nicholson, S. "Reduction of oxygen in lithium-potassium carbonate melt," *Journal of Electroanalytical Chemistry*, 112: 71 (1980).

Agrawal, R., Singh, N., Ribeiro, F., and Delgas, N. "Sustainable fuel for the transportation sector," *Proceedings of the National Academy of Science of the USA* 104(12): 4828–4833 (2007).

Armor, J. "Multiple roles for catalysts in the production of hydrogen," *Applied Catalysis A: General* 176: 159 (1999).

Avgouropoulos, G., Ioannides, T., Matralis, H., Batista, J., and Hocevar, S. "CuO-CeO2 mixed oxide catalysts for the selective oxidation of carbon monoxide in excess hydrogen," *Catalysis Letters* 73(1): 33 (2001).

Bartholomew, C., and Farrauto, R. "Fuel cells: A path toward the hydrogen economy," in *Fundamentals of Industrial Catalytic Processes*. Second Edition. Wiley and Sons, New York (2006).

Bethe G., and Kung, H. "Selective Co oxidation in hydrogen-rich stream over Au/γ Al$_2$O$_3$ catalysts," *Applied Catalysis A: General* 194: 43 (2000).

Bonacci, J., Otchy, T., and Ackerman, T. "Ammonia manufacturing process," U.S. Patent 4,238,468 (1980).

Brdar, D., and Farooque, M. "Materials shape up for MCFC success," *Fuel Cell Review* (2005).

Brown, L. "A comparative study of fuels for on-board hydrogen production for fuel-cell-powered automobiles," *International Journal of Hydrogen Energy* 26: 381 (2001a).

Brown, M., Green, A., Cohn, G., and anderson, H. "Purifying hydrogen by selective oxidation of carbon monoxide," *Industrial Engineering and Chemistry Research* 52: 841 (1960).

Brown, S. "Gearing up to make fuel cells," *Fortune* 168 (2001b).

Castellano, C. Liu, Y., Moini, A., Koermer, G., and Farrauto, R. "Catalyst composition for alcohol reforming," U.S. Patent Application 20070258882, (2007).

Chambers, T., Park, Y., Baker, T., and Rodriguez, N. "Hydrogen storage in graphite nanofibers," *Journal of Physical Chemistry B* 102: 22 (1998).

Chemical and Engineering News. "Conductive polymer bi-polar plates for fuel cells," 37: (2007).

Chin, P., Sun, X., Roberts, G., and Spivey, J. "Preferential oxidation of carbon monoxide with iron-promoted platinum catalysts supported on metal foams," *Applied Catalysis A: General* 302(1): 22–31 (2006).

Clapper, T. *Encyclopedia of Chemical Technology*. p 633. Editors Kirk, R., and Othmer, D. Wiley and Sons, New York (1980).

Cohn, G. "Process for selectively removing carbon monoxide from hydrogen containing gases," U.S. Patent 3, 216,783 (1965).

Cooper, H. "Fuel cells, the hydrogen economy and you," *Chemical Engineering Progress* 37: (2007).

Cornils, A., Hermann, W., Schlogl, R., and Wong, C. *Catalysis from A to Z*, pp 238–239. Wiley and Sons, New York (2000).

Cuzens, J., Durai, K., Woods, R., Farrauto, R., Hwang, S., and Korotkikh, O. "Catalytic auto-thermal fuel processing for fuel cells," AIChE meeting Atlanta, GA, (2000).

Dartt, C., and Davis, M. "Catalysis for environmentally benign processing," *Industrial, Engineering, and Chemistry Research*, 33: 2887–2889 (1994).

Dicks, A. "Advances in catalysis for internal reforming in high temperature fuel cells," *Journal of Power Sources* 71(1–2): 111 (1998).

Dicks, A. "Hydrogen generation from natural gas for the fuel cell system," *Journal of Power Sources* 61: 113 (1996).

Doyle, M., and Rajendran, G. "Perfluorinated membranes," in *Handbook of Fuel Cells*, pp 351–395. Editor Vielstich, W., Lamm, A., and Gasteiger, H. Vol 3 (Part 1), Wiley and Sons, New York (2003).

Farrauto, R. J., Hwang, S., Shore, L., Ruettinger,W., Lampert, J., Giroux, T., Liu. Y., Illinich, O. "New material needs for hydrocarbon fuel processing: Generating hydrogen for the PEM fuel cell," *Annual Review of Materials Research* 33: 1–27 (2003).

Farrauto, R. J. "Introduction to solid polymer membrane fuel cells and reforming natural gas for production of hydrogen," *Applied Catalysis B: Environmental* 56: 3–7 (2005).

Farrauto, R. J., Liu, Y., Ruettinger, W., Ilinich, O., Shore, L., and Giroux, T. "Precious metal catalysts supported on ceramic and metal monolithic structures for the hydrogen economy," *Catalysis Reviews* 49: 141–196 (2007).

Funck, R. "High pressure storage," in *Handbook of Fuel Cells*, pp 83–86. Editors Vielstich, W., Lamm, A., and Gasteiger, H. Wiley and Sons, New York (2003).

Gasteiger, H., Kocha, S., Sompalli, B., and Wagner, F. "Activity benchmarks for Pt, Pt alloys, And non-Pt oxygen reduction catalysts for PEMFCs," *Applied Catalysis B: Environmental* 56(1, 2): 9 (2005).

Gottsefeld, S., and Pafford, J. "A new approach to the problem of carbon monoxide poising in fuel cells at low temperature," *Journal of Electrochemical Society* 135: 2651 (1988).

Gottesfeld, S., and Zawodzinski, T., in R. Alkire, et al. Editors, *Advances in Electrochemical Science and Engineering* 5: 197 (1998).

GM-Quantum At15,000 PSIG, Press Release PR Newswire, (2001).

Gray, P., and Petch, M. "Advances with Hotspot™ fuel processor," *Precious Metal Rev.*, 44(3): 108 (2000).

Grot, W. *Encyclopedia of Polymer Science and Engineering.* Second Edition 16 (1989).

Heck, R., Gulati, S., and Farrauto, R. "The application of monoliths for gas phase catalytic reactions," *Chemical Engineering Journal* 82: 149 (2001).

Hetland, J., and Mulder, G. "In search of a sustainable hydrogen economy: How a large-scale transition to hydrogen may affect the primary energy demand and greenhouse gas emissions," *International Journal of Hydrogen Energy* 32: 736 (2007).

Hirschenhofer, S., Stauffer, D., Englemann, R., and Klett, M. *Fuel Cell Handbook Fourth edition.* (1998).

Hochmuth, J. "The catalytic partial oxidation of methane over a monolith supported catalyst," *Applied Catalysis B: Environmental* 1: 89 (1992).

Hoffmann, J., Yuh, Y., and Jopek, A. "Molten carbonate fuel cells and systems: Electrolyte and material challenges," in *Handbook of Fuel Cells.* pp 921–941. Editors Vielstich, W., Lamm, A., and Gasteiger, H. Wiley and Sons, New York (2003).

Hsu, W., and Gierke, T. "Elastic theory for ionic clustering in perfluorinated ionomers," *Macromolecules* 15: 101 (1982).

Hwang, S., Heck, R., and Yarrington, R. "Fuel cell electric power production," U.S. Patent 4,522,894 (1985).

Hwang, H., and Farrauto, R. "Process for generating hydrogen-rich gas, " U.S. Patent 6,3436,363 (2002).

Hwang, H., and Farrauto, R. "Process for generating a hydrogen-rich gas," U.S. Patent 6,849,572 (2005).

Jones, D., and Roziere, J. "Inorganic/organic composite membranes," in *Handbook of Fuel Cells.* pp 447–455. Editors Vielstich, W., Lamm, A., and Gasteiger, H., Wiley and Sons, New York (2003).

Kahlich, M., Gasteiger, A., and Behm, R. "Kinetics of the selective CO oxidation in H_2 rich gas on Pt/Al_2O_3," *Journal Catalysis* 171: 93 (1997).

Kahlich, M., Gasteiger, A., and Behm, R. "Kinetics for the selective low temperature oxidation of CO in H_2 rich gas over $Au/a\text{-}Al_2O_3$," *Journal of Catalysis* 430: (1999).

Kawada, T., and Mizusaki, J. "Solid oxide fuels cells and systems: Current electrolytes and catalysts," in *Handbook of Fuel Cells*, pp 978–1001, Editors Vielstich, W., Lamm, A., and Gasteiger, H. Wiley and Sons, New York (2003).

Kinoshita, K. "Oxygen electrochemistry," in *Electrochemical O_2 Technology*, pp 19–112. Wiley and Sons, New York (1992).

Kordesch, K., and Simader, G. "Fuel cells and their application," VCH, Germany (1996).

Korotkikh, O., and Farrauto, R. "Selective catalytic oxidation of CO in H_2: Fuel cell applications," *Catalysis Today* 62: 2 (2000).

Korotkikh, O., Farrauto, R., and McFarland, A. "Method for preparation of catalytic material for selective oxidation and catalyst members thereof," U.S. Patent 6,559,094 (2003).

Kreuer, K. "Hydrocarbon membranes," in *Handbook of Fuel Cells*, pp 420–435. Editors Vielstich, W., Lamm, A., and Gasteiger, H., Wiley and Sons, New York (2003).

Krumpelt, M., Ahmed, S., Kumar, R., and Doshi, R. "Method for making hydrogen rich gas from hydrocarbon fuel," U.S. Patent 5,929,286 (1999).

Ladebeck, J., and Wagner, J. "Catalyst development in water gas shift," in *Handbook of Fuel Cells*, pp 190–201. Editors Vielstich, W., Lamm, A., and Gasteiger, H., Wiley and Sons, New York (2003).

Li, Y., Fu, Q., and Stephanopoulis, M. "Low temperature water gas shift reaction over Cu and Ni loaded cerium oxide catalysts," *Applied Catalysis B: Environmental* 27: 179 (2000).

Lin, J., Kunz, R., and Fenton, M. "Membrane/electrode additives for low-humidification operation," in *Handbook of Fuel Cells*, pp 456–464. Editors Vielstich, W., Lamm, A., and Gasteiger, H. Wiley and Sons, New York (2003).

Liu, W., and Stephanopoulis, M. "Total oxidation of carbon monoxide and methane over transition metal fluorite oxide composite catalysts," *Journal of Catalysis* 153: 304, 307 (1995).

Liu, X., Korotkikh, O., and Farrauto, R. J. "Selective catalytic oxidation of CO in H_2: Structural study of Fe oxide promoted Pt catalyst," *Applied Catalysis A: General* 226: 293 (2003).

Ma, X., Velu, S., Kim, J., and Song S. "Deep desulfurization of gasoline by selective adsorption over solid adsorbents and impact of analytical methods on ppm-level sulfur quantification for fuel cell applications," *Applied Catalysis B: Environmental* 56: 137–148 (2005).

Makkus, R., Hemmes, K., and deWitt, J. "Molten carbonate fuel cell reactions," *Electrochemical Society* 141(12): 157 (1992).

Maldonado, A., Yang, F., Qi, G., and Yang R. "Desulfurization of transportation fuels by pi complexation sorbents," *Applied Catalysis B: Environmental* 56: 111–126 (2005).

Malhotra, S., and Datta, R. "Membrane supported nonvolatile acidic electrolytes allow higher temperature operation of proton-exchange membrane fuel cells," *Journal of Electrochemical Society* 144: L-23 (1997).

Mauritz, K., and Moore, R. "State of understanding nafion®," *Chemical Reviews* 104: 4535–4585 (2004).

Metkemeijer, R., and Achard, P. "Comparison of ammonia and methanol applied indirectly in a hydrogen fuel cell", *International Journal of H_2 Energy* 19 (6): 535 (1994).

Mueller, U., Schubert, M, Teich, F., Puetter, H., Arndt, K., and Pastre, J. "Metal-organic frameworks-prospective industrial applications," *Journal of Materials Chemistry* 16: 626–636 (2006).

Nakao, M., and Yoshitake, M. "Composite perfluorinate membranes," in *Handbook of Fuel Cells*, pp 412–419. Editors Vielstich, W., Lamm, A., and Gasteiger, H. Wiley and Sons, New York (2003).

Ng, F., Rahman, A., Ohasi, T., and Jiang, M. "A study of the adsorption of thiophenic sulfur compounds using flow calorimetry," *Applied Catalysis B: Environmental* 56: 127–136 (2005).

Nguyen, K. Krause, C., Balasubramanian, B., Rienke, M., and Valensa, J. "Chicken—Meet the Egg: A cost effective hydrogen supply system," National Hydrogen Association, Washington D.C. (2005).

Nielsen, J. "Production of synthesis gas," *Catalysis Today*, 18(4), 305 (1993).

Oh, S., and Sinkevitch, R. "Carbon monoxide removal from hydrogen rich fuel cell feed streams by selective catalytic oxidation," *Journal of Catalysis*, 142: 254 (1993).

Paganin, V., Ticianelli, E., and Gonzalez, E. "Development and electrochemical studies of gas diffusion electrodes for polymer electrolytic fuel cells," *Journal of Applied Electrochemistry*, 26: 297 (1996).

Passalaqua, E., Freni, S., Barone, F., and Patti, A., "Porous ceramic membranes for internal reforming in molten carbonate fuel cells," *Material Letters* 29: 177 (1996).

Petrow, H., and Allen, R. "Catalytic platinum metal particles on a substrate and method of preparing the catalyst," U.S. Patent 3,992,331 (1976).

Petrow, H., and Allen, R. "Finely particulated colloidal platinum compound and sol for producing same, and method of separation," U.S. Patent 3,992,512 (1976).

Pettersson, L., and Westerholm, P. "State of the art Multi-fuel Reformers for fuel cell vehicles," *J. Hydrogen*, 26: 243 (2001).

Press Release PR Newswire, July 17 (2001).

Ogden, J. "Alternative fuels and prospects-Overview," in *Handbook of Fuel Cell*, pp 3–24. Editors Vielstich, W., Lamm, A., and Gasteiger, H. Wiley and Sons, New York (2003).

Raistrick, D. "Proceedings of the Symposium on diaphragms, Separators, and Ion Exchange Membranes," Editors Van Zee, J., et al. The Electrochemical Society, Pennington, NJ (1986).

Ralph, R. "Clean fuel cell energy for today," *Platinum Metals Review* 43(1): 14–17 (1999).

Ranganathan, E., Bej, S., and Thompson, L. "Methanol steam reforming over Pd/ZnO and Pd/CeO2 catalysts," *Applied Catalysis A: General* 289(2): 153–162 (2005).

Rifkin, J. *The Hydrogen Economy*. Penguin, New York (2002).

Roberts, G., Chin, P., Sun, X., and Spivey, J. "Preferential oxidation of carbon monoxide with Pt/Fe monolithic catalysts: Interactions between external transport and the reverse water gas shift reaction," *Applied Catalysis B: Environmental* 46: 601–611 (2003).

Rostrup-Nielsen, J. "Production of synthesis gas," *Catal. Today* 18(4): 305 (1993).

Sanchez, R., Uneda, A., Tanaka, K., and Haruta, M. "Selective oxidation of CO on hydrogen over gold supported on manganese oxide," *Journal of Catalysis* 168: 125 (1997).

Sandrock, G. "Hydride storage," in *Handbook of Fuel Cells*, pp 101–112. Editors Vielstich, W., Lamm, A., and Gasteiger, H. Wiley and Sons, New York (2003).

Satokawa, S., Kobayashi, Y., and Fujiki, H. "Adsorptive removal of dimethylsulfide and t-butlymercaptan from pipeline natural gas on Ag-zeolite under ambient conditions," *Applied Catalysis B: Environmental* 56: 51–56 (2005).

Schutz, W. "Mannesmann Pilotentwicklung," Munich, Germany private communication (2000).

Shore, L., and Farrauto, R. "Prox catalysts," in *Handbook of Fuel Cells*, pp 211–218. Editors Vielstich, W., Lamm, A., and Gasteiger, H. Wiley and Sons, New York (2003).

Singh, K., and Shaki, V. "Electrochemical studies on Nafion® Membranes," *Journal of Membrane Science* 140(1): 51–56 (1998).

Stone, Y., Ekdunge, P., and Simonsson, S. "Proton conductivity of Nafion 117 as measured by a four-electron AC impedance method," *Journal of Electrochemical Society* 143: 1254 (1996).

Stonehart, P. "Development of advanced noble metal-alloy electrocatalysts for phosphoric acid fuel cells (PAFC)," *Berichte der Bunsengesellschaft für Physikalische Chemie* 94: 913 (1990).

St. Pierre, J., and Wilkinson, D. "Fuel Cells: A new, efficient and cleaner power source," *AIChE Journal* 47: 1482 (2001).

Summer, J., Craeger, E., Ma, J., and DesMarteau, D. "Proton conductivity of Nafion 117 and in a novel bis ((perfluoroalkyl) sulfonyl) imide ionomer membrane," *Journal of Electrochemical Society* 145: 107 (1998).

Teng, Y., Sakurai, H., Ueda, K., and Kobayashi, T. "Oxidative removal of CO contained in hydrogen by using base metal oxide catalysts," *International Journal of Hydrogen Energy* 24: 355 (1999).

Thampan, T., Malhotra, S., Tang, H., and Datta, R. "Modeling of conductive transport proton-exchange membranes for fuel cells," *Journal of Electrochemical Society* 147: 102 (2000).

Tonkovich, J., Zilka, La Mont, M., Wang, Y., and Wegeng, R. "Microchannel reactors for fuel processing applications. 1. Water gas shift reactions," *Chemical Engineering Science*, 54: 2947 (1999).

Trimm, D., and Onsan, Z. "On-board fuel conversion for hydrogen fuel-cell-driven vehicles," *Catalysis Reviews* 43(1,2): 31 (2001).

Voss, D. "Company aims to give fuel cells a little backbone," *Science* 285: 683 (1999).

Wainright, J., Wang, J., Weng, D., Savinell, R., and Litt, M. "Acid doped polybenzimidazole: A new polymer membrane," *Journal of the Electrochemical Society* 142(7): L-121–123 (1995).

Wainright, J., Lii, M., and Savinell, F. "High-temperature membranes," in *Handbook of Fuel Cells*. pp 436–446. Editors Vielstich, W., Lamm, A., and Gasteiger, H. Wiley and Sons, New York (2003).

Wieland, S., Boumann, F., and Stary, K. "New powerful catalysts for autothermal reforming of hydrocarbons and water gas shift reactions for on-bound hydrogen generation in automotive," PEMFC Society of Automotive Engineers (SAE) 2001-01-0234 (2001).

Wolf, J. "Liquid hydrogen technology for vehicles," in *Handbook of Fuel Cells*. pp 89–100. Editors Vielstich, W., Lamm, A., and Gasteiger, H. Wiley and Sons, New York (2003).

Xiao, L., Zhang, H., Jana, E., Scanlon, R., Choe, E., Ramanathan, L., and Benicewicz, B. "Synthesis and characterization of pyridine-based polybenzimidazoles for high temperature polymer electrolyte membranes," *Fuel Cells* 5(2): 287–295 (2005a).

Xiao, L. L., Zhang, H., Scanlon, R., Ramanathan, L., Choe, E., Rogers, D., Apple, T., and Benicewicz, B. "High temperature polybenzimidazoles for fuel cell membranes via a sol gel process," *Chemistry of Materials* 17: 5328–5333 (2005b).

Vielstich, W., Lamm, A., and Gasteiger, H., Editors. *Handbook of Fuel Cells*. Wiley and Sons, New York (2003).

CHAPTER 16 QUESTIONS

1. Consider that your own home (apartment) is powered by a PEM fuel cell that gets its hydrogen from a reformer of natural gas.

   a. Sketch a plot of electricity and your hot water needs versus time of the day for a typical 24-hr day (plot your electrical and heat duty cycle). Demands can be stated as low, medium, and high. Start the day at 5:00 AM and end 24 hr later.

2. What would be the key variable process parameters for the fuel processor and the fuel cell that would have to meet the demands for varying demands of both electricity and hot water?

3. Consider the preferential oxidation reaction (PROX) for selectively removing 0.3% CO (3,000 ppm) from a fuel processing stream containing 75% H_2 by adding O_2/CO at a ratio of 1.5. The inlet temperature is 75 °C. How would you expect the rate-limiting step to change as the CO is progressively oxidized as the reaction proceeds axially down the ceramic monolith bed to an exit concentration of 2-ppm CO?

4. Consider the steam reforming of natural gas (i.e., methane) using 20% Ni/Al_2O_3 solid pellets (10 mm in diameter) located in metal tubes 4 in in diameter all in a box furnace.

   a. What will likely limit the rate of reaction?

   b. Predict how the rate will change if you increase the Ni to 35% for a condition in which catalysis within the bed is controlled by heat transfer.

   c. Washcoat a ceramic monolith with the same amount of total Ni as in b. How will this change the rate of reaction?

   d. Replace the Ni with a highly active Rh/Al_2O_3 washcoat (known to be an excellent steam-reforming catalyst) on the inner walls of the tubular reactor. How will the reaction rate change if the reaction is controlled by heat transfer:

   1) Through the tubular wall?

   2) Heat transfer through the wall is relatively fast?

   e. Recommend an overall better design?

5. a. What are the advantages of the fuel cell versus other means to generate power?

   b. What are the unit operations for making H_2 from natural gas (NG)?

   c. Ideally how will we generate H_2?

   d. Explain the reasons (rate-limiting steps) for three zones in the voltage versus current plot.

Index

Catalytic Air Pollution Control: Commercial Technology, by Ronald M. Heck and Robert J. Farrauto, with Suresh T. Gulati.
Copyright © 2009 John Wiley & Sons, Inc.

518